Relativistic Quantum Mechanics

Theoretical and Mathematical Physics

The series founded in 1975 and formerly (until 2005) entitled *Texts and Monographs in Physics* (TMP) publishes high-level monographs in theoretical and mathematical physics. The change of title to *Theoretical and Mathematical Physics* (TMP) signals that the series is a suitable publication platform for both the mathematical and the theoretical physicist. The wider scope of the series is reflected by the composition of the editorial board, comprising both physicists and mathematicians.

The books, written in a didactic style and containing a certain amount of elementary background material, bridge the gap between advanced textbooks and research monographs. They can thus serve as basis for advanced studies, not only for lectures and seminars at graduate level, but also for scientists entering a field of research.

For further volumes:
http://www.springer.com/series/720

Armin Wachter

Relativistic Quantum Mechanics

Springer

Dr. Armin Wachter
awachter@wachter-hoeber.com

ISSN 1864-5879 e-ISSN 1864-5887
ISBN 978-90-481-3644-5 e-ISBN 978-90-481-3645-2
DOI 10.1007/978-90-481-3645-2

Library of Congress Control Number: 2010928392

Printed on acid-free paper

Springer is part of Springer Science+Business Media (www.springer.com)

Preface

It is more important to repair errors than to prevent them. This is the quintessence of the philosophy of human cognition known as critical rationalism which is perhaps at its most dominant in modern natural sciences. According to it insights are gained through a series of presumptions and refutations, through preliminary solutions that are continuously, rigorously, and thoroughly tested. Here it is of vital importance that insights are never verifiable but, at most, falsifiable. In other words: a natural scientific theory can at most be regarded as "not being demonstrably false" until it can be proven wrong. By contrast, a sufficient criterion to prove its correctness does not exist.

Newtonian mechanics, for example, could be regarded as "not being demonstrably false" until experiments with the velocity of light were performed at the end of the 19th century that were contradictory to the predictions of Newton's theory. Since, so far, Albert Einstein's theory of special relativity does not contradict physical reality (and this theory being simple in terms of its underlying assumptions), relativistic mechanics is nowadays regarded as the legitimate successor of Newtonian mechanics. This does not mean that Newton's mechanics has to be abandoned. It has merely lost its fundamental character as its range of validity is demonstrably restricted to the domain of small velocities compared to that of light.

In the first decade of the 20th century the range of validity of Newtonian mechanics was also restricted with regard to the size of the physical objects being described. At this time, experiments were carried out showing that the behavior of microscopic objects such as atoms and molecules is totally different from the predictions of Newton's theory. The theory more capable of describing these new phenomena is nonrelativistic quantum mechanics and was developed in the subsequent decade. However, already at the time of its formulation, it was clear that the validity of this theory is also restricted as it does not respect the principles of special relativity.

Today, about one hundred years after the advent of nonrelativistic quantum mechanics, it is *quantum field theories* that are regarded as "not being demonstrably false" for the description of microscopic natural phenomena.

They are characterized by the facts that

- they can be Lorentz-covariantly formulated, thus being in agreement with special relativity
- they are many-particle theories with infinitely many degrees of freedom and account very precisely for particle creation and annihilation processes.

Naturally, the way toward these modern theories proceeded through some intermediate steps. One began with nonrelativistic quantum mechanics – in conjunction with its one-particle interpretation – and tried to extend this theory in such a way that it becomes Lorentz-covariant. This initially led to the *Klein-Gordon equation* as a relativistic description of spin-0 particles. However, this equation contains a basic flaw because it leads to solutions with negative energy. Apart from the fact that they seem to have no reasonable interpretation, their existence implies quantum mechanically that stable atoms are not possible as an atomic electron would fall deeper and deeper within the unbounded negative energy spectrum via continuous radiative transitions. Another problem of this equation is the absence of a positive definite probability density which is of fundamental importance for the usual quantum mechanical statistical interpretation. These obstacles are the reason that for a long time, the Klein-Gordon equation was not believed to be physically meaningful.

In his efforts to adhere to a positive definite probability density, Dirac developed an equation for the description of electrons (more generally: spin-1/2 particles) which, however, also yields solutions with negative energy. Due to the very good accordance of Dirac's predictions with experimental results in the low energy regime where negative energy solutions can be ignored (e.g. energy spectrum of the hydrogen atom or gyromagnetic ratio of the electron), it was hardly possible to negate the physical meaning of this theory completely.

In order to prevent electrons from falling into negative energy states, Dirac introduced a trick, the so-called *hole theory*. It claims that the vacuum consists of a completely occupied "sea" of electrons with negative energy which, due to Pauli's exclusion principle, cannot be filled further by a particle. Additionally, this novel assumption allows for an (at least qualitatively acceptable) explanation of processes with changing particle numbers. According to this, an electron with negative energy can absorb radiation, thus being excited into an observable state of positive energy. In addition, this electron leaves a hole in the sea of negative energies indicating the absence of an electron with negative energy. An observer relative to the vacuum interprets this as the presence of a particle with an opposite charge and opposite (i.e. positive) energy. Obviously, this process of *pair creation* implies that, besides the electron, there must exist another particle which distinguishes itself from the electron just by its charge. This particle, the so-called *positron*, was indeed found a short time later and provided an impressive confirmation of Dirac's ideas. Today it is well-known that for each particle there exists an antiparticle with opposite (not necessarily electric) charge quantum numbers.

The problem of the absence of a positive definite probability density could finally be circumvented in the Klein-Gordon theory by interpreting the quantities ρ and $\boldsymbol{j}$ as charge density and charge current density (*charge interpretation*). However, in this case, the transition from positive into negative energy states could not be eliminated in terms of the hole theory, since Pauli's exclusion principle does not apply here and, therefore, a completely filled sea of spin-0 particles with negative energy cannot exist.

The Klein-Gordon as well as the Dirac theory provides experimentally verifiable predictions as long as they are restricted to low energy phenomena where particle creation and annihilation processes do not play any role. However, as soon as one attempts to include high energy processes both theories exhibit deficiencies and contradictions. Today the most successful resort is – due to the absence of contradictions with experimental results – the transition to quantized fields, i.e. to quantum field theories.

This book picks out a certain piece of the cognitive process just described and deals with the theories of Klein, Gordon, and Dirac for the relativistic description of massive, electromagnetically interacting spin-0 and spin-1/2 particles excluding quantum field theoretical aspects as far as possible (relativistic quantum mechanics "in the narrow sense"). Here the focus is on answering the following questions:

- How far can the concepts of nonrelativistic quantum mechanics be applied to relativistic quantum theories?
- Where are the limits of a relativistic one-particle interpretation?
- What similarities and differences exist between the Klein-Gordon and Dirac theories?
- How can relativistic scattering processes, particularly those with pair creation and annihilation effects, be described using the Klein-Gordon and Dirac theories without resorting to the formalism of quantum field theory and where are the limits of this approach?

Unlike many books where the "pure theories" of Klein, Gordon, and Dirac are treated very quickly in favor of an early introduction of field quantization, the book in hand emphasizes this particular viewpoint in order to convey a deeper understanding of the accompanying problems and thus to explicate the necessity of quantum field theories.

This textbook is aimed at students of physics who are interested in a concisely structured presentation of relativistic quantum mechanics "in the narrow sense" and its separation from quantum field theory. With an emphasis on comprehensibility and physical classification, this book ranges on a middle mathematical level and can be read by anybody who has attended theoretical courses of classical mechanics, classical electrodynamics, and nonrelativistic quantum mechanics.

This book is divided into three chapters and an appendix. The first chapter presents the Klein-Gordon theory for the relativistic description of spin-0 particles. As mentioned above, the focus lies on the possibilities and limits of its one-particle interpretation in the usual nonrelativistic quantum mechanical sense. Additionally, extensive symmetry considerations of the Klein-Gordon theory are made, its nonrelativistic approximation is developed systematically in powers of v/c, and, finally, some simple one-particle systems are discussed.

In the second chapter we consider the Dirac theory for the relativistic description of spin-1/2 particles where, again, emphasis is on its one-particle interpretation. Both theories, emanating from certain enhancements of nonrelativistic quantum mechanics, allow for a very direct one-to-one comparison of their properties. This is reflected in the way that the individual sections of this chapter are structured like those of the first chapter – of course, apart from Dirac-specific issues, e.g. the hole theory or spin that are considered separately.

The third chapter covers the description of relativistic scattering processes within the framework of the Dirac and, later on, Klein-Gordon theory. In analogy to nonrelativistic quantum mechanics, relativistic propagator techniques are developed and considered together with the well-known concepts of scattering amplitudes and cross sections. In this way, a scattering formalism is created which enables one-particle scatterings in the presence of electromagnetic background fields as well as two-particle scatterings to be described approximately. Considering concrete scattering processes to lowest orders, the Feynman rules are developed putting all necessary calculations onto a common ground and formalizing them graphically. However, it is to be emphasized that these rules do not, in general, follow naturally from our scattering formalism. Rather, to higher orders they contain solely quantum field theoretical aspects. It is exactly here where this book goes for the first time beyond relativistic quantum mechanics "in the narrow sense". The subsequent discussion of quantum field theoretical corrections (admittedly without their deeper explanation) along with their excellent agreement with experimental results may perhaps provide the strongest motivation in this book to consider quantum field theories as the theoretical fundament of the Feynman rules.

Important equations and relationships are summarized in boxes to allow the reader a well-structured understanding and easy reference. Furthermore, after each section there are a short summary as well as some exercises for checking the understanding of the subject matter. The appendix contains a short compilation of important formulae and concepts.

Finally, we hope that this book helps to bridge over the gap between nonrelativistic quantum mechanics and modern quantum field theories, and explains comprehensibly the necessity for quantized fields by exposing relativistic quantum mechanics "in the narrow sense".

Cologne, March 2010 *Armin Wachter*

Table of Contents

List of Exercises

1. Relativistic Description of Spin-0 Particles

In this chapter, we deal with the relativistic description of spin-0 particles in the "narrow sense" as mentioned in the preface, i.e. on the basis of an adequate enhancement of nonrelativistic quantum mechanics. In doing so, we will adhere to the one-particle interpretation of the nonrelativistic theory to the greatest possible extent. Before we start our discussion, the principles underlying this interpretation are summarized as follows:

Theorem 1.1: Principles of nonrelativistic quantum mechanics

1) The quantum mechanical state of a physical system is described by a state vector $|\psi(t)\rangle$ in a complex unitary Hilbert space $\mathcal{H}$. In this space a positive definite scalar product $\langle\psi|\varphi\rangle$ is defined with the following properties:

- $\langle\psi|\psi\rangle \geq 0$
- $\langle\psi|\varphi\rangle = \langle\varphi|\psi\rangle^*$
- $$\begin{aligned} \langle\psi|\left(\lambda_1|\varphi_1\rangle + \lambda_2|\varphi_2\rangle\right) &= \lambda_1\langle\psi|\varphi_1\rangle + \lambda_2\langle\psi|\varphi_2\rangle \\ \left(\langle\psi_1|\lambda_1 + \langle\psi_2|\lambda_2\right)|\varphi\rangle &= \lambda_1^*\langle\psi_1|\varphi\rangle + \lambda_2^*\langle\psi_2|\varphi\rangle\ , \end{aligned}$$
  with $|\psi_{1,2}\rangle\ , |\varphi_{1,2}\rangle \in \mathcal{H}\ ,\ \lambda_{1,2} \in \mathbb{C}$.

2) Physical observables are quantities that can be measured experimentally. They are described by Hermitean operators with real eigenvalues and a complete orthogonal eigenbasis. The quantum mechanical counterparts to the independent classical quantities "position" x_i and "momentum" p_i are the operators $\hat{x}_i$ and $\hat{p}_i$, for which the following commutation relations hold:

$$[\hat{x}_i, \hat{x}_j] = [\hat{p}_i, \hat{p}_j] = 0\ ,\ [\hat{x}_i, \hat{p}_j] = \mathrm{i}\hbar\delta_{ij}\ ,\ i, j = 1, 2, 3\ .$$

The Hermitean operators corresponding to the classical dynamical variables $\Omega(x_i, p_i)$ are obtained from the mapping

$$\hat{\Omega} = \Omega(x_i \to \hat{x}_i, p_i \to \hat{p}_i)\ .$$

▷

A. Wachter, Relativistic Description of Spin-0 Particles. In: A. Wachter, Relativistic Quantum Mechanics, Theoretical and Mathematical Physics, pp. 1–84 (2011)
DOI 10.1007/978-90-481-3645-2_1

However, there also exist observables without classical analogons such as the particle spin.

3) Every state vector $|\psi\rangle$ can be expanded in the orthonormal eigenbasis $\{|\omega_i\rangle\}$ of an observable $\hat{\Omega}$:

$$|\psi\rangle = \sum_i |\omega_i\rangle \langle\omega_i|\psi\rangle \ , \ \hat{\Omega}|\omega_i\rangle = \omega_i|\omega_i\rangle \ , \ \langle\omega_i|\omega_j\rangle = \delta_{ij} \ .$$

A measurement of a dynamical variable corresponding to the operator $\hat{\Omega}$ yields one of its eigenvalues ω_i with probability

$$W(\omega_i) = \frac{|\langle\omega_i|\psi\rangle|^2}{\langle\psi|\psi\rangle} \ .$$

The statistical average (expectation value) of an observable $\hat{\Omega}$, resulting from a large number of similar measurements on identical systems, is (assuming $|\psi\rangle$ is normalized such that $\langle\psi|\psi\rangle = 1$)

$$\langle\hat{\Omega}\rangle = \langle\psi|\hat{\Omega}\psi\rangle = \langle\psi|\hat{\Omega}|\psi\rangle \ .$$

4) The state vector $|\psi(t)\rangle$ satisfies the Schrödinger equation

$$\mathrm{i}\hbar\frac{\mathrm{d}|\psi(t)\rangle}{\mathrm{d}t} = \hat{H}|\psi(t)\rangle \ ,$$

where $\hat{H}$ denotes the Hermitean operator of total energy (the Hamilton operator). In the simplest case it is obtained from the Hamilton function of the corresponding classical system:

$$\hat{H} = H(x_i \to \hat{x}_i, p_i \to \hat{p}_i) \ .$$

The Hermitecity of $\hat{H}$ leads to the conservation law $\mathrm{d}\langle\psi|\psi\rangle/\mathrm{d}t = 0$.

These basic laws or axioms formulated in the Schrödinger picture can be concretized further by choosing a particular representation (or basis). In the coordinate or position representation which we will mostly use in this book, the state vector $|\psi(t)\rangle$ is represented by a wave function $\psi(\boldsymbol{x}, t)$ encompassing all space-time (and other) information of the physical system. The quantity $|\psi(\boldsymbol{x}, t)|^2$ is interpreted as a probability measure for finding the physical system at the space-time point $(\boldsymbol{x}, t)$. In this representation the position and momentum operators are given by

$$\hat{x}_i = x_i \ , \ \hat{p}_i = -\mathrm{i}\hbar\frac{\partial}{\partial x_i} \ .$$

The corresponding expressions for the scalar product and the expectation value of an observable $\hat{\Omega}$ are

$$\langle\psi|\varphi\rangle = \int \mathrm{d}^3x\psi^\dagger\varphi \ , \ \langle\psi|\hat{\Omega}|\psi\rangle = \int \mathrm{d}^3x\psi^\dagger\hat{\Omega}\psi \ .$$

From this and from the above mentioned 4th axiom follows the conservation of total probability,

$$\frac{\mathrm{d}}{\mathrm{d}t}\int \mathrm{d}^3x|\psi(\boldsymbol{x},t)|^2 = 0 \ ,$$

which is necessary for the statistical one-particle interpretation. On the basis of these principles, particularly the last relation that expresses particle number conservation – or, rather, conservation of the single considered particle – we can already now make some statements about to what extent a relativistic enhancement of the one-particle concept is at all possible.

- Due to the possibility of particle creation at interaction energies that are at least equal to the rest energy of the particle, the range of validity of the one-particle view is restricted to particle energies E, particle momenta $\boldsymbol{p}$, and electromagnetic interaction potentials A^μ, for which

  $$|E - m_0c^2| < m_0c^2 \ , \ |\boldsymbol{p}|, \left|\frac{e}{c}A^\mu\right| < m_0c \ , \ \Delta E \ll m_0c^2 \ , \ \Delta p \ll m_0c \ ,$$

  where m_0 denotes the rest mass of the particle. This is precisely the domain of the *nonrelativistic approximation.*

- Given these restrictions and Heisenberg's uncertainty relation, it follows that

  $$\Delta x \geq \frac{\hbar}{\Delta p} \gg \frac{\hbar}{m_0c} \ .$$

  This means that a relativistic particle cannot be localized more precisely than to an area whose linear extent is large compared to the particle's *Compton wave length* $\lambda_c = \hbar/(m_0c)$.

In the subsequent discussion of the Klein-Gordon theory (as well as of the Dirac theory in the next chapter) these points will be especially taken into account and further concretized.

The main features of the Klein-Gordon theory for the relativistic description of spin-0 particles are developed in the first section of this chapter. Here we will particularly be confronted with negative energy states, which can, however, be related to *antiparticles* using the transformation of *charge conjugation.* The second section deals with the symmetry properties of the Klein-Gordon theory. In addition to continuous symmetries, discrete symmetry transformations are of particular interest as they will lead us to a deeper understanding of the negative eigensolutions. In the third section we extend and complete the one-particle picture of the Klein-Gordon theory. Introducing a *generalized scalar product,* we modify the nonrelativistic quantum mechanical framework in such a way that a consistent one-particle interpretation becomes possible. Furthermore, we discuss the range of validity of the Klein-Gordon one-particle picture and show some interpretational problems outside this range. The fourth section considers the nonrelativistic approximation of the Klein-Gordon theory. First, the nonrelativistic limit is discussed, which leads, as expected, to the laws of nonrelativistic quantum mechanics. Subsequently (higher) relativistic corrections are incorporated by expanding the Klein-Gordon equation in powers of v/c using the *Fouldy-Wouthuysen technique*. This chapter ends with the fifth section, where some simple one-particle systems are considered, particularly with a view to a consistent one-particle interpretation.

Note. To avoid misunderstandings, the terms "wave function", "solution", and "state" are used synonymously in the following. They all refer to the functions that solve the Klein-Gordon equation. In contrast, observable states realized in nature are termed (anti)particles. From now on, the tag " ^ " for quantum mechanical operators is suppressed.

1.1 Klein-Gordon Equation

We start our discussion of the Klein-Gordon theory by writing the Klein-Gordon equation in canonical form. In doing so, we immediately come across two new phenomena, which have no reasonable interpretation within the usual quantum mechanical framework: the existence of negative energy solutions and the absence of a positive definite probability density. Following this, we bring the canonical equation into Hamilton or Schrödinger form, which will turn out to be very useful for subsequent considerations. At the end, we return to the above mentioned two phenomena and develop a physically acceptable interpretation for them using the transformation of charge conjugation.

1.1.1 Canonical and Lorentz-covariant Formulations of the Klein-Gordon Equation

In nonrelativistic quantum mechanics the starting point is the energy-momentum relation

$$E = \frac{\boldsymbol{p}^2}{2m} ,$$

which, using the correspondence rule

$$E \longrightarrow \mathrm{i}\hbar\frac{\partial}{\partial t} , \ \boldsymbol{p} \longrightarrow -\mathrm{i}\hbar\boldsymbol{\nabla} \Longleftrightarrow p^\mu \longrightarrow \mathrm{i}\hbar\partial^\mu \quad \text{(four-momentum)} ,$$

leads to the Schrödinger equation for free particles,

$$\mathrm{i}\hbar\frac{\partial\psi(\boldsymbol{x},t)}{\partial t} = -\frac{\hbar^2}{2m}\boldsymbol{\nabla}^2\psi(\boldsymbol{x},t) .$$

Due to the different orders of its temporal and spatial derivatives, this equation is not Lorentz-covariant (see footnote 1 on page 352 in the Appendix A.1). This means that, passing from one inertial system to another, the equation changes its structure, thus contradicting the principle of relativity. Therefore, in order to arrive at a relativistic quantum mechanical wave equation, it is appropriate to start from the corresponding relativistic energy-momentum relation for free particles,

$$E = \sqrt{c^2 \boldsymbol{p}^2 + m_0^2 c^4}\ , \tag{1.1}$$

where m_0 denotes the rest mass of the particle. Using the above replacement, this leads to

$$\mathrm{i}\hbar \frac{\partial \phi(x)}{\partial t} = \left(-c^2\hbar^2 \boldsymbol{\nabla}^2 + m_0^2 c^4\right)^{1/2} \phi(x)\ ,\ x = (x^\mu)\ .$$

However, this equation has two grave flaws. On the one hand, due to the unsymmetrical appearance of space and time derivatives, the relativistic form invariance of this equation is not apparent. On the other hand, the operator on the right hand side is a square root whose expansion leads to a highly nonlocal theory.

Free Klein-Gordon equation. Both problems can be circumvented by starting with the quadratic form of (1.1), i.e.

$$E^2 = c^2\boldsymbol{p}^2 + m_0^2 c^4 \Longleftrightarrow p_0^2 - \boldsymbol{p}^2 = p_\mu p^\mu = m_0^2 c^2\ .$$

In this case, using the above correspondence rule, one obtains the *free Klein-Gordon equation in canonical form*

$$-\hbar^2 \frac{\partial^2 \phi(x)}{\partial t^2} = \left(-c^2\hbar^2 \boldsymbol{\nabla}^2 + m_0^2 c^4\right)\phi(x)\ ,\ x = (x^\mu)\ . \tag{1.2}$$

This can immediately be brought into Lorentz-covariant form,

$$\left(p_\mu p^\mu - m_0^2 c^2\right)\phi(x) = 0\ , \tag{1.3}$$

so that, for example, the transformational behavior of the wave function ϕ is easy to anticipate when changing the reference system. This equation was suggested by Erwin Schrödinger in 1926 as a relativistic generalization of the Schrödinger equation. Later it was studied in more detail by Oskar Benjamin Klein and Walter Gordon.

First it is to be asserted that, contrary to Schrödinger's equation, the Klein-Gordon equation is a partial differential equation of second order in time. So, to uniquely specify a Klein-Gordon state, one needs two initial values, $\phi(x)$ and $\partial\phi(x)/\partial t$. Furthermore, the Klein-Gordon equation seems to be suited for the description of spin-0 particles (spinless *bosons*), since ϕ is a scalar function and does not possess any internal degrees of freedom or, put differently, the operator in (1.3) only acts on the external degrees of freedom (space-time coordinates) of ϕ.

The free solutions to (1.2) or (1.3) with definite momentum can be easily found. They are

$$\begin{aligned}\phi_{\boldsymbol{p}}^{(1)}(x) &= \mathrm{e}^{-\mathrm{i}(cp_0t-\boldsymbol{px})/\hbar}\ ,\ p_0=+\sqrt{\boldsymbol{p}^2+m_0^2c^2}>0\\ \phi_{\boldsymbol{p}}^{(2)}(x) &= \mathrm{e}^{+\mathrm{i}(cp_0t-\boldsymbol{px})/\hbar}\end{aligned}$$

or

$$\phi_{\boldsymbol{p}}^{(r)}(x)=\mathrm{e}^{-\mathrm{i}\epsilon_r p_\mu x^\mu/\hbar}\ ,\ \epsilon_r=\begin{cases}+1 \text{ for } r=1\\ -1 \text{ for } r=2\ .\end{cases}$$

Note that here and in the following, p_0 is always meant to be the positive square root. Obviously, the Klein-Gordon equation leads to solutions with positive energy eigenvalues $E = +cp_0$ and negative energy eigenvalues $E = -cp_0$ that are separated by the "forbidden" energy interval $]-m_0c^2; m_0c^2[$.[1] While the positive solutions can be interpreted as particle wave functions, the physical meaning of the negative solutions is not clear a priori. This makes the Klein-Gordon theory seem unattractive as a relativistic generalization of Schrödinger's theory. However, as we will see later on, negative solutions can be related to *antiparticles* that are experimentally observable so that the Klein-Gordon theory indeed provides a valuable generalization of Schrödinger's theory. Incidentally, this is why we consider $\phi_{\boldsymbol{p}}^{(2)}(x)$ to be a negative solution with momentum index $\boldsymbol{p}$, although it has the momentum eigenvalue $-\boldsymbol{p}$.

We will return to the interpretational problem of negative solutions later and investigate next some further properties of the Klein-Gordon equation.

Interaction with electromagnetic fields, gauge invariance. In the Klein-Gordon equation, the interaction of a relativistic spin-0 particle with an electromagnetic field can, as in the Schrödinger theory, be taken into account by the following operator replacement, the so-called *minimal coupling*:

$$\mathrm{i}\hbar\frac{\partial}{\partial t}\longrightarrow \mathrm{i}\hbar\frac{\partial}{\partial t}-eA^0\ ,\ \frac{\hbar}{\mathrm{i}}\boldsymbol{\nabla}\longrightarrow\frac{\hbar}{\mathrm{i}}\boldsymbol{\nabla}-\frac{e}{c}\boldsymbol{A}\Longleftrightarrow p^\mu\longrightarrow p^\mu-\frac{e}{c}A^\mu\ ,$$

where $(A^\mu)=\begin{pmatrix}A^0\\ \boldsymbol{A}\end{pmatrix}$ denotes the electromagnetic four-potential and e the electric charge of the particle. With this, (1.2) and (1.3) become[2]

$$\left[\left(\mathrm{i}\hbar\frac{\partial}{\partial t}-eA^0\right)^2-c^2\left(\frac{\hbar}{\mathrm{i}}\boldsymbol{\nabla}-\frac{e}{c}\boldsymbol{A}\right)^2-m_0^2c^4\right]\phi=0 \tag{1.4}$$

and

$$\left[\left(p_\mu-\frac{e}{c}A_\mu\right)\left(p^\mu-\frac{e}{c}A^\mu\right)-m_0^2c^2\right]\phi=0\ . \tag{1.5}$$

---

[1] In the following, the solutions whose energy eigenvalues lie above the forbidden interval (limited from below) are termed *positive solutions* and those with energy eigenvalues below the forbidden interval (limited from above) *negative solutions.*

[2] The minimal coupling is at most correct for structureless point particles which, however, have not been observed so far. Therefore, in (1.5) additional (phenomenologically based) terms of the form $\lambda F_{\mu\nu}F^{\mu\nu}\phi$ with $F^{\mu\nu}=\partial^\mu A^\nu-\partial^\nu A^\mu$ have to be, in principle, taken into consideration.

As is well-known, the Maxwell equations are invariant under local gauge transformations of the kind

$$A^0 \longrightarrow A'^0 = A^0 - \frac{1}{c}\frac{\partial \chi}{\partial t} \ , \quad \boldsymbol{A} \longrightarrow \boldsymbol{A}' = \boldsymbol{A} + \boldsymbol{\nabla}\chi$$

or

$$A^\mu \longrightarrow A'^\mu = A^\mu - \partial^\mu \chi \ , \tag{1.6}$$

where $\chi = \chi(x)$ is an arbitrary real scalar function of the space-time coordinates. As in the nonrelativistic theory, this local gauge invariance can be carried over to the Klein-Gordon equation (1.4) or (1.5) by multiplying the wave function ϕ by a suitably chosen phase:

$$\phi(x) \longrightarrow \phi'(x) = \mathrm{e}^{\mathrm{i}\Lambda(x)}\phi(x) \ . \tag{1.7}$$

In order to find the function Λ, we express (1.5) in terms of the primed quantities and calculate as follows:

$$\begin{aligned} 0 &= \left[\left(p_\mu - \frac{e}{c}A'_\mu - \frac{e}{c}\partial_\mu\chi\right)\left(p^\mu - \frac{e}{c}A'^\mu - \frac{e}{c}\partial^\mu\chi\right) - m_0^2c^2\right]\phi'\mathrm{e}^{-\mathrm{i}\Lambda} \\ &= \left[\left(p_\mu - \frac{e}{c}A'_\mu - \frac{e}{c}\partial_\mu\chi\right)\mathrm{e}^{-\mathrm{i}\Lambda}\left(p^\mu - \frac{e}{c}A'^\mu - \frac{e}{c}\partial^\mu\chi + \hbar\partial^\mu\Lambda\right)\right. \\ &\quad \left. - m_0^2c^2\mathrm{e}^{-\mathrm{i}\Lambda}\right]\phi' \\ &= \mathrm{e}^{-\mathrm{i}\Lambda}\left[\left(p_\mu - \frac{e}{c}A'_\mu - \frac{e}{c}\partial_\mu\chi + \hbar\partial_\mu\Lambda\right)\left(p^\mu - \frac{e}{c}A'^\mu - \frac{e}{c}\partial^\mu\chi + \hbar\partial^\mu\Lambda\right)\right. \\ &\quad \left. - m_0^2c^2\right]\phi' \ . \end{aligned} \tag{1.8}$$

Choosing

$$\Lambda(x) = \frac{e}{\hbar c}\chi(x) \ , \tag{1.9}$$

(1.8) becomes

$$\left[\left(p_\mu - \frac{e}{c}A'_\mu\right)\left(p^\mu - \frac{e}{c}A'^\mu\right) - m_0^2c^2\right]\phi' = 0 \ ,$$

which is formally identical to the Klein-Gordon equation (1.5). Since physical observables are represented by bilinear forms of the kind $\langle\phi^*|\dots|\phi\rangle$, a common equal phase factor does not play any role. Therefore, the Klein-Gordon equation with minimal coupling is invariant under local gauge transformations of the electromagnetic field.[3]

[3] Remarkably, the transformation (1.7) along with (1.9) is the same as the transformation that leads to local gauge invariance in the nonrelativistic theory.

Continuity equation. Multiplying (1.4) or (1.5) by ϕ^* from the left and subsequently subtracting the complex conjugate, one obtains a continuity equation of the form

$$\frac{\partial \rho(x)}{\partial t} + \boldsymbol{\nabla} \boldsymbol{j}(x) = 0 \ , \tag{1.10}$$

with

$$\begin{aligned} \rho(x) &= \frac{\mathrm{i}\hbar}{2m_0c^2}\left[\phi^*\frac{\partial \phi}{\partial t} - \left(\frac{\partial \phi^*}{\partial t}\right)\phi\right] - \frac{e}{m_0c^2}A^0\phi^*\phi \\ \boldsymbol{j}(x) &= -\frac{\mathrm{i}\hbar}{2m_0}\left[\phi^*\boldsymbol{\nabla}\phi - (\boldsymbol{\nabla}\phi^*)\phi\right] - \frac{e}{m_0c}\boldsymbol{A}\phi^*\phi \end{aligned}$$

or, in Lorentz-covariant notation,

$$\partial_\mu j^\mu(x) = 0 \ , \ j^\mu = \frac{\mathrm{i}\hbar}{2m_0}\left(\phi^*\partial^\mu\phi - \phi\partial^\mu\phi^*\right) - \frac{e}{m_0c}A^\mu\phi^*\phi \ , \ (j^\mu) = \begin{pmatrix} c\rho \\ \boldsymbol{j} \end{pmatrix} .$$

Note that an overall factor was introduced in ρ and $\boldsymbol{j}$ due to analogy with nonrelativistic quantum mechanics. As usual, spatial integration of (1.10) yields the conservation law

$$Q = \int \mathrm{d}^3x\rho(x) = \text{const} \ .$$

Obviously, $\rho(x)$ is not positive definite since, at a given time t, ϕ and $\partial\phi/\partial t$ can take on arbitrary values. Therefore, ρ and $\boldsymbol{j}$ cannot be interpreted as probability quantities. This problem, in conjunction with the existence of negative solutions, was the reason that the Klein-Gordon equation was initially rejected and that attempts were made to find a relativistic wave equation of first order in time and with a positive definite probability density. This equation was indeed found by Dirac. However, as we see in Chapter 2, the Dirac equation also yields solutions with negative energy eigenvalues.

To summarize:

Theorem 1.2: Klein-Gordon equation in canonical and Lorentz-covariant forms

The Klein-Gordon equation is the relativistic generalization of Schrödinger's equation for spin-0 particles. For a minimal coupled electromagnetic field, it is

$$\left[\left(\mathrm{i}\hbar\frac{\partial}{\partial t} - eA^0\right)^2 - c^2\left(\frac{\hbar}{\mathrm{i}}\boldsymbol{\nabla} - \frac{e}{c}\boldsymbol{A}\right)^2 - m_0^2c^4\right]\phi(x) = 0 \tag{1.11}$$

or, in manifestly covariant notation,

$$\left[\left(p_\mu - \frac{e}{c}A_\mu\right)\left(p^\mu - \frac{e}{c}A^\mu\right) - m_0^2c^2\right]\phi(x) = 0 \ , \tag{1.12}$$

▷

where m_0 is the rest mass and e the electric charge of the particle. These equations are invariant under local gauge transformations of the electromagnetic field. From the Klein-Gordon equation follows the continuity equation

$$\partial_\mu j^\mu = 0\ ,\ (j^\mu) = \begin{pmatrix} c\rho \\ \boldsymbol{j} \end{pmatrix}\ ,$$

with

$$\left.\begin{aligned} \rho(x) &= \frac{\mathrm{i}\hbar}{2m_0c^2}\left[\phi^*\frac{\partial\phi}{\partial t} - \left(\frac{\partial\phi^*}{\partial t}\right)\phi\right] - \frac{e}{m_0c^2}A^0\phi^*\phi \\ \boldsymbol{j}(x) &= -\frac{\mathrm{i}\hbar}{2m_0}\left[\phi^*\boldsymbol{\nabla}\phi - (\boldsymbol{\nabla}\phi^*)\phi\right] - \frac{e}{m_0c}\boldsymbol{A}\phi^*\phi\ , \end{aligned}\right\} \quad (1.13)$$

as well as the conservation law

$$Q = \int \mathrm{d}^3x\rho(x) = \text{const}\ .$$

The solutions to the free Klein-Gordon equation ($A^\mu = 0$) are

$$\phi_{\boldsymbol{p}}^{(r)}(x) = \frac{1}{(2\pi\hbar)^{3/2}}\sqrt{\frac{m_0c}{p_0}}\mathrm{e}^{-\mathrm{i}\epsilon_r p_\mu x^\mu/\hbar}\ ,\ p_0 = +\sqrt{\boldsymbol{p}^2 + m_0^2c^2}\ ,$$

with momentum eigenvalue $+\boldsymbol{p}$ (for $r = 1$) or $-\boldsymbol{p}$ (for $r = 2$). These solutions are normalized such that

$$\frac{\mathrm{i}\hbar}{2m_0c^2}\int \mathrm{d}^3x\left[\phi_{\boldsymbol{p}}^{(r)*}\frac{\partial\phi_{\boldsymbol{p}'}^{(r')}}{\partial t} - \left(\frac{\partial\phi_{\boldsymbol{p}}^{(r)*}}{\partial t}\right)\phi_{\boldsymbol{p}'}^{(r')}\right] = \epsilon_r\delta_{rr'}\delta(\boldsymbol{p} - \boldsymbol{p}')\ .$$

1.1.2 Hamilton Formulation of the Klein-Gordon Equation

The Klein-Gordon equation from Theorem 1.2 is a differential equation of second order in time. For our subsequent discussion, it is useful to convert it into a system of coupled differential equations of first temporal order. In this way, it acquires a Schrödinger-like form, in which a Hamilton operator can be identified just as in the nonrelativistic theory. Introducing two new fields via

$$\phi = \varphi + \chi\ ,\ \left(\mathrm{i}\hbar\frac{\partial}{\partial t} - eA^0\right)\phi = m_0c^2(\varphi - \chi) \quad (1.14)$$

$$\Longrightarrow \left\{\begin{aligned} \varphi &= \frac{1}{2m_0c^2}\left(m_0c^2 + \mathrm{i}\hbar\frac{\partial}{\partial t} - eA^0\right)\phi \\ \chi &= \frac{1}{2m_0c^2}\left(m_0c^2 - \mathrm{i}\hbar\frac{\partial}{\partial t} + eA^0\right)\phi\ , \end{aligned}\right. \quad (1.15)$$

(1.11) can be rewritten as

$$\left(\mathrm{i}\hbar\frac{\partial}{\partial t} - eA^0\right)(\varphi + \chi) = m_0c^2(\varphi - \chi)$$
$$\left(\mathrm{i}\hbar\frac{\partial}{\partial t} - eA^0\right)(\varphi - \chi) = \left[\frac{1}{m_0}\left(\boldsymbol{p} - \frac{e}{c}\boldsymbol{A}\right)^2 + m_0c^2\right](\varphi + \chi) \, .$$

Addition and subtraction of these two equations lead to the system of coupled differential equations of first order in time,

$$\mathrm{i}\hbar\frac{\partial\varphi}{\partial t} = \frac{1}{2m_0}\left(\boldsymbol{p} - \frac{e}{c}\boldsymbol{A}\right)^2(\varphi + \chi) + (m_0c^2 + eA^0)\varphi$$
$$\mathrm{i}\hbar\frac{\partial\chi}{\partial t} = -\frac{1}{2m_0}\left(\boldsymbol{p} - \frac{e}{c}\boldsymbol{A}\right)^2(\varphi + \chi) - (m_0c^2 - eA^0)\chi \, ,$$

which is equivalent to (1.11). Finally pooling φ and χ into

$$\psi = \begin{pmatrix} \varphi \\ \chi \end{pmatrix}$$

leads to the *Klein-Gordon equation in Hamilton form*

$$\mathrm{i}\hbar\frac{\partial\psi}{\partial t} = H\psi \, , \; H = \frac{\tau_3 + \mathrm{i}\tau_2}{2m_0}\left(\boldsymbol{p} - \frac{e}{c}\boldsymbol{A}\right)^2 + \tau_3 m_0c^2 + eA^0 \, .$$

Here τ_i denote the *Pauli matrices*

$$\tau_1 = \begin{pmatrix} 0 & 1 \\ 1 & 0 \end{pmatrix} , \; \tau_2 = \begin{pmatrix} 0 & -\mathrm{i} \\ \mathrm{i} & 0 \end{pmatrix} , \; \tau_3 = \begin{pmatrix} 1 & 0 \\ 0 & -1 \end{pmatrix} ,$$

which satisfy the following algebra:

$$\tau_i\tau_j = i\epsilon_{ijk}\tau_k + \delta_{ij} \, , \; [\tau_i, \tau_j] = 2i\epsilon_{ijk}\tau_k \, , \; \{\tau_i, \tau_j\} = 2\delta_{ij} \, .$$

The solutions to the free Klein-Gordon equation

$$\mathrm{i}\hbar\frac{\partial\psi}{\partial t} = H^{(0)}\psi \, , \; H^{(0)} = \frac{(\tau_3 + \mathrm{i}\tau_2)\boldsymbol{p}^2}{2m_0} + \tau_3 m_0c^2 \tag{1.16}$$

are given by (see Exercise 1)

$$\psi_{\boldsymbol{p}}^{(1)}(x) = \begin{pmatrix} m_0c + p_0 \\ m_0c - p_0 \end{pmatrix} \mathrm{e}^{-\mathrm{i}p_\mu x^\mu/\hbar}$$
$$\psi_{\boldsymbol{p}}^{(2)}(x) = \begin{pmatrix} m_0c - p_0 \\ m_0c + p_0 \end{pmatrix} \mathrm{e}^{+\mathrm{i}p_\mu x^\mu/\hbar} \, .$$

To calculate ρ and $\boldsymbol{j}$ in the Hamilton formulation, we insert (1.14) and (1.15) into (1.13) and obtain

$$\rho(x) = \psi^\dagger(x)\tau_3\psi(x) = \varphi^*\varphi - \chi^*\chi$$
$$\boldsymbol{j}(x) = -\frac{\mathrm{i}\hbar}{2m_0}\left[\psi^\dagger\tau_3(\tau_3 + \mathrm{i}\tau_2)\boldsymbol{\nabla}\psi - (\boldsymbol{\nabla}\psi^\dagger)\tau_3(\tau_3 + \mathrm{i}\tau_2)\psi\right]$$
$$-\frac{e}{m_0c}\boldsymbol{A}\psi^\dagger\tau_3(\tau_3 + \mathrm{i}\tau_2)\psi \, .$$

Overall, we arrive at the following theorem equivalent to Theorem 1.2:

Theorem 1.3: Klein-Gordon equation in Hamilton form

With the replacements

$$\phi = \varphi + \chi\ ,\ \left(\mathrm{i}\hbar\frac{\partial}{\partial t} - eA^0\right)\phi = m_0c^2(\varphi - \chi)\ ,\ \psi = \begin{pmatrix}\varphi\\ \chi\end{pmatrix}$$

in (1.11), the Klein-Gordon equation in Hamilton form follows as

$$\mathrm{i}\hbar\frac{\partial\psi}{\partial t} = H\psi\ ,\ H = \frac{\tau_3 + \mathrm{i}\tau_2}{2m_0}\left(\boldsymbol{p} - \frac{e}{c}\boldsymbol{A}\right)^2 + \tau_3 m_0c^2 + eA^0\ , \tag{1.17}$$

where τ_i denote the Pauli matrices. The corresponding expressions for ρ and $\boldsymbol{j}$ are

$$\begin{aligned}
\rho(x) &= \psi^\dagger(x)\tau_3\psi(x) = \varphi^*\varphi - \chi^*\chi\\
\boldsymbol{j}(x) &= -\frac{\mathrm{i}\hbar}{2m_0}\left[\psi^\dagger\tau_3(\tau_3 + \mathrm{i}\tau_2)\boldsymbol{\nabla}\psi - (\boldsymbol{\nabla}\psi^\dagger)\tau_3(\tau_3 + \mathrm{i}\tau_2)\psi\right]\\
&\quad -\frac{e}{m_0c}\boldsymbol{A}\psi^\dagger\tau_3(\tau_3 + \mathrm{i}\tau_2)\psi\\
Q &= \int \mathrm{d}^3x\rho(x) = \int \mathrm{d}^3x\psi^\dagger(x)\tau_3\psi(x)\ .
\end{aligned}$$

In the Hamilton formulation, the solutions to the free Klein-Gordon equations are

$$\left.\begin{aligned}
\psi_{\boldsymbol{p}}^{(r)}(x) &= \frac{1}{(2\pi\hbar)^{3/2}}\Psi^{(r)}(\boldsymbol{p})\mathrm{e}^{-\mathrm{i}\epsilon_r p_\mu x^\mu/\hbar}\\
\Psi^{(r)}(\boldsymbol{p}) &= \frac{1}{2\sqrt{m_0cp_0}}\begin{pmatrix} m_0c + \epsilon_r p_0\\ m_0c - \epsilon_r p_0\end{pmatrix}\ ,
\end{aligned}\right\} \tag{1.18}$$

with momentum eigenvalue $+\boldsymbol{p}$ (for $r = 1$) or $-\boldsymbol{p}$ (for $r = 2$). These solutions are normalized such that

$$\begin{aligned}
&\int \mathrm{d}^3x\psi_{\boldsymbol{p}}^{(r)\dagger}(x)\tau_3\psi_{\boldsymbol{p}'}^{(r')}(x) = \epsilon_r\delta_{rr'}\delta(\boldsymbol{p} - \boldsymbol{p}')\\
&\Psi^{(r)\dagger}(\boldsymbol{p})\tau_3\Psi^{(r')}(\boldsymbol{p}) = \epsilon_r\delta_{rr'}\ ,\ \Psi^{(r)}(\boldsymbol{p}) = \Psi^{(r)}(-\boldsymbol{p})\ .
\end{aligned} \tag{1.19}$$

It is important to note that in (1.17) the Hamilton operator H is not Hermitean (since $\mathrm{i}\tau_2$ is not Hermitean). From this it immediately becomes apparent, why it is impossible to find a positive definite probability density (including total probability conservation): using the nonrelativistic scalar product

$$\langle\psi|\,\phi\rangle = \int \mathrm{d}^3x\psi^\dagger\phi\ ,\ \langle\psi|\,\mathcal{O}\,|\phi\rangle = \int \mathrm{d}^3x\psi^\dagger\mathcal{O}\phi \tag{1.20}$$

and the adjunction relation

$$\langle\psi|\,\mathcal{O}\,|\phi\rangle = \langle\phi|\,\mathcal{O}^\dagger\,|\psi\rangle^* \qquad (\mathcal{O}\text{ linear operator})\ , \tag{1.21}$$

we have

$$\mathrm{i}\hbar\frac{\partial\psi}{\partial t} = H\psi \Longrightarrow \mathrm{i}\hbar\psi^\dagger\frac{\partial\psi}{\partial t} = \psi^\dagger H\psi \;,\; -\mathrm{i}\hbar\frac{\partial\psi^\dagger}{\partial t}\psi = (H\psi)^\dagger\psi = (\psi^\dagger H\psi)^*$$

$$\Longrightarrow \mathrm{i}\hbar\frac{\partial}{\partial t}\langle\psi|\,\psi\rangle = \langle\psi|\,H\,|\psi\rangle - \langle\psi|\,H\,|\psi\rangle^* = \langle\psi|\,H - H^\dagger\,|\psi\rangle \neq 0\;.$$

Furthermore, the non-Hermitecity of H is the reason that its eigenstates are generally not orthogonal with respect to (1.20).

Another important consequence of the non-Hermitecity of H is that $\mathrm{e}^{\mathrm{i}H}$ is not unitary. This is one indication that in the Klein-Gordon theory the usage of the scalar product (1.20) seems to be unsuitable as it leads to different results in different pictures (for example, the Schrödinger picture used here, or the Heisenberg picture). We will tackle this problem in Subsection 1.3.1.

1.1.3 Interpretation of Negative Solutions, Antiparticles

So far, we have written down the Klein-Gordon equation in canonical, Lorentz-covariant, and Hamilton forms and looked at some of its formal properties. Now we turn to the negative Klein-Gordon solutions which we have so far ignored. Our aim is to find a physically meaningful interpretation for them as well as for the quantities Q, ρ, and $\boldsymbol{j}$.

Charge conjugation *C*. We again consider the canonical Klein-Gordon equation

$$\left[\left(\mathrm{i}\hbar\frac{\partial}{\partial t} - eA^0\right)^2 - c^2\left(\boldsymbol{p} - \frac{e}{c}\boldsymbol{A}\right)^2 - m_0^2c^4\right]\phi^{(-)}(x) = 0\;, \tag{1.22}$$

where $\phi^{(-)}$ denotes a solution with negative energy. Transforming this equation by taking its complex conjugate, one obtains the mathematically equivalent relation

$$\left[\left(\mathrm{i}\hbar\frac{\partial}{\partial t} + eA^0\right)^2 - c^2\left(\boldsymbol{p} + \frac{e}{c}\boldsymbol{A}\right)^2 - m_0^2c^4\right]\phi_C^{(-)}(x) = 0\;, \tag{1.23}$$

with

$$\phi_C^{(-)}(x) = \phi^{(-)*}(x)\;.$$

The consequences of this become even clearer if we start from the eigenvalue equation of a negative eigenstate $\Psi^{(-)}$ in Hamilton form,

$$\left[\frac{\tau_3 + \mathrm{i}\tau_2}{2m_0}\left(\boldsymbol{p} - \frac{e}{c}\boldsymbol{A}\right)^2 + \tau_3 m_0c^2 + eA^0\right]\Psi^{(-)}(\boldsymbol{x}) = -|E|\Psi^{(-)}(\boldsymbol{x})\;, \tag{1.24}$$

and apply complex conjugation to it. This yields

$$\left[\frac{\tau_3 + \mathrm{i}\tau_2}{2m_0}\left(\boldsymbol{p} + \frac{e}{c}\boldsymbol{A}\right)^2 + \tau_3 m_0c^2 - eA^0\right]\Psi_C^{(-)}(\boldsymbol{x}) = +|E|\Psi_C^{(-)}(\boldsymbol{x})\;, \tag{1.25}$$

with

$$\Psi_C^{(-)}(x) = \tau_1 \Psi^{(-)*}(x) .$$

All in all, if $\phi^{(-)}$ or $\psi^{(-)}$ describes a negative Klein-Gordon state with charge $+e$ within the potential A^μ, then $\phi_C^{(-)} = \phi^{(-)*}$ or $\psi_C^{(-)} = \tau_1\psi^{(-)*}$ describes a positive Klein-Gordon state with charge $-e$ within the same potential A^μ. Correspondingly, the above transformation is called *charge conjugation*. Obviously, it is a reciprocal transformation since its twofold application leads back to the original equation. Furthermore, it is antilinear [4], since, going from (1.22) to (1.23), the relative sign between the differential and potential terms is changed. Therefore, charge conjugation opens us a way to a physical interpretation of negative Klein-Gordon solutions whose charge conjugates are to be regarded as the quantum mechanical wave functions of antiparticles with charge $-e$.

As regards the free Klein-Gordon solutions, charge conjugation yields

$$\phi_{\boldsymbol{p},C}^{(1,2)}(x) = \phi_{\boldsymbol{p}}^{(2,1)}(x) \; , \; \psi_{\boldsymbol{p},C}^{(1,2)}(x) = \psi_{\boldsymbol{p}}^{(2,1)}(x) .$$

In this case the original as well as the charge conjugated wave functions are solutions to the same equation, because the distinction between free states with different charges is not possible.

Charge density, charge current density. We are now in a position to give the quantities Q, ρ, and $\boldsymbol{j}$ physically meaningful interpretations. As we have seen above, the quantity

$$\rho = \psi^\dagger \tau_3 \psi = \varphi^*\varphi - \chi^*\chi \; , \; \int \mathrm{d}^3x \rho(x) = Q = \text{const}$$

cannot generally be taken as a probability density, since it is not positive definite. However, if we restrict ourselves to the validity range of the one-particle interpretation (to be more accurately defined later), i.e. to the nonrelativistic approximation mentioned at the beginning of this chapter, ρ becomes positive definite for positive Klein-Gordon solutions, $|\varphi| \gg |\chi|$, and negative for negative solutions, $|\varphi| \ll |\chi|$ (see Subsection 1.4.1). Since positive solutions belong to particles with charge $+e$ and the charge conjugates of negative solutions belong to antiparticles with charge $-e$, we can interpret the expressions $\rho^{(\pm)}$ (built by $\psi^{(\pm)}$) as *electric charge density* and $\boldsymbol{j}^{(\pm)}$ as *electric charge current density* of a particle or an antiparticle. Consequently, $Q^{(\pm)} = \pm 1$ is the (conserved) total charge of the particle or antiparticle (*charge interpretation*).[5]

---

[4] An operator $\mathcal{O}$ is called antilinear if $\mathcal{O}(\alpha_1\psi_1 + \alpha_2\psi_2) = \alpha_1^*\mathcal{O}\psi_1 + \alpha_2^*\mathcal{O}\psi_2$.

[5] This interpretation can also be maintained outside the validity range of the one-particle picture. In this case Q denotes the conserved total charge of all particles and antiparticles. Consequently, the charge density ρ may take on different signs at different space-time points.

Theorem 1.4: Charge conjugation C and charge interpretation in the Klein-Gordon theory

- In the Klein-Gordon theory the charge conjugation C is defined by the transformation

$$\phi(x) \longrightarrow \phi_C(x) = \phi^*(x) \quad \text{(canonical form)}$$

$$\psi(x) \longrightarrow \psi_C(x) = \tau_1\psi^*(x) \text{ (Hamilton form).}$$

  It turns a positive [negative] Klein-Gordon solution of charge $+e$ [$-e$] into a negative [positive] Klein-Gordon solution of charge $-e$ [$+e$].

- A positive Klein-Gordon solution $\phi^{(+)}$ or $\psi^{(+)}$ represents a physical spin-0 particle of charge $+e$ within the potential A^μ, while the charge conjugate of the negative solution $\phi_C^{(-)}$ or $\psi_C^{(-)}$ (and not the original negative solution) describes the physical antiparticle with opposite charge $-e$ within the same potential A^μ.

- The quantities Q, ρ, and $\boldsymbol{j}$ that are composed of $\phi^{(+)}$ or $\psi^{(+)}$ [$\phi^{(-)}$ or $\psi^{(-)}$] can be interpreted as the electric charge, charge density, and charge current density of the physical particle [antiparticle] (charge interpretation).

While the wave function of an antiparticle is described by the charge conjugated negative solution, one obtains its charge quantities Q, ρ, and $\boldsymbol{j}$ using the original negative solutions. In Section 1.3, we extend this principle to the definition of picture-independent scalar products and expectation values.

Now it becomes clear why we have assigned the negative free Klein-Gordon solution $\phi_{\boldsymbol{p}}^{(2)}$ [$\psi_{\boldsymbol{p}}^{(2)}$] the index $\boldsymbol{p}$, although it possesses the momentum eigenvalue $-\boldsymbol{p}$. This is because this solution should be associated with the corresponding antiparticle (with opposite momentum and energy eigenvalue).

That the statements of Theorem 1.4 do in fact agree with nature is confirmed, on the one hand, by the experimental fact that, for each known spin-0 particle, a corresponding antiparticle has been found. On the other hand, as we see in Chapter 3, they are in accordance with experimentally verifiable predictions from scattering theory.

Overall, we see that the relativistic generalization of Schrödinger's theory to the Klein-Gordon theory leads to a new degree of freedom, the electric charge, whereas the nonrelativistic theory describes states with only one charge sign.[6] In this context it is also important to note that in our considerations we could have equally started with the Klein-Gordon equation for states of charge $-e$, since the sign of the charge does not play a decisive role at any stage. Consequently, particles would carry the charge $-e$ described

[6] This is a characteristic of all relativistic enhancements.

by positive solutions, and antiparticles would have the charge $+e$ described by the charge conjugated negative solutions.

Interpretation of the negative solutions. Although we were able to give the charge conjugated negative Klein-Gordon solutions a physically meaningful interpretation, there are still two serious points open, namely:

- the physical implications stemming from the mere existence of negative solutions and
- the physical interpretation of the negative solutions.

In our previous considerations the existence of solutions with negative energy leads to problems and physical nonsense. Think, for example, of a pion atom consisting of a positively charged nucleus and a circuiting negatively charged pion (spin-0 particle). The corresponding energy spectrum can be calculated, for example, by incorporating the Coulomb potential into the Klein-Gordon equation (see Subsection 1.5.4). It is depicted qualitatively in Figure 1.1.

The bound states directly below the positive energy continuum with $E < m_0c^2$ generally agree with experimental results. So there is no doubt that these are the true bound states of the pion atom. On the other hand, the existence of the negative energy continuum implies that a ground state pion could fall deeper and deeper through continuous radiation transitions.

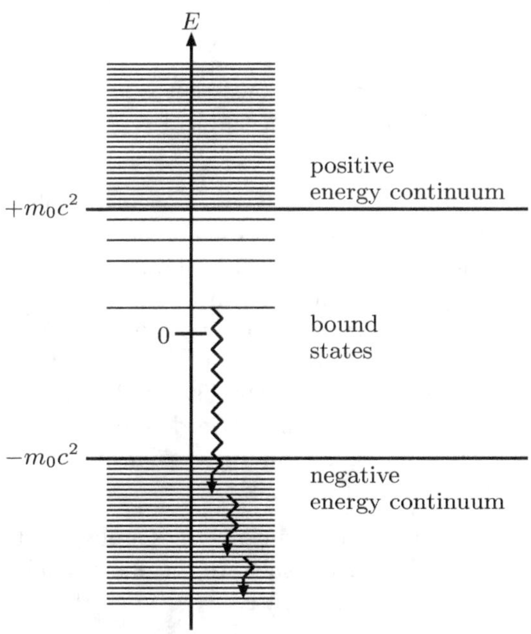

Fig. 1.1. Qualitative energy spectrum of a pion atom. Due to the existence of negative energy states, the pion could fall deeper and deeper through continuous radiation transitions.

Accordingly, the atom would be unstable and a *radiation catastrophe* would occur due to the continuous emission of light.[7] Nevertheless, it is clear that none of these effects are observed; our world could not exist if this decay was present.

As we see later, the same problem exists in the Dirac theory for describing spin-1/2 particles. But there Dirac introduced a trick known as the *hole theory* in order to avoid the radiation catastrophe. According to this model, the vacuum is regarded as a "sea" completely occupied by spin-1/2 particles with negative energies, which, due to Pauli's exclusion principle, cannot be filled by any further particles. Apart from the fact that the radiation catastrophe is now avoided, the negative states acquire a direct physical meaning with physical consequences, for example, the creation and annihilation of particle-antiparticle pairs or the *vacuum polarization.*

It is clear that the hole theory cannot be transferred to the spin-0 case in hand, since the Pauli principle does not apply here. However, even if the hole theory could be applied here in some way, it is to be kept in mind that, in any case, it would mean turning away from the one-particle concept toward a many-particle theory (with infinitely many degrees of freedom). Therefore, within the framework of the targeted one-particle interpretation, we have to leave the physical interpretation of the negative solutions open.

Résumé. All in all, it can be ascertained that using the concepts of charge conjugation and charge interpretation, we can give the positive and the charge conjugated negative energy solutions as well as Q, ρ, and $\boldsymbol{j}$ physically meaningful interpretations as particle, antiparticle, charge, charge density, and charge current density. However, with a view to a consistent one-particle interpretation in the usual nonrelativistic quantum mechanical sense, three points are still open:

[1] The one-particle interpretation requires that positive and negative solutions can be completely decoupled from one another, i.e. that each charged Klein-Gordon state can be represented by a superposition of pure negative or pure positive solutions. However, in general, a Klein-Gordon state is composed of the complete system of positive and negative solutions. We therefore have to clarify under which conditions or within which limits a complete decoupling of positive and negative solutions is possible. Such a splitting leads simultaneously to a positive or negative definite charge density so that a quantum mechanical statistical interpretation becomes possible.

[2] A complete decoupling of positive and negative solutions also implies that not all relativistic operators are applicable with respect to the one-particle concept since they generally mix positive and negative solutions. Hence,

[7] Strictly speaking, the pion atom is unstable due to other effects. However, these effects happen much more slowly than the atom's life time as predicted according to the radiation transitions into negative energy levels.

the question arises: what are meaningful *one-particle operators* and how can they be constructed?

[3] In order to be able to make quantum mechanical probability statements about the state of spin-0 particles, we need a physically meaningful definition of scalar products and expectation values that are independent from the used picture, for example the Schrödinger or the Dirac picture etc. (see the remarks after Theorem 1.3).

As we see in the following pages, these (and other) points can be resolved, so that we finally get a reasonably consistent one-particle picture within certain limits.

At the end of this section we point out that the charge through which a boson distinguishes itself from its antiboson does not necessarily need to be electric. Whereas the pion π^- and antipion π^+ differ indeed in the sign of the electric charge, there also exist bosons such as the kaon K_0 and antikaon $\bar{K}_0$ that are both electrically neutral but possess different signs of the so-called *strangeness charge*. Furthermore, a boson may also carry no charge at all. In this case the corresponding wave function must obey

$$\phi = \phi_C[\psi = \psi_C] \Longrightarrow Q = 0\ ,\ \rho = 0\ ,\ \boldsymbol{j} = 0\ .$$

However, within a consistent one-particle interpretation, considering such neutral particles is problematic, since in this case a complete decoupling of positive and negative solutions is not possible (see [1]).

Summary

- The **Klein-Gordon theory** is the relativistic generalization of nonrelativistic quantum mechanics for the description of spin-0 particles. Starting from the **canonical** or **Lorentz-covariant** representation, this theory can be transferred into **Hamilton form**.
- The Klein-Gordon theory differs from the nonrelativistic theory in two important points: firstly, the Klein-Gordon equation leads to solutions with positive and negative energy. Secondly, due to the non-Hermitecity of the Klein-Gordon-Hamilton operator, j^0 is not positive definite and can therefore not be interpreted as a probability density.
- With the help of the **charge conjugation** and the **charge interpretation**, these two phenomena can be interpreted in a physically meaningful way: **particles** of charge $+e$ are described by positive Klein-Gordon solutions and **antiparticles** of charge $-e$ by the **charge conjugates** of negative solutions. j^0 is the electric **charge density** of the particle and $\boldsymbol{j}$ the corresponding **charge current density**.

▷

- The problems associated with negative solutions (interpretation, **radiation catastrophe**) cannot be solved within the one-particle concept.
- With a view to the most consistent probabilistic one-particle interpretation of the Klein-Gordon theory, two issues are still open: to what extent is a complete decoupling of positive and negative solutions possible, and how can we define a physically sensible as well as picture-independent scalar product?

Exercises

1. Solutions of the free Klein-Gordon equation. Show that the solutions to the free Klein-Gordon equation (1.16) with sharp momentum are given by (1.18).

Solution. To solve (1.16) we make the ansatz

$$\psi(x) = \begin{pmatrix} \varphi_0 \\ \chi_0 \end{pmatrix} e^{i(\boldsymbol{p}\boldsymbol{x}-Et)/\hbar} ,$$

leading to the equation system

$$\left.\begin{aligned} \left(E - \frac{\boldsymbol{p}^2}{2m_0} - m_0c^2\right)\varphi_0 - \frac{\boldsymbol{p}^2}{2m_0}\chi_0 &= 0 \\ \frac{\boldsymbol{p}^2}{2m_0}\varphi_0 + \left(E + \frac{\boldsymbol{p}^2}{2m_0} + m_0c^2\right)\chi_0 &= 0 \,. \end{aligned}\right\} \quad (1.26)$$

It is a necessary condition for the existence of nontrivial solutions that the coefficient determinant vanishes:

$$\begin{vmatrix} E - \frac{\boldsymbol{p}^2}{2m_0} - m_0c^2 & -\frac{\boldsymbol{p}^2}{2m_0} \\ \frac{\boldsymbol{p}^2}{2m_0} & E + \frac{\boldsymbol{p}^2}{2m_0} + m_0c^2 \end{vmatrix} = 0$$

$$\Longleftrightarrow E^2 - \left(\frac{\boldsymbol{p}^2}{2m_0} + m_0c^2\right)^2 + \left(\frac{\boldsymbol{p}^2}{2m_0}\right)^2 = 0 \,.$$

As expected, this again leads to the relativistic energy-momentum relation for free particles:

$$E^2 = \boldsymbol{p}^2c^2 + m_0^2c^4 \Longrightarrow \begin{cases} E^{(+)} = +c\sqrt{\boldsymbol{p}^2 + m_0^2c^2} = +cp_0 \\ E^{(-)} = -c\sqrt{\boldsymbol{p}^2 + m_0^2c^2} = -cp_0 \,. \end{cases}$$

The (unnormalized) solutions corresponding to $E^{(+)}$ and $E^{(-)}$ are finally calculated from (1.26) as

$$E^{(+)} : \psi^{(+)}(x) = \begin{pmatrix} m_0c + p_0 \\ m_0c - p_0 \end{pmatrix} \mathrm{e}^{-\mathrm{i}(cp_0 - \boldsymbol{px})\hbar} \sim \psi_{\boldsymbol{p}}^{(1)}(x)$$

$$E^{(-)} : \psi^{(-)}(x) = \begin{pmatrix} m_0c - p_0 \\ m_0c + p_0 \end{pmatrix} \mathrm{e}^{+\mathrm{i}(cp_0 + \boldsymbol{px})/\hbar} \sim \psi_{-\boldsymbol{p}}^{(2)}(x) \ .$$

2. Lagrange density and energy-momentum tensor of the free Klein-Gordon field. Determine the Lagrange density of the free Klein-Gordon field in the Hamilton formulation. Using the energy-momentum tensor, show that the energy is given by

$$E = \int \mathrm{d}^3x \psi^\dagger \tau_3 H^{(0)} \psi \ , \ H^{(0)} = \frac{\boldsymbol{p}^2}{2m_0}(\tau_3 + \mathrm{i}\tau_2) + \tau_3 m_0 c^2 \ .$$

Solution. In the Hamilton formulation the equation of motion of the free Klein-Gordon field is

$$\mathrm{i}\hbar \frac{\partial \psi}{\partial t} = H^{(0)} \psi \ . \tag{1.27}$$

ψ is a two-component complex field and can be expressed as

$$\psi = \psi_1 + \mathrm{i}\psi_2 \ ,$$

where $\psi_{1,2}$ denote two real fields. Therefore, the Lagrange density can be presented as a function of these two fields and their derivatives. Equivalently, the Lagrange density can also be formulated as a function of ψ, $\bar{\psi} = \psi^\dagger \tau_3$, and their derivatives, leading to

$$\mathcal{L} = \mathrm{i}\hbar \bar{\psi} \frac{\partial \psi}{\partial t} - \frac{\hbar^2}{2m_0} (\boldsymbol{\nabla} \bar{\psi})(\tau_3 + \mathrm{i}\tau_2) \boldsymbol{\nabla} \psi - m_0 c^2 \bar{\psi} \tau_3 \psi \ .$$

In order to see this, we look at the Lagrange equations of the action functional

$$I = \int \mathrm{d}^4x \mathcal{L} \ . \tag{1.28}$$

Its variation with respect to the components of $\bar{\psi}$ leads directly to (1.27), since we have ($\alpha = 1, 2$)

$$\frac{\partial I}{\partial \bar{\psi}_\alpha} = 0 \Longrightarrow \frac{\partial \mathcal{L}}{\partial \bar{\psi}_\alpha} - \partial_\mu \frac{\partial \mathcal{L}}{\partial(\partial_\mu \bar{\psi}_\alpha)} = 0 \quad \text{(Lagrange equation)}$$

$$\Longleftrightarrow \frac{\partial \mathcal{L}}{\partial \bar{\psi}_\alpha} - \frac{\partial}{\partial t} \frac{\partial \mathcal{L}}{\partial(\partial \bar{\psi}_\alpha / \partial t)} - \boldsymbol{\nabla} \frac{\partial \mathcal{L}}{\partial(\boldsymbol{\nabla} \bar{\psi}_\alpha)} = 0$$

$$\frac{\partial \mathcal{L}}{\partial \bar{\psi}_\alpha} = -m_0 c^2 [\tau_3 \psi]_\alpha + \mathrm{i}\hbar \frac{\partial \psi_\alpha}{\partial t} \ , \ \frac{\partial \mathcal{L}}{\partial(\partial \bar{\psi}_\alpha / \partial t)} = 0$$

$$\boldsymbol{\nabla} \frac{\partial \mathcal{L}}{\partial(\boldsymbol{\nabla} \bar{\psi}_\alpha)} = -\frac{\hbar^2 \boldsymbol{\nabla}^2}{2m_0} [(\tau_3 + \mathrm{i}\tau_2)\psi]_\alpha$$

and therefore

$$\mathrm{i}\hbar\frac{\partial\psi_\alpha}{\partial t} = -\frac{\hbar^2\nabla^2}{2m_0}[(\tau_3+\mathrm{i}\tau_2)\psi]_\alpha + m_0c^2[\tau_3\psi]_\alpha \quad \begin{pmatrix}\text{free}\\ \text{Klein-Gordon}\\ \text{equation}\end{pmatrix}. \quad (1.29)$$

Varying (1.28) with respect to the components of ψ, one obtains the corresponding equation of motion for $\bar{\psi}$:

$$\frac{\partial I}{\partial\psi_\alpha} = 0 \Longrightarrow \frac{\partial\mathcal{L}}{\partial\psi_\alpha} - \partial_\mu\frac{\partial\mathcal{L}}{\partial(\partial_\mu\psi_\alpha)} = 0 \quad \text{(Lagrange equation)}$$

$$\Longleftrightarrow \frac{\partial\mathcal{L}}{\partial\psi_\alpha} - \frac{\partial}{\partial t}\frac{\partial\mathcal{L}}{\partial(\partial\psi_\alpha/\partial t)} - \nabla\frac{\partial\mathcal{L}}{\partial(\nabla\psi_\alpha)} = 0$$

$$\frac{\partial\mathcal{L}}{\partial\psi_\alpha} = -m_0c^2[\bar{\psi}\tau_3]_\alpha\,, \quad \frac{\partial\mathcal{L}}{\partial(\partial\psi_\alpha/\partial t)} = \mathrm{i}\hbar\bar{\psi}_\alpha$$

$$\nabla\frac{\partial\mathcal{L}}{\partial(\nabla\psi_\alpha)} = -\frac{\hbar^2\nabla^2}{2m_0}[\bar{\psi}(\tau_3+\mathrm{i}\tau_2)]_\alpha$$

$$\Longrightarrow -\mathrm{i}\hbar\frac{\partial\bar{\psi}_\alpha}{\partial t} = -\frac{\hbar^2\nabla^2}{2m_0}[\bar{\psi}(\tau_3+\mathrm{i}\tau_2)]_\alpha + m_0c^2[\bar{\psi}\tau_3]_\alpha \quad \begin{pmatrix}\text{free}\\ \text{adjoint}\\ \text{Klein-Gordon}\\ \text{equation}\end{pmatrix}.$$

Next we use the energy-momentum tensor

$$T^{\mu\nu} = \frac{\partial\mathcal{L}}{\partial(\partial_\mu\psi_\alpha)}\partial^\nu\psi_\alpha + \frac{\partial\mathcal{L}}{\partial(\partial_\mu\bar{\psi}_\alpha)}\partial^\nu\bar{\psi}_\alpha - g^{\mu\nu}\mathcal{L}$$

to calculate the energy density T^{00}:

$$\begin{aligned}T^{00} &= \frac{\partial\mathcal{L}}{\partial(\partial\psi_\alpha/\partial t)}\frac{\partial\psi_\alpha}{\partial t} + \frac{\partial\mathcal{L}}{\partial(\partial\bar{\psi}_\alpha/\partial t)}\frac{\partial\bar{\psi}_\alpha}{\partial t} - \mathcal{L}\\ &= \frac{\hbar^2}{2m_0}(\nabla\bar{\psi})(\tau_3+\mathrm{i}\tau_2)\nabla\psi + m_0c^2\bar{\psi}\tau_3\psi\,.\end{aligned}$$

This finally leads to the energy

$$\begin{aligned}E &= \int \mathrm{d}^3x T^{00}\\ &= \int \mathrm{d}^3x\left[\frac{\hbar^2}{2m_0}(\nabla\bar{\psi})(\tau_3+\mathrm{i}\tau_2)\nabla\psi + m_0c^2\bar{\psi}\tau_3\psi\right]\\ &\qquad\qquad \text{(partial integration)}\\ &= \int \mathrm{d}^3x\bar{\psi}\left[-\frac{\hbar^2\nabla^2}{2m_0}(\tau_3+\mathrm{i}\tau_2) + m_0c^2\tau_3\right]\psi\\ &= \int \mathrm{d}^3x\bar{\psi}H^{(0)}\psi = \int \mathrm{d}^3x\psi^\dagger\tau_3H^{(0)}\psi\,,\end{aligned}$$

which is positive for positive as well as for negative Klein-Gordon fields. The interpretation of this result becomes apparent in Subsection 1.3.1, where we define a *generalized scalar product* for spin-0 particles and antiparticles consistent with the one-particle concept.

Note: if we had formulated the Lagrange density $\mathcal{L}$ with $\psi^\dagger$ instead of $\bar{\psi}$, we would have arrived at the same equation of motion (1.29). However, we require the action I to be real which, in the case of $\bar{\psi}$, leads to the condition

$$\begin{aligned} I &= \int \mathrm{d}^3x\mathrm{d}t \left[\mathrm{i}\hbar\bar{\psi}\frac{\partial\psi}{\partial t} - \frac{\hbar^2}{2m_0}(\boldsymbol{\nabla}\bar{\psi})(\tau_3 + \mathrm{i}\tau_2)\boldsymbol{\nabla}\psi - m_0c^2\bar{\psi}\tau_3\psi \right] \\ &\qquad\qquad\qquad\qquad\qquad\qquad \text{(partial integration)} \\ &= \int \mathrm{d}^3x\mathrm{d}t\bar{\psi} \left[\mathrm{i}\hbar\frac{\partial}{\partial t} + \frac{\hbar^2\boldsymbol{\nabla}^2}{2m_0}(\tau_3 + \mathrm{i}\tau_2) - m_0c^2\tau_3 \right]\psi \\ &= \int \mathrm{d}^3x\mathrm{d}t\psi^\dagger\tau_3 \left[\mathrm{i}\hbar\frac{\partial}{\partial t} - H^{(0)} \right]\psi = \mathrm{Re}(I) \ . \end{aligned}$$

As one can easily show, this is indeed the case, since $\mathrm{i}\hbar\tau_3\partial/\partial t$ as well as $\tau_3 H^{(0)}$ are Hermitean. By contrast, the integral I will not be real if $\mathcal{L}$ is constructed using the field $\psi^\dagger$ instead of $\bar{\psi}$.

1.2 Symmetry Transformations

In this section we initially postpone our efforts for a physically consistent one-particle interpretation of the Klein-Gordon theory and, instead, begin by considering some further formal properties of the Klein-Gordon equation, namely its symmetry properties. We first define more precisely the terms "transformation" and "symmetry transformation". Then we consider the continuous and discrete symmetries of the Klein-Gordon equation. In doing so, the latter, along with the charge conjugation C from Subsection 1.1.3, will lead us to a better understanding of the negative solutions, especially with a view to the aspired one-particle interpretation.

1.2.1 Active and Passive Transformations

In principle, one distinguishes between two classes of transformations. The first class consists of *active transformations* where the physical state is transformed, while the original and the transformed states are observed from the same reference frame. An example of this kind are the gauge transformations of the electromagnetic field [see (1.6) and (1.7)], which we have already identified as symmetry transformations of the Klein-Gordon theory, as they leave the form of the Klein-Gordon equation unchanged.

The second class is *passive transformations*. Here, not the physical state but the reference frame (or the basis system) is transformed, so that only the perspective changes, from which the same state is observed. These transformations are also called *coordinate transformations* as they always imply a change in the space-time coordinates. In this book we are dealing with relativistic theories, therefore the interesting transformations are the Lorentz transformations.

It is clear that to each passive transformation we can assign an active transformation leading to the same consequences with respect to the appearance or the description of a physical state. In other words, transforming the observer's reference frame or, instead, the physical state in the "opposite direction" leads to the same appearance of a physical state.

In order to clarify the general relationship between a passive and the corresponding active transformation, consider an observer linked to his reference frame and looking at a space point, whose position he denotes by $\boldsymbol{x}$. There he sees a physical state (e.g. spin-0 particle, electromagnetic field) and calls it $z(\boldsymbol{x})$. We initially imagine that a transformation (translation or rotation) of the reference frame is performed and that the observer is told the corresponding transformation law.[8] From this he derives the coordinate vector $\boldsymbol{x}'$ of his original observation point in the transformed system, looks at the original state from the new perspective and denotes its pattern by $z'(\boldsymbol{x}')$. Obviously, this procedure is equivalent to the passive transformation (see Figure 1.2 top)

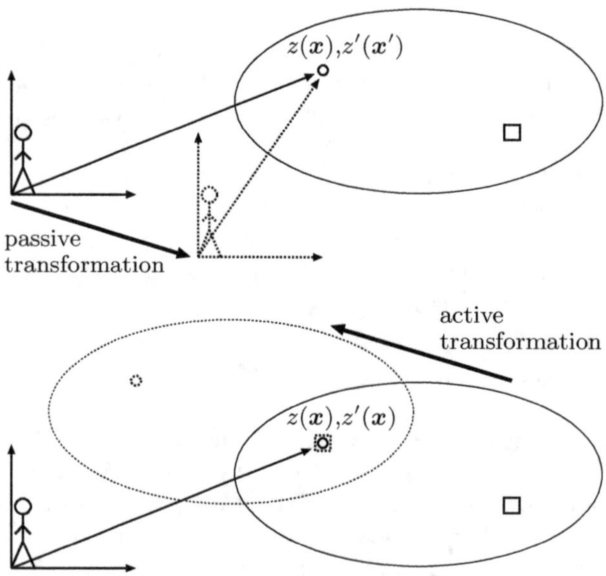

Fig. 1.2. Passive and active transformations. *Upper picture:* from the passive view point the reference frame is shifted to the right and down. *Lower picture:* from the active view point the physical state is shifted in the opposite direction.

[8] Without restricting generality, it is assumed that the physical state is time-independent and the transformation purely spatial.

$$z(\boldsymbol{x}) \longrightarrow z'(\boldsymbol{x}') \ .$$

Now we assume instead that the observer closes his eyes and does not notice the change in the reference frame. Once he opens his eyes again, he sees a pattern different from $z(\boldsymbol{x})$, which he denotes by $z'(\boldsymbol{x})$, because he thinks that he is still looking in the same direction at the same point. Obviously, from the observer's view point, the active transformation

$$z(\boldsymbol{x}) \longrightarrow z'(\boldsymbol{x})$$

has taken place, and he concludes that the physical state itself has been transformed (see Figure 1.2 bottom).

Once the transformation law of a passive transformation, e.g. of the form

$$z'(x') = f_K([z(x)] \ , \ x' = K(x) \ , \ x = K^{-1}(x') \ , \tag{1.30}$$

is known, the corresponding active transformation follows as

$$z'(x) = z'\left[K^{-1}(x')\right] = f_K\left(z\left[K^{-1}(x)\right]\right) \ . \tag{1.31}$$

The mapping passive transformation $\longrightarrow$ active transformation is in principle always possible, whereas the mapping in the opposite direction is generally not possible. This means that there exist active transformations that cannot (or can only partially) be connected to passive transformations. This is particularly apparent, for example, in the charge conjugation transformation (see Theorem 1.4).

With this background, the term "symmetry transformation" can now be specified as follows: *a symmetry transformation leads to formally identical equations of motion and, therefore, to physically equivalent situations, either at the transition from the original to the transformed reference frame (passive case) or at the transition from the original to the transformed physical state (active case).*

Keeping these considerations in mind, we now turn to the symmetry operations of the Klein-Gordon theory.

1.2.2 Lorentz Transformations

The fundamental motivation for the Klein-Gordon equation was that it should obey the principles of special relativity. This implies the form invariance of the Klein-Gordon equation (1.12) under Lorentz transformations (see Appendix A.1) or, rather, under proper Lorentz transformations. However, it is easy to show that the Klein-Gordon equation is form invariant even under general transformations of the kind

$$x^\mu \longrightarrow x'^\mu = \Lambda^\mu{}_\nu x^\nu + a^\mu \ . \tag{1.32}$$

On the one hand, this is because the scalar character of the Klein-Gordon wave function implies that under (1.32) it is changed at most by a phase, which means in the passive case:

$$\phi(x) \longrightarrow \phi'(x') = \lambda\phi(x) \ , \ |\lambda| = 1 \ .$$

On the other hand, in (1.12) the operator acting on ϕ is a Lorentz scalar due to

$$p^\mu \longrightarrow p'^\mu = \Lambda^\mu{}_\nu p^\nu \ , \ A^\mu(x) \longrightarrow A'^\mu(x') = \Lambda^\mu{}_\nu A^\nu(x) \ .$$

From this immediately follows the form invariance of the Klein-Gordon equation under the whole Poincaré group. If we consider only proper Lorentz transformations, i.e. boosts and rotations, $\Lambda^\mu{}_\nu$ depends on continuous parameters whose possible values include the identity transformation. Therefore, in this case, the phase must be $\lambda = 1$.[9]

1.2.3 Discrete Transformations

Parity transformation *P*. As an example of improper (discrete) Lorentz-like symmetry transformations, we consider the orthochronous transformation of space reflection, also called *parity transformation*, which is defined via

$$(\Lambda^\mu{}_\nu) = \begin{pmatrix} 1 & 0 & 0 & 0 \\ 0 & -1 & 0 & 0 \\ 0 & 0 & -1 & 0 \\ 0 & 0 & 0 & -1 \end{pmatrix} \ , \ \Lambda^0{}_0 > 0 \ , \ \det(\Lambda^\mu{}_\nu) = -1 \ .$$

It reverses the sign of the spatial coordinates and leaves the temporal component unchanged. Obviously, it must hold that $\lambda^2 = 1$, since a twofold application of the parity transformation is the identity transformation ($\Lambda^2 = 1$). Thus, in the passive case we have

$$\left.\begin{array}{rcl} \boldsymbol{x} & \longrightarrow & \boldsymbol{x}' = -\boldsymbol{x} \ , \ t \longrightarrow t' = t \\ \phi(\boldsymbol{x},t) & \longrightarrow & \phi_P(\boldsymbol{x}',t') = \lambda_P\phi(\boldsymbol{x},t) \ , \ \lambda_P = \pm 1 \\ A^0(\boldsymbol{x},t) & \longrightarrow & A^0_P(\boldsymbol{x}',t') = A^0(\boldsymbol{x},t) \\ \boldsymbol{A}(\boldsymbol{x},t) & \longrightarrow & \boldsymbol{A}_P(\boldsymbol{x}',t') = -\boldsymbol{A}(\boldsymbol{x},t) \end{array}\right\} \begin{array}{c}\text{passive space} \\ \text{reversal } P,\end{array}$$

where P stands for the parity transformation. This means that, under the parity transformation, ϕ behaves either as a *scalar* (+) or as a *pseudo-scalar* (−). Using the general scheme from (1.30) and (1.31), the corresponding active transformation is obtained as

$$\left.\begin{array}{rcl} \phi(\boldsymbol{x},t) & \longrightarrow & \phi_P(\boldsymbol{x},t) = \lambda_P\phi(-\boldsymbol{x},t) \\ A^0(\boldsymbol{x},t) & \longrightarrow & A^0_P(\boldsymbol{x},t) = A^0(-\boldsymbol{x},t) \\ \boldsymbol{A}(\boldsymbol{x},t) & \longrightarrow & \boldsymbol{A}_P(\boldsymbol{x},t) = -\boldsymbol{A}(-\boldsymbol{x},t) \end{array}\right\} \begin{array}{c}\text{active space} \\ \text{reversal } P.\end{array} \quad (1.33)$$

[9] Wave functions that are not changed under spatial rotations describe, by definition, particles with spin 0. Thus, we have a group or transformation theoretical argument, that the Klein-Gordon equation describes spin-0 particles. In Subsection 2.2.2 we give a transformation theoretical argument for the fact that the Dirac equation describes spin-1/2 particles.

The invariance of the Klein-Gordon equation under this transformation means that the mirror image of a physical Klein-Gordon process also represents a process that can be described by the Klein-Gordon equation.[10] Applying the active parity transformation to the free Klein-Gordon states $\phi_{\boldsymbol{p}}^{(1,2)}(\boldsymbol{x},t)$ yields

$$\phi_{\boldsymbol{p}}^{(1,2)}(\boldsymbol{x},t) \longrightarrow \lambda_P \phi_{\boldsymbol{p}}^{(1,2)}(-\boldsymbol{x},t) = \lambda_P \phi_{-\boldsymbol{p}}^{(1,2)}(\boldsymbol{x},t) \ .$$

On the particle level, this is in accordance with our expectation: the parity transformation reverses the momentum of a spin-0 particle.

Time reversal transformation *T*. Apart from Lorentz-like transformations there also exist non-Lorentz-like discrete symmetry operations, for example the *time reversal transformation*. The physical content of the time reversal can be explained best using the idea of a film. If a physical Klein-Gordon process is recorded by a camera, time reversal means that the film played backward also represents a series of physically realizable events.

The time reversal transformation, which should be better termed "motion reversal transformation", reverses all directions of motions and that of time, and therefore all spatial components of the four-momentum. In contrast, its 0th component remains unchanged due to $p_0 = p_0(\boldsymbol{p}^2)$. The same holds for the four-potential, since $\boldsymbol{A}$ is generated by moving currents and A^0 by charges. Thus, from the passive point of view, time reversal[11] (indicated by the symbol T) means

$$\left.\begin{array}{rcl} \boldsymbol{x} & \longrightarrow & \boldsymbol{x}' = \boldsymbol{x} \ , \ t \longrightarrow t' = -t \\[2mm] A^0(\boldsymbol{x},t) & \longrightarrow & A_T^0(\boldsymbol{x}',t') = A^0(\boldsymbol{x},t) \\[2mm] \boldsymbol{A}(\boldsymbol{x},t) & \longrightarrow & \boldsymbol{A}_T(\boldsymbol{x}',t') = -\boldsymbol{A}(\boldsymbol{x},t) \end{array}\right\} \begin{array}{c}\text{passive time}\\ \text{reversal } T\end{array} \qquad (1.34)$$

and

$$\mathrm{i}\hbar\partial^0 \longrightarrow \mathrm{i}\hbar\partial'^0 = -\mathrm{i}\hbar\partial^0 \ , \ \mathrm{i}\hbar\partial^i \longrightarrow \mathrm{i}\hbar\partial'^i = \mathrm{i}\hbar\partial^i \ .$$

In order to see how the wave function ϕ is transformed under time reversal, we start with the Klein-Gordon equation in the transformed (primed) system,

$$\left[\left(\mathrm{i}\hbar\partial'_\mu - \frac{e}{c}A_{T,\mu}\right)\left(\mathrm{i}\hbar\partial'^\mu - \frac{e}{c}A_T^\mu\right) - m_0^2c^2\right]\phi_T(x') = 0 \ , \qquad (1.35)$$

and express the differential and potential terms by the original quantities

---

[10] This analogy is not yet complete since a mirror merely reverses the component perpendicular to its plane. Only after an additional rotation by π around this vertical line one arrives at the parity transformation. However, the rotation is a proper Lorentz transformation and was already discussed above.

[11] We emphasize again that time reversal is not a Lorentz-like transformation. Strictly speaking, it is therefore not justified to call it a "passive transformation". The time reversal is not to be confused with the nonorthochronous Lorentz transformation of *time reflection*, to which we return in Exercise 3.

$$0 = \left[\left(-i\hbar\partial_0 - \frac{e}{c}A_0\right)\left(-i\hbar\partial^0 - \frac{e}{c}A^0\right) + \left(i\hbar\partial_i + \frac{e}{c}A_i\right)\left(i\hbar\partial^i + \frac{e}{c}A^i\right) - m_0^2c^2\right]\phi_T(x') .$$

As can easily be seen, this relationship leads to the Klein-Gordon equation in the unprimed system, formally identical to (1.35), if ϕ is transformed as

$$\left.\phi(\boldsymbol{x},t) \longrightarrow \phi_T(\boldsymbol{x}',t') = \lambda_T\phi^*(\boldsymbol{x},t) \ , \ |\lambda_T| = 1 \right\} \begin{array}{l}\text{passive time}\\ \text{reversal } T.\end{array} \tag{1.36}$$

Here the condition for λ_T reflects the fact that the twofold application of time reversal leads back to the original state. The active time reversal transformation follows from (1.34) and (1.36) as

$$\left.\begin{array}{rl}\phi(\boldsymbol{x},t) & \longrightarrow \phi_T(\boldsymbol{x},t) = \lambda_T\phi^*(\boldsymbol{x},-t)\\ A^0(\boldsymbol{x},t) & \longrightarrow A_T^0(\boldsymbol{x},t) = A^0(\boldsymbol{x},-t)\\ \boldsymbol{A}(\boldsymbol{x},t) & \longrightarrow \boldsymbol{A}_T(\boldsymbol{x},t) = -\boldsymbol{A}(\boldsymbol{x},-t)\end{array}\right\} \begin{array}{l}\text{active time}\\ \text{reversal } T.\end{array} \tag{1.37}$$

Applied to the free Klein-Gordon solutions, active time reversal yields

$$\phi_{\boldsymbol{p}}^{(1,2)}(\boldsymbol{x},t) \longrightarrow \lambda_T\phi_{\boldsymbol{p}}^{(1,2)*}(\boldsymbol{x},-t) = \lambda_T\phi_{-\boldsymbol{p}}^{(1,2)}(\boldsymbol{x},t) .$$

Like the parity transformation, the time reversal reverses the momentum of a spin-0 particle.

***PCT*-transformation (no symmetry transformation).** We now come to a central point, which is particularly important for the further development of both the one-particle interpretation in the next section as well as the relativistic scattering theory in Chapter 3. On the basis of the charge conjugation discussed in Subsection 1.1.3, the wave function of a physical spin-0 particle of charge $-e$ is obtained by starting with the negative Klein-Gordon solution $\phi^{(-)}$ with charge $+e$ and taking the charge conjugate $\phi_C^{(-)}$ of it. However, since space reversal P and time reversal T are symmetry transformations, we can equally use them for the construction of antiparticle wave functions. Thus, following Theorem 1.4 as well as (1.33) and (1.37), the combination of the three transformations P, T, and C, and their application to the negative solution $\phi^{(-)}$ yields in the active case (ignoring any phases)

$$\phi^{(-)}(x) \longrightarrow \phi_{PCT}^{(-)}(x) = \phi^{(-)}(-x) .$$

Since $\phi_{PCT}^{(-)}$ must be the wave function of an antiparticle because of the C-operation, an important statement, known as *Feynman-Stückelberg interpretation*, follows:

Theorem 1.5: Feynman-Stückelberg interpretation in the Klein-Gordon theory

Due to the *PCT*-transformation, the wave function of a physical spin-0 antiparticle of charge $-e$ can be interpreted as a negative Klein-Gordon solution of charge $+e$ moving backward in space and time.

We can easily prove the correctness of this interpretation by taking the eigenvalue equation for a negative Klein-Gordon state of charge $+e$ in the Hamilton formulation,

$$\left[\frac{\tau_3 + \mathrm{i}\tau_2}{2m_0}\left(\frac{\hbar}{\mathrm{i}}\boldsymbol{\nabla}_{\boldsymbol{x}} - \frac{e}{c}\boldsymbol{A}(x)\right)^2 + \tau_3 m_0 c^2 + eA^0(x)\right]\Psi^{(-)}(x)$$
$$= -|E|\Psi^{(-)}(x) ,$$

and applying the passive PCT-transformation to it. Taking into account

$$x'^\mu = -x^\mu , \ \boldsymbol{\nabla}_{\boldsymbol{x}'} = -\boldsymbol{\nabla}_{\boldsymbol{x}} , \ \Psi_{PCT}(x') = \tau_1\Psi(x) , \ A^\mu_{PCT}(x') = A^\mu(x) ,$$

this yields

$$\left[\frac{\tau_3 + \mathrm{i}\tau_2}{2m_0}\left(\frac{\hbar}{\mathrm{i}}\boldsymbol{\nabla}_{\boldsymbol{x}'} + \frac{e}{c}\boldsymbol{A}_{PCT}(x')\right)^2 + \tau_3 m_0 c^2 - eA^0_{PCT}(x')\right]\Psi^{(-)}_{PCT}(x')$$
$$= +|E|\Psi^{(-)}_{PCT}(x') ,$$

which is the eigenvalue equation for a positive Klein-Gordon state of charge $-e$ with reversed direction of motion in space and time.

The Feynman-Stückelberg interpretation has two important consequences. First, it offers a way to describe antiparticles and particularly their most likely quantum states by the original negative solutions (and not only by their charge conjugates). This fact is utilized in the next section for the definition of physically meaningful expectation values in the sense of the one-particle interpretation. The other consequence is that its application gives a great advantage in describing relativistic scattering processes (Chapter 3) and leads to experimentally verifiable results.

Extended charge conjugation $\mathcal{C}$. The charge conjugation C is a mathematical equivalence operation but not a symmetry transformation as it changes the formal shape of the Klein-Gordon equation [compare (1.22) with (1.23) and (1.24) with (1.25)]. However, we can extend it to a non-Lorentz-like symmetry transformation, if we additionally change the sign of the electromagnetic potentials:

$$\left.\begin{array}{l} \phi(\boldsymbol{x},t) \longrightarrow \phi_{\mathcal{C}}(\boldsymbol{x},t) = \lambda_{\mathcal{C}}\phi^*(\boldsymbol{x},t) , \ |\lambda_{\mathcal{C}}| = 1 \\ A^0(\boldsymbol{x},t) \longrightarrow A^0_{\mathcal{C}}(\boldsymbol{x},t) = -A^0(\boldsymbol{x},t) \\ \boldsymbol{A}(\boldsymbol{x},t) \longrightarrow \boldsymbol{A}_{\mathcal{C}}(\boldsymbol{x},t) = -\boldsymbol{A}(\boldsymbol{x},t) \end{array}\right\} \begin{array}{c}\text{active charge}\\ \text{conjugation } \mathcal{C}.\end{array} \qquad (1.38)$$

We also call this extended transformation *charge conjugation* and introduce the new symbol $\mathcal{C}$ to distinguish it from the former C-transformation. As before, the constraint of $\lambda_{\mathcal{C}}$ honors the fact that twofold application of $\mathcal{C}$ leads back to the original state.

On the level of wave functions the effect of $\mathcal{C}$ is, for example, that the Klein-Gordon equation for a positive solution $\phi^{(+)}$ of charge $+e$ within the potential $+A^\mu$,

$$\left[\left(\mathrm{i}\hbar\frac{\partial}{\partial t} - eA^0\right)^2 - c^2\left(\boldsymbol{p} - \frac{e}{c}\boldsymbol{A}\right)^2 - m_0^2c^4\right]\phi^{(+)}(x) = 0 ,$$

is transformed into the Klein-Gordon equation for a negative solution $\phi_C^{(+)} = \phi^{(+)*}$ with the same charge $+e$ within the potential $A_C^\mu = -A^\mu$, i.e.

$$\left[\left(\mathrm{i}\hbar\frac{\partial}{\partial t} - eA_C^0\right)^2 - c^2\left(\boldsymbol{p} - \frac{e}{c}\boldsymbol{A}_C\right)^2 - m_0^2c^4\right]\phi_C^{(+)}(x) = 0 .$$

However, due to the original C transformation, the latter can be identified with the Klein-Gordon equation for a positive solution $\phi_{CC}^{(+)}$ of charge $-e$ within the potential $-A^\mu$ having the same quantum numbers as the original solution $\phi^{(+)}$.

Therefore, on the physical particle level, the charge conjugation $\mathcal{C}$ means the conversion of a boson into an antiboson with opposite charge and otherwise identical quantum numbers. Consequently, the physical content of the symmetry transformation $\mathcal{C}$ can be described by the classically comprehensible fact that the dynamics of a boson with charge $+e$ within the potential $+A^\mu$ is exactly the same as that of the corresponding antiboson of charge $-e$ within the potential $-A^\mu$.

Further symmetry considerations. We have now discussed all fundamental symmetry transformations of the Klein-Gordon equation. They all mean that the original and the corresponding transformed situations are equivalent. Therefore, arbitrary combinations of symmetry transformations always lead to physically equivalent constellations provided, of course, that the underlying theory yields a correct description of the physics under consideration.

As experiments have shown, the discrete symmetries P, $\mathcal{C}$, and T are indeed realized in nature, both with respect to electromagnetic and strong interaction phenomena which supports the correctness of the Klein-Gordon theory for the description of spin-0 particles. However, this is no longer true for physical processes containing weak interactions where each of the three symmetries is violated. On the other hand, within modern quantum field theories (with any type of interaction), there follows from Lorentz invariance and the usual relationship between spin and statistics the so-called *PCT-theorem.* It states that the three-way combination of P, $\mathcal{C}$, and T is always a symmetry transformation (see Exercise 3). The *PCT*-theorem implies (among other things) that particle and antiparticle have the same life time.

Summary

- Transformations can be classified into **active** and **passive transformations.** At active transformations the physical state, and at passive transformations the underlying basis system is transformed.

▷

- **Symmetry transformations** lead to formally equivalent equations of motions and thus to physically equivalent situations.
- The Klein-Gordon theory is invariant under the full Poincaré group. Discrete symmetry transformations of the theory are the improper Lorentz transformations of **space reflection** P, the non-Lorentz-like transformations of **time reversal** T, and the extended **charge conjugation** $\mathcal{C}$.
- The wave function of a spin-0 antiparticle of charge $-e$ can be interpreted as a negative Klein-Gordon solution of charge $+e$, moving backward in space and time (**Feynman-Stückelberg interpretation**).

Exercises

3. Lorentz behavior of the *PCT*-symmetry transformation (I). Show that the PCT-transformation[12] is a Lorentz-like symmetry operation by considering the improper and nonorthochronous Lorentz transformation of *time reflection.* The latter is also called *Racah time reflection* and is defined via

$$(\Lambda^{\mu}{}_{\nu}) = \begin{pmatrix} -1 & 0 & 0 & 0 \\ 0 & 1 & 0 & 0 \\ 0 & 0 & 1 & 0 \\ 0 & 0 & 0 & 1 \end{pmatrix} \;,\; \Lambda^0{}_0 < 0 \;,\; \det(\Lambda^{\mu}{}_{\nu}) = -1 \;.$$

Solution. For the passive and active transformation laws of the Racah time reflection (denoted by the symbol R), one obtains

$$\left.\begin{aligned} \boldsymbol{x} &\longrightarrow \boldsymbol{x}' = \boldsymbol{x} \;,\; t \longrightarrow t' = -t \\ \phi(\boldsymbol{x},t) &\longrightarrow \phi_R(\boldsymbol{x}',t') = \lambda_R \phi(\boldsymbol{x},t) \;,\; \lambda_R = \pm 1 \\ A^0(\boldsymbol{x},t) &\longrightarrow A^0_R(\boldsymbol{x}',t') = -A^0(\boldsymbol{x},t) \\ \boldsymbol{A}(\boldsymbol{x},t) &\longrightarrow \boldsymbol{A}_R(\boldsymbol{x}',t') = \boldsymbol{A}(\boldsymbol{x},t) \end{aligned}\right\} \begin{array}{c}\text{passive}\\ \text{time}\\ \text{reflection } R\end{array}$$

and

$$\left.\begin{aligned} \phi(\boldsymbol{x},t) &\longrightarrow \phi_R(\boldsymbol{x},t) = \lambda_R \phi(\boldsymbol{x},-t) \\ A^0(\boldsymbol{x},t) &\longrightarrow A^0_R(\boldsymbol{x},t) = -A^0(\boldsymbol{x},-t) \\ \boldsymbol{A}(\boldsymbol{x},t) &\longrightarrow \boldsymbol{A}_R(\boldsymbol{x},t) = \boldsymbol{A}(\boldsymbol{x},-t) \end{aligned}\right\} \begin{array}{c}\text{active}\\ \text{time}\\ \text{reflection } R.\end{array}$$

Comparing the last relations with (1.37) and (1.38), i.e. the active transformation laws of time reversal T and charge conjugation $\mathcal{C}$, it becomes apparent that the combination of $\mathcal{C}$ and T is identical to the Racah time reflection:

[12] Not to be confused with the PCT-transformation.

$$CT = R \Longrightarrow PCT = PR \ .$$

Since P and R are Lorentz-like, it follows that the PCT-transformation is Lorentz-like, too.

1.3 One-Particle Interpretation of the Klein-Gordon Theory

Based on the preceding results, particularly those in Subsections 1.1.3 and 1.2.3, we are now in a position to complete the one-particle interpretation of the Klein-Gordon theory. The essential points that need to be clarified have already been mentioned in Subsection 1.1.3, namely,

[1] under which conditions is a decoupling of the Klein-Gordon theory into two one-particle theories possible,

[2] which operators are appropriate for the one-particle concept and how can they be constructed,

[3] how can physically meaningful and picture- as well as representation-independent one-particle expectation values be defined?

In our subsequent discussion, we first concentrate on point [3] and define a *generalized scalar product* that complies with the findings in Theorem 1.5. This definition also requires a generalization of the terms "Hermitecity" and "unitarity" in order to ensure the picture-independence of the generalized scalar product. Based on this formalism, we then turn to question [2]. Finally, we carry out a thorough discussion of point [1], trace out the range of validity of the one-particle concept and outline inconsistencies outside this range.

For clarity and simplicity and particularly on account of the close affinity to nonrelativistic quantum mechanics, we preferentially use the Hamilton formulation of the Klein-Gordon theory.

1.3.1 Generalized Scalar Product

As already pointed out in Subsection 1.1.3, in the nonrelativistic approximation we have for positive Klein-Gordon solutions $|\varphi| \gg |\chi|$ (and for negative solutions $|\varphi| \ll |\chi|$; see Subsection 1.4.1). At least in this validity range of the one-particle concept (to be defined later in more detail) the charge density for positive solutions is positive definite and can therefore be interpreted as a probability density:

$$\rho = \psi^\dagger \tau_3 \psi \overset{|\varphi| \gg |\chi|}{\approx} \psi^\dagger \psi \ , \quad \int \mathrm{d}^3 x \rho(x) = +1 \ .$$

From this follows that the most likely quantum state of a spin-0 (anti)particle with respect to an observable $\mathcal{O}$ can be described by the usual expectation value of the nonrelativistic theory. For particles with charge $+e$, we have

$$\left.\begin{aligned}\left\langle \psi^{(+)} \middle| \mathcal{O} \middle| \psi^{(+)} \right\rangle &= \int \mathrm{d}^3x \psi^{(+)\dagger}(x) \mathcal{O} \psi^{(+)}(x) \\ \left\langle \psi^{(+)} \middle| \psi^{(+)} \right\rangle &= +1 \ .\end{aligned}\right\} \tag{1.39}$$

In the case of antiparticles of charge $-e$, it follows, due to the charge conjugation C and Theorem 1.5, that

$$\begin{aligned}\left\langle \psi_C^{(-)} \middle| \mathcal{O} \middle| \psi_C^{(-)} \right\rangle &= \int \mathrm{d}^3x \psi_C^{(-)\dagger}(x) \mathcal{O} \psi_C^{(-)}(x) \ , \left\langle \psi_C^{(-)} \middle| \psi_C^{(-)} \right\rangle = +1 \\ &= \int \mathrm{d}^3x \psi^{(-)\dagger}(-x) \mathcal{O} \psi^{(-)}(-x) \\ &= -\int \mathrm{d}^3x \psi^{(-)\dagger}(x) \mathcal{O} \psi^{(-)}(x) \\ &= -\left\langle \psi^{(-)} \middle| \mathcal{O} \middle| \psi^{(-)} \right\rangle \ , \ \left\langle \psi^{(-)} \middle| \psi^{(-)} \right\rangle = -1 \ .\end{aligned} \tag{1.40}$$

The third relation reflects the fact that the reversal of all space and time directions (PCT-transformation) causes the reversal of all eigenvalues and consequently of all expectation values of a wave function. Combining (1.39) and (1.40) finally leads to the *generalized expectation value*

$$\langle \psi | \mathcal{O} | \psi \rangle_{\mathrm{G}} = \int \mathrm{d}^3x \psi(x) \tau_3 \mathcal{O} \psi(x) \ , \tag{1.41}$$

where, in the case of particles with charge $+e$, positive solutions $\psi^{(+)}$ with $\langle \psi^{(+)} | \psi^{(+)} \rangle_{\mathrm{G}} = Q^{(+)} = +1$ and, in the case of antiparticles with charge $-e$, negative solutions $\psi^{(-)}$ with $\langle \psi^{(-)} | \psi^{(-)} \rangle_{\mathrm{G}} = Q^{(-)} = -1$ are to be inserted.

Obviously, this result explains and generalizes the third point from Theorem 1.4, namely, that quantum states of antiparticles can be described by negative Klein-Gordon solutions and not only by their charge conjugates. In this way, we arrive at a desirable symmetry in the description of particles and antiparticles enabling us to regard – with certain reservations – the negative solutions $\psi^{(-)}$ as antiparticle wave functions.[13]

All in all, (1.41) leads to a novel definition of the scalar product that follows, like in the Schrödinger theory, the form of ρ and which has already been anticipated in Theorems 1.2 and 1.3 for the case of free particles (compare to Exercise 2).

---

[13] However, from this another asymmetry follows. For a positive eigensolution of an operator its eigenvalues and expectation values are identical, whereas for a negative eigensolution these quantities are opposites. Therefore, when using (1.41), the third principle of Theorem 1.1 can no longer be sustained. This weak point can be circumvented by adhering to the definition of the nonrelativistic expectation value for both positive and negative solutions and, in return, redefining physical observables in an appropriate way. We do not address this possibility in the following.

Definition: Generalized scalar product

The *generalized scalar product (G-scalar product)* of the Klein-Gordon theory is defined as

$$\langle\psi|\,\phi\rangle_{\mathrm{G}} = \int \mathrm{d}^3x\psi^\dagger(x)\tau_3\phi(x) \ .$$

Two states ψ and ϕ are called *generalized orthogonal* (*G-orthogonal*) if $\langle\psi|\,\phi\rangle_{\mathrm{G}} = 0$. The expression

$$\langle\mathcal{O}\rangle_{\mathrm{G}} = \langle\psi|\,\mathcal{O}\,|\psi\rangle_{\mathrm{G}} = \int \mathrm{d}^3x\psi^{(\pm)\dagger}(x)\tau_3\mathcal{O}\psi(x) \ , \ Q = \langle\psi|\,\psi\rangle_{\mathrm{G}} = \pm 1$$

is the *generalized expectation value (G-expectation value)* of the observable $\mathcal{O}$ denoting the statistical average of many similar measurements of $\mathcal{O}$ on identical spin-0-[anti]particle systems of charge $+e$ $[-e]$. Positive Klein-Gordon solutions $\psi^{(+)}$ with $Q^{(+)} = +1$ are inserted for particles, and negative solutions $\psi^{(-)}$ with $Q^{(-)} = -1$ for antiparticles.

As can easily be shown, $\langle\psi|\,\phi\rangle_{\mathrm{G}}$ has the same properties as $\langle\psi|\,\phi\rangle$ except for positive definiteness, namely

- $\langle\psi|\,\phi+\chi\rangle_{\mathrm{G}} = \langle\psi|\,\phi\rangle_{\mathrm{G}} + \langle\psi|\,\chi\rangle_{\mathrm{G}}$
- $\langle\psi|\,a\phi\rangle_{\mathrm{G}} = a\,\langle\psi|\,\phi\rangle_{\mathrm{G}}$
- $\langle\psi|\,\phi\rangle_{\mathrm{G}} = \langle\phi|\,\psi\rangle_{\mathrm{G}}^*$.

Relating $\langle\psi|\,\mathcal{O}\,|\phi\rangle_{\mathrm{G}}$ back to $\langle\psi|\,\mathcal{O}\,|\phi\rangle$ gives the following adjunction relation corresponding to (1.21):

$$\langle\psi|\,\mathcal{O}\,|\phi\rangle_{\mathrm{G}} = \langle\psi|\,\tau_3\mathcal{O}\,|\phi\rangle = \langle\phi|\,\mathcal{O}^\dagger\tau_3\,|\psi\rangle^* = \langle\phi|\,\tau_3\mathcal{O}^\dagger\tau_3\,|\psi\rangle_{\mathrm{G}}^* \ .$$

From this follows immediately that an operator $\mathcal{O}$ with $\mathcal{O} = \tau_3\mathcal{O}^\dagger\tau_3$ has real G-expectation values. In Exercise 4 we show that such an operator also has the following properties:

- The eigenvalues of charged eigenstates ψ (with $\langle\psi|\,\psi\rangle_{\mathrm{G}} \neq 0$) are real.
- Charged eigenstates corresponding to different eigenvalues are G-orthogonal.

Thus, with a view to the Hermitean operators in nonrelativistic quantum mechanics, we can formulate the following:

Definition: Generalized Hermitean operator

A linear operator $\mathcal{O}$ is called *generalized Hermitean (G-Hermitean)* if

$$\mathcal{O} = \tau_3\mathcal{O}^\dagger\tau_3 \Longleftrightarrow \tau_3\mathcal{O} = (\tau_3\mathcal{O})^\dagger \ ,$$

i.e. if $\tau_3\mathcal{O}$ is an Hermitean operator. Such an operator has real G-expectation values.

▷

With respect to the one-particle interpretation, only those G-Hermitean operators represent physically meaningful quantities whose eigenstates form a complete system, i.e. in which each Klein-Gordon state with a finite G-norm can be expanded (*one-particle operators*).

From this follows generally (not only for the free case) that charged eigenstates of H and $\boldsymbol{p}$ are G-orthogonal since H and $\boldsymbol{p}$ are G-Hermitean operators.

Apart from G-Hermitean operators, the G-scalar product leads to another important class of operators, namely the transformation operators which leave the scalar product invariant. To this we consider the operator U as well as the transformation

$$\psi' = U\psi \ , \ \phi' = U\phi$$

and require that

$$\langle \psi' | \phi' \rangle_{\mathrm{G}} = \langle \psi' | \tau_3 | \phi' \rangle = \langle \psi | U^\dagger \tau_3 U | \phi \rangle = \langle \psi | \tau_3 U^\dagger \tau_3 U | \phi \rangle_{\mathrm{G}} \stackrel{!}{=} \langle \psi | \phi \rangle_{\mathrm{G}} \, .$$

Therefore, we define:

Definition: Generalized unitary operator

A linear transformation operator U is called *generalized unitary (G-unitary)* if

$$\tau_3 U^\dagger \tau_3 = U^{-1} \, .$$

Such an operator leaves the G-scalar product invariant.

G-unitary operators also possess the following properties analogous to those of unitary operators (see Exercise 4):

- The product of two G-unitary operators is also a G-unitary operator.
- If U describes an infinitesimal G-unitary transformation $U = 1 + \mathrm{i}\epsilon\mathcal{O}$ with $|\epsilon| \ll 1$, then, $\mathcal{O}$ is G-Hermitean.
- If $\mathcal{O}$ is G-Hermitean, then, $\mathrm{e}^{\mathrm{i}\mathcal{O}}$ is G-unitary.

Because of the last property, $\mathrm{e}^{\mathrm{i}H}$ is a G-unitary operator, since H is G-Hermitean. This means that the chosen definition of the G-scalar product also ensures its picture-independence; using $\mathrm{e}^{\mathrm{i}H}$ (or $\mathrm{e}^{\mathrm{i}Ht/\hbar}$) we can, as usual, switch between different pictures (Schrödinger picture, Heisenberg picture etc.), all of them giving equivalent descriptions of the Klein-Gordon theory with respect to the generalizations discussed.

1.3.2 One-particle Operators and Feshbach-Villars Representation

Having solved problem [3] through the definition of the G-scalar product, we now turn to the generalized form of the Ehrenfest theorem in order to tackle the question which G-Hermitean operators represent physically meaningful one-particle operators in the sense of the definition of page 32 and how they can be constructed (point [2]).

Generalized Ehrenfest theorem. Our starting point is the Heisenberg equation known from nonrelativistic quantum mechanics and clearly also valid in the relativistic case:

$$\mathrm{i}\hbar\frac{\mathrm{d}\mathcal{O}_{\mathrm{H}}}{\mathrm{d}t} = [\mathcal{O}_{\mathrm{H}}, H_{\mathrm{H}}] + \mathrm{i}\hbar\frac{\partial\mathcal{O}_{\mathrm{H}}}{\partial t} \;.$$

It represents an operator equation in the Heisenberg picture which corresponds to the Hamiltonian Klein-Gordon equation. The states and operators in the Schrödinger picture (without index) and in the Heisenberg picture (with index H) are connected via the representation-independent relations

$$|\psi_{\mathrm{H}}\rangle = \mathrm{e}^{-\mathrm{i}\hbar H(t-t_0)}\,|\psi(t)\rangle = |\psi(t_0)\rangle \;,\; \mathcal{O}_{\mathrm{H}} = \mathrm{e}^{-\mathrm{i}\hbar H(t-t_0)}\mathcal{O}\mathrm{e}^{\mathrm{i}\hbar H(t-t_0)} \;.$$

Multiplying the Heisenberg equation by τ_3 from the left and taking into account $\mathrm{d}\,|\psi_{\mathrm{H}}\rangle\,/\mathrm{d}t = 0$, we obtain the (picture- and representation-independent) *Heisenberg equation for G-expectation values*

$$\frac{\mathrm{d}\,\langle\mathcal{O}\rangle_{\mathrm{G}}}{\mathrm{d}t} = \frac{1}{\mathrm{i}\hbar}\,\langle[\mathcal{O},H]\rangle_{\mathrm{G}} + \left\langle\frac{\partial\mathcal{O}}{\partial t}\right\rangle_{\mathrm{G}}$$

as well as – for explicitly time-independent operators ($\partial\mathcal{O}/\partial t = 0$) – the *generalized Ehrenfest theorem*

$$\frac{\mathrm{d}\,\langle\mathcal{O}\rangle_{\mathrm{G}}}{\mathrm{d}t} = \frac{1}{\mathrm{i}\hbar}\,\langle[\mathcal{O},H]\rangle_{\mathrm{G}} \;. \tag{1.42}$$

Without the index G, both relations are also valid in the nonrelativistic theory. There the Ehrenfest theorem also implies the formal equivalence with the Hamilton equations of classical mechanics where the classical quantities are replaced by their mean values. Examples of this kind are [with $H = \boldsymbol{p}^2/2m_0 + V(\boldsymbol{x})$]

$$\frac{\mathrm{d}\,\langle\boldsymbol{p}\rangle}{\mathrm{d}t} = \frac{1}{\mathrm{i}\hbar}\,\langle[\boldsymbol{p},H]\rangle = -\,\langle\boldsymbol{\nabla}V\rangle \longleftrightarrow \frac{\mathrm{d}\boldsymbol{p}}{\mathrm{d}t} = -\frac{\partial H}{\partial\boldsymbol{x}} = -\boldsymbol{\nabla}V$$

$$\frac{\mathrm{d}\,\langle\boldsymbol{x}\rangle}{\mathrm{d}t} = \frac{1}{\mathrm{i}\hbar}\,\langle[\boldsymbol{x},H]\rangle = \left\langle\frac{\boldsymbol{p}}{m_0}\right\rangle \longleftrightarrow \frac{\mathrm{d}\boldsymbol{x}}{\mathrm{d}t} = \frac{\partial H}{\partial\boldsymbol{p}} = \frac{\boldsymbol{p}}{m_0} \;.$$

In the relativistic Klein-Gordon theory this correspondence principle does not hold generally. For example, in the free case, we have the operator equation

$$\frac{\mathrm{d}\,\langle\boldsymbol{x}\rangle_{\mathrm{G}}}{\mathrm{d}t} = \frac{1}{\mathrm{i}\hbar}\left\langle[\boldsymbol{x},H^{(0)}]\right\rangle_{\mathrm{G}} = \left\langle\frac{(\tau_3+\mathrm{i}\tau_2)\boldsymbol{p}}{m_0}\right\rangle_{\mathrm{G}} \;,\; H^{(0)} \text{ from (1.16)} \;,$$

whereas the classical relationship is given by

$$\frac{\mathrm{d}\boldsymbol{x}}{\mathrm{d}t} = \frac{\boldsymbol{p}}{m} = \frac{c\boldsymbol{p}}{p_0} \;,\; m = m_0/\sqrt{1-v^2/c^2} \;. \tag{1.43}$$

As one can see, the right hand sides of both equations are different. Another important difference to the nonrelativistic theory is that the "velocity operator"

$$\boldsymbol{v} = \frac{(\tau_3 + \mathrm{i}\tau_2)\boldsymbol{p}}{m_0} \tag{1.44}$$

is G-Hermitean but does not represent an observable or a one-particle operator in the sense of the definition of page 32 since the eigenvalues of the matrix $\tau_3 + \mathrm{i}\tau_2$ are zero. Furthermore and against our expectation, the operator $\boldsymbol{v}$ is not constant for free particles , due to $[\boldsymbol{v}, H^{(0)}] \neq 0$. From all of this we must conclude that, in the Klein-Gordon theory, not all operators (e.g. constructed via the generalized Ehrenfest theorem) are physically sensible.

The reason for this phenomenon lies in the one-particle concept: according to the definition of one-particle operators, only even operators are allowed that do not mix positive and negative states since only these may possess a complete basis system.[14] Since each operator $\mathcal{O}$ is divisible into an even and an odd operator,

$$\mathcal{O} = [\mathcal{O}] + \{\mathcal{O}\} \;,\; [\mathcal{O}] = \text{even} \;,\; \{\mathcal{O}\} = \text{odd} \;,$$

its even part, i.e. the sought one-particle operator, can be separated.

Obviously, the free Hamilton operator $H^{(0)}$ and the momentum operator $\boldsymbol{p}$ are one-particle operators, since they possess the positive and negative states (1.18) as a (common) eigenbasis. In contrast, the position operator $\boldsymbol{x}$ is not an even operator, as $[\boldsymbol{x}, H^{(0)}]$ is not even. As discussed in the next section, the full Hamilton operator H from (1.17) is not even, either.

Feshbach-Villars representation. The general investigation of even and odd operators becomes much simplified in a specific representation (in a specific basis system), where the positive and negative states are of the form

$$\psi^{(+)} \sim \begin{pmatrix} 1 \\ 0 \end{pmatrix} \;,\; \psi^{(-)} \sim \begin{pmatrix} 0 \\ 1 \end{pmatrix} \;.$$

This is because here the even operator $[\mathcal{O}]$ of an operator $\mathcal{O}$ is its diagonal part:

$$\mathcal{O} = \begin{pmatrix} \mathcal{O}_{11} & \mathcal{O}_{12} \\ \mathcal{O}_{21} & \mathcal{O}_{22} \end{pmatrix} \;,\; [\mathcal{O}] = \begin{pmatrix} \mathcal{O}_{11} & 0 \\ 0 & \mathcal{O}_{22} \end{pmatrix} \;,\; \{\mathcal{O}\} = \begin{pmatrix} 0 & \mathcal{O}_{12} \\ \mathcal{O}_{21} & 0 \end{pmatrix} \;.$$

The transition to such a representation is equivalent to the diagonalization of the Hamilton operator which, in general, is only possible approximately (see the next section). However, in the free case the diagonalization can be carried

[14] An operator $\mathcal{O}$ is called even if $\mathcal{O}\psi^{(\pm)} = \psi'^{(\pm)}$, where $\psi^{(\pm)}$ and $\psi'^{(\pm)}$ denote arbitrary positive (+) and negative (−) states. If, on the other hand, $\mathcal{O}\psi^{(\pm)} = \psi'^{(\mp)}$, the operator $\mathcal{O}$ is called odd.

out exactly and leads to the so-called *Feshbach-Villars representation (FV-representation)*. The transition from the Schrödinger momentum representation (in which $\boldsymbol{p}$ and p_0 are $\mathbb{C}$-numbers) to the corresponding Feshbach-Villars representation is facilitated by the G-unitary transformation operator

$$U = \frac{(m_0c + p_0) - \tau_1(m_0c - p_0)}{2\sqrt{m_0cp_0}} \;,\; U^{-1} = \frac{(m_0c + p_0) + \tau_1(m_0c - p_0)}{2\sqrt{m_0cp_0}} \;,$$

since we have

$$\tilde{\Psi}^{(1)} = U\Psi^{(1)}(\boldsymbol{p}) = \begin{pmatrix} 1 \\ 0 \end{pmatrix} \;,\; \tilde{\Psi}^{(2)} = U\Psi^{(2)}(\boldsymbol{p}) = \begin{pmatrix} 0 \\ 1 \end{pmatrix}$$

and (see Exercise 5)

$$\tilde{H}^{(0)} = UH^{(0)}U^{-1} = cp_0\tau_3 \;,\; \tilde{H}^{(0)}\tilde{\Psi}^{(1,2)} = \pm cp_0\tilde{\Psi}^{(1,2)} \tag{1.45}$$

as well as

$$\tilde{\boldsymbol{p}} = U\boldsymbol{p}U^{-1} = UU^{-1}\boldsymbol{p} = \boldsymbol{p} \;.$$

The last relations show, once again, that $H^{(0)}$ and $\boldsymbol{p}$ are even operators, i.e. $H^{(0)} = [H^{(0)}]$, $\boldsymbol{p} = [\boldsymbol{p}]$. Furthermore, contrary to $H^{(0)}$, the operator $\tilde{H}^{(0)}$ is Hermitean.

One-particle operators for position and velocity. With the help of the Feshbach-Villars representation and the transformation operator U, we are now able to determine the *one-particle position operator* from $\boldsymbol{x}$ as well as the *one-particle velocity operator* from (1.44) for the free case. To this end, it is useful to distinguish explicitly between the following representations (in the Schrödinger picture):

- Position representation: in this representation (which we have mostly been working in so far), the position operator $\boldsymbol{x}$ is given by the $\mathbb{C}$-number $\boldsymbol{x}$, the momentum operator $\boldsymbol{p}$ by $-\mathrm{i}\hbar\boldsymbol{\nabla}$, and the Hamilton operator $H^{(0)}$ by (1.16).
- Momentum representation: here $\boldsymbol{x}$ is given by $\mathrm{i}\hbar\boldsymbol{\nabla}_{\boldsymbol{p}}$ and $\boldsymbol{p}$ by the $\mathbb{C}$-number $\boldsymbol{p}$.
- FV-momentum representation: this is obtained by diagonalizing $H^{(0)}$ in momentum representation as shown in (1.45).[15]

Let us now consider the position operator in momentum representation,

$$\boldsymbol{x} = \mathrm{i}\hbar\boldsymbol{\nabla}_{\boldsymbol{p}} \;.$$

In FV-momentum representation, it becomes (see Exercise 5)

[15] Note that we are dealing with two different categories of representations that are combined. In the first category $\boldsymbol{x}$ or $\boldsymbol{p}$ are diagonal (position or momentum representation). In the second one the Hamilton operator $H^{(0)}$ is diagonal and results essentially from a rotation in τ-space (Feshbach-Villars representation).

$$\tilde{\boldsymbol{x}} = \mathrm{i}\hbar U \boldsymbol{\nabla}_{\boldsymbol{p}} U^{-1} = \mathrm{i}\hbar \boldsymbol{\nabla}_{\boldsymbol{p}} - \mathrm{i}\hbar \frac{\tau_1 \boldsymbol{p}}{2p_0^2} . \tag{1.46}$$

Since τ_1 is nondiagonal, the one-particle position operator follows as

$$[\tilde{\boldsymbol{x}}] = \mathrm{i}\hbar \boldsymbol{\nabla}_{\boldsymbol{p}} . \tag{1.47}$$

It is the canonical conjugate to the momentum operator since we have

$$[[\tilde{\boldsymbol{x}}]_i, [\tilde{\boldsymbol{p}}]_j] = \mathrm{i}\hbar \left[\frac{\partial}{\partial p_i}, p_j\right] = \mathrm{i}\hbar\delta_{ij} ,$$

as in nonrelativistic quantum mechanics. Inverting the transformation U, one obtains from (1.47) the one-particle position operator in momentum representation (see Exercise 5):

$$[\boldsymbol{x}] = \mathrm{i}\hbar U^{-1} \boldsymbol{\nabla}_{\boldsymbol{p}} U = \mathrm{i}\hbar \boldsymbol{\nabla}_{\boldsymbol{p}} + \mathrm{i}\hbar \frac{\tau_1 \boldsymbol{p}}{2p_0^2} , \; \boldsymbol{p} = \mathbb{C}\text{-number} , \tag{1.48}$$

and from this the one-particle position operator in position representation:

$$[\boldsymbol{x}] = \boldsymbol{x} + \mathrm{i}\hbar \frac{\tau_1 \boldsymbol{p}}{2p_0^2} , \; \boldsymbol{p} = -\mathrm{i}\hbar \boldsymbol{\nabla} .$$

To determine the one-particle velocity operator $[\tilde{\boldsymbol{v}}]$ in FV-momentum representation, we use the generalized Ehrenfest theorem (1.42) with $[\tilde{\boldsymbol{x}}]$ from (1.47) and $[\tilde{H}^{(0)}] = \tilde{H}^{(0)} = cp_0\tau_3$ to find

$$\langle [\tilde{\boldsymbol{v}}] \rangle_{\mathrm{G}} = \frac{\mathrm{d}\langle [\tilde{\boldsymbol{x}}] \rangle_{\mathrm{G}}}{\mathrm{d}t} = \frac{1}{\mathrm{i}\hbar} \left\langle \left[[\tilde{\boldsymbol{x}}], \tilde{H}^{(0)} \right] \right\rangle_{\mathrm{G}} = \left\langle \tau_3 \frac{c\boldsymbol{p}}{p_0} \right\rangle_{\mathrm{G}}$$

$$\Longrightarrow [\tilde{\boldsymbol{v}}] = \frac{c\boldsymbol{p}\tau_3}{p_0} , \; \boldsymbol{p} = \mathbb{C}\text{-number} .$$

Thus, in FV-momentum representation and in the case of positive states, we have the same relations between the one-particle velocity and momentum operators as in classical relativistic mechanics. For negative states, this is only true with respect to the absolute amounts. The one-particle velocity operator in momentum representation is calculated from the last relation as (see Exercise 5)

$$[\boldsymbol{v}] = U^{-1} \frac{c\boldsymbol{p}\tau_3}{p_0} U = \frac{(\tau_3 + \mathrm{i}\tau_2)\boldsymbol{p}}{2m_0} + \frac{(\tau_3 - \mathrm{i}\tau_2) m_0 c^2 \boldsymbol{p}}{2p_0^2} , \; \boldsymbol{p} = \mathbb{C}\text{-number} . \tag{1.49}$$

In position representation this finally yields

$$[\boldsymbol{v}] = \frac{(\tau_3 + \mathrm{i}\tau_2)\boldsymbol{p}}{2m_0} + \frac{(\tau_3 - \mathrm{i}\tau_2) m_0 c^2 \boldsymbol{p}}{2p_0^2} , \; \boldsymbol{p} = -\mathrm{i}\hbar \boldsymbol{\nabla} .$$

This operator has a common eigenbasis with $H^{(0)}$ and $\boldsymbol{p}$, namely, the positive and negative Klein-Gordon solutions $\psi_{\boldsymbol{p}}^{(r)}$ from (1.18) with the energy eigenvalues $\epsilon_r c p_0$, the momentum eigenvalues $\epsilon_r \boldsymbol{p}$, and the one-particle velocity eigenvalue $c\boldsymbol{p}/p_0$.

Theorem 1.6: One-particle operators and FV-representation in the Klein-Gordon theory

In the one-particle interpretation of the Klein-Gordon theory only those G-Hermitean operators that do not mix positive and negative states (compare to the definition of page 32) are sensible descriptions of physical quantities. In the free case the corresponding Hamilton operator $H^{(0)}$ can be diagonalized using the G-unitary Feshbach-Villars transformation

$$U = \frac{(m_0c + p_0) - \tau_1(m_0c - p_0)}{2\sqrt{m_0cp_0}} ,$$

which leads to the Feshbach-Villars representation. Here the even part of an operator can be determined very easily as it is simply given by its diagonal part.

Contrary to $H^{(0)}$ and $\boldsymbol{p}$, the position operator $\boldsymbol{x}$ and the velocity operator $\boldsymbol{v}$ are not even. Transforming them into the Feshbach-Villars representation, separating their diagonal parts, and subsequently transforming the latter back, one finds the one-particle position operator $[\boldsymbol{x}]$ and the one-particle velocity operator $[\boldsymbol{v}]$ in the usual position or momentum representation to be

$$[\boldsymbol{x}] = \boldsymbol{x} + \mathrm{i}\hbar\frac{\tau_1\boldsymbol{p}}{2p_0^2} , \ [\boldsymbol{v}] = \frac{(\tau_3 + \mathrm{i}\tau_2)\boldsymbol{p}}{2m_0} + \frac{(\tau_3 - \mathrm{i}\tau_2)m_0c^2\boldsymbol{p}}{2p_0^2} .$$

In this context note the following: although the operator $[\boldsymbol{v}]$ seems to be acceptable within the one-particle interpretation, it contains a flaw, since in position representation, we have:

$$[\boldsymbol{p}]\psi_{\boldsymbol{p}}^{(1,2)}(x) = \pm\boldsymbol{p}\psi_{\boldsymbol{p}}^{(1,2)}(x) , \ [\boldsymbol{v}]\psi_{\boldsymbol{p}}^{(1,2)}(x) = +\frac{c\boldsymbol{p}}{p_0}\psi_{\boldsymbol{p}}^{(1,2)}(x) .$$

This means that, for negative solutions $\psi_{\boldsymbol{p}}^{(2)}$, the eigenvalue (or G-expectation value) of $[\boldsymbol{v}]$ is opposed to the eigenvalue (or G-expectation value) of $[\boldsymbol{p}]$. The reason for this unphysical behavior is related to the fact that according to Theorem 1.5, negative solutions are propagating backward in time.

Smearing of position wave functions. We now come to an important consequence resulting from the difference between the usual position operator $\boldsymbol{x}$ and the one-particle position operator $[\boldsymbol{x}]$ or, equally, from the noncommutativity of $\boldsymbol{x}$ and the Feshbach-Villars transformation U. In FV-momentum representation the common eigenstates of the *charge operator* τ_3 and the one-particle position operator $[\tilde{\boldsymbol{x}}]$ are given by

$$\tilde{\phi}_{\boldsymbol{x}'}^{(1)}(\boldsymbol{p}) = \frac{1}{(2\pi\hbar)^{3/2}}\begin{pmatrix}1\\0\end{pmatrix}e^{-\mathrm{i}\boldsymbol{p}\boldsymbol{x}'} , \ \tilde{\phi}_{\boldsymbol{x}'}^{(2)}(\boldsymbol{p}) = \frac{1}{(2\pi\hbar)^{3/2}}\begin{pmatrix}0\\1\end{pmatrix}e^{-\mathrm{i}\boldsymbol{p}\boldsymbol{x}'} ,$$

with

$$[\tilde{\boldsymbol{x}}]\tilde{\phi}_{\boldsymbol{x}'}^{(r)}(\boldsymbol{p}) = \mathrm{i}\hbar\boldsymbol{\nabla}_{\boldsymbol{p}}\tilde{\phi}_{\boldsymbol{x}'}^{(r)}(\boldsymbol{p}) = \boldsymbol{x}'\tilde{\phi}_{\boldsymbol{x}'}^{(r)}(\boldsymbol{p}) , \ \tau_3\tilde{\phi}_{\boldsymbol{x}'}^{(r)}(\boldsymbol{p}) = \epsilon_r\tilde{\phi}_{\boldsymbol{x}'}^{(r)}(\boldsymbol{p}) .$$

The corresponding eigenstates of the one-particle position operator $[\boldsymbol{x}]$ in momentum representation follow from this as

$$\phi_{\boldsymbol{x}'}^{(r)}(\boldsymbol{p}) = U^{-1}\tilde{\phi}_{\boldsymbol{x}'}^{(r)}(\boldsymbol{p}) = \frac{1}{(2\pi\hbar)^{3/2}}\frac{1}{2\sqrt{m_0cp_0}}\begin{pmatrix} m_0c+\epsilon_r p_0 \\ m_0c-\epsilon_r p_0\end{pmatrix}\mathrm{e}^{-\mathrm{i}\boldsymbol{p}\boldsymbol{x}'/\hbar} .$$

As known from nonrelativistic quantum mechanics, the transition from momentum to position representation is carried out via

$$\psi_{\boldsymbol{x}'}^{(r)}(\boldsymbol{x}) = \frac{1}{(2\pi\hbar)^{3/2}}\int \mathrm{d}^3\boldsymbol{p}\mathrm{e}^{\mathrm{i}\boldsymbol{p}\boldsymbol{x}/\hbar}\phi_{\boldsymbol{x}'}^{(r)}(\boldsymbol{p}) .$$

After some lengthy calculations which we do not present here, one finally finds

$$\psi_{\boldsymbol{x}'}^{(1)}(\boldsymbol{x}) \overset{z\gg 1}{\approx} \begin{pmatrix} z^{-7/4}+z^{-9/4} \\ z^{-7/4}-z^{-9/4}\end{pmatrix}\mathrm{e}^{-z} , \quad \psi_{\boldsymbol{x}'}^{(2)}(\boldsymbol{x}) \overset{z\gg 1}{\approx} \begin{pmatrix} z^{-7/4}-z^{-9/4} \\ z^{-7/4}+z^{-9/4}\end{pmatrix}\mathrm{e}^{-z} ,$$

with $z = m_0c|\boldsymbol{x}-\boldsymbol{x}'|/\hbar$. Hence, in position representation, the eigenfunctions of the one-particle position operator $[\boldsymbol{x}]$ are not strict $\delta(\boldsymbol{x}-\boldsymbol{x}')$-functions, but exhibit a certain smearing over an area of magnitude

$$z \sim 1 \Longrightarrow |\boldsymbol{x}-\boldsymbol{x}'| \sim \frac{\hbar}{m_0c} .$$

These considerations lead us to the following conclusions: due to $[\boldsymbol{x}, U] \neq 0$, the Feshbach-Villars transformation U is a nonlocal transformation. The transformed wave function $\tilde{\psi}(x)$ is obtained from the original wave function $\psi(x)$ by averaging its position argument $\boldsymbol{x}$ over an area whose linear extent is comparable to the Compton wave length of the particle. As already mentioned in the introduction of this chapter, with respect to the one-particle interpretation, only those Klein-Gordon wave packets are physically sensible whose extent is large compared to the corresponding Compton wave length. Therefore, averaging effects stemming from nonlocal changes of representations can essentially be ignored.

1.3.3 Validity Range of the One-particle Concept

Using the Hamilton form of the Klein-Gordon equation and introducing the G-scalar product, we have so far developed a formalism which references nonrelativistic quantum mechanics in many respects. Within it the positive and (charge conjugated) negative solutions can be interpreted as two different one-particle systems with opposite charge signs and positive G-expectation values. In order to complete the one-particle concept, we still have to investigate the conditions of its physical consistency and thus clarify the still open point [1].

First of all, from the introductory remarks of this chapter, it is clear that the one-particle interpretation of the Klein-Gordon theory can only be applied to physical situations where processes changing the particle number (particle creation and annihilation) do not play any role. This is the case only for particle energies, particle momenta, and electromagnetic potentials for which

$$|E - m_0c^2| < m_0c^2 \ , \ |\boldsymbol{p}|, \left|\frac{e}{c}A^\mu\right| < m_0c \ , \ \Delta p \ll m_0c \ .$$

This, in turn, implies, due to Heisenberg's uncertainty relation, a necessary position uncertainty of

$$\Delta x \gg \lambda_{\rm c} = \frac{\hbar}{m_0c} \tag{1.50}$$

for the wave packet of the considered particle. Now we examine what additional constraints need to be satisfied for the decoupling of the Klein-Gordon theory into two one-particle theories with pure positive or negative solutions (point [1]) so that

- charged particles and antiparticles can be reasonably described
- the charge density is either positive or negative allowing a quantum mechanical statistical interpretation of G-expectation values.

To this end, we consider a free Klein-Gordon wave packet of charge $+e$ located around the origin at $t = 0$ according to

$$\psi(\boldsymbol{x}, t=0) = (\pi\Delta^2)^{-3/4}\mathrm{e}^{-\boldsymbol{x}^2/(2\Delta^2)}\begin{pmatrix}1\\0\end{pmatrix} \ , \ Q = \langle\psi|\,\psi\rangle_{\rm G} = +1$$

and ask, in which circumstances negative parts with significant amplitude are to be expected? To do this, we decompose the wave packet into its Fourier components,[16]

$$\psi(\boldsymbol{x}, t=0) = \left(\frac{\Delta^2}{\pi\hbar^2}\right)^{3/4}\int\frac{\mathrm{d}^3p'}{(2\pi\hbar)^{3/2}}\mathrm{e}^{-\boldsymbol{p}'^2\Delta^2/(2\hbar^2)}\mathrm{e}^{\mathrm{i}\boldsymbol{p}'\boldsymbol{x}/\hbar}\begin{pmatrix}1\\0\end{pmatrix} \ , \tag{1.51}$$

and compare this expression with the general solution for $t = 0$,

$$\psi(\boldsymbol{x}, t=0) = \int \mathrm{d}^3p' \sum_{r=1}^{2} a^{(r)}(\boldsymbol{p}')\psi^{(r)}_{\boldsymbol{p}'}(\boldsymbol{x}, t=0) \ .$$

This yields

$$\left(\frac{\Delta^2}{\pi\hbar^2}\right)^{3/4}\mathrm{e}^{-\boldsymbol{p}'^2\Delta^2/(2\hbar^2)}\begin{pmatrix}1\\0\end{pmatrix} = a^{(1)}(\boldsymbol{p}')\Psi^{(1)}(\boldsymbol{p}') + a^{(2)}(-\boldsymbol{p}')\Psi^{(2)}(-\boldsymbol{p}') \ .$$

Multiplying this equation by $\Psi^{(r)\dagger}(\boldsymbol{p}')\tau_3$ from the left and using (1.19) we obtain the series coefficients $a^{(r)}(\boldsymbol{p}')$ as

[16] To avoid confusion, we denote the Fourier momentum by $\boldsymbol{p}'$. The group momentum of the wave packet is $\boldsymbol{p} = \boldsymbol{0}$.

$$a^{(1)}(\boldsymbol{p}') = \left(\frac{\Delta^2}{\pi\hbar^2}\right)^{3/4} \frac{m_0c + p_0'}{2\sqrt{m_0cp_0'}} \mathrm{e}^{-\boldsymbol{p}'^2\Delta^2/(2\hbar^2)}$$

$$a^{(2)}(-\boldsymbol{p}') = \left(\frac{\Delta^2}{\pi\hbar^2}\right)^{3/4} \frac{m_0c - p_0'}{2\sqrt{m_0cp_0'}} \mathrm{e}^{-\boldsymbol{p}'^2\Delta^2/(2\hbar^2)} .$$

We see that the amplitudes $a^{(2)}(\boldsymbol{p}')$ of the negative solution $\psi^{(2)}_{\boldsymbol{p}'}(\boldsymbol{x})$ in the wave packet are nonzero, which reflects the fact that only the positive and negative solutions together form a complete system. The ratio of the amplitudes of positive and negative solutions is

$$\frac{a^{(2)}(-\boldsymbol{p}')}{a^{(1)}(\boldsymbol{p}')} = \frac{m_0c - p_0'}{m_0c + p_0'} = \frac{m_0c - \sqrt{m_0^2c^2 + \boldsymbol{p}'^2}}{m_0c + \sqrt{m_0^2c^2 + \boldsymbol{p}'^2}} .$$

From this follows that only the amplitudes of negative solutions with Fourier momenta $|\boldsymbol{p}'| \gtrsim m_0c$ contribute significantly to the wave packet. On the other hand, the Fourier transformation (1.51) shows that only momenta with $|\boldsymbol{p}'| \lesssim \hbar/\Delta$ are predominantly present in the wave packet. We therefore conclude that for a significant contribution of negative solutions, the wave packet must be localized within an area whose extent is comparable to the Compton wave length of the spin-0 particle: $\Delta \lesssim \hbar/m_0c$. Put differently: demanding the Klein-Gordon theory to be completely decoupleable into two one-particle theories leads again to the constraint (1.50), i.e. to Klein-Gordon wave packets whose extent is large compared to the corresponding Compton wave length.

For the sake of clarity, we summarize the limits of the one-particle interpretation of the Klein-Gordon theory as follows:

Theorem 1.7: Validity range of the one-particle concept

A consistent one-particle interpretation of the Klein-Gordon theory is possible only in those cases where

- the particle's energy and momentum as well as the electromagnetic potentials obey the conditions

$$|E - m_0c^2| < m_0c^2 , \ |\boldsymbol{p}|, \left|\frac{e}{c}A^\mu\right| < m_0c , \ \Delta E \ll m_0c^2 , \ \Delta p \ll m_0c ,$$

- the particle's wave packet has a spatial extent Δ, which is large compared to the corresponding wave length:

$$\Delta \gg \lambda_{\mathrm{c}} = \frac{\hbar}{m_0c} .$$

The second condition follows, on the one hand, from the first condition in conjunction with Heisenberg's uncertainty relation and, on the other hand, from the requirement that the charged spin-0 particle or antiparticle must be representable by sole positive or negative Klein-Gordon solutions. With ▷

these prerequisites, the Klein-Gordon theory decouples into two one-particle theories with positive and negative definite charge densities, allowing a statistical one-particle interpretation. Beyond these limits, the one-particle concept leads to contradictions that can be solved satisfactorily only within quantum field theoretical considerations.

Such contradictions appear, for example, in the *shaky movement* (see Exercise 7) and in the *Klein paradox*, to which we now turn.

1.3.4 Klein Paradox

We consider a onedimensional spin-0 particle, that, coming from the left, is scattered against a potential step of the form (see Figure 1.3)

$$eA^0(z) = V(z) = \left\{ \begin{array}{l} 0 \ \text{ for } z < 0 \text{ (area I)} \\ V_0 \ \text{ for } z > 0 \text{ (area II)} \end{array} \right\} , \ V_0 > 0 \ , \ \boldsymbol{A} = \boldsymbol{0} \ .$$

The corresponding canonical Klein-Gordon equation is

$$\left(\mathrm{i}\hbar \frac{\partial}{\partial t} - V(z)\right)^2 \phi(z,t) + \left(c^2\hbar^2 \frac{\mathrm{d}^2}{\mathrm{d}z^2} - m_0^2 c^4\right) \phi(z,t) = 0 \ .$$

Separating the time-dependent part via

$$\phi(z,t) = \Phi(z)\mathrm{e}^{-\mathrm{i}Et/\hbar} \ ,$$

we obtain the stationary equation

$$\frac{\mathrm{d}^2\Phi(z)}{\mathrm{d}z^2} = \frac{1}{c^2\hbar^2} \left\{m_0^2 c^4 - [E - V(z)]^2\right\} \Phi(z) \ . \qquad (1.52)$$

For its general solution in the areas I ($z < 0$) and II ($z > 0$), we set

$$\Phi_{\mathrm{I}}(z) = \Phi_{\mathrm{in}}(z) + \Phi_{\mathrm{ref}}(z) \ , \ \Phi_{\mathrm{II}}(z) = \Phi_{\mathrm{trans}}(z) \ ,$$

with

$$\Phi_{\mathrm{in}}(z) = A\mathrm{e}^{\mathrm{i}k_1 z} \ , \ \Phi_{\mathrm{ref}}(z) = B\mathrm{e}^{-\mathrm{i}k_1 z} \ , \ \Phi_{\mathrm{trans}}(z) = C\mathrm{e}^{\mathrm{i}k_2 z}$$

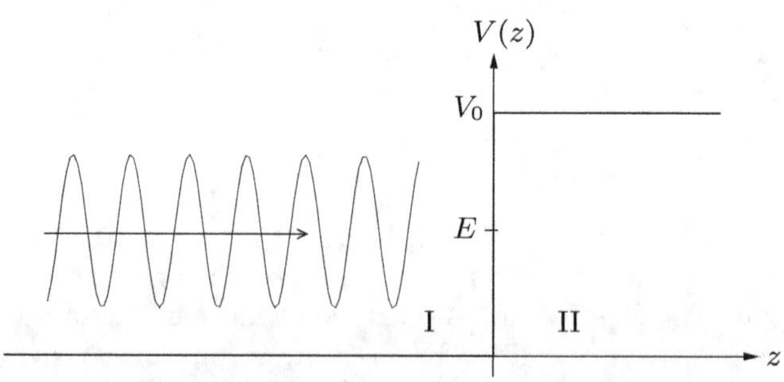

Fig. 1.3. Onedimensional potential step.

$$k_1 = \sqrt{\frac{E^2 - m_0^2 c^4}{c^2 \hbar^2}} \;,\; k_2 = \sqrt{\frac{(E - V_0)^2 - m_0^2 c^4}{c^2 \hbar^2}} \;,$$

where $\Phi_{\rm in}$, $\Phi_{\rm ref}$, $\Phi_{\rm trans}$ denote the incoming, reflected, and transmitted parts, respectively. The integration constants A, B, and C follow from the continuity conditions of $\Phi(z)$ and $\Phi'(z)$ at $z = 0$ as[17]

$$B = \frac{1-r}{1+r} A \;,\; C = \frac{2A}{1+r} \;,\; r = \frac{k_2}{k_1} \;.$$

Depending on the choice of V_0 or E, we initially distinguish three cases:

1. case: $E > V_0 + m_0 c^2$. In this case, the wave number k_2 is real implying that the transmitted wave in area II is oscillating, and we have $r > 0$. The current densities of the incoming, reflected, and transmitted parts in the z-direction are calculated as

$$T = \frac{j_{\rm trans}}{j_{\rm in}} = \frac{4r}{(1+r)^2} \;,\; R = -\frac{j_{\rm ref}}{j_{\rm in}} = \frac{(1-r)^2}{(1+r)^2} = 1 - T \;.$$

For each value $r > 0$, the reflection and transmission coefficients obey $0 < R, T < 1$, in accordance with our expectations.

2. case: $V_0 - m_0 c^2 < E < V_0 + m_0 c^2 \;,\; E > m_0 c^2$. Here k_2 is imaginary, and the transmitted wave is exponentially damped down.

3. case: $m_0 c^2 < E < V_0 - m_0 c^2 \Longrightarrow V_0 > 2 m_0 c^2$. As in the first case, the wave number k_2 is real yielding an oscillating transmitted wave in area II.

Obviously, in the first two cases, the Klein-Gordon solutions behave similarly to those of nonrelativistic quantum mechanics and can be interpreted as the scattering of a particle of charge $+e$ at the (from its view) repulsive potential barrier. However, the third case is starkly contradictory to our expectations with respect to the one-particle interpretation, since the potential step should be inpenetrable for a quantum mechanical particle with $E < V_0$. Another contradiction – in the second case in parts and in the whole third case – is the different current density signs of the incoming and transmitted waves for $E - V_0 < 0$,

$$\begin{aligned} \rho_{\rm in}(z) &= \frac{E}{m_0 c^2} |\Phi_{\rm in}(z)|^2 > 0 \qquad , \; z < 0 \\ \rho_{\rm trans}(z) &= \frac{E - V_0}{m_0 c^2} |\Phi_{\rm trans}(z)|^2 < 0 \;\;,\; z > 0 \;. \end{aligned}$$

[17] The continuity of $\Phi(z)$ at $z = 0$ is a consequence of charge current conservation. Moreover, for potentials $V(z)$ with a finite step at $z = 0$, we have

$$\Phi'(+\delta) - \Phi'(-\delta) = \int\limits_{-\delta}^{+\delta} \mathrm{d}z \frac{\mathrm{d}}{\mathrm{d}z} \Phi'(z) \sim \int\limits_{-\delta}^{+\delta} \mathrm{d}z \left[m_0^2 c^4 - (E - V(z))^2 \right] \Phi(z) \stackrel{\delta \to 0}{\longrightarrow} 0 \;.$$

According to this, the transmitted part must be regarded as a negative Klein-Gordon solution whose energy $E - V_0$ relative to the potential V_0 is indeed less than zero (see Figure 1.4).

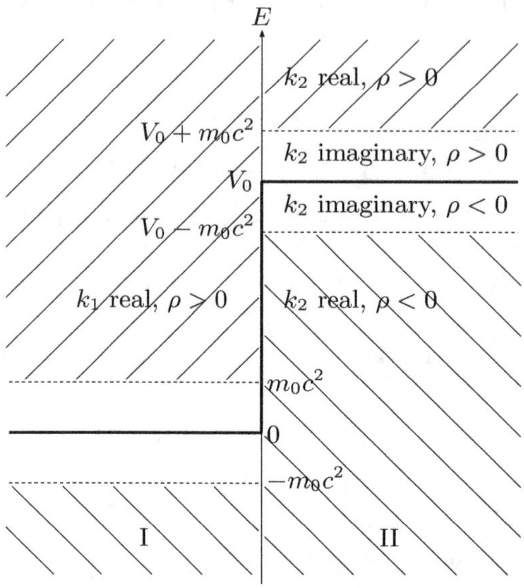

Fig. 1.4. Energy intervals of the onedimensional potential step. In area I the positive solutions lie within the interval $E > m_0c^2$ and the negative solutions within the interval $E < -m_0c^2$. In area II the positive solutions (with energy $E - V_0$ relative to the potential V_0) lie within the interval $E > V_0 + m_0c^2$ and the negative ones within the interval $E < V_0 - m_0c^2$. In between there are the solutions of the "forbidden" energy intervals.

The reasons for these circumstances are rooted in the fact that the increase of V_0 up to a value around E corresponds to a decrease of the wave packet's penetration depth in area II down to $1/k_2 \approx \hbar/m_0c$, i.e. down to the Compton wave length of the incoming particle. Choosing, instead, a potential with an increase less than m_0c^2 per Compton wave length, one can show that the paradoxes tend to vanish. In other words: the above mentioned difficulties result from too-strong a localization of the particle (see Theorem 1.7).

It is interesting that, when leaving the level of the one-particle interpretation, the third case allows a physically sensible (but at best qualitatively acceptable) interpretation in terms of *pair creation*. Here we have to keep in mind that the negative wave function $\Phi_{\rm trans}$ with momentum eigenvalue $+\hbar k_2$ and energy $E - V_0 < -m_0c^2$ corresponds to an antiparticle of charge $-e$ flying with momentum $-\hbar k_2$ from the right toward the potential step. However, since we are assuming incoming movements from the left to the right, an incoming antiparticle from the right does not make any sense. On the other hand, we have the freedom to choose the sign of k_2. So, replacing

k_2 by $-k_2$, the transmitted wave Φ_{trans} now corresponds to an antiparticle moving to the right with charge $-e$ and momentum $+\hbar|k_2|$. Moreover, we then have

$$r < 0 \Longrightarrow R > 1 \ , \ T < 0 \ .$$

These relations can finally be interpreted as particle-antiparticle creation in the following way: all particles, coming from the left are totally reflected at the potential step. Additionally, particle-antiparticle pairs are created with the particles moving to the left ($R > 1$) and the antiparticles to the right ($T < 0$).

Let us, for completeness, also consider the remaining two energy intervals:

4. case: $-m_0c^2 < E < m_0c^2$. In this case, a solution does not exist as long as we adhere to a rightward directed incoming movement.

5. case: $E < -m_0c^2$. k_2 is real, and we have again an oscillating wave in area II. Replacing k_1 and k_2 by $k_1 = -|k_1|$ and $k_2 = -|k_2|$, this can be interpreted within the one-particle picture, similarly to the above considerations, as follows: a rightward incoming antiparticle with charge $-e$ is scattered at the (from its point of view) attractive potential barrier. Compared to the first case, the charge current densities j_{in}, j_{ref}, and j_{trans} have opposite signs, and we again obtain $r > 0 \Longrightarrow 0 < R, T < 1$.

Summary

- With the help of the **G-scalar product** and the **G-expectation value**, statistical mean measurement values of spin-0 [anti]particle systems can be described symmetrically using positive [negative] Klein-Gordon solutions.
- The G-scalar product leads to the definition of **G-Hermitean** and **G-unitary operators** that correspond to Hermitean and unitary operators of nonrelativistic quantum mechanics.
- The G-Hermitecity of the Klein-Gordon-Hamilton operator ensures the picture-independence of the G-scalar product.
- Within the one-particle interpretation, only those G-Hermitean operators can be regarded as observables that are **even operators**, i.e. that do not mix positive and negative Klein-Gordon solutions (**one-particle interpretation**). The even part of an operator can be determined best in a representation where the Hamilton operator is diagonal. In the free case this is given by the **Feshbach-Villars representation**.
- The **Feshbach-Villars transformation** is a nonlocal transformation. The position argument $\boldsymbol{x}$ of a wave function $\psi(x)$ is averaged or smeared over a region whose extent is equal to the Compton wave length of the particle.

- The validity range of the one-particle probabilistic interpretation is restricted, on the one hand, to small energies where particle creation processes can be ignored and, on the other hand, to Klein-Gordon wave packets whose extent is large compared to the corresponding wave length.
- The **Klein paradox** is a simple example of interpretational difficulties of the one-particle concept stemming from too-strong a localization of Klein-Gordon wave packets. Beyond the one-particle concept, this phenomenon can be qualitatively interpreted as **pair creation**.

Exercises

4. Properties of G-Hermitean and G-unitary operators. Verify the following properties of G-Hermitean and G-unitary operators:

a) For charged eigenstates the eigenvalues of a G-Hermitean operator are real.

b) Charged eigenstates of a G-Hermitean operator to different eigenvalues are G-orthogonal.

c) The product of two G-unitary operators is a G-unitary operator, too.

d) If U describes the infinitesimal G-unitary transformation $U = 1 + \mathrm{i}\epsilon\mathcal{O}$, $|\epsilon| \ll 1$, then $\mathcal{O}$ is G-Hermitean.

e) If $\mathcal{O}$ is G-Hermitean, then $\mathrm{e}^{\mathrm{i}\mathcal{O}}$ is G-unitary.

Solution.

To a) Let $\mathcal{O}$ be a G-Hermitean operator and

$$\mathcal{O}\,|\psi\rangle = a\,|\psi\rangle \;\;,\;\; \langle\psi|\,\psi\rangle_{\mathrm{G}} \neq 0\;.$$

Then

$$a\,\langle\psi|\,\psi\rangle_{\mathrm{G}} = \langle\psi|\,\mathcal{O}\,|\psi\rangle_{\mathrm{G}} = \langle\psi|\,\mathcal{O}\,|\psi\rangle_{\mathrm{G}}^* = a^*\,\langle\psi|\,\psi\rangle_{\mathrm{G}}^* \Longrightarrow a = a^*\;.$$

To b) Let at least $|\,\phi\rangle$ be charged and

$$\mathcal{O}\,|\psi\rangle = a\,|\psi\rangle \;\;,\;\; \mathcal{O}\,|\phi\rangle = b\,|\phi\rangle \;\;,\;\; a \neq b\;,\;\; b = b^*\;.$$

Then, we have

$$\langle\phi|\,\mathcal{O}\,|\psi\rangle_{\mathrm{G}} = a\,\langle\phi|\,\psi\rangle_{\mathrm{G}} \tag{1.53}$$

and

$$\begin{aligned}\langle\psi|\,\mathcal{O}\,|\phi\rangle_{\mathrm{G}} = b\,\langle\psi|\,\phi\rangle_{\mathrm{G}} &\Longrightarrow \langle\phi|\,\mathcal{O}\,|\psi\rangle_{\mathrm{G}}^* = b\,\langle\phi|\,\psi\rangle_{\mathrm{G}}^* \\ &\Longrightarrow \langle\phi|\,\mathcal{O}\,|\psi\rangle_{\mathrm{G}} = b\,\langle\phi|\,\psi\rangle_{\mathrm{G}}\;.\end{aligned} \tag{1.54}$$

The difference between (1.53) and (1.54) leads to

$$0 = (a-b)\,\langle\phi|\,\psi\rangle_{\mathrm{G}} \Longrightarrow \langle\phi|\,\psi\rangle_{\mathrm{G}} = 0\;.$$

To c) Let

$$\tau_3 U^\dagger \tau_3 = U^{-1} \;,\; \tau_3 V^\dagger \tau_3 = V^{-1} \;.$$

Then

$$\tau_3 [UV]^\dagger \tau_3 UV = \tau_3 V^\dagger U^\dagger \tau_3 UV = \tau_3 V^\dagger \tau_3 \tau_3 U^\dagger \tau_3 UV = V^{-1}U^{-1}UV = 1$$
$$\Longrightarrow \tau_3 [UV]^\dagger \tau_3 = [UV]^{-1} \;.$$

To d)

$$1 = \tau_3(1 - \mathrm{i}\epsilon\mathcal{O}^\dagger)\tau_3(1 + \mathrm{i}\epsilon\mathcal{O}) = 1 - \mathrm{i}\epsilon\tau_3\mathcal{O}^\dagger\tau_3 + \mathrm{i}\epsilon\mathcal{O} + \mathcal{O}\left(\epsilon^2\right)$$
$$\Longrightarrow \tau_3\mathcal{O}^\dagger\tau_3 = \mathcal{O} \;.$$

To e) Due to

$$\tau_3\left(\mathcal{O}^\dagger\right)^n\tau_3 = \tau_3\mathcal{O}^\dagger\mathcal{O}^\dagger\ldots\mathcal{O}^\dagger\tau_3 = \tau_3\mathcal{O}^\dagger\tau_3\tau_3\mathcal{O}^\dagger\tau_3\ldots\tau_3\mathcal{O}^\dagger\tau_3 = \mathcal{O}^n \;,$$

it follows that

$$\begin{aligned}\tau_3\left(\mathrm{e}^{\mathrm{i}\mathcal{O}}\right)^\dagger\tau_3 &= \tau_3\mathrm{e}^{-\mathrm{i}\mathcal{O}^\dagger}\tau_3 = \sum_n \frac{(-\mathrm{i})^n}{n!}\tau_3\left(\mathcal{O}^\dagger\right)^n\tau_3 = \sum_n \frac{(-\mathrm{i})^n}{n!}\mathcal{O}^n \\ &= \mathrm{e}^{-\mathrm{i}\mathcal{O}} = \left(\mathrm{e}^{\mathrm{i}\mathcal{O}}\right)^{-1} \;.\end{aligned}$$

5. Feshbach-Villars transformation (I). Prove the relations (1.45), (1.46), (1.48), (1.49).

Solution.

To (1.45) and (1.49). Taking into account

$$(1 \pm \tau_1)(\tau_3 \pm \mathrm{i}\tau_2) = 0 \;,\; (1 \pm \tau_1)(\tau_3 \mp \mathrm{i}\tau_2) = 2(\tau_3 \mp \mathrm{i}\tau_2)$$

and

$$\begin{aligned}U^2 &= \frac{(1+\tau_1)p_0}{2m_0c} + \frac{(1-\tau_1)m_0c}{2p_0} \\ H^{(0)} &= \frac{(\tau_3 + \mathrm{i}\tau_2)p_0^2}{2m_0} + \frac{(\tau_3 - \mathrm{i}\tau_2)m_0c^2}{2} \;,\end{aligned}$$

it follows for $\tilde{H}^{(0)}$ in the FV-momentum and coordinate representation that

$$\tilde{H}^{(0)} = UH^{(0)}U^{-1} = U^2H^{(0)} = \frac{(\tau_3 + \mathrm{i}\tau_2)cp_0}{2} + \frac{(\tau_3 - \mathrm{i}\tau_2)cp_0}{2} = cp_0\tau_3$$

and for the one-particle velocity operator $[\boldsymbol{v}]$ in the momentum and coordinate representation that

$$[\boldsymbol{v}] = U^{-1}\frac{c\boldsymbol{p}\tau_3}{p_0}U = \frac{c\boldsymbol{p}\tau_3}{p_0}U^2 = \frac{(\tau_3 + \mathrm{i}\tau_2)\boldsymbol{p}}{2m_0} + \frac{(\tau_3 - \mathrm{i}\tau_2)m_0c^2\boldsymbol{p}}{2p_0^2} \;.$$

Of course, the last relationship is also obtained using the generalized Ehrenfest theorem

$$[\boldsymbol{v}] = \left[\boldsymbol{\nabla}_{\boldsymbol{p}} + \frac{\tau_1 \boldsymbol{p}}{2p_0^2}, H^{(0)}\right] ,$$

where the expression on the left hand side of the commutator is the one-particle position operator in the momentum representation.

To (1.46) and (1.48). In the FV-momentum representation the position operator obeys

$$\tilde{\boldsymbol{x}} = \mathrm{i}\hbar U \boldsymbol{\nabla}_{\boldsymbol{p}} U^{-1} = \mathrm{i}\hbar \boldsymbol{\nabla}_{\boldsymbol{p}} + \mathrm{i}\hbar U\left(\boldsymbol{\nabla}_{\boldsymbol{p}} U^{-1}\right) ,$$

and in the momentum representation the one-particle position operator fulfills

$$[\boldsymbol{x}] = \mathrm{i}\hbar U^{-1} \boldsymbol{\nabla}_{\boldsymbol{p}} U = \mathrm{i}\hbar \boldsymbol{\nabla}_{\boldsymbol{p}} + \mathrm{i}\hbar U^{-1}\left(\boldsymbol{\nabla}_{\boldsymbol{p}} U\right) .$$

Moreover, we have

$$\begin{aligned}
\boldsymbol{\nabla}_{\boldsymbol{p}} p_0 &= \boldsymbol{\nabla}_{\boldsymbol{p}} \sqrt{\boldsymbol{p}^2 + m_0^2 c^2} = \frac{\boldsymbol{p}}{\sqrt{\boldsymbol{p}^2 + m_0^2 c^2}} = \frac{\boldsymbol{p}}{p_0} \\
\boldsymbol{\nabla}_{\boldsymbol{p}} \frac{p_0}{2\sqrt{m_0 c p_0}} &= \frac{1}{2\sqrt{m_0 c}} \boldsymbol{\nabla}_{\boldsymbol{p}} \sqrt{p_0} = \frac{1}{4\sqrt{m_0 c p_0}} \boldsymbol{\nabla}_{\boldsymbol{p}} p_0 = \frac{\boldsymbol{p} p_0}{4\sqrt{m_0 c p_0^5}} \\
\boldsymbol{\nabla}_{\boldsymbol{p}} \frac{m_0 c}{2\sqrt{m_0 c p_0}} &= \frac{m_0 c}{2\sqrt{m_0 c}} \boldsymbol{\nabla}_{\boldsymbol{p}} \frac{1}{\sqrt{p_0}} = -\frac{m_0 c}{4\sqrt{m_0 c p_0^3}} \boldsymbol{\nabla}_{\boldsymbol{p}} p_0 = -\frac{m_0 c \boldsymbol{p}}{4\sqrt{m_0 c p_0^5}} .
\end{aligned}$$

From this follows

$$\begin{aligned}
\left(\boldsymbol{\nabla}_{\boldsymbol{p}} U^{-1}\right) &= \frac{(p_0 - m_0 c) - \tau_1 (p_0 + m_0 c)}{4\sqrt{m_0 c p_0^5}} \boldsymbol{p} \\
\left(\boldsymbol{\nabla}_{\boldsymbol{p}} U\right) &= \frac{(p_0 - m_0 c) + \tau_1 (p_0 + m_0 c)}{4\sqrt{m_0 c p_0^5}} \boldsymbol{p}
\end{aligned}$$

and, therefore,

$$U\left(\boldsymbol{\nabla}_{\boldsymbol{p}} U^{-1}\right) = -\frac{\tau_1 \boldsymbol{p}}{2p_0^2} , \quad U^{-1}\left(\boldsymbol{\nabla}_{\boldsymbol{p}} U\right) = +\frac{\tau_1 \boldsymbol{p}}{2p_0^2} .$$

6. Construction of one-particle operators using the sign operator (I). Construct the one-particle operators $[\boldsymbol{x}]$ and $[\boldsymbol{v}]$ using the G-Hermitean *sign operator*

$$\Lambda = \frac{H^{(0)}}{\sqrt{H^{(0)2}}} = \frac{\frac{\boldsymbol{p}^2}{2m_0}(\tau_3 + \mathrm{i}\tau_2) + m_0 c^2 \tau_3}{c p_0} .$$

Solution. Evidently, the operator Λ possesses the eigenfunctions $\psi_{\boldsymbol{p}}^{(1,2)}(x)$ with the eigenvalues (energy signs) ± 1. We can use this property for the construction of one-particle operators in the following way: let $[\mathcal{O}]$ and $\{\mathcal{O}\}$ be the even and odd parts of the operator

$$\mathcal{O} = [\mathcal{O}] + \{\mathcal{O}\} .$$

Then, for arbitrary wave packets $\psi^{(\pm)}(x)$ built by pure positive or negative free Klein-Gordon solutions, we have

$$\begin{aligned}
\mathcal{O}\psi^{(+)} &= [\mathcal{O}]\psi^{(+)} + \{\mathcal{O}\}\psi^{(+)} \\
\mathcal{O}\psi^{(-)} &= [\mathcal{O}]\psi^{(-)} + \{\mathcal{O}\}\psi^{(-)} \\
\Lambda\mathcal{O}\Lambda\psi^{(+)} &= \Lambda\mathcal{O}\psi^{(+)} = [\mathcal{O}]\psi^{(+)} - \{\mathcal{O}\}\psi^{(+)} \\
\Lambda\mathcal{O}\Lambda\psi^{(-)} &= -\Lambda\mathcal{O}\psi^{(-)} = [\mathcal{O}]\psi^{(-)} - \{\mathcal{O}\}\psi^{(-)} \;.
\end{aligned}$$

Adding the first and the third, and subtracting the second and the fourth relation, it follows that

$$[\mathcal{O}] = \frac{1}{2}(\mathcal{O} + \Lambda\mathcal{O}\Lambda) \;,\; \{\mathcal{O}\} = \frac{1}{2}(\mathcal{O} - \Lambda\mathcal{O}\Lambda) \;.$$

Clearly, if $\mathcal{O}$ is a G-Hermitean operator, the operators $[\mathcal{O}]$, $\{\mathcal{O}\}$, and $\Lambda\mathcal{O}\Lambda$ are G-Hermitean, too:

$$\begin{aligned}
(\tau_3\Lambda\mathcal{O}\Lambda)^\dagger &= \Lambda^\dagger\mathcal{O}^\dagger\Lambda^\dagger\tau_3 = \Lambda^\dagger\mathcal{O}^\dagger(\tau_3\Lambda)^\dagger = \Lambda^\dagger\mathcal{O}^\dagger\tau_3\Lambda = \Lambda^\dagger(\tau_3\mathcal{O})^\dagger\Lambda \\
&= \Lambda^\dagger\tau_3\mathcal{O}\Lambda = (\tau_3\Lambda)^\dagger\mathcal{O}\Lambda = \tau_3\Lambda\mathcal{O}\Lambda \;.
\end{aligned}$$

To determine the one-particle position operator, we change to the momentum representation ($\boldsymbol{x} = \mathrm{i}\hbar\boldsymbol{\nabla}_{\boldsymbol{p}}$, $\boldsymbol{p} = \mathbb{C}$-number) and calculate as follows:

$$(\boldsymbol{\nabla}_{\boldsymbol{p}}\Lambda) = \frac{\boldsymbol{p}(\tau_3 + \mathrm{i}\tau_2)}{m_0 c p_0} - \frac{\boldsymbol{p}H^{(0)}}{c p_0^3} \;,\; \Lambda(\boldsymbol{\nabla}_{\boldsymbol{p}}\Lambda) = \frac{\boldsymbol{p}\tau_3(\tau_3 + \mathrm{i}\tau_2)}{p_0^2} - \frac{\boldsymbol{p}}{p_0^2} = \frac{\tau_1\boldsymbol{p}}{p_0^2}$$

$$\Longrightarrow \Lambda\boldsymbol{x}\Lambda = \mathrm{i}\hbar\boldsymbol{\nabla}_{\boldsymbol{p}} + \mathrm{i}\hbar\Lambda(\boldsymbol{\nabla}_{\boldsymbol{p}}\Lambda) = \mathrm{i}\hbar\boldsymbol{\nabla}_{\boldsymbol{p}} + \frac{\mathrm{i}\hbar\tau_1\boldsymbol{p}}{p_0^2} \;.$$

From this we obtain in the momentum and coordinate representation

$$[\boldsymbol{x}] = \frac{1}{2}(\boldsymbol{x} + \Lambda\boldsymbol{x}\Lambda) = \boldsymbol{x} + \frac{\mathrm{i}\hbar\tau_1\boldsymbol{p}}{2p_0^2}$$

in accordance with Theorem 1.6. For the one-particle velocity operator, a similar calculation yields

$$\boldsymbol{v}\Lambda = \frac{\boldsymbol{p}(\tau_3 + \mathrm{i}\tau_2)\tau_3}{p_0^2} \;,\; \Lambda\boldsymbol{v}\Lambda = \frac{(\tau_3 - \mathrm{i}\tau_2)m_0c^2\boldsymbol{p}}{p_0^2}$$

$$\Longrightarrow [\boldsymbol{v}] = \frac{1}{2}(\boldsymbol{v} + \Lambda\boldsymbol{v}\Lambda) = \frac{(\tau_3 + \mathrm{i}\tau_2)\boldsymbol{p}}{2m_0} + \frac{(\tau_3 - \mathrm{i}\tau_2)m_0c^2\boldsymbol{p}}{2p_0^2} \;,$$

again complying with Theorem 1.6.

7. Shaky movement (I). Show that the mean current $\langle \boldsymbol{j} \rangle$ of an arbitrary Klein-Gordon wave packet includes a temporally oscillating movement, if it is composed of positive and negative components. What is the interpretation of this fact?

Solution. In the canonical formulation the wave packet may be

$$\phi(x) = \phi^{(+)}(x) + \phi^{(-)}(x) \ , \ \phi^{(\pm)}(x) = \int \mathrm{d}^3 p a^{(1,2)}(\boldsymbol{p})\phi_{\boldsymbol{p}}^{(1,2)}(x) \ .$$

Using the adjunction relation $\langle \phi | \boldsymbol{A} | \psi \rangle = \langle \psi | \boldsymbol{A}^\dagger | \phi \rangle^*$, it follows that

$$\begin{aligned}
\langle \boldsymbol{j} \rangle &= \frac{1}{2m_0}\left[\langle \phi | \boldsymbol{p} | \phi \rangle - \langle \phi^* | \boldsymbol{p} | \phi^* \rangle\right] \\
&= \frac{1}{2m_0}\left[\langle \phi | \boldsymbol{p} | \phi \rangle + \langle \phi | \boldsymbol{p} | \phi \rangle^*\right] \\
&= \frac{1}{m_0}\langle \phi | \boldsymbol{p} | \phi \rangle \\
&= \frac{1}{m_0}\left\langle \phi^{(+)} + \phi^{(-)} \middle| \boldsymbol{p} \middle| \phi^{(+)} + \phi^{(-)} \right\rangle \\
&= \frac{1}{m_0}\left[\left\langle \phi^{(+)} \middle| \boldsymbol{p} \middle| \phi^{(+)} \right\rangle + \left\langle \phi^{(-)} \middle| \boldsymbol{p} \middle| \phi^{(-)} \right\rangle\right. \\
&\quad \left. + \left\langle \phi^{(+)} \middle| \boldsymbol{p} \middle| \phi^{(-)} \right\rangle + \left\langle \phi^{(-)} \middle| \boldsymbol{p} \middle| \phi^{(+)} \right\rangle\right] \\
&= \frac{1}{m_0}\left[\left\langle \phi^{(+)} \middle| \boldsymbol{p} \middle| \phi^{(+)} \right\rangle + \left\langle \phi^{(-)} \middle| \boldsymbol{p} \middle| \phi^{(-)} \right\rangle\right] \\
&\quad + \frac{2}{m_0}\mathrm{Re}\left[\left\langle \phi^{(+)} \middle| \boldsymbol{p} \middle| \phi^{(-)} \right\rangle\right] \\
&= \underbrace{\int \mathrm{d}^3 p \frac{c\boldsymbol{p}}{p_0}\left|a^{(1)}(\boldsymbol{p})\right|^2}_{\langle \boldsymbol{j} \rangle^{(+)}} \ \underbrace{- \int \mathrm{d}^3 p \frac{c\boldsymbol{p}}{p_0}\left||a^{(2)}(\boldsymbol{p})\right|^2}_{\langle \boldsymbol{j} \rangle^{(-)}} \\
&\quad + 2\mathrm{Re}\left(\int \mathrm{d}^3 p \frac{c\boldsymbol{p}}{p_0} \mathrm{e}^{2\mathrm{i}p_0 x^0/\hbar} a^{(1)*}(\boldsymbol{p}) a^{(2)}(-\boldsymbol{p})\right) \ .
\end{aligned}$$

Besides the time-independent mean currents of the positive and negative parts, there are also mixed terms oscillating very fast in time. The frequency of this shaky movement (German: *Zitterbewegung*) is of order of $2m_0c^2/\hbar$. Interpreting this phenomenon within the enforced one-particle picture, we must conclude that the "particle" described by ϕ performs a periodic oscillatory movement around its (classical) trajectory. This example shows, once again, that the description of neutral spin-0 particles by real Klein-Gordon wave packets

$$\phi^{(0)}(x) = \phi^{(+)}(x) + \phi^{(-)}(x) \ , \ \phi^{(-)*} = \phi^{(+)}$$

is problematic with respect to the one-particle interpretation, since it inevitably involves a shaky movement.

1.4 Nonrelativistic Approximation of the Klein-Gordon Theory

It is a necessary condition for the correctness of the Klein-Gordon theory that, in the nonrelativistic limit, it passes over to the laws of nonrelativistic quantum mechanics. This limiting process is the subject of this section. First, we discuss the nonrelativistic approximation to leading order of v/c that will lead us to the well-known nonrelativistic Schrödinger equation for spin-0 particles. Afterward, we include relativistic corrections of higher orders. For this we use the *Fouldy-Wouthuysen transformation* by which the Klein-Gordon-Hamilton operator cannot be diagonalized exactly but (in principle) to every required finite order of v/c.

1.4.1 Nonrelativistic Limit

Dealing with the nonrelativistic limit of the Klein-Gordon theory, the substitutions (1.15) prove to be very useful as, in this limit, we have for a positive Klein-Gordon solution[18]

$$\varphi = \left[1 + \mathcal{O}\left(\frac{v^2}{c^2}\right)\right]\phi^{(+)} \;,\; \chi = \mathcal{O}\left(\frac{v^2}{c^2}\right)\phi^{(+)}$$

and for a negative solution

$$\varphi = \mathcal{O}\left(\frac{v^2}{c^2}\right)\phi^{(-)} \;,\; \chi = \left[1 + \mathcal{O}\left(\frac{v^2}{c^2}\right)\right]\phi^{(-)} \;.$$

This means that for positive solutions, the lower component of ψ is suppressed by a factor of v^2/c^2 compared to the upper one, whereas for negative solutions the reverse is true. Consequently, the term $(\boldsymbol{p} - e\boldsymbol{A}/c)^2\chi/2m_0$ in the upper part of the Klein-Gordon equation (1.17) can be ignored up to order $\mathcal{O}\left(v^2/c^2\right)$ for positive solutions leading to

$$\left.\begin{aligned} \psi &= \begin{pmatrix} 1 \\ \mathcal{O}\left(v^2/c^2\right) \end{pmatrix}\varphi \\ \mathrm{i}\hbar\frac{\partial\varphi}{\partial t} &= \left[\frac{1}{2m_0}\left(\boldsymbol{p} - \frac{e}{c}\boldsymbol{A}\right)^2 + m_0c^2 + eA^0 + \mathcal{O}\left(\frac{v^4}{c^4}\right)\right]\varphi \;. \end{aligned}\right\} \quad (1.55)$$

---

[18] In the nonrelativistic limit we can assume that the fields eA^0 and $e\boldsymbol{A}/c$ have at most a magnitude in the range of the particle's energy and momentum, i.e. $|eA^0| \approx m_0v^2/2 \ll m_0c^2$ and $|e\boldsymbol{A}/c| \approx m_0v \ll m_0c$. Thus, it follows that

$$(\boldsymbol{p} - e\boldsymbol{A}/c)^2\phi^{(\pm)}/2m_0 = m_0c^2\mathcal{O}\left(v^2/c^2\right)\phi^{(\pm)}$$

and

$$\begin{aligned} (\mathrm{i}\hbar\partial/\partial t - eA^0)\phi^{(+)} &= m_0c^2\left[+1 + \mathcal{O}\left(v^2/c^2\right)\right]\phi^{(+)} \text{ for positive states} \\ (\mathrm{i}\hbar\partial/\partial t - eA^0)\phi^{(-)} &= m_0c^2\left[-1 + \mathcal{O}\left(v^2/c^2\right)\right]\phi^{(-)} \text{ for negative states.} \end{aligned}$$

Up to the rest energy term m_0c^2, this equation coincides with the well-known nonrelativistic Schrödinger equation for spinless particles within an electromagnetic field. Correspondingly, in the case of negative solutions we obtain from (1.17) the equation

$$\left.\begin{aligned}\psi &= \begin{pmatrix}\mathcal{O}\left(v^2/c^2\right) \\ 1\end{pmatrix}\chi \\ \mathrm{i}\hbar\frac{\partial\chi}{\partial t} &= \left[-\frac{1}{2m_0}\left(\boldsymbol{p}-\frac{e}{c}\boldsymbol{A}\right)^2 - m_0c^2 + eA^0 + \mathcal{O}\left(\frac{v^4}{c^4}\right)\right]\chi\,.\end{aligned}\right\} \quad (1.56)$$

Combining equations (1.55) and (1.56) finally yields the Hamiltonian Klein-Gordon equation

$$\left.\begin{aligned}\mathrm{i}\hbar\frac{\partial\psi}{\partial t} &= H^{\mathrm{nr}}\psi \\ H^{\mathrm{nr}} &= \tau_3\left[m_0c^2 + \frac{1}{2m_0}\left(\boldsymbol{p}-\frac{e}{c}\boldsymbol{A}\right)^2\right] + eA^0 + \mathcal{O}\left(\frac{v^4}{c^4}\right)\,,\end{aligned}\right\} \quad (1.57)$$

correct up to order $\mathcal{O}\left(v^2/c^2\right)$ with the diagonal, G-Hermitean, and Hermitean Hamilton operator H^{nr}. Multiplying (1.57) from the left by $\psi^\dagger\tau_3$ and subsequently subtracting the adjoint equation leads to the continuity equation

$$\frac{\partial\rho(x)}{\partial t} + \boldsymbol{\nabla}\boldsymbol{j}(x) = 0\,,$$

with the positive or negative definite charge density

$$\rho = \psi^\dagger\tau_3\psi \approx \begin{cases}\varphi^*\varphi \geq 0 \text{ for positive states} \\ -\chi^*\chi \leq 0 \text{ for negative states}\,,\end{cases}$$

correct up to order $\mathcal{O}\left(v^2/c^2\right)$, and the charge current density

$$\boldsymbol{j} = \frac{\hbar}{2\mathrm{i}m}\left[\psi^\dagger\boldsymbol{\nabla}\psi - (\boldsymbol{\nabla}\psi^\dagger)\psi - \frac{2\mathrm{i}e}{\hbar c}\boldsymbol{A}\psi^\dagger\psi\right]\,.$$

The last expression is formally identical to the probability current of nonrelativistic quantum mechanics.[19]

---

[19] Since H^{nr} in (1.57) is Hermitean, one could define a positive definite probability density $\rho = \psi^\dagger\psi$ for particle and antiparticle (associated with a conserved total probability) that is connected to the corresponding probability current density $\boldsymbol{j}$ via $\partial\rho/\partial t + \boldsymbol{\nabla}\boldsymbol{j} = 0$. However, this relationship would hold only in the present representation. It would get lost if we changed the representation through a G-unitary transformation, because then the Hamilton operator would still be G-Hermitean but not Hermitean any more. The same argument holds for the diagonalization of the free canonical Klein-Gordon equation (Subsection 1.3.2) as well as for the general equation (next subsection).

1.4.2 Relativistic Corrections

In the nonrelativistic Klein-Gordon equation (1.57), the positive and negative solutions are completely decoupled, which is reflected by the diagonal form of the Hamilton operator H^{nr}. Here we can restrict ourselves to the upper or lower component to obtain a theory for particles or antiparticles that can be interpreted with respect to our generalized quantum mechanical formalism.

From Subsection 1.3.2 we know that, in the free case, the Hamiltonian Klein-Gordon equation can be exactly diagonalized by resorting to the Feshbach-Villars representation. Therefore, the question naturally arises whether there exists a method for the exact diagonalization of the general Klein-Gordon equation where positive and negative solutions are explicitly decoupled in all orders of v/c. As it turns out, this is not possible (see Exercise 8) due to a quantum field theoretical effect, the so-called *vacuum polarization*.[20]

On the other hand, with the help of the *Fouldy-Wouthuysen transformation*, it is always possible to diagonalize the Klein-Gordon-Hamilton operator to any desired order of v/c. Using this method, wave functions and operators are successively transformed (carried over into a new representation) in such a way that, in the new representation, the Hamilton operator is even (diagonal) up to the respective order of v/c. By disregarding its odd (antidiagonal) part, we again obtain two explicitly decoupled one-particle theories for particle and antiparticle that can be interpreted, as before, up to this order. Hence, this method can be regarded as a generalization of the Feshbach-Villars transformation for the free case discussed in Subsection 1.3.2.

In order to illustrate the use of the Fouldy-Wouthuysen transformation, we consider the Klein-Gordon equation (1.17) in the form[21]

$$m_0c^2K\psi = 0 \ , \ K = \tau_3 + \epsilon + \omega \ ,$$

where

$$\epsilon = -\frac{1}{m_0c^2}\left(\mathrm{i}\hbar\frac{\partial}{\partial t} - eA^0\right) + \frac{\tau_3}{2m_0^2c^2}\left(\boldsymbol{p} - \frac{e}{c}\boldsymbol{A}\right)^2 = \mathcal{O}(1) + \mathcal{O}\left(\frac{v^2}{c^2}\right) \quad (1.58)$$

or

$$\tau_3 + \epsilon = \mathcal{O}\left(\frac{v^2}{c^2}\right)$$

is an even (diagonal) operator and

$$\omega = \frac{\mathrm{i}\tau_2}{2m_0^2c^2}\left(\boldsymbol{p} - \frac{e}{c}\boldsymbol{A}\right)^2 = \mathcal{O}\left(\frac{v^2}{c^2}\right) \quad (1.59)$$

---

20 Vacuum polarization denotes the creation of charged particle-antiparticle pairs in strong electromagnetic fields, for example, in the close vicinity of the sources. These *virtual particles* interact with the considered particle – in addition to the external field – thus leading to a polarization of the vacuum (see Section 3.4).

21 Here we introduce dimensionless operators to facilitate the treatment of orders in v/c.

is odd (antidiagonal). Passing over to a new representation via the transformation

$$U = e^{iS} ,$$

the Klein-Gordon equation becomes

$$m_0 c^2 K' \psi' = 0 , \; \psi' = U\psi , \; K' = UKU^{-1} .$$

The Fouldy-Wouthuysen transformation is now characterized by the fact that the transformation U or S is chosen in such a way that K' can also be split into an even and an odd operator according to

$$K' = \tau_3 + \epsilon' + \omega' , \; \tau_3 + \epsilon' = \mathcal{O}\left(\frac{v^2}{c^2}\right) , \; \omega' = \mathcal{O}\left(\frac{v^4}{c^4}\right) \text{ (or higher)} ,$$

where ω' is suppressed (at least) by a factor of v^2/c^2 compared to ω and hence is (at least) two orders higher than $\tau_3 + \epsilon'$ or $\tau_3 + \epsilon$. Coming from this representation, we can again seek a transformation U' so that, in the corresponding new representation, we have

$$m_0 c^2 K'' \psi'' = 0 , \; \psi'' = U' \psi' , \; K'' = U' K' U'^{-1} = \tau_3 + \epsilon'' + \omega'' ,$$

with

$$\tau_3 + \epsilon'' = \mathcal{O}\left(\frac{v^2}{c^2}\right) , \; \omega'' = \mathcal{O}\left(\frac{v^6}{c^6}\right) \text{ (or higher)} .$$

This method can be iterated infinitely so that the order of the odd operator can be raised arbitrarily. Once the desired order of the odd operator is reached, the corresponding even operator yields the relativistic corrections to both one-particle theories with an error of this order.[22]

Starting from

$$K = \tau_3 + \epsilon + \omega , \; \epsilon \text{ from (1.58)} , \; \omega \text{ from (1.59)} ,$$

we now concretize our considerations and show how to obtain the diagonal Klein-Gordon-Hamilton operator, correct up to order $\mathcal{O}\left(v^4/c^4\right)$. To this end, we need the *Baker-Hausdorff expansion*[23]

---

[22] As each power of v/c corresponds to a factor of $v/c \sim p/m_0 c$, the Fouldy-Wouthuysen transformation can also be regarded as a series expansion in powers of $1/m_0$.

[23] This is easy to verify when considering the operator function

$$F(\lambda) = e^{i\lambda S} K e^{-i\lambda S} = \sum_{n=0}^{\infty} \frac{\lambda^n}{n!} \left.\frac{d^n F}{d\lambda^n}\right|_{\lambda=0} ,$$

with

$$\frac{d^n F}{d\lambda^n} = i^n e^{i\lambda S} [S, [S, \cdots, [S, K]] \cdots] e^{-i\lambda S} .$$

$$\begin{aligned} K' &= e^{iS} K e^{-iS} \\ &= K + i[S,K] + \frac{i^2}{2!}[S,[S,K]] + \frac{i^3}{3!}[S,[S,[S,K]]] + \ldots \ . \end{aligned} \tag{1.60}$$

Since the transformed operator K' contains the original operator K, we must choose S so that ω is eliminated in K'. As we see shortly, this can be achieved by the choice

$$U = e^{iS} \ , \ S = -\frac{i\tau_3\omega}{2} \ . \tag{1.61}$$

Taking into account $\tau_3\omega = -\omega\tau_3$ and $\tau_3\epsilon = \epsilon\tau_3$, we now calculate as follows:

$$\left.\begin{aligned} i[S,K] &= \frac{1}{2}\left[\tau_3\omega, \tau_3 + \epsilon + \omega\right] = -\omega + \tau_3\omega^2 + \frac{\tau_3}{2}[\omega,\epsilon] \\ \frac{i^2}{2}[S,[S,K]] &= \frac{1}{4}\left[\tau_3\omega, -\omega + \tau_3\omega^2 + \frac{\tau_3}{2}[\omega,\epsilon]\right] \\ &= -\frac{\tau_3\omega^2}{2} - \frac{\omega^3}{2} - \frac{1}{8}[\omega,[\omega,\epsilon]] \\ \frac{i^3}{6}[S,[S,[S,K]]] &= \frac{1}{6}\left[\tau_3\omega, -\frac{\tau_3\omega^2}{2} - \frac{\omega^3}{2} - \frac{1}{8}[\omega,[\omega,\epsilon]]\right] \\ &= \frac{\omega^3}{6} - \frac{\tau_3\omega^4}{6} + \frac{\tau_3}{48}[\omega,[\omega,[\omega,\epsilon]]] \ . \end{aligned}\right\} \tag{1.62}$$

Inserting these expressions into (1.60) yields

$$K' = \tau_3 + \epsilon' + \omega' \ ,$$

with[24]

$$\begin{array}{ccccccc} & & & \mathcal{O}\left(\frac{v^2}{c^2}\right) & \mathcal{O}\left(\frac{v^4}{c^4}\right) & \mathcal{O}\left(\frac{v^8}{c^8}\right) & \mathcal{O}\left(\frac{v^6}{c^6}\right) \\ & & & \downarrow & \downarrow & \downarrow & \downarrow \\ \tau_3 + \epsilon' & = & \tau_3 + \epsilon & +\frac{\tau_3\omega^2}{2} & -\frac{\tau_3\omega^4}{8} & -\frac{1}{8}[\omega,[\omega,\epsilon]] & + \ldots = \mathcal{O}\left(\frac{v^2}{c^2}\right) \end{array}$$

and

$$\omega' = -\frac{\omega^3}{3} + \frac{\tau_3}{2}[\omega,\epsilon] + \frac{\tau_3}{48}[\omega,[\omega,[\omega,\epsilon]]] + \ldots = \mathcal{O}\left(\frac{v^4}{c^4}\right) \ .$$

As desired, ω is eliminated from K' due to the first row of (1.62), and ω' is raised up by two orders.[25] It should be clear that all subsequent transformations have the same structure. Applying the transformation

$$U' = e^{iS'} \ , \ S' = -\frac{i\tau_3\omega'}{2}$$

[24] As long as the operator ϵ appears in commutators of the form $[\ldots,[\omega,\epsilon]\ldots]$, it holds that $\epsilon = \mathcal{O}\left(v^2/c^2\right)$.

[25] Note that the last equation leads to the even operator $K' = \tau_3 + \epsilon$, correct up to order $\mathcal{O}\left(v^2/c^2\right)$, as well as to the equation $m_0c^2(\tau_3 + \epsilon)\psi' = 0$, in accordance with (1.57).

to K', we find for K''

$$K'' = \tau_3 + \epsilon'' + \omega'' \;, \tag{1.63}$$

with

$$\begin{array}{ccccccc} & & \mathcal{O}\left(\frac{v^2}{c^2}\right) & \mathcal{O}\left(\frac{v^8}{c^8}\right) & \mathcal{O}\left(\frac{v^{16}}{c^{16}}\right) & \mathcal{O}\left(\frac{v^{10}}{c^{10}}\right) & \\ & & \downarrow & \downarrow & \downarrow & \downarrow & \\ \tau_3 + \epsilon'' & = & \tau_3 + \epsilon' & +\frac{\tau_3\omega'^2}{2} & -\frac{\tau_3\omega'^4}{8} & -\frac{1}{8}[\omega', [\omega', \epsilon']] & + \ldots = \mathcal{O}\left(\frac{v^2}{c^2}\right) \end{array}$$

and

$$\omega'' = -\frac{\omega'^3}{3} + \frac{\tau_3}{2}[\omega', \epsilon'] + \frac{\tau_3}{48}[\omega', [\omega', [\omega', \epsilon']]] + \ldots = \mathcal{O}\left(\frac{v^6}{c^6}\right) \;.$$

Ignoring all terms of order $\mathcal{O}\left(v^6/c^6\right)$ (and higher), K'' becomes an even operator and is given by

$$\begin{aligned} K'' &= \tau_3 + \epsilon' = \tau_3 + \epsilon + \frac{\tau_3\omega^2}{2} \\ &= \tau_3\left[1 + \frac{1}{2m_0^2c^2}\left(\boldsymbol{p} - \frac{e}{c}\boldsymbol{A}\right)^2 - \frac{1}{8m_0^4c^4}\left(\boldsymbol{p} - \frac{e}{c}\boldsymbol{A}\right)^4\right] \\ &\quad - \frac{1}{m_0c^2}\left(\mathrm{i}\hbar\frac{\partial}{\partial t} - eA^0\right) \;. \end{aligned}$$

The Hamiltonian Klein-Gordon equation follows from this as

$$\mathrm{i}\hbar\frac{\partial\psi''}{\partial t} = H''\psi'' \;,$$

with the diagonal, G-Hermitean, and Hermitean Hamilton operator

$$H'' = \tau_3\left[m_0c^2 + \frac{1}{2m_0}\left(\boldsymbol{p} - \frac{e}{c}\boldsymbol{A}\right)^2 - \frac{1}{8m_0^3c^2}\left(\boldsymbol{p} - \frac{e}{c}\boldsymbol{A}\right)^4\right] + eA^0$$

and the wave function

$$\psi''(x) = \mathrm{e}^{-\mathrm{i}\tau_3\omega'/2}\mathrm{e}^{-\mathrm{i}\tau_3\omega/2}\psi(x) \;.$$

As desired, we now have two explicitly decoupled $\mathcal{O}\left(v^4/c^4\right)$-one-particle theories, whose relativistic corrections result solely from the expansion of the relativistic kinetic energy

$$\sqrt{c^2\left(\boldsymbol{p} - \frac{e}{c}\boldsymbol{A}\right)^2 + m_0^2c^4} \;.$$

As we show in Exercise 9, electric interaction corrections do not come into play until a precision of order $\mathcal{O}\left(v^6/c^6\right)$.

Keep in mind that the Fouldy-Wouthuysen transformation U in (1.61) and all subsequent transformations U', ... are G-unitary operators since S, S', ... are G-Hermitean. From this follows the invariance of G-expectation values for operators transforming as $U[\cdot]U^{-1}$. Thus, it is indeed justified to speak of "Fouldy-Wouthuysen representations". However, there is an important restriction for the Hamilton operator itself, because the transition

$$K\psi = 0 \longrightarrow K'\psi' = 0 \ , \ K' = UKU^{-1} \ , \ \psi' = U\psi$$

implies the transformation

$$\mathrm{i}\hbar\frac{\partial\psi}{\partial t} = H\psi \longrightarrow \mathrm{i}\hbar\frac{\partial\psi'}{\partial t} = H'\psi' \ , \ H' = U\left(H - \mathrm{i}\hbar\frac{\partial}{\partial t}\right)U^{-1}$$

of the Klein-Gordon equation. This means that the G-expectation values of the original and transformed Hamilton operators coincide only if $\partial \boldsymbol{A}/\partial t = \boldsymbol{0}$.

Theorem 1.8: Fouldy-Wouthuysen transformation in the Klein-Gordon theory

The Fouldy-Wouthuysen transformation provides a systematic procedure for the diagonalization of the Klein-Gordon-Hamilton operator up to any (finite) order of v/c. Writing the Klein-Gordon equation (1.17) in the form

$$m_0c^2K^{(0)}\psi^{(0)} = 0 \ , \ K^{(0)} = \tau_3 + \epsilon^{(0)} + \omega^{(0)} \ ,$$

with the dimensionless even operators $\epsilon^{(0)}$, $\tau_3 + \epsilon^{(0)} = \mathcal{O}\left(v^2/c^2\right)$ and the odd operator $\omega^{(0)} = \mathcal{O}\left(v^2/c^2\right)$, and iterating the relations according to

$$\begin{aligned} K^{(n)} &= \tau_3 + \epsilon^{(n)} + \omega^{(n)} = U^{(n-1)}K^{(n-1)}U^{(n-1)^{-1}} \\ \psi^{(n)}(x) &= U^{(n-1)}\psi^{(n-1)}(x) \\ U^{(n)} &= \exp\left(-\frac{\mathrm{i}\tau_3\omega^{(n)}}{2}\right) \qquad \text{(G-unitary)} \ , \end{aligned}$$

one obtains new representations of the Klein-Gordon theory where

$$\tau_3 + \epsilon^{(n)} = \mathcal{O}\left(\frac{v^2}{c^2}\right) \ , \ \omega^{(n)} = \mathcal{O}\left(\frac{v^{2n+2}}{c^{2n+2}}\right) \ .$$

Ignoring the odd operator, the even part of $K^{(n)}$ leads to two explicitly decoupled one-particle theories for particle and antiparticle, correct up to order $\mathcal{O}\left(v^{2n}/c^{2n}\right)$. They can be interpreted with respect to our generalized quantum mechanical formalism.

As in the Feshbach-Villars representation, the one-particle operators of the Fouldy-Wouthuysen representations can be constructed by transforming the original (relativistic) operators appropriately, and subsequently separating its even (diagonal) part. Here the one-particle position operator exhibits similar properties as in the case of the Feshbach-Villars transformation. This means that, due to $[\boldsymbol{x}, U] \neq 0$, the Fouldy-Wouthuysen transformation is nonlocal, too, and leads to a smearing of the coordinate wave function or, rather, the position argument, with a magnitude comparable to the Compton wave length of the particle (see Exercise 9).

At the end of this section, it should be emphasized, once again, that the Fouldy-Wouthuysen method makes sense only in those cases where, firstly, the one-particle interpretation is applicable in the sense of Theorem 1.7 and, secondly, the Fouldy-Wouthuysen expansion converges. It is particularly not applicable to physical problems with strong or rapidly changing fields where particle creation and annihilation processes need to be taken into account.

Summary

- To lowest order, the nonrelativistic approximation of the Klein-Gordon theory leads to a diagonal, G-Hermitean, and Hermitean Hamilton operator (**nonrelativistic limit**). From this follow two explicitly decoupled one-particle theories for particle and antiparticle, the former coinciding with the laws of the nonrelativistic Schrödinger theory.
- In the general case, as opposed to the field-free case, the Klein-Gordon-Hamilton operator cannot be diagonalized exactly but only approximately. For this the **Fouldy-Wouthuysen method** can be used, where the Hamilton operator is successively diagonalized to increasing orders of v/c. Ignoring the odd part, one obtains a diagonal, G-Hermitean, and Hermitean Hamilton operator, correct up to the considered order of v/c, from which two explicitly decoupled one-particle theories can be derived.
- Like the Feshbach-Villars transformation, the **Fouldy-Wouthuysen transformation** is also nonlocal, leading to a smearing of the position argument over a range comparable to the Compton wave length of the particle.
- The Fouldy-Wouthuysen method is physically sensible only in those cases where the v/c-expansion converges and the one-particle interpretation is applicable.

Exercises

8. Diagonalizability of the Hamiltonian Klein-Gordon equation.

a) Show that only in the free case do the two components φ and χ of ψ each obey the canonical Klein-Gordon equation.

b) Evaluate the commutator

$$\left[\left(\boldsymbol{p}-\frac{e}{c}\boldsymbol{A}\right)^2, \mathrm{i}\hbar\frac{\partial}{\partial t}-eA^0\right] .$$

Solution.

To a) Our starting point is the Hamiltonian Klein-Gordon equation (1.17) in the form

$$\left[\left(\mathrm{i}\hbar\frac{\partial}{\partial t} - eA^0\right) - \frac{\tau_3 + \mathrm{i}\tau_2}{2m_0}\left(\boldsymbol{p} - \frac{e}{c}\boldsymbol{A}\right)^2 - \tau_3 m_0 c^2\right]\psi = 0 \ .$$

Multiplying this from the left by

$$\left[\left(\mathrm{i}\hbar\frac{\partial}{\partial t} - eA^0\right) + \frac{\tau_3 + \mathrm{i}\tau_2}{2m_0}\left(\boldsymbol{p} - \frac{e}{c}\boldsymbol{A}\right)^2 + \tau_3 m_0 c^2\right]$$

yields

$$\begin{aligned}
0 = &\left[\left(\mathrm{i}\hbar\frac{\partial}{\partial t} - eA^0\right)^2 - \frac{\tau_3 + \mathrm{i}\tau_2}{2m_0}\left(\mathrm{i}\hbar\frac{\partial}{\partial t} - eA^0\right)\left(\boldsymbol{p} - \frac{e}{c}\boldsymbol{A}\right)^2\right.\\
&-\tau_3 m_0 c^2\left(\mathrm{i}\hbar\frac{\partial}{\partial t} - eA^0\right) + \frac{\tau_3 + \mathrm{i}\tau_2}{2m_0}\left(\boldsymbol{p} - \frac{e}{c}\boldsymbol{A}\right)^2\left(\mathrm{i}\hbar\frac{\partial}{\partial t} - eA^0\right)\\
&-\frac{\tau_3 c^2}{2}(\tau_3 - \mathrm{i}\tau_2)\left(\boldsymbol{p} - \frac{e}{c}\boldsymbol{A}\right)^2 + \tau_3 m_0 c^2\left(\mathrm{i}\hbar\frac{\partial}{\partial t} - eA^0\right)\\
&\left.-\frac{\tau_3 c^2}{2}(\tau_3 + \mathrm{i}\tau_2)\left(\boldsymbol{p} - \frac{e}{c}\boldsymbol{A}\right)^2 - m_0^2 c^4\right]\psi\\
= &\left[\left(\mathrm{i}\hbar\frac{\partial}{\partial t} - eA^0\right)^2 - c^2\left(\boldsymbol{p} - \frac{e}{c}\boldsymbol{A}\right)^2 - m_0^2 c^4\right]\psi\\
&+\frac{\tau_3 + \mathrm{i}\tau_2}{2m_0}\left[\left(\boldsymbol{p} - \frac{e}{c}\boldsymbol{A}\right)^2, \left(\mathrm{i}\hbar\frac{\partial}{\partial t} - eA^0\right)\right]\psi \ .
\end{aligned}$$

From this follows that each component of ψ fulfills the Klein-Gordon equation only for the free case, since only then the commutator term vanishes exactly. Put differently, only in the free case is the Hamiltonian Klein-Gordon equation exactly diagonalizable. However, at weak or weakly varying fields, the commutator term is small compared to the others so that the canonical equation is solved by φ and χ at least approximately. In this case, an approximate diagonalization of the Hamiltonian equation seems possible.

To b) Taking into account

$$\left(\boldsymbol{p} - \frac{e}{c}\boldsymbol{A}\right)^2 = \boldsymbol{p}^2 - \frac{e}{c}(\boldsymbol{p}\boldsymbol{A}) - \frac{2e}{c}\boldsymbol{A}\boldsymbol{p} + \frac{e^2}{c^2}\boldsymbol{A}^2 \ ,$$

we have

$$\begin{aligned}
\left[\left(\boldsymbol{p} - \frac{e}{c}\boldsymbol{A}\right)^2, \mathrm{i}\hbar\frac{\partial}{\partial t} - eA^0\right] = &-e[\boldsymbol{p}^2, A^0] - \frac{\mathrm{i}e\hbar}{c}\left[(\boldsymbol{p}\boldsymbol{A}), \frac{\partial}{\partial t}\right]\\
&-\frac{2\mathrm{i}e\hbar}{c}\left[\boldsymbol{A}\boldsymbol{p}, \frac{\partial}{\partial t}\right] + \frac{e^2}{c}[\boldsymbol{A}\boldsymbol{p}, A^0]\\
&+\frac{\mathrm{i}e^2\hbar}{c^2}\left[\boldsymbol{A}^2, \frac{\partial}{\partial t}\right] \ .
\end{aligned}$$

The commutators on the right hand side are calculated as

$$[\boldsymbol{p}^2, A^0] = (\boldsymbol{p}^2 A^0) + 2(\boldsymbol{p}A^0)\boldsymbol{p}\;,\;\left[(\boldsymbol{p}\boldsymbol{A}), \frac{\partial}{\partial t}\right] = -\left(\boldsymbol{p}\frac{\partial \boldsymbol{A}}{\partial t}\right)$$

$$\left[\boldsymbol{A}\boldsymbol{p}, \frac{\partial}{\partial t}\right] = -\frac{\partial \boldsymbol{A}}{\partial t}\boldsymbol{p}\;,\;[\boldsymbol{A}\boldsymbol{p}, A^0] = \boldsymbol{A}(\boldsymbol{p}A^0)\;,\;\left[\boldsymbol{A}^2, \frac{\partial}{\partial t}\right] = -2\boldsymbol{A}\frac{\partial \boldsymbol{A}}{\partial t}\;.$$

Thus, it follows that

$$\begin{aligned}\left[\left(\boldsymbol{p}-\frac{e}{c}\boldsymbol{A}\right)^2, \mathrm{i}\hbar\frac{\partial}{\partial t} - eA^0\right] &= -2e\left[(\boldsymbol{p}A^0) - \frac{\mathrm{i}\hbar}{c}\frac{\partial \boldsymbol{A}}{\partial t}\right]\left(\boldsymbol{p}-\frac{e}{c}\boldsymbol{A}\right)\\ &\quad -e\left\{\boldsymbol{p}\left[(\boldsymbol{p}A^0) - \frac{\mathrm{i}\hbar}{c}\frac{\partial \boldsymbol{A}}{\partial t}\right]\right\}\\ &= -2\mathrm{i}e\hbar\boldsymbol{E}\left(\boldsymbol{p}-\frac{e}{c}\boldsymbol{A}\right) - \mathrm{i}e\hbar(\boldsymbol{p}\boldsymbol{E})\;, \end{aligned}\qquad (1.64)$$

with the electric field $\boldsymbol{E} = -(\boldsymbol{\nabla}A^0) - e\partial\boldsymbol{A}/c\partial t$.

9. Diagonal Hamiltonian Klein-Gordon equation up to $\mathcal{O}\left(v^6/c^6\right)$. Diagonalize the Hamiltonian Klein-Gordon equation (1.17) up to order $\mathcal{O}\left(v^6/c^6\right)$.

Solution. If we apply the Fouldy-Wouthuysen transformation

$$U'' = \mathrm{e}^{\mathrm{i}S''}\;,\;S'' = -\frac{\mathrm{i}\tau_3\omega''}{2}$$

to K'' from (1.63), we obtain the expression

$$K''' = \tau_3 + \epsilon''' + \omega'''\;,$$

with

$$\begin{array}{ccccccc} & & \mathcal{O}\left(\frac{v^2}{c^2}\right) & \mathcal{O}\left(\frac{v^{12}}{c^{12}}\right) & \mathcal{O}\left(\frac{v^{24}}{c^{24}}\right) & \mathcal{O}\left(\frac{v^{14}}{c^{14}}\right) & \\ & & \downarrow & \downarrow & \downarrow & \downarrow & \\ \tau_3 + \epsilon''' & = \tau_3 + \epsilon'' & +\frac{\tau_3\omega''^2}{2} & -\frac{\tau_3\omega''^4}{8} & & -\frac{1}{8}[\omega'', [\omega'', \epsilon'']] & + \ldots = \mathcal{O}\left(\frac{v^2}{c^2}\right)\end{array}$$

and

$$\omega''' = -\frac{\omega''^3}{3} + \frac{\tau_3}{2}[\omega'', \epsilon''] + \frac{\tau_3}{48}[\omega'', [\omega'', [\omega'', \epsilon'']]] + \ldots = \mathcal{O}\left(\frac{v^8}{c^8}\right)\;.$$

Ignoring all terms of order $\mathcal{O}\left(v^8/c^8\right)$ (and higher) leads to the even operator

$$K''' = \tau_3 + \epsilon'' = \tau_3 + \epsilon' = \tau_3 + \epsilon + \frac{\tau_3\omega^2}{2} - \frac{1}{8}[\omega, [\omega, \epsilon]]\;.$$

Because of (1.64), the last term is simplified to

$$\begin{aligned}[\omega, \epsilon] &= \frac{\mathrm{i}\tau_2}{2m_0^3c^4}\left[\left(\boldsymbol{p}-\frac{e}{c}\boldsymbol{A}\right)^2, \mathrm{i}\hbar\frac{\partial}{\partial t} - eA^0\right] + \frac{\mathrm{i}}{4m_0^4c^4}\left(\boldsymbol{p}-\frac{e}{c}\boldsymbol{A}\right)^4[\tau_2, \tau_3]\\ &= \frac{e\hbar\tau_2}{m_0^3c^4}\boldsymbol{E}\left(\boldsymbol{p}-\frac{e}{c}\boldsymbol{A}\right) + \frac{e\hbar\tau_2}{2m_0^3c^4}(\boldsymbol{p}\boldsymbol{E}) - \frac{\tau_1}{2m_0^4c^4}\left(\boldsymbol{p}-\frac{e}{c}\boldsymbol{A}\right)^4\end{aligned}$$

$$\Longrightarrow [\omega,[\omega,\epsilon]] = \frac{\mathrm{i}e\hbar}{2m_0^5c^6}\left[\left(\boldsymbol{p}-\frac{e}{c}\boldsymbol{A}\right)^2, \boldsymbol{E}\left(\boldsymbol{p}-\frac{e}{c}\boldsymbol{A}\right)\right] + \frac{\mathrm{i}e\hbar}{4m_0^5c^6}\left[\left(\boldsymbol{p}-\frac{e}{c}\boldsymbol{A}\right)^2, (\boldsymbol{p}\boldsymbol{E})\right] - \frac{\tau_3}{2m_0^6c^6}\left(\boldsymbol{p}-\frac{e}{c}\boldsymbol{A}\right)^6 .$$

All in all, the Hamiltonian Klein-Gordon equation

$$\mathrm{i}\hbar\frac{\partial\psi'''}{\partial t} = H'''\psi''' ,$$

follows, correct up to order $\mathcal{O}\left(v^6/c^6\right)$, with the diagonal, G-Hermitean, and Hermitean Hamilton operator

$$\begin{aligned} H''' = \tau_3 \Bigg[& m_0c^2 \\ & + \frac{1}{2m_0}\left(\boldsymbol{p}-\frac{e}{c}\boldsymbol{A}\right)^2 - \frac{1}{8m_0^3c^2}\left(\boldsymbol{p}-\frac{e}{c}\boldsymbol{A}\right)^4 + \frac{1}{16m_0^5c^4}\left(\boldsymbol{p}-\frac{e}{c}\boldsymbol{A}\right)^6\Bigg] \\ & + eA^0 - \frac{\mathrm{i}e\hbar}{16m_0^4c^5}\left[\left(\boldsymbol{p}-\frac{e}{c}\boldsymbol{A}\right)^2, \boldsymbol{E}\left(\boldsymbol{p}-\frac{e}{c}\boldsymbol{A}\right)\right] \\ & - \frac{\mathrm{i}e\hbar}{32m_0^4c^5}\left[\left(\boldsymbol{p}-\frac{e}{c}\boldsymbol{A}\right)^2, (\boldsymbol{p}\boldsymbol{E})\right] \end{aligned} \qquad (1.65)$$

and the wave function

$$\psi'''(x) = \mathrm{e}^{-\mathrm{i}\tau_3\omega''/2}\mathrm{e}^{-\mathrm{i}\tau_3\omega'/2}\mathrm{e}^{-\mathrm{i}\tau_3\omega/2}\psi(x) .$$

The appearance of the relativistic corrections in (1.65) can be understood as a consequence of the nonlocality of the Fouldy-Wouthuysen transformation and the accompanying smearing of the position argument $\boldsymbol{x}$: the effective potential acting on the wave function at $\boldsymbol{x}$ in a certain Fouldy-Wouthuysen representation is composed of the contributions of the original potential averaged over an area around $\boldsymbol{x}$. Therefore, the whole potential has the form of a multipole expansion of the original potential.

1.5 Simple One-Particle Systems

The last section of this chapter deals with some simple Klein-Gordon one-particle systems where the canonical representation is used throughout. As an extension of the Klein paradox from Subsection 1.3.4, we initially consider the onedimensional potential well and discuss its types of solutions particularly with respect to their one-particle interpretation. Then we turn to the problem of spherically symmetric potentials that can be, analogously to the nonrelativistic case, related back to a radial equation by separating the angular-dependent part. Examples of this *radial Klein-Gordon equation* are the free particle, the spherically symmetric potential well, and the Coulomb potential, the last providing a simple description of pion atoms. At the end we modify the Coulomb potential by an oscillator-like potential term in order to take into account the finite extent of atomic nuclei. In so doing, the fundamental limits of the one-particle concept will become apparent once again.

1.5.1 Potential Well

We start our investigation of simple one-particle systems with a spin-0 particle bound within or coming from the left and scattering against a onedimensional potential well of the form

$$eA^{(0)}(z) = V(z) = \left\{ \begin{array}{lll} 0 & \text{for } -a < z < a & \text{(area II)} \\ V_0 & \text{else} & \text{(area I, III)} \end{array} \right\} , \quad V_0 > 0 \, . \tag{1.66}$$

To solve the corresponding Klein-Gordon equation, we initially separate the time-dependent part, as in the case of the potential step in Subsection 1.3.4:

$$\phi(z,t) = \Phi(z)\mathrm{e}^{-\mathrm{i}Et/\hbar} \, .$$

This leads us to the stationary equation (1.52) with $V(z)$ from (1.66). Before tackling this equation in detail, we can try, as in the case of the Klein paradox, to get a qualitative overview of all possible configurations (see Figure 1.5).

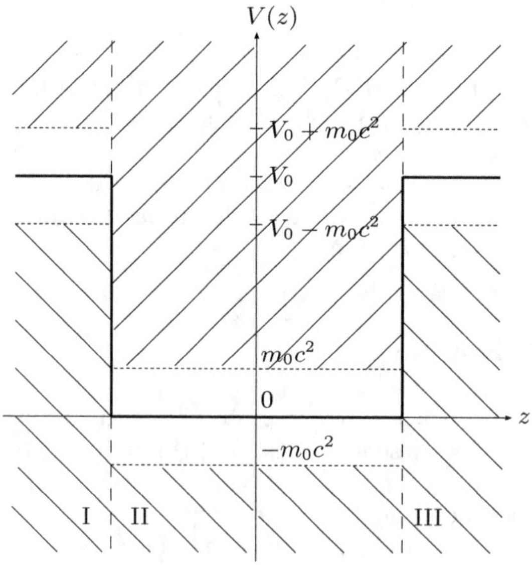

Fig. 1.5. Energy intervals of the onedimensional potential well.

1. case: $E > V_0 + m_0c^2$. In all three areas I ($z < -a$), II ($-a < z < a$), and III ($z > a$), we expect to have oscillating Klein-Gordon solutions with a positive charge density. This can be interpreted as a normal scattering of a particle with charge $+e$ against the (from its point of view) attractive potential barrier, since at the area borders, only "allowed" positive energy regimes are in contact.

2. case: $V_0 - m_0c^2 < E < V_0 + m_0c^2$, $E > m_0c^2$. In this case, the solutions of the "allowed" positive energy regime of II dive into the "forbidden" positive ($0 < E - V_0 < m_0c^2$) or negative ($-m_0c^2 < E - V_0 < 0$) energy regimes of I and III. Thus, the solutions in I and III are expected to fall off exponentially with positive or negative charge density, while the solutions in II are oscillating with positive charge density. This situation corresponds to a bound particle of charge $+e$.

3. case: $m_0c^2 < E < V_0 - m_0c^2 \Longrightarrow V_0 > 2m_0c^2$. Here the "allowed" positive energy regime of II turns into the "allowed" negative energy regimes of I and III. The solutions should therefore oscillate in all three areas with positive charge density in II and negative charge densities in I and III. Beyond the one-particle picture, this can be viewed as the scattering of an antiparticle of charge $-e$ against the (from its point of view) repulsive potential barrier, whereas *quasi-bound particle resonances* appear in area II.

4. case: $-m_0c^2 < E < m_0c^2$. This energy interval encompasses the "forbidden" positive ($0 < E < m_0c^2$) and negative ($-m_0c^2 < E < 0$) energy regimes in II that turn into the "allowed" negative energy regimes of I and III. We therefore expect exponentially decreasing solutions with positive and negative charge density in area II and oscillating solutions with negative charge density in I and III. This case corresponds to the tunneling of an antiparticle of charge $-e$ through the (from its point of view) repulsive potential barrier.

5. case: $E < -m_0c^2$. Only "allowed" negative energies are involved here implying oscillating solutions with negative charge densities in all three areas. This is the case of a normal scattering of an antiparticle of charge $-e$ at the (from its point of view) repulsive potential barrier.

Let us now consider the above cases in detail:

1., 3., and 5. case in detail. In these three scattering cases, we assume a particle or an antiparticle coming from the left. For the solution of (1.52) in the areas I, II, and III, we make the ansatz

$$\left.\begin{aligned} \Phi_{\rm I}(z) &= \Phi_{\rm in}(z) + \Phi_{\rm ref}(z) \\ &\qquad \Phi_{\rm in}(z) = A\mathrm{e}^{\mathrm{i}k_1 z} \ , \ \Phi_{\rm ref}(z) = B\mathrm{e}^{-\mathrm{i}k_1 z} \\ \Phi_{\rm II}(z) &= C\mathrm{e}^{\mathrm{i}k_2 z} + D\mathrm{e}^{-\mathrm{i}k_2 z} \\ \Phi_{\rm III}(z) &= \Phi_{\rm trans}(z) = E\mathrm{e}^{\mathrm{i}k_1 z} \ , \end{aligned}\right\} \quad (1.67)$$

with

$$k_1 = \pm\sqrt{\frac{(E-V_0)^2 - m_0^2c^4}{c^2\hbar^2}} \;,\; k_2 = \pm\sqrt{\frac{E^2 - m_0^2c^4}{c^2\hbar^2}}$$

and $k_1 = +|k_1|, k_2 = +|k_2|$ in the 1. case, $k_1 = -|k_1|, k_2 = +|k_2|$ in the 3. case, and $k_1 = -|k_1|, k_2 = -|k_2|$ in the 5. case. The continuity conditions of $\Phi(z)$ and $\Phi'(z)$ at the area borders $z = \pm a$ lead to the following conditional equations for the integration constants A, B, C, D, and E:

$$\left.\begin{aligned}
A\mathrm{e}^{-\mathrm{i}k_1a} + B\mathrm{e}^{\mathrm{i}k_1a} &= C\mathrm{e}^{-\mathrm{i}k_1a} + D\mathrm{e}^{\mathrm{i}k_1a} \\
k_1\left(A\mathrm{e}^{-\mathrm{i}k_1a} - B^{\mathrm{i}k_1a}\right) &= k_2\left(C\mathrm{e}^{-\mathrm{i}k_2a} - D\mathrm{e}^{\mathrm{i}k_2a}\right) \\
C\mathrm{e}^{\mathrm{i}k_2a} + D\mathrm{e}^{-\mathrm{i}k_2a} &= E\mathrm{e}^{\mathrm{i}k_1a} \\
k_2\left(C\mathrm{e}^{\mathrm{i}k_2a} - D\mathrm{e}^{-\mathrm{i}k_2a}\right) &= k_1E\mathrm{e}^{\mathrm{i}k_1a} \;.
\end{aligned}\right\} \quad (1.68)$$

After some calculations, one finds from this the reflection and transmission coefficients

$$\left.\begin{aligned}
R &= -\frac{j_{\mathrm{ref}}}{j_{\mathrm{in}}} = \frac{(k_1^2-k_2^2)^2\sin^2 2k_2a}{4k_1^2k_2^2 + (k_1^2-k_2^2)^2\sin^2 2k_2a} \\
T &= \frac{j_{\mathrm{trans}}}{j_{\mathrm{in}}} = \frac{4k_1^2k_2^2}{4k_1^2k_2^2 + (k_1^2-k_2^2)^2\sin^2 2k_2a} = 1 - R \;.
\end{aligned}\right\} \quad (1.69)$$

Both coefficients oscillate between zero and one depending on k_2 or E. An interesting special case is $\sin 2k_2a = 0$, i.e.

$$E^2 = n^2\frac{c^2\hbar^2\pi^2}{4a^2} + m_0^2c^4 \;,\; n = 1, 2, \ldots \;,$$

where the reflection coefficient vanishes exactly.

4. case in detail. In this tunneling case, we also assume an antiparticle coming from the left. For the Klein-Gordon solution, we can therefore apply ansatz (1.67), where k_1 and k_2 are chosen as

$$k_1 = -\sqrt{\frac{(E-V_0)^2 - m_0^2c^4}{c^2\hbar^2}} \;,\; k_2 = \mathrm{i}\kappa_2 \;,\; \kappa_2 = \sqrt{\frac{m_0^2c^4 - E^2}{c^2\hbar^2}} \;.$$

Taking into account $k_2^2 = -\kappa_2^2$ and $\sin^2 2k_2a = -\sinh^2 2\kappa_2a$, the reflection and transmission coefficients follow immediately from (1.69) as

$$\begin{aligned}
R &= \frac{(k_1^2+\kappa_2^2)^2\sinh^2 2\kappa_2a}{4k_1^2\kappa_2^2 + (k_1^2+\kappa_2^2)^2\sinh^2 2\kappa_2a} \\
T &= \frac{4k_1^2\kappa_2^2}{4k_1^2\kappa_2^2 + (k_1^2+\kappa_2^2)^2\sinh^2 2\kappa_2a} = 1 - R \;.
\end{aligned}$$

As before, both coefficients range between zero and one. However, the transition coefficient and, with it, the penetration probability of the potential barrier decreases exponentially with a and increases exponentially with E.

2. case in detail. Contrary to all the other cases, we expect bound states here. Therefore, our solution ansatz for (1.52) in the areas I, II, and III is

$$\begin{aligned} \Phi_{\mathrm{I}}(z) &= A\mathrm{e}^{\kappa_1 z} \\ \Phi_{\mathrm{II}}(z) &= B\cos k_2 z + C\sin k_2 z \\ \Phi_{\mathrm{III}}(z) &= D\mathrm{e}^{-\kappa_1 z} \end{aligned}$$

$$\kappa_1 = \sqrt{\frac{m_0^2c^4 - (E-V_0)^2}{c^2\hbar^2}} \ , \ k_2 = \sqrt{\frac{E^2 - m_0^2c^4}{c^2\hbar^2}} \ ,$$

where Φ_{II} is written in trigonometric form for convenience. The continuity conditions at the area borders yield the equations

$$\begin{aligned} A\mathrm{e}^{-\kappa_1 a} &= B\cos k_2 a - C\sin k_2 a \\ \kappa_1 A\mathrm{e}^{-\kappa_1 a} &= k_2 B\sin k_2 a + k_2 C\cos k_2 a \\ D\mathrm{e}^{-\kappa_1 a} &= B\cos k_2 a + C\sin k_2 a \\ -\kappa_1 D\mathrm{e}^{-\kappa_1 a} &= -k_2 B\sin k_2 a + k_2 C\cos k_2 a \ . \end{aligned}$$

Combining the first two and the last two equations leads to

$$\kappa_1 = k_2\frac{B\sin k_2 a + C\cos k_2 a}{B\cos k_2 a - C\sin k_2 a} = k_2\frac{B\sin k_2 a - C\cos k_2 a}{B\cos k_2 a + C\sin k_2 a} \ ,$$

that, in turn, implies $BC = 0$. This means that we have to distinguish between two cases from which follow two different quantization conditions for the energy E:[26]

2.a: $C = 0 \Longrightarrow A = D$.

$$\tan k_2 a = \frac{\kappa_1}{k_2} \ .$$

2.b: $B = 0 \Longrightarrow A = -D$.

$$-\cot k_2 a = \tan\left(k_2 a + \frac{\pi}{2}\right) = \frac{\kappa_1}{k_2} \ .$$

By (numerically) solving these two equations, we finally obtain the possible energy values of the bound states in area II.

The transmission coefficient of the 1., 3., 4., and 5. case is depicted as a function of the energy E/m_0c^2 in Figure 1.6. A potential well of width $a = 6\hbar/m_0c$ and of height $V_0 = 3m_0c^2$ is chosen here (so that the condition $V_0 > 2m_0c^2$ for the occurrence of the 3. case is fulfilled). For the "proper"

[26] Note that the Klein-Gordon equation (1.52) is parity invariant due to the symmetric form of the potential (1.66). This means that it is invariant under the replacement $z \longrightarrow -z$. From this follows that, with $\Phi(z)$, the wave function $\Phi(-z)$ is a Klein-Gordon solution to the same energy E. Due to the linearity of (1.52), these solutions can be combined to give the new solutions

$$\Phi_{\pm}(z) = \Phi(z) \pm \Phi(-z) \ , \ \Phi_{\pm}(z) = \pm\Phi_{\pm}(-z)$$

with a defined parity. The case 2.a corresponds to solutions with positive and the case 2.b to solutions with negative parity.

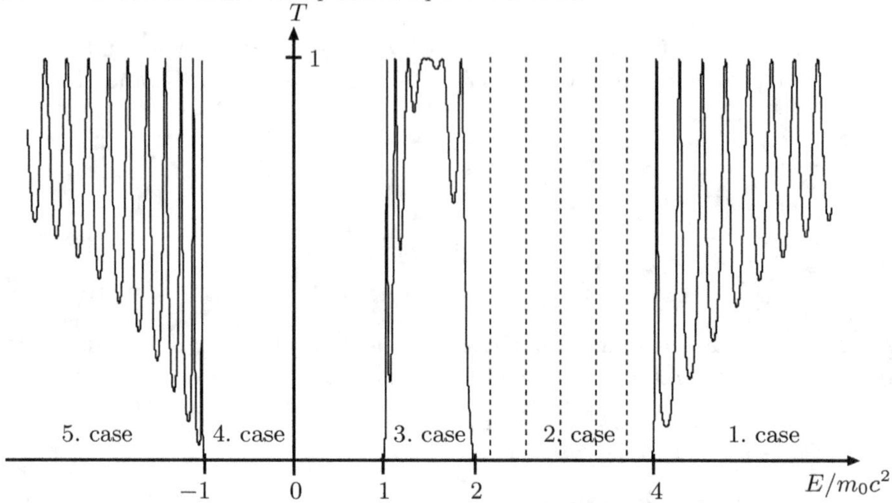

Fig. 1.6. Transmission coefficient of a onedimensional potential well of height $V_0 = 3m_0c^2$ and width $a = 6\hbar/m_0c$ as a function of E. The dashed lines indicate the energy values of the bound states.

scattering cases (1. and 5. case), T shows the typical oscillatory behavior approaching one with decreasing amplitude at the outskirts of the energy range. Since the transmission coefficient increases exponentially with the width a in the 4. case (tunneling case), it is practically zero there. Between the tunneling case and the 2. case lies the resonance case 3, where T is oscillating, too, and which cannot be interpreted within the one-particle picture in the strict sense; in the vicinity of the transmission maxima, an incoming antiparticle hits against a quasi-bound state appearing as particle resonance so that the antiparticle can penetrate the (from its point of view) repulsive potential barrier almost without resistance.

1.5.2 Radial Klein-Gordon Equation

If the Klein-Gordon equation contains a centrally symmetric potential $eA^0(x) = V(\boldsymbol{x}) = V(|\boldsymbol{x}|)$, $\boldsymbol{A} = \boldsymbol{0}$, it possesses a central symmetry. As in non-relativistic quantum mechanics, it is then appropriate to rewrite the Klein-Gordon equation in terms of spherical coordinates,

$$x = r\cos\varphi\sin\theta\ ,\ \ y = r\sin\varphi\sin\theta\ ,\ \ z = r\cos\theta\ ,\ \ r = |\boldsymbol{x}|\ ,$$

in order to separate the angular and radial parts. Our starting point is the Klein-Gordon equation

$$\left(\mathrm{i}\hbar\frac{\partial}{\partial t} - V\right)^2 \phi(x) + \left(c^2\hbar^2\boldsymbol{\nabla}^2 - m_0^2c^4\right)\phi(x) = 0\ ,$$

which can be immediately cast into the time-independent Klein-Gordon equation

$$\left[(E-V)^2+c^2\hbar^2\boldsymbol{\nabla}^2-m_0^2c^4\right]\Phi(\boldsymbol{x})=0\;,\;\;\phi(x)=\Phi(\boldsymbol{x})\mathrm{e}^{-\mathrm{i}Et/\hbar}\;. \tag{1.70}$$

Next we write the momentum term as

$$\hbar^2\boldsymbol{\nabla}^2=-p_r^2-\frac{\boldsymbol{L}^2}{r^2}\;,$$

where

$$p_r=-\mathrm{i}\hbar\left(\frac{\partial}{\partial r}+\frac{1}{r}\right)\;,\;\;p_r^2=\frac{\partial^2}{\partial r^2}+\frac{2}{r}\frac{\partial}{\partial r}$$

denotes the *radial momentum* and $\boldsymbol{L}=\boldsymbol{x}\times\boldsymbol{p}$ the angular momentum whose eigenfunctions are given by the spherical harmonics $Y_{l,m}(\theta,\varphi)$ (see Appendix A.3):

$$\begin{aligned}\boldsymbol{L}^2Y_{l,m}&=\hbar^2l(l+1)Y_{l,m}\;,\;\;l=0,1,2,\ldots\\ L_zY_{l,m}&=\hbar mY_{l,m}\;,\;\;m=-l,\ldots,l\;.\end{aligned}$$

With this (1.70) turns into

$$\left[(E-V)^2+c^2\hbar^2\left(\frac{\partial^2}{\partial r^2}+\frac{2}{r}\frac{\partial}{\partial r}\right)-\frac{c^2\boldsymbol{L}^2}{r^2}-m_0^2c^4\right]\Phi(\boldsymbol{x})=0\;.$$

Using the ansatz

$$\Phi(\boldsymbol{x})=g_l(r)Y_{l,m}(\theta,\varphi)\;,$$

the angular-dependent part can now be separated so that we are finally led to

Theorem 1.9: Radial Klein-Gordon equation for centrally symmetric potentials

The solutions to the time-independent Klein-Gordon equation with a centrally symmetric potential,

$$\left\{[E-V(r)]^2+c^2\hbar^2\boldsymbol{\nabla}^2-m_0^2c^4\right\}\Phi(\boldsymbol{x})=0\;,$$

can be written in the spherical representation as

$$\Phi_{l,m}(r,\theta,\varphi)=g_l(r)Y_{l,m}(\theta,\varphi)\;.$$

The functions g_l obey the *radial Klein-Gordon equation*

$$\left[\frac{\mathrm{d}^2}{\mathrm{d}r^2}+\frac{2}{r}\frac{\mathrm{d}}{\mathrm{d}r}-\frac{l(l+1)}{r^2}+k^2\right]g_l(r)=0\;,\;\;k^2=\frac{(E-V)^2-m_0^2c^4}{c^2\hbar^2} \tag{1.71}$$

or, using $g_l(r)=u_l(r)/r$,

$$\left[\frac{\mathrm{d}^2}{\mathrm{d}r^2}-\frac{l(l+1)}{r^2}+k^2\right]u_l(r)=0\;. \tag{1.72}$$

▷

These solutions also satisfy the relations

$$\begin{aligned} \boldsymbol{L}^2\Phi_{l,m}(r,\theta,\varphi) &= \hbar^2 l(l+1)\Phi_{l,m}(r,\theta,\varphi)\ ,\ l=0,1,2,\ldots \\ L_z\Phi_{l,m}(r,\theta,\varphi) &= \hbar m\Phi_{l,m}(r,\theta,\varphi)\ ,\ m=-l,\ldots,l \\ \left[\Phi_{l,m}\right]_P(r,\theta,\varphi) &= (-1)^l\Phi_{l,m}(r,\theta,\varphi)\ . \end{aligned}$$

The last equation follows from $Y_{l,m}(\pi-\theta,\varphi+\pi) = (-1)^l Y_{l,m}(\theta,\varphi)$ [P=active parity transformation].

Note that (1.71) and (1.72) are formally identical to the corresponding radial equations of the Schrödinger theory, with $k^2 = 2m_0(E-V)/\hbar^2$. Furthermore, the following points should be kept in mind:

- Of the solutions g_l and u_l, only those are physically sensible that are integrable with respect to the G-scalar product. Contrary to the nonrelativistic case, these integrability conditions depend on the potential, and the following integrals must exist:

$$\int_0^\infty \mathrm{d}r r^2 g_l^2(r)\ ,\ \int_0^\infty \mathrm{d}r r^2 g_l^2(r)V(r) \quad \text{or} \quad \int_0^\infty \mathrm{d}r u_l^2(r)\ ,\ \int_0^\infty \mathrm{d}r u_l^2(r)V(r)\ .$$

- If the potential diverges more slowly than $1/r^2$ at the origin: $\lim_{r\to 0} r^2V(r) = 0$, the equation

$$\frac{\mathrm{d}^2u_l}{\mathrm{d}r^2} - \frac{l(l+1)}{r^2}u_l = 0$$

holds around the origin whose solutions are $u_l(r) \sim r^{l+1}$ (regular solution) and $u_l(r) \sim r^{-l}$.

- If, for $r \to \infty$, the potential falls off faster than $1/r$: $\lim_{r\to\infty} rV(r) = 0$, we have for large r

$$\frac{\mathrm{d}^2u}{\mathrm{d}r^2} + \frac{E^2 - m_0^2c^4}{c^2\hbar^2}u = 0\ .$$

The solutions to this equation behave asymptotically as

$$\begin{aligned} |E| < m_0c^2 &: \quad u(r) \sim \mathrm{e}^{-kr}, \mathrm{e}^{kr} \\ |E| > m_0c^2 &: \quad u(r) \sim \mathrm{e}^{\mathrm{i}kr}, \mathrm{e}^{-\mathrm{i}kr}\ ,\ k^2 = \left|\frac{E^2 - m_0^2c^4}{c^2\hbar^2}\right|\ . \end{aligned}$$

1.5.3 Free Particle and Spherically Symmetric Potential Well

As an application of centrally symmetric potential problems, we first consider the easiest case of a free spin-0 particle ($V = 0$). The discussion can be carried out analogously to the nonrelativistic case. Using the substitutions

$$\rho = kr \ , \ g_l(r) = \hat{g}_l(\rho) \ , \ k^2 = \frac{E^2 - m_0^2c^4}{c^2\hbar^2} \ ,$$

the radial Klein-Gordon equation (1.71) becomes the spherical Bessel equation (see Appendix A.2)

$$\left[\frac{\mathrm{d}^2}{\mathrm{d}\rho^2} + \frac{2}{\rho}\frac{\mathrm{d}}{\mathrm{d}\rho} + 1 - \frac{l(l+1)}{\rho^2}\right]\hat{g}_l(\rho) = 0 \ . \tag{1.73}$$

Its solutions are the spherical Bessel functions whose form and asymptotic behavior are given by

$$j_l(\rho) = (-\rho)^l \left(\frac{1}{\rho}\frac{\mathrm{d}}{\mathrm{d}\rho}\right)^l \frac{\sin\rho}{\rho} \sim \begin{cases} \dfrac{\rho^l}{(2l+1)!!} & \text{for } \rho \to 0 \\ \dfrac{\sin(\rho - l\pi/2)}{\rho} & \text{for } \rho \to \infty \end{cases}$$

$$n_l(\rho) = (-\rho)^l \left(\frac{1}{\rho}\frac{\mathrm{d}}{\mathrm{d}\rho}\right)^l \frac{\cos\rho}{\rho} \sim \begin{cases} \dfrac{(2l-1)!!}{\rho^{l+1}} & \text{for } \rho \to 0 \\ \dfrac{\cos(\rho - l\pi/2)}{\rho} & \text{for } \rho \to \infty \ . \end{cases}$$

Special combinations of these functions called Hankel functions are of particular interest:

$$\begin{aligned} h_l^{(+)}(\rho) &= n_l(\rho) + \mathrm{i}j_l(\rho) \stackrel{\rho\to\infty}{\longrightarrow} \frac{\mathrm{e}^{\mathrm{i}(\rho - l\pi/2)}}{\rho} \\ h_l^{(-)}(\rho) &= n_l(\rho) - \mathrm{i}j_l(\rho) \stackrel{\rho\to\infty}{\longrightarrow} \frac{\mathrm{e}^{-\mathrm{i}(\rho - l\pi/2)}}{\rho} \ . \end{aligned}$$

Their asymptotic behavior for $k^2 > 0$ corresponds to outgoing and incoming spherical waves, respectively. Depending on E, we now have to distinguish between two cases:

- $|E| < m_0c^2$: here $\hat{g}_l(\rho) = h_l^{(+)}(\rho)$ is the only bounded solution to (1.73). However, it has a pole of order $l+1$ at the origin. Therefore, the eigenvalue problem has no solution; in accordance with our expectations, there are no free (anti)particles with an energy E within the "forbidden" interval $-m_0c^2 < E < m_0c^2$.

- $|E| > m_0c^2$: in this case the equation (1.73) has exactly one solution bounded everywhere, namely, $\hat{g}_l(\rho) = j_l(\rho)$. The physically possible solution to the Klein-Gordon equation (1.71) is, therefore,

  $$g_l(r) = j_l(kr) \ .$$

Note that the above considerations can easily be carried over to the case of a potential $V(r)$ that can be split into regions of constant potential values V_i. In this case E needs to be replaced by $E - V_i$ in each region.

Spherically symmetric potential well. Keeping the above results in mind, we now turn to the problem of a spin-0 particle within a spherically symmetric potential well of the form

$$eA^0(r) = V(r) = \left\{ \begin{array}{l} -V_0 \text{ for } r < a \text{ (area I)} \\ \\ \quad 0 \quad \text{for } r > a \text{ (area II)} \end{array} \right\} , \quad V_0 > 0 \, .$$

We first find the possible solutions to (1.71) within the areas I and II. In the inner area I, the regular solution at the origin is

- $|E + V_0| > m_0c^2$: $g_l^{(\mathrm{I})}(r) = A_l j_l(k_1 r)$, $k_1 = \sqrt{\dfrac{(E+V_0)^2 - m_0^2c^4}{c^2\hbar^2}}$

or

- $|E + V_0| < m_0c^2$: $g_l^{(\mathrm{I})}(r) = A_l j_l(\mathrm{i}\kappa_1 r)$, $\kappa_1 = \sqrt{\dfrac{m_0^2c^4 - (E+V_0)^2}{c^2\hbar^2}}$.

In the outer area II, two cases must be distinguished:

- $|E| < m_0c^2$ (bound states): here

$$g_l^{(\mathrm{II})}(r) = B_l h_l^{(+)}(\mathrm{i}\kappa_2 r) \, , \quad \kappa_2 = \sqrt{\frac{m_0^2c^4 - E^2}{c^2\hbar^2}}$$

is the only solution bounded at infinity. The continuity conditions at $r = a$,

$$g_l^{(\mathrm{I})}(a) = g_l^{(\mathrm{II})}(a) \, , \quad \frac{\mathrm{d}}{\mathrm{d}r} g_l^{(\mathrm{I})}(r)\bigg|_{r=a} = \frac{\mathrm{d}}{\mathrm{d}r} g_l^{(\mathrm{II})}(r)\bigg|_{r=a} \, ,$$

determine the ratio of the integration constants A_l and B_l. Both conditions can be satisfied simultaneously only for certain discrete values of E; they determine the energy levels of the bound states. For l=0 states and additionally assuming $E + V_0 > m_0c^2$ (2. case, see below), follows the condition

$$\tan k_1 a = -\frac{k_1}{\kappa_2} \, . \tag{1.74}$$

- $|E| > m_0c^2$ (unbound states): the general solution is a linear combination of the spherical Bessel functions that can be written as

$$g_l^{(\mathrm{II})}(r) = B_l \left[j_l(k_2 r)\cos\delta_l + n_l(k_2 r)\sin\delta_l \right] \, , \quad k_2 = \sqrt{\frac{E^2 - m_0^2c^4}{c^2\hbar^2}} \, .$$

For $l = 0$ states the corresponding continuity condition yields for the phase δ_0

$$\tan(k_2 a + \delta_0) = \frac{k_2}{k_1}\tan k_1 a \, , \tag{1.75}$$

if $|E + V_0| > m_0c^2$ (1., 3., and 5. case, see below), or

$$\tan(k_2 a + \delta_0) = \frac{k_2}{\kappa_1}\tanh \kappa_1 a \, , \tag{1.76}$$

if $|E + V_0| < m_0c^2$ (4. case, see below).

As in Subsection 1.5.1, these solutions can be classified and interpreted as follows (see Figure 1.7):

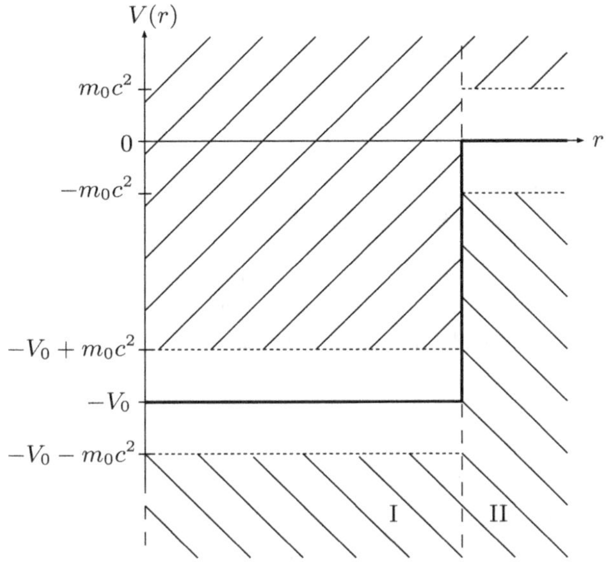

Fig. 1.7. Energy intervals of the spherically symmetric potential well.

1. case: $E > m_0c^2$. In both areas I and II only "allowed" positive energies are involved. This corresponds to a normal scattering of a particle with charge $+e$ against the (from its point of view) attractive potential well (compare to 1. case from Subsection 1.5.1).

2. case: $-m_0c^2 < E < m_0c^2$, $E + V_0 > m_0c^2$. At this energy interval the "allowed" positive energies of I adjoin the "forbidden" positive and negative energy regimes of II. We therefore have a bound particle of charge $+e$ (compare to 2. case in Subsection 1.5.1).

3. case: $-V_0 + m_0c^2 < E < -m_0c^2 \Longrightarrow V_0 > 2m_0c^2$. Here the "allowed" positive energy regime of I touches the "allowed" negative energy regime of II. Beyond the one-particle picture, this can be interpreted as the scattering of an antiparticle with charge $-e$ against the (from its point of view) repulsive potential well in the presence of particle resonances (compare to 3. case in Subsection 1.5.1).

4. case: $-V_0 - m_0c^2 < E < -V_0 + m_0c^2$. The "allowed" negative energy regime of II adjoins the "forbidden" positive and negative energy regimes of I. This corresponds to a normal scattering of an antiparticle with charge $-e$ against the (from its point of view) repulsive potential well while the penetration depth in area I decreases exponentially (compare to 4. case in Subsection 1.5.1).

5. case: $E < -V_0 - m_0c^2$. Only "allowed" negative energies are present in I and II. Thus, we have a normal scattering of an antiparticle with charge $-e$ against the (from its point of view) repulsive potential well (compare to 5. case in Subsection 1.5.1).

2. case in detail. Let us now consider the bound case in more detail. We set $V_0 = Ze^2/a$ and take the potential well as a naive model for the electrostatic binding of a pion atom consisting of a nucleus with charge $-Ze = +Z|e|$ and an orbiting pion of charge $+e = -|e|$. Figure 1.8 shows the corresponding energies for 1s-states as a function of the "nucleus charge number" Z and the "nucleus radius" a following from (1.74). As we can see, for each a-value there exists a Z-interval, only in which bound 1s-states are possible. At the lower Z-border the solutions of the upper energy continuum dive into the bound region and approach the lower energy continuum with increasing Z. However, we have to bear in mind that, below the zero-energy, the one-

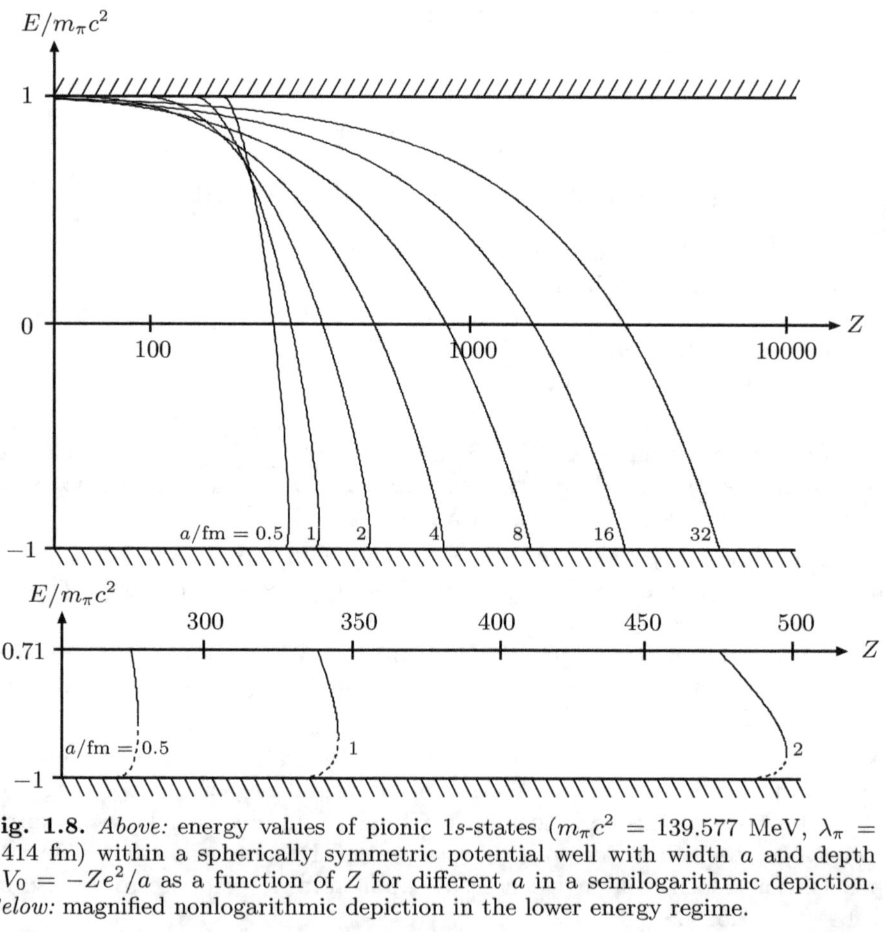

Fig. 1.8. *Above:* energy values of pionic 1s-states ($m_\pi c^2 = 139.577$ MeV, $\lambda_\pi = 1.414$ fm) within a spherically symmetric potential well with width a and depth $-V_0 = -Ze^2/a$ as a function of Z for different a in a semilogarithmic depiction. *Below:* magnified nonlogarithmic depiction in the lower energy regime.

particle interpretation breaks down more and more due to the high binding energy $|E_{\mathrm{B}}| = |E - m_\pi c^2| > m_\pi c^2$.

It is also striking that for small a values (small potential reach and large potential depth), i.e. far beyond the validity range of the one-particle concept, the energy curves exhibit a strong left bend in the lower energy regime (two solutions for a given Z). This left bend results from the antiparticle states entering the bound region from the lower energy continuum and joining the particle states at the points of infinite slope. Remarkably, this potential can therefore bind particles and antiparticles simultaneously (*Schiff-Snyder effect*).

1.5.4 Coulomb Potential

We now refine our description of pion atoms and consider the problem of a spin-0 particle within a Coulomb potential of the form

$$eA^0(r) = V(r) = -\frac{Ze^2}{r} = -\frac{Z\hbar c\alpha_e}{r} \;,\; \alpha_e = \frac{e^2}{\hbar c} = 1/137.03602 \;,$$

where α_e denotes the *fine structure constant.* In this case the radial Klein-Gordon equation (1.72) is

$$\left[\frac{\mathrm{d}^2}{\mathrm{d}r^2} - \frac{l(l+1) - (Z\alpha_e)^2}{r^2} + \frac{2EZ\alpha_e}{\hbar cr} - \frac{m_0^2c^4 - E^2}{\hbar^2c^2}\right] u_l(r) = 0 \;. \tag{1.77}$$

Since we wish to restrict our discussion to bound states, i.e. to the energy interval $-m_0c^2 < E < m_0c^2$, we can introduce the quantities

$$\rho = \beta r \;,\; u_l(r) = \hat{u}_l(\rho) \;,\; \beta = 2\sqrt{\frac{m_0^2c^4 - E^2}{\hbar^2c^2}} \;,\; \lambda = \frac{2EZ\alpha_e}{\beta\hbar c}$$

in order to rewrite (1.77) as

$$\left[\frac{\mathrm{d}^2}{\mathrm{d}\rho^2} - \frac{l'(l'+1)}{\rho^2} + \frac{\lambda}{\rho} - \frac{1}{4}\right] \hat{u}_l(\rho) = 0 \;, \tag{1.78}$$

with

$$l'(l'+1) = l(l+1) - (Z\alpha_e)^2$$

$$\Longrightarrow l' = -\frac{1}{2} \pm \sqrt{\left(l + \frac{1}{2}\right)^2 - (Z\alpha_e)^2} \;. \tag{1.79}$$

Obviously, the equation (1.78) is formally identical to the radial Schrödinger equation of the nonrelativistic Coulomb problem. We can therefore follow the way of solving the nonrelativistic problem by first considering (1.78) for the asymptotic regions $\rho \to 0, \infty$.

$\rho \to 0$: in this case (1.78) reduces to the equation

$$\left[\frac{\mathrm{d}^2}{\mathrm{d}\rho^2} - \frac{l'(l'+1)}{\rho^2}\right]\hat{u}_l(\rho) = 0 , \quad (1.80)$$

having the two solutions $\hat{u}_l(\rho) = \rho^{l'+1}$ and $\rho^{-l'}$. Due to (1.79), they can be summarized in the one solution $\hat{u}_l(\rho) = \rho^{l'+1}$, whereas the sign in (1.79) remains to be clarified. Now the following points have to be taken into account:

- Physical solutions to our problem exist only for

  $$l + \frac{1}{2} > Z\alpha_e .$$

  Otherwise, the quantity $l' = -1/2 \pm \mathrm{i}\sigma$, $\sigma = \sqrt{(Z\alpha_e)^2 - (l+1/2)^2}$ would be complex so that, near the origin, we would obtain wave functions of the form $\hat{u}_l(\rho) \sim \rho^{1/2}\exp(\pm\mathrm{i}\sigma\ln\rho)$ that oscillate infinitely often for $\rho \to 0$ and hence yield divergent expectation values of the kinetic energy.

- Because of the $1/r$-behavior of the Coulomb potential, the integrability of the wave function near the origin implies the constraint

  $$l' + 1 > 0 ,$$

  which is always true for the positive sign in (1.79). If, on the other hand, Z is sufficiently large for a given l, this constraint is also fulfilled by the negative sign in (1.79). However, in this case we can find another constraint that finally rules out the negative sign, for example, that the expectation value of the kinetic energy must exist.

All in all, the expression

$$\hat{u}_l(\rho) = \rho^{l'+1} , \ l' = -\frac{1}{2} + s , \ s = \sqrt{\left(l+\frac{1}{2}\right)^2 - (Z\alpha_e)^2}$$

remains as the only physically sensible solution to (1.80).

$\rho \to \infty$: here (1.78) turns into the equation

$$\left(\frac{\mathrm{d}^2}{\mathrm{d}\rho^2} - \frac{1}{4}\right)\hat{u}(\rho) = 0 .$$

Its bounded solution at infinity is

$$\hat{u}(\rho) = \mathrm{e}^{-\rho/2} .$$

Putting both asymptotic regions together we are led to the ansatz

$$\hat{u}_l(\rho) = \rho^{l'+1}\mathrm{e}^{-\rho/2}f(\rho)$$

for the solution of (1.78), from which the differential equation

$$\rho f''(\rho) + (2l' + 2 - \rho)f'(\rho) + (\lambda - l' - 1)f(\rho) = 0$$

follows. Inserting the power expansion

$$f(\rho) = \sum_{k=0}^{\infty} a_k \rho^k$$

yields

$$\sum_{k=0}^{\infty} \left[(k+1)(k+2l'+2)a_{k+1} + (\lambda - l' - 1 - k)a_k\right] \rho^k = 0 \ ,$$

giving the following recursion formula for the expansion coefficients a_i:

$$a_{k+1} = \frac{k+l'+1-\lambda}{(k+1)(k+2l'+2)} a_k \ .$$

For the wave function u_l to obey the integrability condition, i.e. to converge at infinity, the power expansion must break at some $k = n'$, i.e.

$$\lambda = n' + l' + 1 \ , \ n' = 0, 1, 2, \ldots \ .$$

This is the quantization condition for λ and thus for the energy levels of the bound spin-0 states. From this it follows that

$$\frac{E^2(Z\alpha_e)^2}{m_0^2c^4 - E^2} = \left[n' + \frac{1}{2} + \sqrt{\left(l + \frac{1}{2}\right)^2 - (Z\alpha_e)^2}\right]^2$$

$$\Longrightarrow E_{n',l} = \frac{m_0c^2}{\sqrt{1 + \frac{(Z\alpha_e)^2}{\left(n' + \frac{1}{2} + \sqrt{\left(l+\frac{1}{2}\right)^2 - (Z\alpha_e)^2}\right)^2}}} \ .$$

As we can see, the above constraint $l + 1/2 \geq Z\alpha_e$ is also necessary for the existence of real (i.e. physical) energy eigenvalues. It is most restrictive for l=0-states and implies: $l = 0 \Longrightarrow Z < 68.5$, $l = 1 \Longrightarrow Z < 205.5$, and so on. Bound states do not exist for larger Z-values. For a comparison with the nonrelativistic spectroscopic notation, we introduce the principal quantum number

$$n = n' + l + 1$$

and finally obtain (see Figure 1.9)

$$E_{n,l} = \frac{m_0c^2}{\sqrt{1 + \frac{(Z\alpha_e)^2}{\left(n - \left(l+\frac{1}{2}\right) + \sqrt{\left(l+\frac{1}{2}\right)^2 - (Z\alpha_e)^2}\right)^2}}} \ , \quad \begin{array}{l} n = 1, 2, \ldots \\ l = 0, 1, \ldots, n-1 \ . \end{array} \tag{1.81}$$

Obviously, the degeneracy with respect to the angular momentum, which we encounter in nonrelativistic quantum mechanics, is now removed. Expanding (1.81) in powers of $Z\alpha_e$ yields

$$E_{n,l} = m_0c^2 \left[1 - \frac{(Z\alpha_e)^2}{2n^2} - \frac{(Z\alpha_e)^4}{2n^4}\left(\frac{n}{l+\frac{1}{2}} - \frac{3}{4}\right) + \ldots\right] \ . \tag{1.82}$$

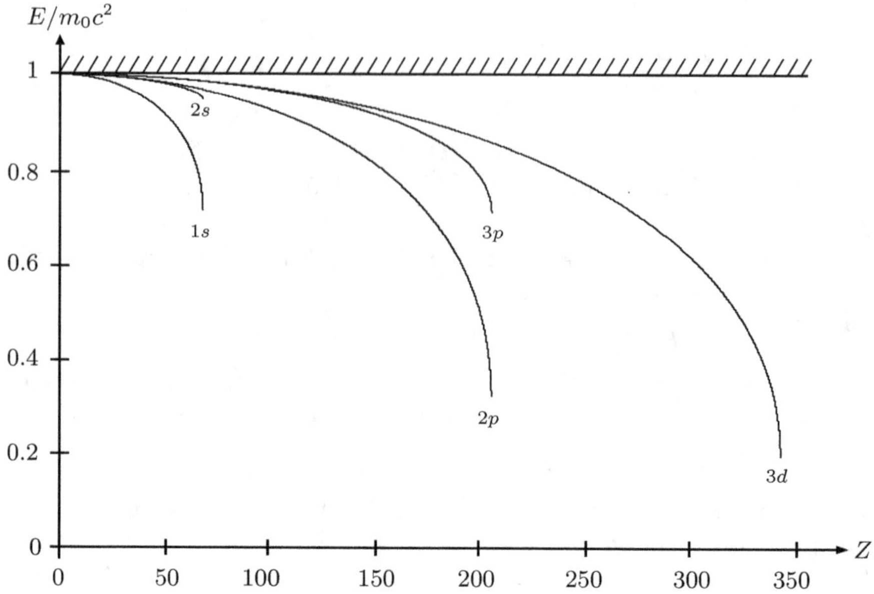

Fig. 1.9. Energy values of bound spin-0 states within a Coulomb potential of the form $V(r) = -Ze^2/r$ as a function of Z. Each curve ends at a certain maximal Z-value. For larger Z we get complex (unphysical) energies.

The first term is the rest energy. The second term is the binding energy of the nonrelativistic Coulomb problem, and the third one is a relativistic correction where the removal of the angular momentum degeneracy is manifest. It is identical to the correction stemming from the inclusion of the perturbation operator $H' = -(\boldsymbol{p}^2)^2/(8m_0^3c^2)$ in the Schrödinger equation.

The following items have to be kept in mind for a better understanding of the above findings:

- In our calculations we have assumed an infinitely heavy atomic nucleus. However, since it has de facto a finite mass, the nucleus and the pion move around a common center of gravity that does not coincide with the nucleus's center. In Newtonian mechanics as well as in nonrelativistic quantum mechanics this center of gravity motion can easily be separated from the relative motion, whereas the latter is related back to an effective one-particle problem with the reduced mass
$$\mu = \frac{m_0 M}{m_0 + M} \ .$$
However, this division is problematic in relativistic quantum mechanics. Firstly, there does not exist a satisfactory relativistic two-center equation and, secondly, the center of gravity system can no longer be defined geometrically but only dynamically. In order to take into account the mutual repulsion of the particles, one usually proceeds, as in the nonrelativistic

case, by adhering to the original one-center equation and assigning the pion the above reduced mass μ. Contrary to the hydrogen atom, this correction is fairly large for the pion atom, since the pion is about 273 times heavier than the electron ($m_\pi c^2 = 139.577$ MeV, $m_e c^2 = 0.511$ MeV).

- The description of the interaction between nucleus and pion using an external static Coulomb field disregards the finite propagational velocity of the force between these two constituents. We therefore expect significant impacts of those *retardation effects* particularly in the inner shells of the pion atom where the pion's velocity is comparable to the velocity of light. Here the *external field approximation* and the above mentioned *reduced mass approximation* are expected to become poor.
- As we have seen, there seem to exist no bound l=0-states for $Z > 68.5$. On the other hand, we know that there definitely exist nuclei with larger nucleus numbers so that, also for those cases, the Klein-Gordon equation should yield an acceptable explanation. The reason for this discrepancy lies in disregarding the nucleus's finite extent. For example, if we compare the Bohr radius of a pion, $R_\pi = 1/(m_\pi c Z \alpha_e) \approx 200$ fm$/Z$, with the nucleus radius $R_{\mathrm{K}} \approx 1.5 \cdot A^{1/3}$ fm (A = nucleus number), we expect, particularly for large Z, a considerable overlap of the pion wave function with the nucleus. In the next subsection, we use a modified Coulomb potential in order to take the nucleus's finite extent into account, and we will see that, in this case, bound states do exist even for large Z and small l.
- Since the probability of finding the pion near the nucleus is considerably large, we have to include strong interaction corrections whose effects can be significant.
- Finally, we must also take vacuum polarization effects into account leading to a screening of the nucleus charge near the nucleus and hence to a further modification of the Coulomb potential.

1.5.5 Oscillator-Coulomb Potential

All the influences mentioned above have to be taken into account for a more realistic description of pion atoms. We now single out the finite extent of the nucleus and (naively) consider it to be a homogeneous charged sphere. The corresponding potential is (R = nucleus radius)

$$eA^0(r) = V(r) = \begin{cases} -\dfrac{Ze^2}{2R}\left(3 - \dfrac{r^2}{R^2}\right) & \text{for } r < R \text{ (area I)} \\ -\dfrac{Ze^2}{r} & \text{for } r > R \text{ (area II)}\,. \end{cases}$$

It is composed of an oscillator-like potential in the inner area I and the Coulomb potential of the preceding subsection in area II. In area I the radial Klein-Gordon equation is

$$\left\{\frac{\mathrm{d}^2}{\mathrm{d}r^2} - \frac{l(l+1)}{r^2} + \frac{\left[E + \frac{Ze^2}{2R}\left(3 - \frac{r^2}{R^2}\right)^2\right] - m_0^2c^4}{\hbar^2c^2}\right\} u_l(r) = 0\ .$$

Introducing the quantities

$$A = \frac{E}{\hbar c}\ ,\ B = A^2 - \frac{m_0^2c^2}{\hbar^2}\ ,\ C = \frac{Z\alpha_e}{2R^3}\ ,$$

it can be simplified to

$$\left[\frac{\mathrm{d}^2}{\mathrm{d}r^2} - \frac{l(l+1)}{r^2} + B - 2ACr^2 + C^2r^4\right] u_l(r) = 0\ .$$

To solve this equation we make the power series expansion ansatz

$$u_l^{(\mathrm{I})}(r) = r^{l+1}\sum_{k=0}^{\infty} c_k r^{2k}\ ,$$

leading to the following conditional equation for the coefficients c_k:

$$\begin{aligned} 0 &= \sum_{k=0}^{\infty} c_k(2k+l+1)(2k+l)r^{2k+l-1} - \sum_{k=0}^{\infty} l(l+1)c_k r^{2k+l-1} \\ &\quad + B\sum_{k=0}^{\infty} c_k r^{2k+l+1} - 2AC\sum_{k=0}^{\infty} c_k r^{2k+l+3} + C^2\sum_{k=0}^{\infty} c_k r^{2k+l+5}\ . \end{aligned}$$

Comparing the coefficients for each single order in r finally yields

$$c_1 = -\frac{Bc_0}{2(2l+3)}\ ,\ c_2 = -\frac{Bc_1 - 2ACc_0}{4(2l+5)}$$

$$c_k = -\frac{Bc_{k-1} - 2ACc_{k-2} + C^2c_{k-3}}{2k(2l+2k+1)}\ ,\ k \geq 3\ .$$

In area II the radial equations (1.77) and (1.78) hold. Their general solution is given by

$$u_l^{(\mathrm{II})}(r) = \hat{u}_l^{(\mathrm{II})}(\rho) = \mathrm{e}^{-\rho/2}\sum_{k=0}^{\infty} a_k\rho^{1/2+s+k} + \mathrm{e}^{-\rho/2}\sum_{k=0}^{\infty} b_k\rho^{1/2-s+k}\ , \quad (1.83)$$

with

$$a_{k+1} = \frac{k+1/2+s-\lambda}{(k+1)(k+1+2s)}a_k\ ,\ b_{k+1} = \frac{k+1/2-s-\lambda}{(k+1)(k+1-2s)}b_k \quad (1.84)$$

and

$$\rho = \beta r\ ,\ \beta = 2\sqrt{\frac{m_0^2c^4 - E^2}{\hbar^2c^2}}\ ,\ \lambda = \frac{2EZ\alpha_e}{\beta\hbar c}\ ,\ s = \sqrt{\left(l+\frac{1}{2}\right)^2 - (Z\alpha_e)^2}.$$

Up to the irrelevant coefficient c_0, our problem involves three quantities to be determined, namely the coefficients a_0 and b_0, as well as λ or the energy E.

On the other side we have the same number of conditions, the two continuity conditions at $r = R$ and the required correct asymptotic behavior of the wave function for $r \to \infty$.

Extracting a_0 and b_0 from the sums in (1.84), the expression (1.83) can be rewritten as

$$u_l^{(\mathrm{II})}(r) = a_0 \Omega(s,\rho) + b_0 \Omega(-s,\rho) ,$$

with

$$\Omega(s,\rho) = \mathrm{e}^{-\rho/2} \sum_{k=0}^{\infty} a'_k \rho^{1/2+s+k} , \; a'_{k+1} = \frac{k+1/2+s-\lambda}{(k+1)(k+1+2s)} a'_k , \; a'_0 = 1.$$

Now we are able to express both continuity conditions by the compact equation system

$$\begin{aligned} u_l^{(\mathrm{I})}(R) &= u_l^{(\mathrm{II})}(R) &&= a_0 \Omega(s,\beta R) &&+ b_0 \Omega(-s,\beta R) \\ \left.\frac{\mathrm{d}u_l^{(\mathrm{I})}(r)}{\mathrm{d}r}\right|_{r=R} &= \left.\frac{\mathrm{d}u_l^{(\mathrm{II})}(r)}{\mathrm{d}r}\right|_{r=R} &&= a_0 \left.\frac{\mathrm{d}\Omega(s,\rho)}{\mathrm{d}r}\right|_{r=R} &&+ b_0 \left.\frac{\mathrm{d}\Omega(-s,\rho)}{\mathrm{d}r}\right|_{r=R} , \end{aligned}$$

which, for given values of m_0, Z, R, l, and E, can be uniquely solved.[27] The sought energy eigenvalues E are determined via the required asymptotic behavior of the radial wave function at infinity.[28] Depending on the number of zeros, the found energies and states can be classified using the usual spectroscopic notation (one zero $\Longrightarrow n = 1$, two zeros $\Longrightarrow n = 2$, and so on).

In Figure 1.10 the energy eigenvalues of bound pionic $1s$-, $2s$-, and $2p$-states are depicted against the nucleus charge number Z, where a nucleus radius of $R = 10$ fm is chosen. As we can see, there is an almost linear relationship between the state energies and the nucleus charge number. Furthermore, we perceive that, compared to the pure Coulomb problem, there also exist bound states for large Z and small l values (compare to Figure 1.9). At $Z \approx 760$ [935, 1025] the energy value for $1s$-[$2s$-, $2p$-]pions reaches $E = 0$. For larger Z it becomes negative (compare to Figure 1.8). At $Z \approx 1450$ [1670, 1785], we finally find the energy $E = -m_\pi c^2$.

At the end of this subsection, we draw on the results of our last example to highlight again the basic interpretational difficulties of the Klein-Gordon theory in relation to the one-particle picture. Table 1.1 shows for two different nucleus charge numbers Z the G-expectation value of the $1s$-pion radius, $\langle r \rangle_\mathrm{G}$, the electrostatic oscillator-Coulomb potential V at the point $\langle r \rangle_\mathrm{G}$, the binding energy $E_\mathrm{B} = E_{1\mathrm{s}} - m_\pi c^2$, as well as the mean quadratic deviation $\Delta r = \sqrt{\langle r^2 \rangle_\mathrm{G} - \langle r \rangle_\mathrm{G}^2}$. Comparing these values with the pion's rest energy

[27] For real s both quantities a_0 and b_0 are real, too. If s is imaginary, we have $\Omega(-s,\rho) = \Omega^*(s,\rho)$ and therefore $b_0 = a_0^*$ so that, also in this case, only two quantities remain, namely $\mathrm{Re}(a_0)$ and $\mathrm{Im}(a_0)$.

[28] To do this, one solves the equation system for different energies within the range $-m_0c^2 < E < m_0c^2$ and checks the asymptotic behavior of the corresponding solutions.

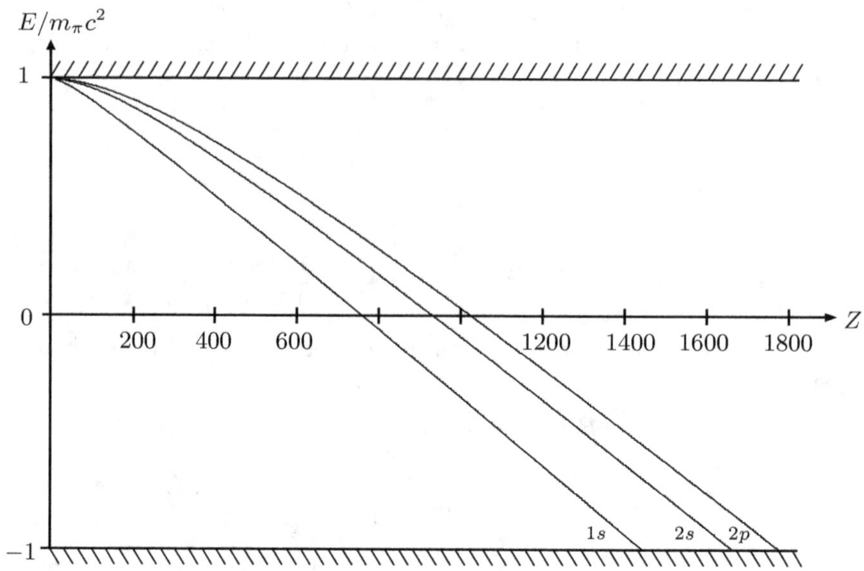

Fig. 1.10. Energy values of bound 1*s*-, 2*s*-, and 2*p*-pion states in the field of a homogeneously charged sphere (oscillator-Coulomb potential) as a function of Z. The sphere's radius (nucleus radius) is $R = 10$ fm.

	$Z = 2$	$Z = 1450$
$\langle r \rangle_{\mathrm{G}}$	146.4 fm	3.7 fm
$V\left(\langle r \rangle_{\mathrm{G}}\right)$	−0.02 MeV	−298.9 MeV
E_{B}	−0.05 MeV	−278.8 MeV
Δr	84.3 fm	1.6 fm

Tab. 1.1. Characteristic values of the bound 1*s*-pion state in the oscillator-Coulomb potential for the weak ($Z = 2$) and the strong ($Z = 1450$) binding cases.

$m_\pi c^2 = 139.577$ MeV and its Compton wave length $\lambda_\pi = 1.414$ fm, we have for the weak binding case

$$Z = 2: \quad |E_{\mathrm{B}}|, |V(\langle r \rangle_{\mathrm{G}})| \ll m_\pi c^2 \ , \ \Delta r \gg \lambda_\pi \ .$$

Here all prerequisites of Theorem 1.7 are fulfilled so that the interpretation of our results within the one-particle concept seems to be justified. However, in the strong binding case we have

$$Z = 1450: \quad |E_{\mathrm{B}}|, |V(\langle r \rangle_{\mathrm{G}})| \approx 2m_\pi c^2 \ , \ \Delta r \approx \lambda_\pi \ .$$

These relations clearly contradict the assumptions in Theorem 1.7, thus showing the impossibility of the one-particle interpretation. A direct confirmation of these conclusions is obtained by considering the radial charge density of the 1s-pion state,

$$r^2\rho(r) = \frac{E - V(r)}{m_\pi c^2} u_l^2(r) \ .$$

In the weak binding case ($Z = 2$) E is positive, and the radial charge density is positive definite. By contrast, in the strong binding case ($Z = 1450$) E is negative so that the radial charge density does not show a uniform course and turns into negative values at $r \approx 15$ fm (see Figure 1.11). This is clearly incompatible with the one-particle concept. The physical meaning of this change of sign for strong fields (as for the onedimensional and spherically symmetric potential wells) can ultimately be understood only within quantum field theories where the number of particles is variable.

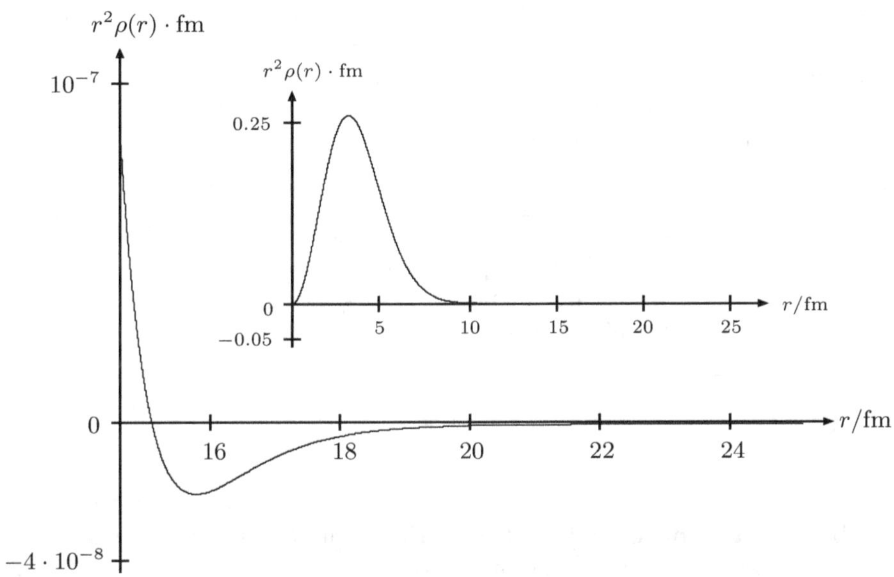

Fig. 1.11. G-normalized radial charge density of the 1s-pion state in the oscillator-Coulomb potential with $Z = 1450$ and $R = 10$ fm. The large picture shows a magnified extract of the small one. At $r \approx 15$ fm the charge density changes its sign.

Summary

- Studying the onedimensional potential well, we encounter different classes of solutions depending on the particle's energy that can be interpreted more or less consistently within the one-particle concept as **scattering** or **binding** of an (anti)particle.
- For centrally symmetric potentials we can separate the angular-dependent part to obtain the **radial Klein-Gordon equation**, which is formally identical to the nonrelativistic radial Schrödinger equation.
- For the free case (and that of a potential with constant regions) this equation turns into the **spherical Bessel differential equation**, whose solutions are given by the **spherical Bessel functions**.
- Exactly like nonrelativistic electron atoms, pion atoms can be described approximately using the Coulomb potential. Contrary to the nonrelativistic case, the angular momentum degeneracy is removed. At small angular momentum values l we find bound pion states only for correspondingly small nucleus charge numbers Z. This results from disregarding the finite extent of the nucleus.
- Taking the nucleus's finite extent into account by using an oscillator-Coulomb potential, we also find bound pion states for large Z-values.
- The basic difficulties of the one-particle interpretation in the presence of strong fields can be clearly demonstrated by means of the solutions to the oscillator-Coulomb problem.

Exercises

10. Exponential potential. Calculate the bound l=0-states of spin-0 particles in an exponential potential of the form

$$eA^0(r) = V(r) = -Z\alpha \mathrm{e}^{-r/a} \ , \ \alpha = m_0 c^2 \alpha_e \ ,$$

where Z denotes the nucleus charge number, α the coupling constant, and a the decline constant.

Solution. Since we consider only s-states, the centrifugal term vanishes and the radial Klein-Gordon equation (1.72) acquires a very simple form:

$$u''(r) + k^2 u(r) = 0 \ , \ k^2 = \frac{[E - V(r)]^2 - m_0^2 c^4}{\hbar^2 c^2} \ .$$

For its solution we make the separation and substitution ansatz

$$u(r) = \mathrm{e}^{r/2a} \omega(t) \ , \ t = 2\mathrm{i} Z\alpha \frac{a}{\hbar c} \mathrm{e}^{-r/a} \ .$$

From this and taking into account

$$\frac{\mathrm{d}t}{\mathrm{d}r} = -\frac{2\mathrm{i}Z\alpha}{\hbar c}\mathrm{e}^{-r/a} \ , \ \mathrm{e}^{-r/a} = -\frac{\mathrm{i}\hbar ct}{2Z\alpha a} \ , \ \mathrm{e}^{-2r/a} = -\frac{\hbar^2 c^2 t^2}{4Z^2\alpha^2 a^2}$$

as well as

$$\begin{aligned} k^2 &= \frac{E^2 - m_0^2c^4}{\hbar^2c^2} - \frac{\mathrm{i}Et}{\hbar ca} - \frac{t^2}{4a^2} \\ u'(r) &= \frac{1}{2a}\mathrm{e}^{r/2a}\omega(t) - \frac{2\mathrm{i}Z\alpha}{\hbar c}\mathrm{e}^{-r/2a}\omega'(t) \\ u''(r) &= \frac{1}{4a^2}\mathrm{e}^{r/2a}\omega(t) - \frac{4Z^2\alpha^2}{\hbar^2c^2}\mathrm{e}^{-3r/2a}\omega''(t) \ , \end{aligned}$$

we obtain the differential equation

$$\omega''(t) + \left(\frac{1/4 - p^2a^2}{t^2} - \frac{\mathrm{i}Ea}{\hbar ct} - \frac{1}{4}\right)\omega(t) = 0 \ , \ p^2 = \frac{m_0^2c^4 - E^2}{\hbar^2c^2} \ , \qquad (1.85)$$

which is formally identical to (1.78), if in the latter

$$\lambda = -\frac{\mathrm{i}Ea}{\hbar c} \ , \ l' = -\frac{1}{2} + pa$$

is chosen. The regular solution to (1.85) at $r \to \infty$ (i.e. $t = 0$) is therefore

$$\omega(t) = t^{1/2+pa}\mathrm{e}^{-t/2}\sum_{k=0}^{\infty} a_k t^k \ , \ a_{k+1} = \frac{k + 1/2 + pa - \lambda}{(k+1)(k+2pa+1)}a_k \ .$$

The discrete energy values follow from the constraint that $u(r)$ [or $\omega(t)$] must vanish at the origin [or at $t_0 = t(r=0)$]:

$$\sum_{k=0}^{\infty} a_k t_0^k = 0 \ , \ t_0 = \frac{2\mathrm{i}Z\alpha a}{\hbar c} \ .$$

From this implicit conditional equation the energy values E for s-states can be determined numerically.

Figure 1.12 shows the energy values of pionic 1s-, 2s-, and 3s-states depending on the nucleus charge number Z, where a decline constant of $a = \lambda_\pi = 1.414$ fm is chosen. Similarly to the spherically symmetric potential well (Figure 1.8) and the oscillator-Coulomb potential (Figure 1.10), we also find here certain Z intervals for the bound states, where the energy values coming from the upper energy continuum decrease with increasing Z and eventually border on the lower energy continuum at $Z = 778$ (1s), $Z = 1278$ (2s), and $Z = 1754$ (3s).

If we now reduce the decline constant to $a = 0.2 \cdot \lambda_\pi$, the lower border point for the 1s-state is increased to $Z = 2158$, as can be seen from Figure 1.13. Additionally and similarly to the spherically symmetric potential well, we see a strong left bend in the lower energy regime stemming again from the antiparticle states entering the bound region from the lower energy continuum. The existence of these antiparticle states is again connected to the very short reach and the very large depth of the potential.

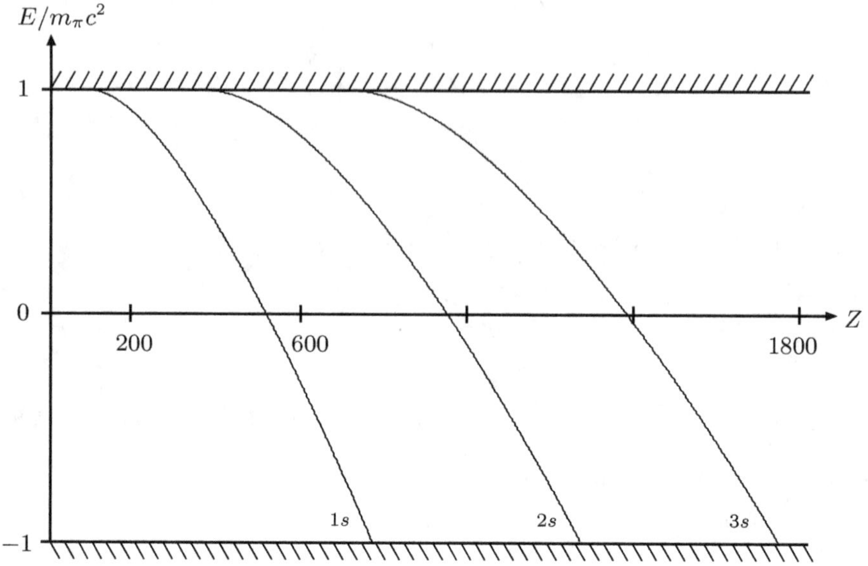

Fig. 1.12. Energy values of bound 1*s*-, 2*s*-, and 3*s*-pion states within an exponential potential with a decline constant $a = \lambda_\pi$ as a function of Z.

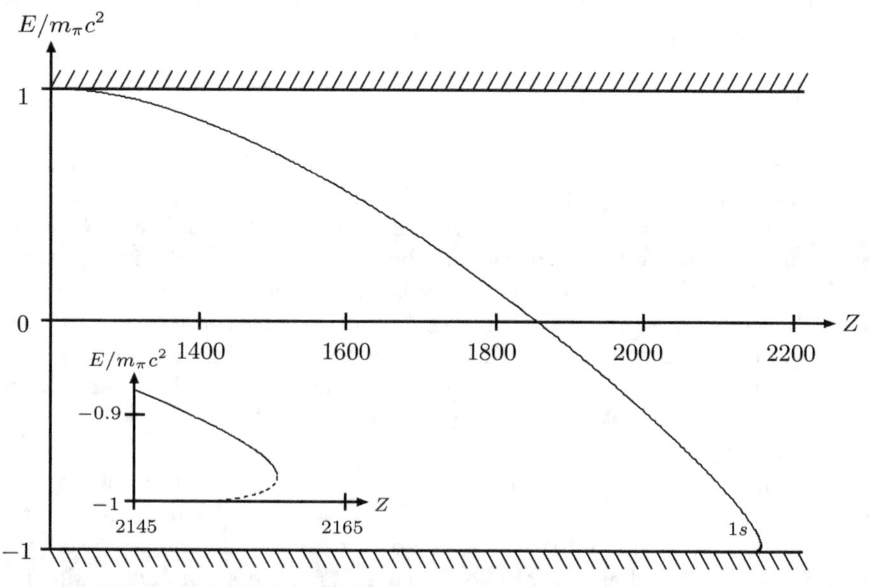

Fig. 1.13. Energy values of bound 1*s*-pion states within an exponential potential with a decline constant of $a = 0.2 \cdot \lambda_\pi$ as a function of Z. The small picture shows a magnified extract in the lower energy regime.

2. Relativistic Description of Spin-1/2 Particles

In the preceding chapter we discussed the Klein-Gordon theory for the description of spin-0 particles as well as its interpretation in relation to the quantum mechanical one-particle concept. There, for didactic reasons, we disregarded the chronological order where a sensible one-particle interpretation of the Klein-Gordon theory was not seriously considered until Dirac established a relativistic quantum mechanics for electrons (more generally: spin-1/2 particles).

In this chapter we turn to the Dirac theory and also pay specific attention to a physically consistent one-particle interpretation and its limits (relativistic quantum mechanics "in the narrow sense"). In practice, this means that we are again guided by the basic principles presented in the introduction of Chapter 1, namely

- the principles of nonrelativistic quantum mechanics (Theorem 1.1),
- the limitation of the one-particle interpretation to small interaction energies compared to the rest energy of the particle as well as to a large position uncertainty of the wave function compared to the Compton wave length.

In order to make the similarities and differences in the Klein-Gordon and Dirac theories as transparent as possible, the following sections – except for Dirac-specific topics – are structured similarly to those of the first chapter.

The first section deals with the foundations of the Dirac theory for the description of spin-1/2 particles, which, as we will see, exhibit similar features as the Klein-Gordon theory. In the second section we look at the continuous and discrete symmetries of the Dirac theory. The third section is dedicated to the extension, completion, and delimitation of its one-particle interpretation. The nonrelativistic approximation of the Dirac theory to different orders of v/c is the subject of the fourth section, where we will also resort to the Fouldy-Wouthuysen method to systematically diagonalize the Dirac-Hamilton operator. As in the Klein-Gordon case, this chapter ends with some detailed considerations of simple one-particle systems in the fifth section.

A. Wachter, Relativistic Description of Spin-1/2 Particles. In: A. Wachter, Relativistic Quantum Mechanics, Theoretical and Mathematical Physics, pp. 85–176 (2011)
DOI 10.1007/978-90-481-3645-2_2

2.1 Dirac Equation

As we saw in Subsection 1.1.1, the relativistic energy momentum relation for free particles,

$$E^2 = \boldsymbol{p}^2 c^2 + m_0^2 c^4 \ , \tag{2.1}$$

leads through the operator replacement

$$E \longrightarrow \mathrm{i}\hbar \frac{\partial}{\partial t} \ , \ \boldsymbol{p} \longrightarrow -\mathrm{i}\hbar \boldsymbol{\nabla}$$

to a scalar wave equation namely the Klein-Gordon equation. This equation does not allow a positive definite probability density in the sense of non-relativistic quantum mechanics due to the presence of the second temporal derivative. Additionally, as a consequence of the quadratic energy momentum dependence, there exist solutions with negative energy that need to be interpreted adequately.

In his efforts to adhere to a positive definite probability density, Paul Dirac sought a relativistic generalization of Schrödinger's equation of first order in time. Indeed he found this equation in 1928, which describes, contrary to the Klein-Gordon equation, spin-1/2 particles. However, as every relativistic wave equation, it also encompasses solutions with negative energy whose physical meaning is a priori not clear.

In this section we derive the Dirac equation and other basic relationships of the Dirac theory in the canonical as well as in the Lorentz-covariant formulation. As it turns out, the solutions to the Dirac equation exhibit an inner degree of freedom which is to be interpreted as a quantum mechanical spin with the quantum number $s = 1/2$. Furthermore, we discuss formal properties of the spin operator, cast it into a Lorentz-covariant form, and utilize it for the construction of projection operators. At the end we turn to the negative solutions and their interpretation where, again, the charge conjugation transformation and, additionally, the *hole theory* will play an important role.

2.1.1 Canonical Formulation of the Dirac Equation

The starting point of Dirac's considerations was a relativistic generalization of the Schrödinger equation for free particles of the form

$$\mathrm{i}\hbar \frac{\partial \psi(x)}{\partial t} = H^{(0)} \psi(x) \ , \ x = (x^\mu) \ , \ H^{(0)} \text{ Hermitean} \ , \tag{2.2}$$

to which he initially assigned three constraints:

- Equation (2.2) must be Lorentz-covariant. Since its temporal derivative is of first order, the spatial derivatives must be of first order, too.
- Equation (2.2) must yield the relativistic energy momentum relation (2.1) in operator form.

- The quantity $\rho = \psi^*\psi$ must be the temporal component of a conserved four-vector j^μ (i.e. there must exist a Lorentz-covariant continuity equation) so that its integral over the whole space is invariant.

Due to the first two constraints, the following ansatz seems to be reasonable for the Hamilton operator in (2.2):

$$H^{(0)} = c\boldsymbol{\alpha}\boldsymbol{p} + \beta m_0 c^2 \;,\; \boldsymbol{p} = -\mathrm{i}\hbar\boldsymbol{\nabla} \;,\; m_0 = \text{rest mass} \;, \tag{2.3}$$

with the secondary condition

$${H^{(0)}}^2 = c^2\boldsymbol{p}^2 + m_0^2c^4 \tag{2.4}$$

or

$$-\hbar^2\frac{\partial^2\psi(x)}{\partial t^2} = (c^2\boldsymbol{p}^2 + m_0^2c^4)\psi(x) \quad \text{(Klein-Gordon equation)} \;.$$

Obviously, condition (2.4) implies that α_i and β cannot be ordinary numbers as there are no mixed terms in $\boldsymbol{\alpha}\boldsymbol{p}$ and β. In order to find out what algebraic structure α_i and β possess, we write (2.2) and (2.3) in the form

$$\mathrm{i}\hbar\frac{\partial\psi}{\partial t} = \left(\frac{\hbar c}{\mathrm{i}}\sum_i \alpha_i\partial_i + \beta m_0 c^2\right)\psi \tag{2.5}$$

and iterate this relation:

$$\begin{aligned}
-\hbar^2\frac{\partial^2\psi}{\partial t^2} &= \mathrm{i}\hbar\frac{\partial}{\partial t}\left(\frac{\hbar c}{\mathrm{i}}\sum_i \alpha_i\partial_i\psi + \beta m_0 c^2\psi\right)\\
&= \frac{\hbar c}{\mathrm{i}}\sum_j \alpha_j\partial_j\left(\frac{\hbar c}{\mathrm{i}}\sum_i \alpha_i\partial_i\psi + \beta m_0 c^2\psi\right)\\
&\quad+\beta m_0 c^2\left(\frac{\hbar c}{\mathrm{i}}\sum_i \alpha_i\partial_i\psi + \beta m_0 c^2\psi\right)\\
&= -\hbar^2c^2\sum_{i,j}\frac{\alpha_i\alpha_j + \alpha_j\alpha_i}{2}\partial_i\partial_j\psi\\
&\quad+\frac{\hbar mc^2}{\mathrm{i}}\sum_i(\alpha_i\beta + \beta\alpha_i)\,\partial_i\psi + \beta^2m_0^2c^4\psi \;.
\end{aligned}$$

From this we perceive that the secondary condition (2.4) can be fulfilled only if α_i and β are matrices obeying the algebra

$$\{\alpha_i, \alpha_j\} = 2\delta_{ij} \;,\; \{\alpha_i, \beta\} = 0 \;,\; \alpha_i^2 = \beta^2 = 1 \;. \tag{2.6}$$

Additionally, these matrices must be Hermitean in order for the Hamilton operator itself to be Hermitean:

$$\alpha_i = \alpha_i^\dagger \;,\; \beta = \beta^\dagger \;.$$

From the last relation in (2.6) it follows further that the eigenvalues of the matrices are confined to the values ± 1. On the other hand, the anticommutators along with the cyclic permutability of the trace, $\mathrm{tr}(ab) = \mathrm{tr}(ba)$, imply that the trace of the matrices vanishes, as we have, for example,

$$\mathrm{tr}(\alpha_i) = +\mathrm{tr}(\beta^2\alpha_i) = +\mathrm{tr}(\beta\alpha_i\beta) = -\mathrm{tr}(\alpha_i) = 0 \ .$$

However, since the trace is just the sum of the eigenvalues, the number of positive and negative eigenvalues must be equal. From this follows that the matrices must be of an even-numbered dimension. The smallest even-numbered dimension $N = 2$ is excluded because here, we are only able to find three anticommuting matrices, namely the Pauli matrices. The smallest dimension, for which the condition (2.6) for α_i and β can be satisfied, is $N = 4$. We will be concentrating exclusively on this in the following.

One of the most common explicit representations of the algebra (2.6) is the *Dirac representation*

$$\alpha_i = \begin{pmatrix} 0 & \sigma_i \\ \sigma_i & 0 \end{pmatrix} \ , \ \beta = \begin{pmatrix} 1 & 0 \\ 0 & -1 \end{pmatrix} \ .$$

Here σ_i denote the Pauli matrices known from Subsection 1.1.2, for which we will use constantly the symbol σ instead of τ in the context of the Dirac equation. Another useful representation is the *Weyl representation.* It is defined via

$$\alpha_i = \begin{pmatrix} \sigma_i & 0 \\ 0 & -\sigma_i \end{pmatrix} \ , \ \beta = \begin{pmatrix} 0 & -1 \\ -1 & 0 \end{pmatrix} \ .$$

Due to $N = 4$, (2.2) and (2.5) now become a fourdimensional matrix equation,

$$\mathrm{i}\hbar\frac{\partial\psi_i(x)}{\partial t} = \sum_{j=1}^{4}\left[c(\boldsymbol{\alpha p})_{ij} + \beta_{ij}m_0c^2\right]\psi_j(x) \ , \ i = 1,2,3,4 \ , \tag{2.7}$$

and the wave function ψ becomes a fourdimensional column vector,

$$\psi(x) = \begin{pmatrix} \psi_1(x) \\ \psi_2(x) \\ \psi_3(x) \\ \psi_4(x) \end{pmatrix} \ ,$$

which is usually called *bispinor.* Equation (2.7) denotes the so-called *free Dirac equation in canonical* or *Hamilton form,* for which we will mostly use the abridging vectorial notation

$$\mathrm{i}\hbar\frac{\partial\psi}{\partial t} = H^{(0)}\psi \ , \ H^{(0)} = c\boldsymbol{\alpha p} + \beta m_0c^2 \ . \tag{2.8}$$

Before we study this equation in more detail, we notice that, as desired, it is a partial differential equation of first order in space and time. Furthermore, its Hamilton operator is Hermitean so that we can hope to find a positive definite probability density. However, whether the Dirac equation is Lorentz-covariant and thus the rest of the above conditions are also fulfilled is not obvious at this stage and remains to be verified.

Solutions of the free Dirac equation. The solutions to the free Dirac equation (2.8) with defined momentum can be found as easily as those of the free Klein-Gordon equation. In the Dirac representation they are (see Exercise 11)

$$\left.\begin{aligned}\psi_{\boldsymbol{p}}^{(1,2)}(x) &= \begin{pmatrix} \chi^{(1,2)} \\ \dfrac{\boldsymbol{\sigma p}\chi^{(1,2)}}{p_0 + m_0 c} \end{pmatrix} \mathrm{e}^{-\mathrm{i}(cp_0 - \boldsymbol{px})\hbar} \\ \psi_{\boldsymbol{p}}^{(3,4)}(x) &= \begin{pmatrix} \dfrac{\boldsymbol{\sigma p}\chi^{(3,4)}}{p_0 + m_0 c} \\ \chi^{(3,4)} \end{pmatrix} \mathrm{e}^{+\mathrm{i}(cp_0 - \boldsymbol{px})/\hbar} \;, \end{aligned}\right\} \tag{2.9}$$

with

$$p_0 = +\sqrt{\boldsymbol{p}^2 + m_0^2 c^2} > 0 \;,$$

where $\chi^{(1,2)}$ and $\chi^{(3,4)}$ respectively denote two linearly independent constant two-column spinors. As in the Klein-Gordon case, we are again confronted with two types of solutions: the ones with positive energy $E = +cp_0$ that can be interpreted as particle wave functions and the others with negative energy $E = -cp_0$, between whom lies the "forbidden" energy interval $]-m_0c^2 : m_0c^2[$ (compare to Subsection 1.1.1). For the same reasons as in the Klein-Gordon case, it is clear that the mere existence of the negative Dirac solutions seems initially incomprehensible so that we must seek a physically sensible interpretation of them. As we see in Subsection 2.1.6 (and as the reader might already suspect), there exists again a relationship between the negative solutions and antiparticles explaining the reversed assignment of momentum eigenvalue and index.

Due to the freedom in choosing the spinors $\chi^{(1,2)}$ and $\chi^{(3,4)}$, the positive and negative Dirac solutions in (2.9) are not yet uniquely specified. Besides the Hamilton operator $H^{(0)}$ and the momentum operator $\boldsymbol{p}$, we therefore expect there to exist another operator acting only on the wave functions' inner degrees of freedom and composing a complete set of commuting observables together with $H^{(0)}$ and $\boldsymbol{p}$. In Subsection 2.1.4 we see that this operator is closely connected to the spin whose quantum number is fixed to 1/2. From this we deduce that the Dirac equation seems to be appropriate for the description of spin-1/2 particles (spin-1/2 *fermions*).

Interaction with electromagnetic fields, gauge invariance. In the Dirac theory (as in all quantum mechanical theories) the interaction of a relativistic spin-1/2 particle with an electromagnetic field can be taken into account by the operator replacement (minimal coupling)[1]

[1] This replacement is valid only for structureless point particles. See Exercises 15 and 27.

$$\mathrm{i}\hbar\frac{\partial}{\partial t} \longrightarrow \mathrm{i}\hbar\frac{\partial}{\partial t} - eA^0 \ , \ \boldsymbol{p} \longrightarrow \boldsymbol{p} - \frac{e}{c}\boldsymbol{A}$$

within the free Dirac equation (2.8). This yields the equation

$$\mathrm{i}\hbar\frac{\partial\psi(x)}{\partial t} = H\psi \ , \ H = c\boldsymbol{\alpha}\left(\boldsymbol{p} - \frac{e}{c}\boldsymbol{A}\right) + eA^0 + \beta m_0 c^2 \ , \tag{2.10}$$

with the Hermitean Hamilton operator H and the electric particle charge e. Performing a calculation similar to the Klein-Gordon case, we can easily confirm that this equation is invariant under the local gauge transformations

$$A^0 \longrightarrow A'^0 = A^0 - \frac{1}{c}\frac{\partial\chi}{\partial t} \ , \ \boldsymbol{A} \longrightarrow \boldsymbol{A}' = \boldsymbol{A} + \boldsymbol{\nabla}\chi$$

of the electric field, if the wave function ψ is simultaneously multiplied by an appropriate phase:

$$\psi \longrightarrow \psi' = \psi\mathrm{e}^{\mathrm{i}\Lambda(x)} \ , \ \Lambda(x) = \frac{e}{\hbar c}\chi(x) \ .$$

Continuity equation. Due to the Hermitecity of the Hamilton operator, we expect that, contrary to the Klein-Gordon equation, the Dirac equation allows the definition of a positive definite probability density in the sense of nonrelativistic quantum mechanics. To show this, we calculate as follows: multiplication of (2.10) from the left with $\psi^\dagger = (\psi_1^*, \psi_2^*, \psi_3^*, \psi_4^*)$ yields

$$\mathrm{i}\hbar\psi^\dagger\frac{\partial\psi}{\partial t} = \frac{\hbar c}{\mathrm{i}}\psi^\dagger\boldsymbol{\alpha}\boldsymbol{\nabla}\psi - e\psi^\dagger\boldsymbol{\alpha}\boldsymbol{A}\psi + eA^0\psi^\dagger\psi + m_0c^2\psi^\dagger\beta\psi \ . \tag{2.11}$$

Taking the adjoint of (2.10) (while taking into account $\boldsymbol{\alpha} = \boldsymbol{\alpha}^\dagger$, $\beta = \beta^\dagger$) and subsequently multiplying this from the right by ψ, we get

$$-\mathrm{i}\hbar\frac{\partial\psi^\dagger}{\partial t}\psi = -\frac{\hbar c}{\mathrm{i}}(\boldsymbol{\nabla}\psi^\dagger)\boldsymbol{\alpha}\psi - e\psi^\dagger\boldsymbol{\alpha}\boldsymbol{A}\psi + eA^0\psi^\dagger\psi + m_0c^2\psi^\dagger\beta\psi \ . \tag{2.12}$$

Subtracting the last two equations leads to a continuity equation of the form

$$\frac{\partial\rho(x)}{\partial t} + \boldsymbol{\nabla}\boldsymbol{j}(x) = 0 \ , \ \text{mit} \quad \rho = \psi^\dagger\psi \ , \ \boldsymbol{j} = \psi^\dagger c\boldsymbol{\alpha}\psi \ . \tag{2.13}$$

Applying Gauss's law finally yields

$$\frac{\partial}{\partial t}\int \mathrm{d}^3x\rho = -\int \mathrm{d}^3x\boldsymbol{\nabla}\boldsymbol{j} = -\oint \mathrm{d}\boldsymbol{F}\boldsymbol{j} = 0 \ .$$

Together with

$$\psi^\dagger\psi = \sum_i \psi_i^*\psi_i = \sum_i |\psi_i|^2 \geq 0 \ ,$$

this indeed justifies the interpretation of ρ as a positive definite probability density and, correspondingly, $\boldsymbol{j}$ as a probability current density. Furthermore, it follows that we can carry over the scalar product from nonrelativistic quantum mechanics (in coordinate representation)

$$\langle \psi | \phi \rangle = \int \mathrm{d}^3 x \psi^\dagger(x)\phi(x)$$

including all consequences. In addition to the items of Theorem 1.1, these consequences are

- the orthogonality of eigenstates of Hermitean operators with different eigenvalues,
- the picture- and representation-independence of the scalar product under unitary transformations.

So, in the Hermitean Dirac case and contrary to the non-Hermitean Klein-Gordon case, we do not need to modify the nonrelativistic quantum mechanical terms "scalar product", "Hermitecity", and "unitarity" due to physical or representational reasons. In the more detailed discussion of particles and antiparticles in Subsection 2.1.6, we see how the nonrelativistic expectation value as well as its physical interpretation [see 3) in Theorem 1.1] can be transferred to the Dirac case.

Theorem 2.1: Dirac equation in canonical form

The Dirac equation is the relativistic generalization of Schrödinger's equation for spin-1/2 particles. For a minimal coupled electromagnetic field its canonical form is

$$\mathrm{i}\hbar\frac{\partial\psi(x)}{\partial t} = H\psi(x) \ , \ H = c\boldsymbol{\alpha}\left(\boldsymbol{p} - \frac{e}{c}\boldsymbol{A}\right) + eA^0 + \beta m_0 c^2 \ , \qquad (2.14)$$

where m_0 denotes the rest mass and e the electric charge of the particle. ψ is a fourdimensional column vector (bispinor), and α_i, β are fourdimensional Hermitean matrices obeying the algebra

$$\{\alpha_j, \alpha_k\} = 2\delta_{jk} \ , \ \{\alpha_j, \beta\} = 0 \ , \ \alpha_j^2 = \beta^2 = 1 \ .$$

In the Dirac representation they are

$$\alpha_i = \begin{pmatrix} 0 & \sigma_i \\ \sigma_i & 0 \end{pmatrix} \ , \ \beta = \begin{pmatrix} 1 & 0 \\ 0 & -1 \end{pmatrix} \ .$$

The Dirac equation is invariant under local gauge transformations of the electromagnetic field. Furthermore, it allows the definition of a positive definite probability density

$$\rho(x) = \psi^\dagger(x)\psi(x) \ , \ \int \mathrm{d}^3 x \rho(x) = \text{const}$$

as well as a probability current density

$$\boldsymbol{j}(x) = \psi^\dagger(x) c\boldsymbol{\alpha}\psi(x)$$

▷

that are connected via the continuity equation

$$\frac{\partial \rho}{\partial t} + \nabla \boldsymbol{j} = 0 \ .$$

As in Schrödinger's theory and contrary to the Klein-Gordon theory, the scalar product is defined by (compare the definition of the G-scalar product, Subsection 1.3.1)

$$\langle \psi | \phi \rangle = \int \mathrm{d}^3 x \psi^\dagger(x) \phi(x) \ .$$

In the Dirac representation the solutions to the free Dirac equation are

$$\psi_{\boldsymbol{p}}^{(r)}(x) = \frac{1}{(2\pi\hbar)^{3/2}} \sqrt{\frac{m_0 c}{p_0}} \mathrm{e}^{-\mathrm{i}\epsilon_r p_\mu x^\mu/\hbar} \omega^{(r)}(\boldsymbol{p}) \ , \ \epsilon_r = \begin{cases} +1 \text{ for } r = 1,2 \\ -1 \text{ for } r = 3,4 \end{cases}$$

$$p_0 = +\sqrt{\boldsymbol{p}^2 + m_0^2 c^2} \ ,$$

with

$$\omega^{(1,2)}(\boldsymbol{p}) = \sqrt{\frac{p_0 + m_0 c}{2 m_0 c}} \begin{pmatrix} \chi^{(1,2)} \\ \frac{\boldsymbol{\sigma}\boldsymbol{p}}{p_0 + m_0 c} \chi^{(1,2)} \end{pmatrix} \ , \ \chi^{(i)\dagger}\chi^{(j)} = \delta_{ij}$$

$$\omega^{(3,4)}(\boldsymbol{p}) = \sqrt{\frac{p_0 + m_0 c}{2 m_0 c}} \begin{pmatrix} \frac{\boldsymbol{\sigma}\boldsymbol{p}}{p_0 + m_0 c} \chi^{(3,4)} \\ \chi^{(3,4)} \end{pmatrix} \ , \ \chi^{(i)\dagger}\chi^{(j)} = \delta_{ij}$$

(momentum eigenvalue $+\boldsymbol{p}$ for $r = 1,2$ and $-\boldsymbol{p}$ for $r = 3,4$), and are normalized such that

$$\left\langle \psi_{\boldsymbol{p}}^{(r)} \middle| \psi_{\boldsymbol{p}'}^{(r')} \right\rangle = \delta_{rr'} \delta(\boldsymbol{p} - \boldsymbol{p}') \ , \ \omega^{(r)\dagger}(\epsilon_r \boldsymbol{p}) \omega^{(r')}(\epsilon_{r'} \boldsymbol{p}) = \frac{p_0}{m_0 c} \delta_{rr'} . \quad (2.15)$$

Due to the freedom in the choice of $\chi^{(r)}$, these solutions are not yet uniquely specified.

Besides (2.15) there also exist the following useful completeness and orthogonality relations, which are all proven in Exercise 18 (Section 2.2):

$$\left.\begin{aligned} \bar{\omega}^{(r)}(\boldsymbol{p}) \omega^{(r')}(\boldsymbol{p}) &= \epsilon_r \delta_{rr'} \\ \sum_{r=1}^{4} \epsilon_r \omega_\alpha^{(r)}(\boldsymbol{p}) \bar{\omega}_\beta^{(r)}(\boldsymbol{p}) &= \delta_{\alpha\beta} \\ \sum_{r=1}^{4} \omega_\alpha^{(r)}(\epsilon_r \boldsymbol{p}) \omega_\beta^{(r)\dagger}(\epsilon_r \boldsymbol{p}) &= \frac{p_0}{m_0 c} \delta_{\alpha\beta} \ . \end{aligned}\right\} \quad (2.16)$$

Here

$$\bar{\psi} = \psi^\dagger \beta \quad (2.17)$$

denotes the *Dirac-adjoint* or simply *adjoint bispinor* in relation to ψ.

2.1.2 Dirac Equation in Lorentz-Covariant Form

Having found a relativistic generalization of the Schrödinger equation of first order in space and time possessing a positive definite probability density, we are left to show that, with respect to Dirac's three constraints presented on page 86, this equation has the same form in all inertial systems, in accordance with the relativity principle. Because of the symmetry between $ct = x^0$ and x^i, it is advantageous to introduce the *γ-matrices*

$$\gamma^0 = \beta \ , \ \gamma^i = \beta\alpha_i \ ,$$

which, due to (2.6), obey the *Clifford algebra*

$$\{\gamma^\mu, \gamma^\nu\} = 2g^{\mu\nu} \ , \ (\gamma^\mu)^2 = g^{\mu\mu} \ . \tag{2.18}$$

Furthermore, following from the Hermitecity of α_i and β, we have the relations

$$\gamma^{\mu\dagger} = g^{\mu\mu}\gamma^\mu \Longleftrightarrow \gamma^{\mu\dagger} = \gamma^0\gamma^\mu\gamma^0 \ . \tag{2.19}$$

Using these matrices as well as the four-notation for the momentum and the electromagnetic field,

$$(p^\mu) = \mathrm{i}\hbar(\partial^\mu) = \mathrm{i}\hbar \begin{pmatrix} \partial/(c\partial t) \\ -\boldsymbol{\nabla} \end{pmatrix} \ , \ (A^\mu) = \begin{pmatrix} A^0 \\ \boldsymbol{A} \end{pmatrix} \ ,$$

the canonical Dirac equation (2.14) can be reformulated as

$$\left[\gamma^\mu \left(p_\mu - \frac{e}{c}A_\mu(x)\right) - m_0 c\right]\psi(x) = 0 \ , \tag{2.20}$$

clearly with the same free solutions as in Theorem 2.1. In the Dirac representation the γ-matrices are given by

$$\gamma^0 = \begin{pmatrix} 1 & 0 \\ 0 & -1 \end{pmatrix} \ , \ \gamma^i = \begin{pmatrix} 0 & \sigma_i \\ -\sigma_i & 0 \end{pmatrix} \ ,$$

and in the Weyl representation they are

$$\gamma^0 = \begin{pmatrix} 0 & -1 \\ -1 & 0 \end{pmatrix} \ , \ \gamma^i = \begin{pmatrix} 0 & \sigma_i \\ -\sigma_i & 0 \end{pmatrix} \ .$$

For the Dirac equation to be Lorentz-covariant (form invariant), the following needs to be shown: given a physical state $\psi(x)$ in the inertial system K, there must exist an explicit description that allows the calculation of the corresponding $\psi'(x')$ in the inertial system K' describing the same physical state (passive transformation, see Subsection 1.2.1). Additionally, $\psi'(x')$ must be a solution to the primed equation

$$\left[\gamma'^\mu \left(p'_\mu - \frac{e}{c}A'_\mu(x')\right) - m_0 c\right]\psi'(x') = 0 \ ,$$

where the primed matrices also obey the relations (2.18) and (2.19). However, one can show that two sets of matrices being in accordance with (2.18) differ only by a unitary transformation:

$$\gamma'^{\mu} = U^{\dagger}\gamma^{\mu}U \ , \ U^{\dagger} = U^{-1} \ .$$

This means that we have just another representation of the γ-matrices in the primed system. Without restricting generality, we therefore assume that the γ-matrices have the same form in both systems so that the primed Dirac equation can be rewritten as

$$\left[\gamma^{\mu}\left(p'_{\mu} - \frac{e}{c}A'_{\mu}(x')\right) - m_0 c\right]\psi'(x') = 0 \ . \tag{2.21}$$

The *bispinor transformation* $D(\Lambda)$ which transfers $\psi(x)$ into $\psi'(x')$ must be linear since the Lorentz transformation Λ of the coordinates is linear, too. This means[2]

$$\psi'(x') = \psi'(\Lambda x) = D(\Lambda)\psi(x) = D(\Lambda)\psi(\Lambda^{-1}x')$$

or, equivalently,

$$\psi(x) = D^{-1}(\Lambda)\psi'(x') = D^{-1}(\Lambda)\psi'(\Lambda x)$$

and

$$\psi(x) = D(\Lambda^{-1})\psi'(x') \ .$$

The last two relations imply the equalization

$$D^{-1}(\Lambda) = D(\Lambda^{-1}) \ .$$

In order to obtain a conditional equation for D, we express the primed quantities in (2.21) by the unprimed ones:

$$\left[\gamma^{\mu}\left(p_{\nu} - \frac{e}{c}A_{\nu}(x)\right)[\Lambda^{-1}]^{\nu}{}_{\mu} - m_0 c\right]D(\Lambda)\psi(x) = 0 \ .$$

Multiplying this equation from the left by $D^{-1}(\Lambda)$ yields

$$\left[D^{-1}(\Lambda)\gamma^{\mu}\left(p_{\nu} - \frac{e}{c}A_{\nu}(x)\right)[\Lambda^{-1}]^{\nu}{}_{\mu}D(\Lambda) - m_0 c\right]\psi(x) = 0 \ .$$

From this we perceive that the Dirac equation is Lorentz-covariant if, and only if, to each Lorentz transformation Λ there exists a matrix $D(\Lambda)$ for which[3]

$$D^{-1}(\Lambda)\gamma^{\mu}[\Lambda^{-1}]^{\nu}{}_{\mu}D(\Lambda) = \gamma^{\nu} \Longleftrightarrow D^{-1}(\Lambda)\gamma^{\mu}D(\Lambda) = \Lambda^{\mu}{}_{\nu}\gamma^{\nu} \ . \tag{2.22}$$

It should be mentioned in advance that the matrix $D(\Lambda)$ is generally not unitary. In Exercise 12 it is shown that most generally the relation

[2] Of all the transformations $x^{\mu} \to x'^{\mu} = \Lambda^{\mu}{}_{\nu}x^{\nu} + a^{\mu}$ of the full Poincaré group, we only consider the homogeneous Lorentz transformations ($a^{\mu} = 0$) since the invariance of the Dirac equation under space-time translations is obvious:

$$x^{\mu} \to x'^{\mu} = x^{\mu} + a^{\mu} \Longrightarrow A'^{\mu}(x') = A^{\mu}(x) \ , \ p'^{\mu} = p^{\mu} \Longrightarrow \psi'(x') = \psi(x) \ .$$

[3] Since we are assuming the same γ-matrices in the unprimed and primed systems, it is not justified to consider γ^{μ} to be four-vectors as suggested by the μ-indices. However, in the next subsection we see that, with respect to Lorentz transformations, bilinear forms of the kind $\bar{\psi}\cdots\psi$ behave as if γ^{μ} was a four-vector.

$$D^\dagger(\Lambda) = b\gamma^0 D^{-1}(\Lambda)\gamma^0 \ , \ b = \frac{\Lambda^0{}_0}{|\Lambda^0{}_0|} = \pm 1 \tag{2.23}$$

holds if $\det(D) = 1$ is presupposed.

So far we have found the conditional equation (2.22) for the bispinor transformation $D(\Lambda)$. To some extent, it represents a necessary criterion for the Lorentz covariance of the Dirac equation. In order to complete our line of argument, we are left to show that such bispinor transformations indeed exist by, for example, explicitly constructing them. We will return to this issue in Section 2.2. However, it should be mentioned here that it is indeed possible to write down the corresponding bispinor transformation not only for proper Lorentz transformations (which would be sufficient) but also for all transformations of the Poincaré group.

Adjoint bispinor, adjoint Dirac equation. In (2.17) we already introduced the adjoint bispinor

$$\bar{\psi} = \psi^\dagger\beta = \psi^\dagger\gamma^0 \ ,$$

whose explicit forms in the Dirac and Weyl representations are given by

$$\bar{\psi} = (\psi_1^*, \psi_2^*, -\psi_3^*, -\psi_4^*) \quad \text{and} \quad \bar{\psi} = (-\psi_3^*, -\psi_4^*, -\psi_1^*, -\psi_2^*) \ .$$

A big advantage of its use is that, under Lorentz transformations, it behaves inversely to ψ since, taking (2.23) into account, we find

$$\begin{aligned}\bar{\psi}'(x') &= \psi'^\dagger(x')\gamma^0 = [D\psi(x)]^\dagger\gamma^0 = \psi^\dagger(x)D^\dagger\gamma^0 = b\psi^\dagger(x)\gamma^0 D^{-1} \\ &= b\bar{\psi}(x)D^{-1} \ .\end{aligned} \tag{2.24}$$

This means that we can combine $\bar{\psi}$ and ψ to form *covariant bilinear forms* with a defined transformational behavior under Lorentz transformations (see below as well as Subsection 2.1.3).

Using the relation

$$(\mathrm{i}\hbar\gamma^\mu\partial_\mu\psi)^\dagger = \left(\mathrm{i}\hbar\gamma^\mu\frac{\partial\psi}{\partial x^\mu}\right)^\dagger = -\mathrm{i}\hbar\frac{\partial\psi^\dagger}{\partial x^\mu}\gamma^{\mu\dagger} = -\mathrm{i}\hbar\partial_\mu\psi^\dagger\gamma^{\mu\dagger} \ ,$$

one obtains

$$\left(-\mathrm{i}\hbar\partial_\mu - \frac{e}{c}A_\mu\right)\psi^\dagger\gamma^{\mu\dagger} - m_0 c\psi^\dagger = 0 \ ,$$

which is the Hermitean conjugate to the Dirac equation (2.20). Multiplying this relation by γ^0 from the right and exploiting (2.19) finally yields the *adjoint Dirac equation*

$$\left(-p_\mu - \frac{e}{c}A_\mu\right)\bar{\psi}\gamma^\mu - m_0 c\bar{\psi} = 0 \ ,$$

which is equivalent to the Dirac equation.

Lorentz covariance of the continuity equation. With the help of the adjoint bispinor, it is now easy to show that the quantity

$$j^\mu = c\psi^\dagger\gamma^0\gamma^\mu\psi = c\bar{\psi}\gamma^\mu\psi \ , \ (j^\mu) = \begin{pmatrix} c\rho \\ \boldsymbol{j} \end{pmatrix}$$

is a contravariant four-vector under orthochronous Lorentz transformations and hence that the continuity equation can be written in the manifestly covariant fashion

$$\partial_\mu j^\mu = 0 \ .$$

This is because, taking into account (2.22) and (2.24) with $b = +1$, it follows that

$$\begin{aligned} j'^\mu(x') &= c\bar{\psi}'(x')\gamma^\mu\psi'(x') = c\bar{\psi}(x)D^{-1}(\Lambda)\gamma^\mu D(\Lambda)\psi(x) \\ &= c\bar{\psi}(x)\Lambda^\mu{}_\nu\gamma^\nu\psi(x) = \Lambda^\mu{}_\nu j^\mu(x) \ . \end{aligned}$$

Theorem 2.2: Dirac equation in Lorentz-covariant form

Using the γ-matrices

$$\gamma^0 = \beta \ , \ \gamma^j = \beta\alpha_j \ ,$$

the Dirac equation for a minimal coupled electromagnetic field is

$$\left[\gamma^\mu\left(p_\mu - \frac{e}{c}A_\mu(x)\right) - m_0c\right]\psi(x) = 0 \ . \tag{2.25}$$

It is Lorentz-covariant if, for each Lorentz transformation

$$\Lambda: \quad x^\mu \longrightarrow x'^\mu = \Lambda^\mu{}_\nu x^\nu \ ,$$

one can find a bispinor transformation

$$D(\Lambda): \quad \psi(x) \longrightarrow \psi'(x') = D(\Lambda)\psi(x)$$

with

$$D^{-1}(\Lambda)\gamma^\mu D(\Lambda) = \Lambda^\mu{}_\nu\gamma^\nu \ . \tag{2.26}$$

With this constraint, the four-current density

$$j^\mu = c\psi^\dagger\gamma^0\gamma^\mu\psi = c\bar{\psi}\gamma^\mu\psi$$

transforms as a contravariant four-vector under orthochronous Lorentz transformations, and the Lorentz-covariant continuity equation

$$\partial_\mu j^\mu = 0$$

holds. The adjoint bispinor

$$\bar{\psi} = \psi^\dagger\gamma^0$$

▷

obeys the adjoint Dirac equation

$$\left(-p_\mu - \frac{e}{c}A_\mu\right)\bar{\psi}\gamma^\mu - m_0 c\bar{\psi} = 0 .$$

Using

$$D^\dagger(\Lambda) = b\gamma^0 D^{-1}(\Lambda)\gamma^0 , \; b = \frac{\Lambda^0{}_0}{|\Lambda^0{}_0|} = \pm 1 , \; \det(D) = 1 ,$$

the transformational behavior of the adjoint bispinor follows as

$$\bar{\psi}(x) \longrightarrow \bar{\psi}'(x') = b\bar{\psi}(x)D^{-1}(\Lambda) .$$

2.1.3 Properties of γ-Matrices and Covariant Bilinear Forms

Before continuing the discussion of Dirac's equation with respect to its solutions, we present some intermediate considerations about formal properties of the γ-matrices and their use for the composition of *covariant bilinear forms.*

Complete basis system. The four matrices γ^μ are linearly independent, but they do not compose a complete basis within the 16-dimensional space of 4×4-matrices. However, by simple matrix multiplications of the γ^μ, it is possible to find 16 linearly independent basis elements of this space. They are listed in Tab. 2.1.

		explicit form	
Notation	Number	$\left(\Gamma^{(n)}\right)^2 = +1$	$\left(\Gamma^{(n)}\right)^2 = -1$
$\Gamma^{(\mathrm{S})}$: 1	1	1	
$\Gamma^{(\mathrm{V})}$: γ^μ	4	γ^0	$\gamma^1, \gamma^2, \gamma^3$
$\Gamma^{(\mathrm{T})}$: $\gamma^\mu\gamma^\nu$, $\mu < \nu$	6	$\gamma^0\gamma^1, \gamma^0\gamma^2, \gamma^0\gamma^3$	$\gamma^1\gamma^2, \gamma^1\gamma^3, \gamma^2\gamma^3$
$\Gamma^{(\mathrm{A})}$: $\gamma^\mu\gamma^5$	4	$\mathrm{i}\gamma^0\gamma^2\gamma^3, \mathrm{i}\gamma^0\gamma^3\gamma^1, \mathrm{i}\gamma^0\gamma^1\gamma^2$	$\mathrm{i}\gamma^1\gamma^2\gamma^3$
$\Gamma^{(\mathrm{P})}$: γ^5	1	$\mathrm{i}\gamma^0\gamma^1\gamma^2\gamma^3$	

Tab. 2.1. Basis elements of the 4×4-matrix space.

As for the γ-matrices themselves, the square of the matrices summarized by the five types $\Gamma^{(n)}$ is +1 or −1. The upper indices stand for scalar (S), vector (V), tensor (T), pseudo-scalar (P), and pseudo- or axial vector (A). They refer to the respective transformational behavior of the corresponding matrices in combination with $\bar{\psi}$ and ψ under Lorentz transformations (see below). The unity matrix alone commutes with all the other matrices. Every other matrix commutes with exactly 8 of the 16 matrices and anticommutes with the remaining 8. It is clear that each product of more than four γ-matrices necessarily contains matrices with a same index so that such a product can be related back to one of the 16 elements by using the anticommutator relation (2.18).

Now, by the following line of argument, we can show that the above listed matrices indeed compose a complete basis within the space of 4×4-matrices:

- For each matrix $\Gamma^{(n)} \neq \Gamma^{(\mathrm{S})}$ there exists a matrix $\Gamma^{(m)}$ so that
$$\Gamma^{(n)}\Gamma^{(m)} = -\Gamma^{(m)}\Gamma^{(n)} .$$
This implies a vanishing trace of $\Gamma^{(n)}$,
$$\begin{aligned}\pm\mathrm{tr}\left[\Gamma^{(n)}\right] &= \mathrm{tr}\left[\Gamma^{(n)}\left(\Gamma^{(m)}\right)^2\right] = -\mathrm{tr}\left[\Gamma^{(m)}\Gamma^{(n)}\Gamma^{(m)}\right] \\ &= -\mathrm{tr}\left[\Gamma^{(n)}\left(\Gamma^{(m)}\right)^2\right] = 0 , \end{aligned} \tag{2.27}$$
where, in the second last relation, the cyclic permutability of the trace has been used.
- For each $\Gamma^{(a)}$ and $\Gamma^{(b)} \neq \Gamma^{(a)}$ there exists an $\Gamma^{(n)} \neq \Gamma^{(\mathrm{S})}$ so that
$$\Gamma^{(a)}\Gamma^{(b)} = \Gamma^{(n)} . \tag{2.28}$$
- Now we assume the existence of numbers a_n with
$$\sum_n a_n\Gamma^{(n)} = 0 . \tag{2.29}$$
Multiplying this relation by $\Gamma^{(m)}$ and subsequently taking the trace yields
$$\sum_n a_n \mathrm{tr}\left[\Gamma^{(n)}\Gamma^{(m)}\right] = 0 .$$
In the case of $\Gamma^{(m)} \neq \Gamma^{(\mathrm{S})}$ we have, due to (2.27) and (2.28), $a_m = 0$. If, on the other hand, $\Gamma^{(m)} = \Gamma^{(\mathrm{S})}$, we find $a_\mathrm{S} = 0$. Therefore, all coefficients in (2.29) are zero, which proves the linear independence of $\Gamma^{(n)}$.

Inverse matrices. As for four-vectors (more generally: Lorentz tensors), we can define the matrix γ_μ corresponding to γ^μ by lowering the index via
$$\gamma_\mu = g_{\mu\nu}\gamma^\nu .$$
Based on (2.18) (no summation over μ) we have
$$\gamma^\mu\gamma_\mu = \gamma^\mu g_{\mu\nu}\gamma^\nu = \left(\gamma^\mu\right)^2 g_{\mu\mu} = g^{\mu\mu}g_{\mu\mu} = 1 .$$
This means that γ_μ is the inverse matrix to γ^μ. Therefore, one generally obtains the inverse matrix to $\Gamma^{(n)}$ by reversing the order of its matrices γ^μ, replacing them by γ_μ and adding a sign to each i-factor. Following this procedure, for example, the inverse matrix of $\mathrm{i}\gamma^1\gamma^2\gamma^3$ reads $-\mathrm{i}\gamma_3\gamma_2\gamma_1$.

Covariant bilinear forms. As already mentioned above, we can use the γ-matrix combinations $\Gamma^{(n)}$ from Tab. 2.1 to build covariant bilinear forms with a well-defined transformational behavior under Lorentz transformations. We now specify such bilinear forms and the corresponding transformational behavior in more detail, restricting ourselves to orthochronous transformations ($b = +1$) and making use of the relations

$$\psi \longrightarrow \psi' = D\psi \;,\; \bar{\psi} \longrightarrow \bar{\psi}' = \bar{\psi}D^{-1} \;,\; D^{-1}\gamma^\mu D = \Lambda^\mu{}_\nu \gamma^\nu \;.$$

- *Scalar bilinear form,* $\Gamma^{(\mathrm{S})}$. This bilinear form is given by $\bar{\psi}1\psi = \bar{\psi}\psi$ and obviously transforms as a Lorentz scalar:

$$\bar{\psi}\psi \longrightarrow \bar{\psi}D^{-1}D\psi = \bar{\psi}\psi \;.$$

- *Vectorial bilinear form,* $\Gamma^{(\mathrm{V})}$. It is composed of $\bar{\psi}\gamma^\mu\psi$ and represents the four-current which we already know to transform as a contravariant vector:

$$\bar{\psi}\gamma^\mu\psi \longrightarrow \bar{\psi}D^{-1}\gamma^\mu D\psi = \bar{\psi}\Lambda^\mu{}_\nu\gamma^\nu\psi = \Lambda^\mu{}_\nu\bar{\psi}\gamma^\nu\psi \;.$$

- *Tensorial bilinear form,* $\Gamma^{(\mathrm{T})}$. This quantity is $\bar{\psi}\gamma^\mu\gamma^\nu\psi$. Its transformational behavior is calculated as

$$\begin{aligned}\bar{\psi}\gamma^\mu\gamma^\nu\psi \longrightarrow \;& \bar{\psi}D^{-1}\gamma^\mu\gamma^\nu D\psi = \bar{\psi}D^{-1}\gamma^\mu DD^{-1}\gamma^\nu D\psi \\ &= \Lambda^\mu{}_\alpha\Lambda^\nu{}_\beta\bar{\psi}\gamma^\alpha\gamma^\beta\psi \;.\end{aligned}$$

  Thus, $\bar{\psi}\gamma^\mu\gamma^\nu\psi$ transforms as a contravariant tensor of rank 2.

- *Pseudo-scalar bilinear form,* $\Gamma^{(\mathrm{P})}$. A corresponding calculation for $\bar{\psi}\gamma^5\psi$ yields

$$\begin{aligned}\bar{\psi}\gamma^5\psi = \mathrm{i}\bar{\psi}\gamma^0\gamma^1\gamma^2\gamma^3\psi \longrightarrow \;& \mathrm{i}\bar{\psi}D^{-1}\gamma^0 DD^{-1}\gamma^1 DD^{-1}\gamma^2 DD^{-1}\gamma^3 D\psi \\ &= \mathrm{i}\Lambda^0{}_\alpha\Lambda^1{}_\beta\Lambda^2{}_\delta\Lambda^3{}_\rho\bar{\psi}\gamma^\alpha\gamma^\beta\gamma^\delta\gamma^\rho\psi \;. \qquad (2.30)\end{aligned}$$

  For further evaluation we have to keep in mind that terms with at least two equal indices do not contribute, as, for example, we have

$$\Lambda^2{}_\mu\Lambda^3{}_\mu\gamma^\mu\gamma^\mu = \Lambda^2{}_\mu g^{\mu\nu}\Lambda^3{}_\nu$$

  that vanishes, due to (A.2) and (A.3) in the Appendix. Furthermore, because of the anticommutativity of the γ-matrices, it follows that

$$\begin{aligned}\Lambda^0{}_\alpha\Lambda^1{}_\beta\Lambda^2{}_\delta\Lambda^3{}_\rho\gamma^\alpha\gamma^\beta\gamma^\delta\gamma^\rho &= \epsilon^{\alpha\beta\delta\rho}\Lambda^0{}_\alpha\Lambda^1{}_\beta\Lambda^2{}_\delta\Lambda^3{}_\rho\gamma^0\gamma^1\gamma^2\gamma^3 \\ &= \det(\Lambda)\gamma^0\gamma^1\gamma^2\gamma^3 \;.\end{aligned}$$

  All in all, (2.30) can therefore be written as

$$\bar{\psi}\gamma^5\psi \longrightarrow \det(\Lambda)\bar{\psi}\gamma^5\psi \;.$$

  Consequently, $\bar{\psi}\gamma^5\psi$ is a pseudo-scalar whose transformational behavior differs from that of a scalar by an additional factor of $\det(\Lambda)$.

- *Pseudo- or axial vector bilinear form,* $\Gamma^{(\mathrm{A})}$. The last bilinear form is $\bar{\psi}\gamma^\mu\gamma^5\psi$. Here a line of argument similar to the previous point leads to the following transformation law:

$$\begin{aligned}\bar{\psi}\gamma^\mu\gamma^5\psi &\longrightarrow \mathrm{i}\Lambda^\mu{}_\nu\Lambda^0{}_\alpha\Lambda^1{}_\beta\Lambda^2{}_\delta\Lambda^3{}_\rho\bar{\psi}\gamma^\mu\gamma^\alpha\gamma^\beta\gamma^\delta\gamma^\rho\psi\\ &= \mathrm{i}\Lambda^\mu{}_\nu\epsilon^{\alpha\beta\delta\rho}\bar{\psi}\gamma^\mu\gamma^0\gamma^1\gamma^2\gamma^3\psi\\ &= \det(\Lambda)\Lambda^\mu{}_\nu\bar{\psi}\gamma^\nu\gamma^5\psi\ .\end{aligned}$$

Therefore, $\bar{\psi}\gamma^\mu\gamma^5\psi$ is a contravariant pseudo-vector. It transforms as a vector up to an additional factor of $\det(\Lambda)$.

2.1.4 Spin Operator

In Subsection 2.1.1 we already pointed out that, besides $H^{(0)}$ and $\boldsymbol{p}$, there must exist another operator acting solely on the inner degrees of freedom of the Dirac wave functions and commuting with $H^{(0)}$ and $\boldsymbol{p}$. In order to find this operator, we initially consider the free Hamilton operator for resting free particles in the Dirac or Weyl representation,

$$\mathcal{H}^{(0)} = \beta m_0 c^2\ ,$$

and notice that the operator

$$\boldsymbol{S} = \frac{h}{2}\hat{\boldsymbol{\sigma}}\ ,\ \hat{\boldsymbol{\sigma}} = \begin{pmatrix}\boldsymbol{\sigma} & 0\\ 0 & \boldsymbol{\sigma}\end{pmatrix} \tag{2.31}$$

commutes with $\mathcal{H}^{(0)}$. Since $\boldsymbol{S}$, on the one hand, exhibits a formal similarity to the spin operator of nonrelativistic quantum mechanics and, on the other hand, obeys the typical commutator relations

$$[S_i, S_j] = \mathrm{i}\hbar\epsilon_{ijk}S_k\ ,\ i,j,k = 1,2,3$$

for quantum mechanical angular momenta, it can be identified with the spin operator of the Dirac theory. Its quantum number follows from the attribution

$$\boldsymbol{S}^2 = \frac{\hbar^2}{4}\hat{\boldsymbol{\sigma}}^2 = \frac{3}{4}\hbar^2 = \hbar^2 s(s+1) \Longrightarrow s = \frac{1}{2}\ ,$$

indicating that the Dirac solutions describe spin-1/2 particles.[4] Furthermore, it follows from the well-known rules of the angular momentum algebra that the projection of the spin onto an arbitrary spatial direction,

$$S_{\boldsymbol{n}^{(0)}} = \boldsymbol{n}^{(0)}\boldsymbol{S}\ ,\ |\boldsymbol{n}^{(0)}| = 1\ ,$$

possesses the eigenvalues or quantum numbers

$$\hbar m_{\boldsymbol{n}^{(0)}}\ ,\ m_{\boldsymbol{n}^{(0)}} = \pm 1/2\ .$$

Therefore, the sought operator is $S_{\boldsymbol{n}^{(0)}}$. Together with $\mathcal{H}^{(0)}$, $\boldsymbol{p}$ (and $\boldsymbol{S}^2$), it composes a complete set of commuting observables. Consequently, in the

[4] A more formal transformation theoretical argument is given in Subsection 2.2.2.

rest case, the Dirac solutions in Theorem 2.1 can be uniquely specified by quoting their eigenvalues of energy, momentum, and the spin projection onto an arbitrary space direction. If the spin is oriented in or opposite to the z-direction ($m_z = \pm 1/2$), the corresponding spinors may be chosen as, for example,

$$\chi^{(1,3)} \sim \begin{pmatrix} 1 \\ 0 \end{pmatrix} \Longrightarrow S_z \omega^{(1,3)}(\mathbf{0}) = +\frac{\hbar}{2}\omega^{(1,3)}(\mathbf{0})$$

$$\chi^{(2,4)} \sim \begin{pmatrix} 0 \\ 1 \end{pmatrix} \Longrightarrow S_z \omega^{(2,4)}(\mathbf{0}) = -\frac{\hbar}{2}\omega^{(2,4)}(\mathbf{0}) ,$$

with

$$S_z = S_{\boldsymbol{n}^{(0)}} , \ \boldsymbol{n}^{(0)} = \begin{pmatrix} 0 \\ 0 \\ 1 \end{pmatrix} .$$

Let us now turn to the more general case of a moving free particle described by the Hamilton operator $H^{(0)}$ from (2.8). Contrary to the nonrelativistic theory, in this case, the spin operator $\boldsymbol{S}$ or $S_{\boldsymbol{n}^{(0)}}$ does not, in general, commute with $H^{(0)}$. However, we can easily extend it to a Lorentz-covariant operator that commutes with $H^{(0)}$ and $\boldsymbol{p}$ in every case. Here we notice the following according to Subsection 2.1.2: if $\Lambda_{\boldsymbol{v}}$ denotes the proper Lorentz transformation which transforms from the rest system of a spin-1/2 particle to a reference frame moving with velocity $\boldsymbol{v}$, it holds that

$$\begin{aligned}
\omega^{(r)}(-\boldsymbol{p}) = D(\Lambda_{\boldsymbol{v}})\omega^{(r)}(\mathbf{0}) &\Longleftrightarrow \omega^{(r)}(\boldsymbol{p}) = D(\Lambda_{-\boldsymbol{v}})\omega^{(r)}(\mathbf{0}) \\
&\Longleftrightarrow \omega^{(r)}(\mathbf{0}) = D^{-1}(\Lambda_{-\boldsymbol{v}})\omega^{(r)}(\boldsymbol{p}) \\
&\Longleftrightarrow \omega^{(r)}(\mathbf{0}) = D(\Lambda_{-\boldsymbol{v}}^{-1})\omega^{(r)}(\boldsymbol{p}) \\
&\Longleftrightarrow \omega^{(r)}(\mathbf{0}) = D(\Lambda_{\boldsymbol{v}})\omega^{(r)}(\boldsymbol{p}) ,
\end{aligned}$$

where $D(\Lambda_{\boldsymbol{v}})$ is the bispinor transformation belonging to $\Lambda_{\boldsymbol{v}} = \Lambda_{-\boldsymbol{v}}^{-1}$, $\omega^{(r)}(\mathbf{0})$ are the bispinors for resting particles, and $\omega^{(r)}(\boldsymbol{p})$ are the bispinors for free particles with momentum $\boldsymbol{p}$. If we now assume that the bispinors $\chi^{(r)}$ are chosen in a way that the bispinors $\omega^{(r)}(\mathbf{0})$ of the rest system are eigenstates of $S_{\boldsymbol{n}^{(0)}}$, i.e.

$$S_{\boldsymbol{n}^{(0)}}\omega^{(r)}(\mathbf{0}) = \hbar m_{\boldsymbol{n}^{(0)}}\omega^{(r)}(\mathbf{0}) , \ m_{\boldsymbol{n}^{(0)}} = \pm\frac{1}{2} ,$$

it follows that

$$\begin{aligned}
&S_{\boldsymbol{n}^{(0)}} D(\Lambda_{\boldsymbol{v}})\omega^{(r)}(\boldsymbol{p}) = \hbar m_{\boldsymbol{n}^{(0)}} D(\Lambda_{\boldsymbol{v}})\omega^{(r)}(\boldsymbol{p}) \\
&\Longleftrightarrow D^{-1}(\Lambda_{\boldsymbol{v}}) S_{\boldsymbol{n}^{(0)}} D(\Lambda_{\boldsymbol{v}})\omega^{(r)}(\boldsymbol{p}) = \hbar m_{\boldsymbol{n}^{(0)}}\omega^{(r)}(\boldsymbol{p}) .
\end{aligned}$$

In other words, if $\omega^{(r)}(\mathbf{0})$ are eigenstates of $S_{\boldsymbol{n}^{(0)}}$, then $\omega^{(r)}(\boldsymbol{p})$ are eigenstates of the *Lorentz-covariant spin operator*

$$S(n,p) = D^{-1}(\Lambda_{\boldsymbol{v}}) S_{\boldsymbol{n}^{(0)}} D(\Lambda_{\boldsymbol{v}}) \tag{2.32}$$

with the same eigenvalues. The desired commutator relations necessarily follow from this:

$$\left[S(n,p), H^{(0)}\right] = [S(n,p), \boldsymbol{p}] = 0 \ .$$

The notation $S(n,p)$ becomes apparent if we cast (2.32) in a slightly different form and eliminate the bispinor transformation D. To do this, we initially rewrite $S_{\boldsymbol{n}^{(0)}}$ in the form

$$S_{\boldsymbol{n}^{(0)}} = \frac{\hbar}{2m_0c}\gamma^5\gamma^\mu n_\mu^{(0)}\gamma^\nu p_\nu^{(0)} \ ,$$

where

$$\left(n^{(0)\mu}\right) = \begin{pmatrix} 0 \\ \boldsymbol{n}^{(0)} \end{pmatrix} \ , \ \left(p^{(0)\mu}\right) = \begin{pmatrix} m_0c \\ \boldsymbol{0} \end{pmatrix}$$

respectively denote the four-vector extension of the direction of spin projection and the four-momentum in the rest system. Using (2.26) the relation (2.32) turns into

$$\begin{aligned} S(n,p) &= D^{-1}(\Lambda_{-\boldsymbol{v}}^{-1})S_{\boldsymbol{n}^{(0)}}D(\Lambda_{-\boldsymbol{v}}^{-1}) \\ &= \frac{\hbar}{2m_0c}\gamma^5 D^{-1}(\Lambda_{-\boldsymbol{v}}^{-1})\gamma^\mu D(\Lambda_{-\boldsymbol{v}}^{-1})n_\mu^{(0)}D^{-1}(\Lambda_{-\boldsymbol{v}}^{-1})\gamma^\nu D(\Lambda_{-\boldsymbol{v}}^{-1})p_\nu^{(0)} \\ &= \frac{\hbar}{2m_0c}\gamma^5[\Lambda_{-\boldsymbol{v}}^{-1}]^\mu{}_\alpha\gamma^\alpha n_\mu^{(0)}[\Lambda_{-\boldsymbol{v}}^{-1}]^\nu{}_\beta\gamma^\beta p_\nu^{(0)} \\ &= \frac{\hbar}{2m_0c}\gamma^5\gamma^\alpha n_\mu^{(0)}[\Lambda_{-\boldsymbol{v}}^{-1}]^\mu{}_\alpha\gamma^\beta p_\nu^{(0)}[\Lambda_{-\boldsymbol{v}}^{-1}]^\nu{}_\beta \\ &= \frac{\hbar}{2m_0c}\gamma^5\gamma^\alpha n_\alpha\gamma^\beta p_\beta \ , \end{aligned} \tag{2.33}$$

with

$$n^\mu = [\Lambda_{-\boldsymbol{v}}]^\mu{}_\nu n^{(0)\nu} \ , \ p^\mu = [\Lambda_{-\boldsymbol{v}}]^\mu{}_\nu p^{(0)\nu} \ .$$

In the second row of (2.33) the relation $[D,\gamma^5] = 0$ is used, which is valid for proper Lorentz transformations.[5] All in all, we can therefore record:

Theorem 2.3: Lorentz-covariant spin operator and four-polarization

The Lorentz-covariant spin operator (more precisely: the Lorentz-covariant projection of the spin) of the Dirac theory is representation-independently

$$S(n,p) = \frac{\hbar}{2m_0c}\gamma^5\gamma^\mu n_\mu\gamma^\nu p_\nu \ , \ \left[S(n,p), H^{(0)}\right] = [S(n,p), \boldsymbol{p}] = 0 \ .$$

▷

[5] $\{\gamma^5, \gamma^\mu\} = 0 \Longrightarrow [\gamma^5, \sigma_{\mu\nu}] = 0$; see Theorem 2.6 in Subsection 2.2.1.

$n = (n^\mu)$ denotes the *four-polarization* and $p = (p^\mu)$ the four-momentum that are connected to the corresponding expressions of the rest system via

$$p^\mu = [\Lambda_{-\boldsymbol{v}}]^\mu{}_\nu p^{(0)\nu} \;,\; (p^{(0)\mu}) = \begin{pmatrix} m_0 c \\ \mathbf{0} \end{pmatrix}$$

$$n^\mu = [\Lambda_{-\boldsymbol{v}}]^\mu{}_\nu n^{(0)\nu} \;,\; (n^{(0)\mu}) = \begin{pmatrix} 0 \\ \boldsymbol{n}^{(0)} \end{pmatrix} .$$

If the free bispinors $\omega^{(r)}(\mathbf{0})$ are eigenstates of $S(n^{(0)}, p^{(0)}) = S_{\boldsymbol{n}^{(0)}}$, then $\omega^{(r)}(\boldsymbol{p}) = D(\Lambda_{-\boldsymbol{v}})\omega^{(r)}(\mathbf{0})$ are eigenstates of $S(n,p)$ with the same eigenvalues.

A special case of the Lorentz-covariant spin operator is the so-called *helicity operator*. It is defined by the projection of the spin $\boldsymbol{S}$ onto the momentum direction, i.e. $\boldsymbol{n}^{(0)} = \boldsymbol{p}/|\boldsymbol{p}|$. Taking into account[6] $[S_{\boldsymbol{p}}, D(\Lambda_{\boldsymbol{v}})] = 0$ and (2.32), it can be written as

$$S(p,p) = S_{\boldsymbol{p}} = \frac{\boldsymbol{p}\boldsymbol{S}}{|\boldsymbol{p}|} = \frac{\hbar}{2|\boldsymbol{p}|}\gamma^5\gamma^0\boldsymbol{\gamma}\boldsymbol{p} \;,$$

where the right hand side expression is independent from the chosen representation. The corresponding quantum number $m_{\boldsymbol{p}}$ is called *helicity*.

2.1.5 Projection Operators

For practical calculations it is often helpful to have operators that project bispinors with a given energy sign and spin orientation out of the general solutions to the free Dirac equation with momentum index $\boldsymbol{p}$.

Energy projection operators. The projectors for solutions with defined energy follow immediately from the free Dirac equation in momentum space:

$$\gamma^\mu p_\mu \omega^{(1,2)}(\boldsymbol{p}) = m_0 c\,\omega^{(1,2)}(\boldsymbol{p}) \;,\; \gamma^\mu p_\mu \omega^{(3,4)}(\boldsymbol{p}) = -m_0 c\,\omega^{(3,4)}(\boldsymbol{p}) \;. \quad (2.34)$$

On this basis, we can define the operators

$$\Lambda_+(p) = \frac{\gamma^\mu p_\mu + m_0 c}{2m_0 c} \;,\; \Lambda_-(p) = \frac{-\gamma^\mu p_\mu + m_0 c}{2m_0 c} \;,$$

with the properties

$$\Lambda_+(p)\omega^{(1,2)}(\boldsymbol{p}) = \omega^{(1,2)}(\boldsymbol{p}) \quad,\quad \Lambda_+(p)\omega^{(3,4)}(\boldsymbol{p}) = 0$$

$$\Lambda_-(p)\omega^{(1,2)}(\boldsymbol{p}) = 0 \quad,\quad \Lambda_-(p)\omega^{(3,4)}(\boldsymbol{p}) = \omega^{(3,4)}(\boldsymbol{p}) \;.$$

Thus, $\Lambda_+(p)$ projects onto positive and $\Lambda_-(p)$ onto negative solutions at arbitrary spin orientation. This complete and Lorentz-covariant projection system satisfies the equations

$$\Lambda_\pm^2(p) = \Lambda_\pm(p) \;,\; \Lambda_\pm(p)\Lambda_\mp(p) = 0 \;,\; \Lambda_+(p) + \Lambda_-(p) = 1 \quad (2.35)$$

that are characteristic for projection operators.

[6] See (2.60) in Subsection 2.2.1.

Spin projection operators. In order to obtain analogous expressions for the spin, we assume that $\chi^{(r)}$ are chosen so that

$$\left.\begin{aligned} S(n,p)\omega^{(1,3)}(\boldsymbol{p}) &= +\frac{\hbar}{2}\omega^{(1,3)}(\boldsymbol{p}) \quad \begin{pmatrix}\text{spin oriented toward}\\ \boldsymbol{n}^{(0)} \text{ in the rest system}\end{pmatrix} \\ S(n,p)\omega^{(2,4)}(\boldsymbol{p}) &= -\frac{\hbar}{2}\omega^{(2,4)}(\boldsymbol{p}) \quad \begin{pmatrix}\text{spin oriented toward}\\ -\boldsymbol{n}^{(0)} \text{ in the rest system}\end{pmatrix} . \end{aligned}\right\} \quad (2.36)$$

Following Theorem 2.3 and (2.34), we then have

$$\begin{aligned} S(n,p)\omega^{(1,2)}(\boldsymbol{p}) &= \frac{\hbar}{2m_0c}\gamma^5\gamma^\mu n_\mu \gamma^\nu p_\nu \omega^{(1,2)}(\boldsymbol{p}) \\ &= \frac{\hbar}{2}\gamma^5\gamma^\mu n_\mu \omega^{(1,2)}(\boldsymbol{p}) \\ &= \pm\frac{\hbar}{2}\omega^{(1,2)}(\boldsymbol{p}) \\ \Longrightarrow \gamma^5\gamma^\mu n_\mu \omega^{(1,2)}(\boldsymbol{p}) &= \pm\omega^{(1,2)}(\boldsymbol{p}) \\ S(n,p)\omega^{(3,4)}(\boldsymbol{p}) &= \frac{\hbar}{2m_0c}\gamma^5\gamma^\mu n_\mu \gamma^\nu p_\nu \omega^{(3,4)}(\boldsymbol{p}) \\ &= -\frac{\hbar}{2}\gamma^5\gamma^\mu n_\mu \omega^{(3,4)}(\boldsymbol{p}) \\ &= \pm\frac{\hbar}{2}\omega^{(3,4)}(\boldsymbol{p}) \\ \Longrightarrow \gamma^5\gamma^\mu n_\mu \omega^{(3,4)}(\boldsymbol{p}) &= \mp\omega^{(3,4)}(\boldsymbol{p}) . \end{aligned}$$

This yields the Lorentz-covariant spin projectors

$$\Sigma(n) = \frac{1}{2}(1+\gamma^5\gamma^\mu n_\mu) \ , \ \Sigma(-n) = \frac{1}{2}(1-\gamma^5\gamma^\mu n_\mu) \ ,$$

acting on $\omega^{(r)}(\boldsymbol{p})$ in the following way:

$$\left.\begin{aligned} \Sigma(n)\omega^{(1,4)}(\boldsymbol{p}) &= \omega^{(1,4)}(\boldsymbol{p}) \quad , \quad \Sigma(n)\omega^{(2,3)}(\boldsymbol{p}) = 0 \\ \Sigma(-n)\omega^{(1,4)}(\boldsymbol{p}) &= 0 \quad , \quad \Sigma(-n)\omega^{(2,3)}(\boldsymbol{p}) = \omega^{(2,3)}(\boldsymbol{p}) . \end{aligned}\right\} \quad (2.37)$$

This means that $\Sigma(n)$ [$\Sigma(-n)$] projects onto bispinors with positive [negative] energy whose spin is oriented toward $+\boldsymbol{n}^{(0)}$ in the rest system, and onto bispinors with negative [positive] energy whose spin is oriented toward $-\boldsymbol{n}^{(0)}$ in the rest system. We can easily see that $\Sigma(\pm n)$ also compose a complete projection system fulfilling the corresponding characteristic equations similar to (2.35). Since the projectors for energy and spin both possess $\omega^{(r)}(\boldsymbol{p})$ as a common system of eigenvectors, we further have $[\Lambda_\pm(p), \Sigma(\pm n)] = 0$. We can therefore immediately write down four additional projection operators,

$$
\begin{aligned}
P_1(\boldsymbol{p}) &= \Lambda_+(p)\Sigma(n) \\
P_2(\boldsymbol{p}) &= \Lambda_+(p)\Sigma(-n) \\
P_3(\boldsymbol{p}) &= \Lambda_-(p)\Sigma(-n) \\
P_4(\boldsymbol{p}) &= \Lambda_-(p)\Sigma(n) \ ,
\end{aligned}
$$

acting as

$$
P_r(\boldsymbol{p})\omega^{(r')}(\boldsymbol{p}) = \delta_{rr'}\omega^{(r)}(\boldsymbol{p}) \ .
$$

In (2.37) it is apparent that the action of the spin projector for negative energy states is reverse to what would be expected based on (2.36). We have already encountered a similar situation in Theorem 2.1 where the solutions with negative energy and momentum index $\boldsymbol{p}$ are eigenfunctions of the momentum operator with eigenvalue $-\boldsymbol{p}$. The physical reason for these seemingly reversed assignments is connected to the interpretation of the negative solutions addressed in the next subsection.

Theorem 2.4: Projection operators for energy and spin

Introducing for $\omega^{(r)}(\boldsymbol{p})$ the commonly used notation

$$
\begin{aligned}
\omega^{(1)}(\boldsymbol{p}) &= u(p,n) \quad , \ \omega^{(2)}(\boldsymbol{p}) = u(p,-n) \\
\omega^{(3)}(\boldsymbol{p}) &= v(p,-n) \ , \ \omega^{(4)}(\boldsymbol{p}) = v(p,n) \ ,
\end{aligned}
$$

with the secondary condition (2.36), the Dirac eigensolutions [to $H^{(0)}$, $\boldsymbol{p}$, and $S(n,p)$] can be specified by their energy sign ϵ, four-momentum index $p = (p^\mu)$, and four-polarization index $n = (n^\mu)$ as

$$
\psi_{\epsilon,p,n}(x) = \frac{1}{(2\pi\hbar)^{3/2}}\sqrt{\frac{m_0 c}{p_0}}\mathrm{e}^{-\mathrm{i}\epsilon p_\mu x^\mu \hbar} \times \left\{ \begin{array}{l} u(p,n) \text{ for } \epsilon = +1 \\ v(p,n) \text{ for } \epsilon = -1 \end{array} \right\}
$$

(compare to Theorem 2.1). For negative solutions the four-momentum and four-polarization indices are opposite to the corresponding eigenvalues. The Lorentz-covariant energy and spin projectors are

$$
\Lambda_\pm(p) = \frac{\pm\gamma^\mu p_\mu + m_0 c}{2m_0 c} \ , \ \Sigma(n) = \frac{1}{2}(1+\gamma^5\gamma^\mu n_\mu) \ .
$$

They act on the bispinors u, v in the following way:

$$
\Lambda_+(p)\left\{ \begin{array}{c} u(p,\pm n) \\ v(p,\pm n) \end{array} \right\} = \left\{ \begin{array}{c} u(p,\pm n) \\ 0 \end{array} \right\}
$$

$$
\Lambda_-(p)\left\{ \begin{array}{c} u(p,\pm n) \\ v(p,\pm n) \end{array} \right\} = \left\{ \begin{array}{c} 0 \\ v(p,\pm n) \end{array} \right\}
$$

▷

$$\Sigma(+n)\left\{\begin{matrix} u, v(p,+n) \\ u, v(p,-n) \end{matrix}\right\} = \left\{\begin{matrix} u, v(p,+n) \\ 0 \end{matrix}\right\}$$

$$\Sigma(-n)\left\{\begin{matrix} u, v(p,+n) \\ u, v(p,-n) \end{matrix}\right\} = \left\{\begin{matrix} 0 \\ u, v(p,-n) \end{matrix}\right\} .$$

Put differently, from a general free Dirac solution with four-momentum index p,

- $\Lambda_{\pm}(p)$ projects onto its portions with energy sign $\epsilon = \pm$
- $\Sigma(\pm n)$ projects onto its portions with four-polarization index $\pm n$.

2.1.6 Interpretation of Negative Solutions, Antiparticles and Hole Theory

So far we have disregarded the negative solutions to the Dirac equation and their interpretation. The problems due to their mere existence are the same as those addressed within the discussion of the negative Klein-Gordon solutions in Subsection 1.1.3.

Charge conjugation *C*. In that subsection we saw that there exists a connection between the negative Klein-Gordon solution $\phi^{(-)}$ of charge $+e$ and its charge conjugate $\phi_C^{(-)}$ with charge $-e$, where the latter was identified with an antiparticle wave function with positive energy. We therefore conjecture that a similar relationship can be established in the case in hand. Therefore, starting from the Dirac equation

$$\left[\gamma^\mu\left(\mathrm{i}\hbar\partial_\mu - \frac{e}{c}A^\mu(x)\right) - m_0 c\right]\psi^{(-)}(x) = 0 \tag{2.38}$$

for a negative solution $\psi^{(-)}$ of charge $+e$ in the potential A^μ, it should be possible to deduce the Dirac equation

$$\left[\gamma^\mu\left(\mathrm{i}\hbar\partial_\mu + \frac{e}{c}A^\mu(x)\right) - m_0 c\right]\psi_C^{(-)}(x) = 0 \tag{2.39}$$

for a positive solution $\psi_C^{(-)}$ of charge $-e$ in the same potential A^μ. Obviously, as in the Klein-Gordon case, this is possible only if the corresponding transformation of $\psi^{(-)}$ is antilinear since moving from (2.38) to (2.39) the relative sign between the differential and potential terms changes. Therefore, we make the following (reciprocal) ansatz:

$$\psi_C^{(-)}(x) = C\psi^{(-)*}(x) \ , \ C^2 = 1 \ , \ C \text{ linear.}$$

Inserting this expression into (2.39), multiplying from the left by C^{-1}, and subsequently taking the complex conjugate, we finally arrive at (2.38), if C satisfies the condition

$$C^{-1}\gamma^\mu C = -\gamma^{*\mu} . \qquad (2.40)$$

The solution to this equation can be found easily. In the Dirac or Weyl representation it is

$$C = \mathrm{i}\gamma^2 .$$

That $\psi_C^{(-)}$ is indeed a solution with positive energy can be seen most easily by considering the eigenvalue equation of a negative energy state $\Psi^{(-)}$ [compare to (1.24), (1.25)]:

$$\left[c\boldsymbol{\alpha}\left(\boldsymbol{p}-\frac{e}{c}\boldsymbol{A}\right)+eA^0+\beta m_0c^2\right]\Psi^{(-)}(\boldsymbol{x}) = -|E|\Psi^{(-)}(\boldsymbol{x}) .$$

Complex conjugation and subsequent multiplication from the left by $\mathrm{i}\gamma^2 = \mathrm{i}\beta\alpha_2$ leads, along with (2.6) and $\alpha_{1,3} = \alpha_{1,3}^*$, $\alpha_2 = -\alpha_2^*$, to the eigenvalue equation

$$\left[c\boldsymbol{\alpha}\left(\boldsymbol{p}+\frac{e}{c}\boldsymbol{A}\right)-eA^0+\beta m_0c^2\right]\Psi_C^{(-)}(\boldsymbol{x}) = +|E|\Psi_C^{(-)}(\boldsymbol{x}) ,$$

with

$$\Psi_C^{(-)}(\boldsymbol{x}) = \mathrm{i}\gamma^2\Psi^{(-)*}(\boldsymbol{x}) .$$

All in all, we can therefore state (compare to Theorem 1.4):

Theorem 2.5: Charge conjugation C in the Dirac theory

- In the Dirac and Weyl representations the charge conjugation C of the Dirac theory is defined by the transformation

  $$\psi(x) \longrightarrow \psi_C(x) = \mathrm{i}\gamma^2\psi^*(x) .$$

  It turns a positive [negative] Dirac solution of charge $+e$ [$-e$] into a negative [positive] Dirac solution of charge $-e$ [$+e$].

- A positive Dirac solution $\psi^{(+)}$ represents a physical spin-1/2 particle of charge $+e$ in the potential A^μ, whereas the charge conjugate of the negative solution $\psi_C^{(-)}$ (and not the original negative solution) describes the physical antiparticle with opposite charge $-e$ in the same potential A^μ.

With respect to the first two points in both Theorems 1.4 and 2.5, the circumstances of the Klein-Gordon and Dirac cases are completely identical. In particular, the Dirac theory, as a relativistic enhancement of Schrödinger's theory, also leads to a new degree of freedom and hence to the prediction of antiparticles which, so far, have been verified experimentally for every known spin-1/2 particle. However, the third point in Theorem 1.4 (charge interpretation) does not have a counterpart in the above Theorem since here, Q, ρ, and $\boldsymbol{j}$ are to be interpreted as probability quantities (see Theorem 2.1).

Based on Theorem 2.5 and the annotations before Theorem 2.1, we can now specify the quantum mechanical expectation value of an observable for particles and antiparticles in the Dirac theory (compare to the definition of the G-expectation value in the Klein-Gordon theory, Subsection 1.3.1):

Definition: Expectation value in the Dirac theory

In the Dirac theory the expectation value of the observable $\mathcal{O}$ is defined by

$$\langle \mathcal{O} \rangle = \langle \psi | \mathcal{O} | \psi \rangle = \int \mathrm{d}^3x \psi^\dagger(x) \mathcal{O} \psi(x) \ , \ \langle \psi | \psi \rangle = +1 \ . \tag{2.41}$$

It denotes the statistical average of many similar measurements of $\mathcal{O}$ at identical spin-1/2 [anti]particle systems of charge $+e$ [$-e$]. For particles positive Dirac solutions $\psi^{(+)}$, and for antiparticles charge conjugated negative solutions $\psi_C^{(-)}$ are to be inserted.

Before we turn to the interpretation of the negative solutions, let us look at the charge conjugation C in more detail. We consider the expectation value of an operator $\mathcal{O}$ in the state ψ_C and calculate in the Dirac or Weyl representation as follows:

$$\begin{aligned} \langle \mathcal{O} \rangle_C &= \langle \psi_C | \mathcal{O} | \psi_C \rangle = \int \mathrm{d}^3x \psi_C^\dagger \mathcal{O} \psi_C = \int \mathrm{d}^3x (\mathrm{i}\gamma^2\psi^*)^\dagger \mathcal{O} \mathrm{i}\gamma^2\psi^* \\ &= \int \mathrm{d}^3x \psi^{*\dagger} \gamma^{2\dagger} \mathcal{O} \gamma^2 \psi^* = \int \mathrm{d}^3x \psi^{*\dagger} \gamma^0 \gamma^2 \gamma^0 \mathcal{O} \gamma^2 \psi^* \\ &= -\int \mathrm{d}^3x \psi^{*\dagger} \gamma^2 \gamma^0 \gamma^0 \mathcal{O} \gamma^2 \psi^* = -\int \mathrm{d}^3x \psi^{*\dagger} \gamma^2 \mathcal{O} \gamma^2 \psi^* \\ &= -\left(\int \mathrm{d}^3x \psi^\dagger \gamma^2 \mathcal{O}^* \gamma^2 \psi \right)^* = -\langle \psi | \gamma^2 \mathcal{O}^* \gamma^2 | \psi \rangle^* \ . \end{aligned} \tag{2.42}$$

With these relations, the following can be derived easily (see Exercise 14):

$$\left. \begin{aligned} \langle \beta \rangle_C &= -\langle \beta \rangle \\ \langle \boldsymbol{x} \rangle_C &= \langle \boldsymbol{x} \rangle \\ \langle \boldsymbol{\alpha} \rangle_C &= \langle \boldsymbol{\alpha} \rangle \\ \langle \boldsymbol{p} \rangle_C &= -\langle \boldsymbol{p} \rangle \\ \langle \boldsymbol{S} \rangle_C &= -\langle \boldsymbol{S} \rangle \\ \langle \boldsymbol{L} \rangle_C &= -\langle \boldsymbol{L} \rangle \ , \ L = \boldsymbol{x} \times \boldsymbol{p} \\ \langle \boldsymbol{J} \rangle_C &= -\langle \boldsymbol{J} \rangle \ , \ \boldsymbol{J} = \boldsymbol{L} + \boldsymbol{S} \end{aligned} \right\} \tag{2.43}$$

$$\langle H(-e) \rangle_C = -\langle H(e) \rangle \ , \ H(e) = c\boldsymbol{\alpha}\left(\boldsymbol{p} - \frac{e}{c}\boldsymbol{A}\right) + eA^0 + \beta m_0 c^2 \ . \tag{2.44}$$

Furthermore, we have

$$\rho_C = \psi_C^\dagger \psi_C = \psi^\dagger \psi = \rho \ , \ \boldsymbol{j}_C = \psi_C^\dagger c\boldsymbol{\alpha} \psi_C = \psi^\dagger c\boldsymbol{\alpha} \psi = \boldsymbol{j} \ . \tag{2.45}$$

According to this, ψ_C and ψ possess the same probability density and probability current density in all space-time points. Therefore, the electric charge and current densities of ψ_C and ψ are opposite to each other. From (2.44) it becomes apparent once again that the solutions to the Dirac equation with negative energy correspond to the charge conjugated solutions with positive energy and vice versa. Finally, (2.43) and (2.44) express the important result that the charge conjugation C reverses the charge, energy, momentum, and spin of a Dirac state.

Applied to free Dirac solutions, the charge conjugation yields, up to a constant phase factor (see Exercise 13),

$$\psi_{\epsilon,p,n}(x) \longleftrightarrow \psi_{-\epsilon,p,n}(x) \tag{2.46}$$

showing that $u(p,n)$ and $v(p,n)$ are the charge conjugates of each other and that the original and charge conjugated states are described by the same four-momentum and four-polarization indices. This is clearly a consequence of the reversed assignment of the eigenvalues and indices for momentum and spin in the negative solutions which we have introduced analogously to the Klein-Gordon case, since the negative solutions should refer to antiparticles described by the charge conjugates of the negative solutions.[7]

Interpretation of negative solutions, hole theory As in the Klein-Gordon case, the mere existence of negative Dirac solutions causes difficulties, namely with respect to their physical consequences and to their interpretation (compare to the discussion in Subsection 1.1.3). In particular, we are again confronted with the problem that the theory seems to allow the transition from positive energy states to deeper and deeper negative energy levels (see Figure 1.1), although those transitions obviously do not occur in nature (stability of matter). As we already noted in Subsection 1.1.3, we must postulate the nonexistence of the radiation catastrophe as long as we restrict ourselves to the one-particle view. However, beyond the one-particle picture, the Dirac theory offers an explanation found by Dirac himself known as *hole theory*.

In this theory the dilemma caused by the negative Dirac solutions is solved by filling the negative energy levels with electrons (more generally: spin-1/2 particles) in agreement with Pauli's exclusion principle. Accordingly, the vacuum state is the one where all negative energy levels are occupied by electrons and all positive energy levels are empty. As a consequence, the transition catastrophe is now eliminated since Pauli's exclusion principle forbids the transition of real electrons (with positive energy) into the completely filled *sea of negative energies* (see Figure 2.1 left).

This novel assumption of a sea occupied by electrons has many consequences. For example, an electron with negative energy can shift into a state of positive energy by absorbing radiation. If this occurs, one observes an electron with charge $+e$ and energy $+E$. Additionally, a hole is created in the sea of negative energies indicating the absence of an electron with charge $+e$

[7] As in the Klein-Gordon case, we note that in the free case the distinction of charged particles is not possible. Therefore, the original and the charge conjugated states are solutions to the same Dirac equation.

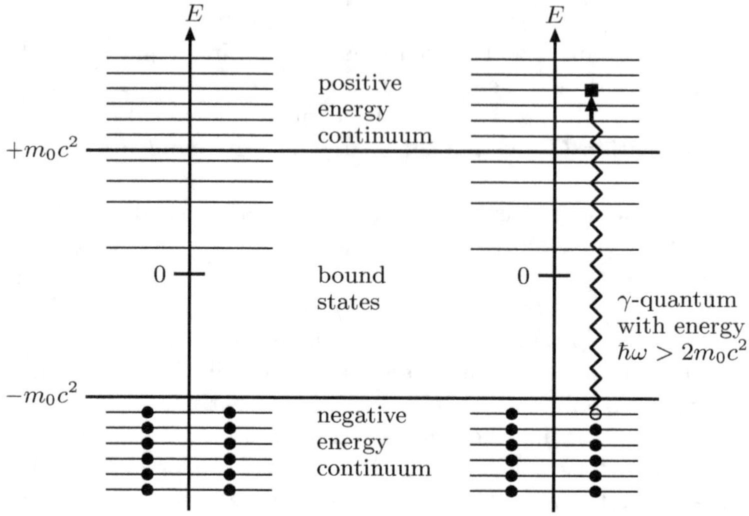

Fig. 2.1. *Left:* within the hole theory, the vacuum is characterized by the fact that each negative energy level is occupied by two electrons (spin up and spin down) (•), whereas all positive energy levels are empty. *Right:* in the process of pair creation an electron of negative energy absorbs radiation with energy $\hbar\omega > 2m_0c^2$ and shifts into a positive electron state (■). This causes a hole in the sea of negative energies (◦) which appears as an additional antiparticle (positron).

and energy $-E$. An observer relative to the vacuum perceives this hole as a particle of charge $-e$ and energy $+E$ (antiparticle) (Figure 2.1 right). Thus, the hole theory also provides an explanation for the creation of particle-antiparticle pairs (*pair creation*). This new perspective implies that there must exist a unique relationship between negative Dirac solutions with charge $+e$ and positive solutions with charge $-e$ which is, as we already know, provided by the charge conjugation C. Accordingly, within the hole picture, an occupied electron state with negative energy is described by $\psi^{(-)}$, whereas its absence, i.e. the corresponding hole, is described by $\psi_C^{(-)}$, which is the wave function of the antielectron, the so-called *positron*. This interpretation becomes clear again once we write down the charge, energy, and momentum balances of the pair creation process:

$$
\begin{aligned}
Q_{\text{photon}} = 0 &= Q_{\text{electron pos. energy}} - Q_{\text{electron neg. energy}} \\
&= Q_{\text{electron}} + Q_{\text{positron}} \\
E_{\text{photon}} = \hbar\omega &= E_{\text{electron pos. energy}} - E_{\text{electron neg. energy}} \\
&= E_{\text{electron}} + E_{\text{positron}} \\
\boldsymbol{p}_{\text{photon}} = \hbar\boldsymbol{k} &= \boldsymbol{p}_{\text{electron pos. energy}} - \boldsymbol{p}_{\text{electron neg. energy}} \\
&= \boldsymbol{p}_{\text{electron}} + \boldsymbol{p}_{\text{positron}} \,.
\end{aligned}
$$

The effect opposite to pair creation, the particle-antiparticle annihilation (*pair annihilation*), can equally be described within the hole theory. Here a light emitting electron falls into an electron hole in the sea of negative energies and hence destroys the positron associated with the hole.

As in the Klein-Gordon case, note that in our discussion of the Dirac theory, the charge sign does not play any role so that we could also have interpreted positrons of charge $-e$ as particles (described by positive Dirac solutions) and electrons of charge $-e$ as antiparticles (described by charge conjugated negative solutions). Consequently, the Dirac sea would have consisted of positrons with negative energy and its holes would have been electrons.

Further consequences of the hole theory. If one tries to examine the conclusions of the hole theory in more detail, insufficiencies and contradictions inevitably arise that can ultimately be solved only within quantum field theoretical considerations. Nevertheless, the hole theory is of great importance since, for the first time, it provided a (naive) model of the vacuum which is not characterized by the absence of everything but possesses an internal structure that is subject to modification. For example, external electromagnetic fields can deform the wave functions of the electrons within the sea of negative energies thus creating a measurable vacuum polarization (displacement charge of the vacuum) compared to the field-free case. For a real electron repelling the electrons of the Dirac sea, this deformation implies that its electric charge is weakened compared to its "bare" charge – an effect which is observed, for example, in the energy spectrum of the hydrogen atom.

A further consequence of the hole theory in the present form is that the vacuum obviously possesses an infinite charge and energy that must be *renormalized* to zero by fixing an appropriate charge and energy zero point. This procedure is principally possible although not very satisfactory (aesthetic). However, the most serious flaws of the hole theory are, firstly, that it contains an asymmetry in the description of particles and antiparticles. Secondly, the question remains unanswered how to treat the mutual interaction of the occupied states within the sea of negative energies.

Finally, we note the following: the hole theory was introduced by Dirac, among others, to provide an interpretation of the negative Dirac solutions as well as to give a plausible explanation for the absence of the radiation catastrophe. However, this theory inevitably exceeds the one-particle concept since it describes particles with both charge signs simultaneously. As we have already stated, we have to accept the fact that in the Dirac as well as in the Klein-Gordon case, the problem of negative energies cannot be solved within a strict one-particle interpretation.

Résumé. As in the discussion of the Klein-Gordon theory at the end of Subsection 1.1.3, we now draw an interim summary by gathering the already clarified as well as still open points relevant to the desired one-particle interpretation.

With respect to the interpretation of the Dirac and Klein-Gordon solutions, the circumstances are completely equal: positive solutions describe particles with charge $+e$, and charge conjugated negative solutions describe antiparticles with charge $-e$. Concerning the quantities ρ and $\boldsymbol{j}$: in the Klein-Gordon case they do not come into question as probability quantities due to the non-Hermitecity of the Hamilton operator. Instead, they can be interpreted as charge quantities where positive Klein-Gordon solutions are to be inserted for particles and negative solutions for antiparticles. In contrast, the Dirac-Hamilton operator is Hermitean so that ρ can be interpreted as a probability density and $\boldsymbol{j}$ as a probability current density. For particles we have to insert positive Dirac solutions, and for antiparticles the charge conjugated negative solutions or – due to $\rho_C = \rho$, $\boldsymbol{j}_C = \boldsymbol{j}$ – the original negative solutions.

For the completion and inner consistency of the one-particle interpretation of the Dirac theory, the following points remain to be clarified that will be tackled in Section 2.3 (compare to the points [1], [2], and [3] in Subsection 1.1.3):

[1] A necessary condition for the one-particle interpretation is that positive and negative solutions are completely decoupled since only then particles and antiparticles can be described sensibly. However, as in the Klein-Gordon case, only the positive solutions together with the negative ones compose a complete function system. Therefore, we have to ask to what extent a complete decoupling is possible.

[2] A complete decoupling also implies that we must strive toward physically sensible one-particle operators that do not mix positive and negative solutions.

Here we do not need to cite a point equivalent to [3] in Subsection 1.1.3, since with (2.41), we have already found a physically sensible definition of quantum mechanical statistical expectation values that are invariant under unitary transformations (picture-independent).

Parallel to the final remarks at the end of Subsection 1.1.3, we note that, besides the electric charge, there exist other types of charge by which some fermions differ from their antifermions. For example, there are quarks carrying a complicated *color charge* in addition to their electric charge.

Summary

- The **Dirac equation** is a fourdimensional system of coupled differential equations of first order in space and time. It represents the relativistic generalization of Schrödinger's equation for spin-1/2 particles.

▷

- The Dirac theory differs from the nonrelativistic theory in one important point: it encompasses solutions (**bispinors**) for both positive and negative energies.
- The Hermitecity of the Dirac-Hamilton operator allows the definitions of a positive definite probability density and a probability current density that are formally identical to those of nonrelativistic quantum mechanics. We can therefore carry over the scalar product and expectation value from the nonrelativistic theory, including the consequences with respect to Hermitean and unitary operators.
- Combining the 16 basis elements of the γ-matrix space with $\bar{\psi}$ and ψ, one can build **covariant bilinear forms** with a well-defined transformational behavior under orthochronous Lorentz transformations.
- The free plane Dirac wave functions are not yet uniquely specified as the eigensolutions to the Hamilton and momentum operators. They possess an additional degree of freedom leading to a spin operator with quantum number $s = 1/2$.
- Due to the **charge conjugation**, the Dirac solutions can be interpreted in the following way: **particles** of charge $+e$ are described by positive Dirac solutions and **antiparticles** of charge $-e$ by the **charge conjugated** negative solutions.
- Within the one-particle concept, the problems connected to negative solutions (interpretation, radiation catastrophe) cannot be solved. Beyond the one-particle picture, the **hole theory** provides a qualitatively acceptable explanation for them. According to it, the vacuum possesses a modifiable inner structure with physical consequences (**pair creation** and **annihilation**, **vacuum polarization**).
- With a view to the most consistent one-particle probability interpretation of the Dirac theory, it remains to be clarified to what extent a complete decoupling of positive and negative solutions is possible.

Exercises

11. Solutions of the free Dirac equation. Show that in the Dirac representation the solutions to the free Dirac equation (2.8) with a sharp momentum are given by (2.9).

Solution. To solve this problem, we can proceed similarly to Exercise 1. Our ansatz is

$$\psi(x) = \begin{pmatrix} \varphi_0 \\ \chi_0 \end{pmatrix} \mathrm{e}^{\mathrm{i}(\boldsymbol{p}\boldsymbol{x}-Et)/\hbar} ,$$

where φ_0 and χ_0 each denote two-component constant spinors. Inserting this into (2.8) leads to the equation system

$$\left.\begin{aligned}(E-m_0c^2)\varphi_0 - c\boldsymbol{\sigma}\boldsymbol{p}\chi_0 &= 0\\ -c\boldsymbol{\sigma}\boldsymbol{p}\varphi_0 + (E+m_0c^2)\chi_0 &= 0\ ,\end{aligned}\right\} \tag{2.47}$$

which possesses nontrivial solutions only if the coefficient determinant vanishes:

$$\begin{vmatrix} E-m_0c^2 & -c\boldsymbol{\sigma}\boldsymbol{p}\\ -c\boldsymbol{\sigma}\boldsymbol{p} & E+m_0c^2\end{vmatrix} = E^2 - m_0c^2c^4 - c^2(\boldsymbol{\sigma}\boldsymbol{p})(\boldsymbol{\sigma}\boldsymbol{p}) = 0\ .$$

From this and taking into account the identity

$$(\boldsymbol{\sigma}\boldsymbol{A})(\boldsymbol{\sigma}\boldsymbol{B}) = \boldsymbol{A}\boldsymbol{B} + \mathrm{i}\boldsymbol{\sigma}(\boldsymbol{A}\times\boldsymbol{B})\ ,$$

we obtain the relativistic energy-momentum relation for free particles,

$$E^2 - m_0^2c^4 - c^2\boldsymbol{p}^2 = 0 \Longrightarrow \begin{cases} E^{(+)} = +c\sqrt{\boldsymbol{p}^2+m_0^2c^2} = +cp_0\\ E^{(-)} = -c\sqrt{\boldsymbol{p}^2+m_0^2c^2} = -cp_0\ ,\end{cases}$$

in accordance with our expectations. The solutions to our problem finally result from the insertion of $E^{(+)}$ and $E^{(-)}$ into (2.47). They can be written in the (unnormalized) form

$$E^{(+)}:\ \psi^{(+)}(x) = \begin{pmatrix}\chi^{(1,2)}\\ \dfrac{\boldsymbol{\sigma}\boldsymbol{p}\chi^{(1,2)}}{p_0+m_0c}\end{pmatrix}\mathrm{e}^{-\mathrm{i}(cp_0-\boldsymbol{p}\boldsymbol{x})\hbar} \sim \psi_{\boldsymbol{p}}^{(1,2)}(x)$$

$$E^{(-)}:\ \psi^{(-)}(x) = \begin{pmatrix}\dfrac{-\boldsymbol{\sigma}\boldsymbol{p}\chi^{(3,4)}}{p_0+m_0c}\\ \chi^{(3,4)}\end{pmatrix}\mathrm{e}^{+\mathrm{i}(cp_0+\boldsymbol{p}\boldsymbol{x})/\hbar} \sim \psi_{-\boldsymbol{p}}^{(3,4)}(x)\ ,$$

where $\chi^{(1,2)}$ and $\chi^{(3,4)}$ respectively denote two linearly independent spinors.

12. Nonunitarity of bispinor transformations (I). Verify (2.23).

Solution. We have the relations [see (2.19) and (2.22)]

$$\Lambda^\mu{}_\nu\gamma^\nu = D^{-1}\gamma^\mu D \Longleftrightarrow \Lambda^\mu{}_\nu\gamma^{\nu\dagger} = D^\dagger\gamma^{\mu\dagger}D^{\dagger-1} \tag{2.48}$$

and

$$\gamma^{\mu\dagger} = \gamma^0\gamma^\mu\gamma^0\ . \tag{2.49}$$

From the second equation of (2.48) and from (2.49) follows that

$$\begin{aligned}&\Lambda^\mu{}_\nu\gamma^0\gamma^\nu\gamma^0 = D^\dagger\gamma^0\gamma^\mu\gamma^0D^{\dagger-1}\\ &\Longrightarrow \gamma^0\Lambda^\mu{}_\nu\gamma^0\gamma^\nu\gamma^0\gamma^0 = \gamma^0D^\dagger\gamma^0\gamma^\mu\gamma^0D^{\dagger-1}\gamma^0 = \gamma^0D^\dagger\gamma^0\gamma^\mu\left(\gamma^0D^\dagger\gamma^0\right)^{-1}\ .\end{aligned}$$

On the other hand, the first relation of (2.48) can be rewritten as

$$\Lambda^{\mu}{}_{\nu}\gamma^{\nu} = \Lambda^{\mu}{}_{\nu}\gamma^{0}\gamma^{\nu\dagger}\gamma^{0} = \gamma^{0}\Lambda^{\mu}{}_{\nu}\gamma^{0}\gamma^{\nu}\gamma^{0}\gamma^{0} = D^{-1}\gamma^{\mu}D \ .$$

Comparing the last two relations leads to

$$D^{-1}\gamma^{\mu}D = \gamma^{0}D^{\dagger}\gamma^{0}\gamma^{\mu}\left(\gamma^{0}D^{\dagger}\gamma^{0}\right)^{-1}$$

or

$$\gamma^{\mu} = D\gamma^{0}D^{\dagger}\gamma^{0}\gamma^{\mu}\left(\gamma^{0}D^{\dagger}\gamma^{0}\right)^{-1}D^{-1} = D\gamma^{0}D^{\dagger}\gamma^{0}\gamma^{\mu}\left(D\gamma^{0}D^{\dagger}\gamma^{0}\right)^{-1} \ .$$

Thus, $D\gamma^{0}D^{\dagger}\gamma^{0}$ commutes with all γ^{μ} and is therefore proportional to the unity matrix:

$$D\gamma^{0}D^{\dagger}\gamma^{0} = b \Longrightarrow D\gamma^{0}D^{\dagger} = b\gamma^{0}. \tag{2.50}$$

From this, we obtain the sought relation

$$D^{\dagger} = b\left(D\gamma^{0}\right)^{-1}\gamma^{0} = b\gamma^{0}D^{-1}\gamma^{0}$$

as well as

$$\left(D\gamma^{0}D^{\dagger}\right)^{\dagger} = D\gamma^{0}D^{\dagger} = b^{*}\gamma^{0} \Longrightarrow b = b^{*} \ .$$

In order to determine the real constant b, we use that $\det(D) = 1$ was presupposed. The calculation of the determinant of (2.50) then yields

$$b^{4} = 1 \Longrightarrow b = \pm 1 \ .$$

Next we consider the equation

$$D^{\dagger}D = b\gamma^{0}D^{-1}\gamma^{0}D = b\gamma^{0}\Lambda^{0}{}_{\nu}\gamma^{\nu} = b\Lambda^{0}{}_{0} + b\sum_{k=1}^{3}\Lambda^{0}{}_{k}\underbrace{\gamma^{0}\gamma^{k}}_{\alpha_k} \tag{2.51}$$

and argue as follows: since $\det(D^{\dagger}D)$ is equal to the product of all eigenvalues, all of them must be nonzero. Furthermore, the operator $D^{\dagger}D$ is Hermitean. For its real eigenvalues, it therefore holds that

$$D^{\dagger}D\psi_{a} = a\psi_{a} \Longrightarrow a\psi_{a}^{\dagger}\psi_{a} = \psi_{a}^{\dagger}D^{\dagger}D\psi_{a} = (D\psi_{a})^{\dagger}D\psi_{a} > 0 \Longrightarrow a > 0 \ .$$

Since the trace of $D^{\dagger}D$ is equal to the sum of all eigenvalues, it follows from (2.51) and $\mathrm{tr}(\alpha_k) = 0$ that

$$0 < \mathrm{tr}(D^{\dagger}D) = 4b\Lambda^{0}{}_{0} \ .$$

So, in total, we have

$$b = \frac{\Lambda^{0}{}_{0}}{|\Lambda^{0}{}_{0}|} = \begin{cases} +1 & \text{for orthochronous Lorentz transformations} \\ & \text{preserving the time direction} \\ -1 & \text{for nonorthochronous Lorentz transformations} \\ & \text{changing the time direction.} \end{cases}$$

13. Charge conjugation of free Dirac states. Verify (2.46).

Solution. Our starting point is a plane free Dirac wave function $\psi_{\epsilon,p,n}$ with energy sign ϵ, four-momentum index p, and four-polarization index n fulfilling the projection relation

$$\psi_{\epsilon,p,n}(x) = \left(\frac{\epsilon\gamma^\mu p_\mu + m_0 c}{2m_0 c}\right)\left(\frac{1+\gamma^5\gamma^\mu n_\mu}{2}\right)\psi_{\epsilon,p,n}(x) \ .$$

On the basis of (2.40), $\{C, \gamma^5\} = 0$, $\gamma^5 = \gamma^{5*}$, and the fact that in the first bracket term $p_\mu = p_\mu^*$ is a real number (and not an operator), the charge conjugated wave follows as

$$\begin{aligned}
[\psi_{\epsilon,p,n}]_C(x) &= C\left(\frac{\epsilon\gamma^{\mu*} p_\mu + m_0 c}{2m_0 c}\right)\left(\frac{1+\gamma^5\gamma^{\mu*} n_\mu}{2}\right)\psi^*_{\epsilon,p,n}(x) \\
&= C\left(\frac{\epsilon\gamma^{\mu*} p_\mu + m_0 c}{2m_0 c}\right)C^{-1}C\left(\frac{1+\gamma^5\gamma^{\mu*} n_\mu}{2}\right)C^{-1}C\psi^*_{\epsilon,p,n}(x) \\
&= \left(\frac{-\epsilon\gamma^{\mu} p_\mu + m_0 c}{2m_0 c}\right)\left(\frac{1+\gamma^5\gamma^{\mu} n_\mu}{2}\right)C\psi^*_{\epsilon,p,n}(x) \\
&= \left(\frac{-\epsilon\gamma^{\mu} p_\mu + m_0 c}{2m_0 c}\right)\left(\frac{1+\gamma^5\gamma^{\mu} n_\mu}{2}\right)[\psi_{\epsilon,p,n}]_C(x) \\
&= \psi_{-\epsilon,p,n}(x) \ .
\end{aligned}$$

14. Expectation values of charge conjugated Dirac states. Show that the relations (2.43), (2.44), and (2.45) are true.

Solution. We provide the proof with the help of (2.42) and the adjunction relation $\langle\psi|\,\mathcal{O}\,|\psi\rangle^* = \langle\psi|\,\mathcal{O}^\dagger\,|\psi\rangle$ in the Dirac or Weyl representation.

To show: $\langle\beta\rangle_C = -\langle\beta\rangle$, with $\beta = \gamma^0$.

$$\begin{aligned}
\langle\beta\rangle_C &= -\langle\psi|\,\gamma^2\gamma^0\gamma^2\,|\psi\rangle^* = \langle\psi|\,\gamma^2\gamma^2\gamma^0\,|\psi\rangle^* = -\langle\psi|\,\gamma^0\,|\psi\rangle^* \\
&= -\langle\psi|\,\gamma^0\,|\psi\rangle = -\langle\beta\rangle \ .
\end{aligned}$$

To show: $\langle\boldsymbol{x}\rangle_C = \langle\boldsymbol{x}\rangle$.

$$\begin{aligned}
\langle\boldsymbol{x}\rangle_C &= -\left(\int \mathrm{d}^3x\psi^\dagger\gamma^2\boldsymbol{x}\gamma^2\psi\right)^* = \left(\int \mathrm{d}^3x\psi^\dagger\boldsymbol{x}\psi\right)^* = \langle\psi|\,\boldsymbol{x}\,|\psi\rangle^* \\
&= \langle\psi|\,\boldsymbol{x}\,|\psi\rangle = \langle\boldsymbol{x}\rangle \ .
\end{aligned}$$

To show: $\langle\boldsymbol{\alpha}\rangle_C = \langle\boldsymbol{\alpha}\rangle$, with $\alpha_i = \gamma^0\gamma^i$.

$$\begin{aligned}
\gamma^2\alpha_i^*\gamma^2 &= \gamma^2\gamma^0\gamma^{i*}\gamma^2 \\
&= \left\{\begin{array}{c} \gamma^2\gamma^0\gamma^i\gamma^2 = \gamma^2\gamma^2\gamma^0\gamma^i = -\gamma^0\gamma^i \text{ for } i=1,3 \\ -\gamma^2\gamma^0\gamma^2\gamma^2 = \gamma^2\gamma^0 = -\gamma^0\gamma^2 \text{ for } i=2 \end{array}\right\} = -\alpha_i
\end{aligned}$$

$$\Longrightarrow \langle\alpha_i\rangle_C = -\langle\psi|\,\gamma^2\alpha_i^*\gamma^2\,|\psi\rangle^* = \langle\psi|\,\alpha_i\,|\psi\rangle^* = \langle\psi|\,\alpha_i\,|\psi\rangle = \langle\alpha_i\rangle \ .$$

To show: $\langle \boldsymbol{p} \rangle_C = -\langle \boldsymbol{p} \rangle$, with $\boldsymbol{p} = -\mathrm{i}\hbar\boldsymbol{\nabla}$.

$$\begin{aligned}\langle \boldsymbol{p} \rangle_C &= -\left[\int \mathrm{d}^3x\psi^\dagger\gamma^2(\mathrm{i}\hbar\boldsymbol{\nabla})\gamma^2\psi\right]^* = -\left[\int \mathrm{d}^3x\psi^\dagger(-\mathrm{i}\hbar\boldsymbol{\nabla})\psi\right]^* \\ &= -\langle\psi|\,\boldsymbol{p}\,|\psi\rangle^* = -\langle\psi|\,\boldsymbol{p}\,|\psi\rangle = -\langle \boldsymbol{p} \rangle \ .\end{aligned}$$

To show: $\langle \boldsymbol{S} \rangle_C = -\langle \boldsymbol{S} \rangle$, with $\boldsymbol{S} = \frac{\hbar}{2}\begin{pmatrix}\boldsymbol{\sigma} & 0\\ 0 & \boldsymbol{\sigma}\end{pmatrix}$.

$$\begin{aligned}\gamma^2 S_i^*\gamma^2 &= \frac{\hbar}{2}\begin{pmatrix}0 & \sigma_2\\ -\sigma_2 & 0\end{pmatrix}\begin{pmatrix}\sigma_i^* & 0\\ 0 & \sigma_i^*\end{pmatrix}\begin{pmatrix}0 & \sigma_2\\ -\sigma_2 & 0\end{pmatrix} \\ &= -\frac{\hbar}{2}\begin{pmatrix}\sigma_2\sigma_i^*\sigma_2 & 0\\ 0 & \sigma_2\sigma_i^*\sigma_2\end{pmatrix} = \frac{\hbar}{2}\begin{pmatrix}\sigma_i^{*T} & 0\\ 0 & \sigma_i^{*T}\end{pmatrix} = S_i^\dagger\end{aligned}$$

$$\Longrightarrow \langle S_i \rangle_C = -\left\langle\psi|\,S_i^\dagger\,|\psi\right\rangle^* = -\langle\psi|\,S_i\,|\psi\rangle = -\langle S_i \rangle \ .$$

To show: $\langle \boldsymbol{L} \rangle_C = -\langle \boldsymbol{L} \rangle$, with $\boldsymbol{L} = \boldsymbol{x}\times(-\mathrm{i}\hbar\boldsymbol{\nabla})$.

$$\begin{aligned}\langle \boldsymbol{L} \rangle_C &= -\left[\int \mathrm{d}^3x\psi^\dagger\gamma^2\boldsymbol{x}\times(\mathrm{i}\hbar\boldsymbol{\nabla})\gamma^2\psi\right]^* = -\left[\int \mathrm{d}^3x\psi^\dagger\boldsymbol{x}\times(-\mathrm{i}\hbar\boldsymbol{\nabla})\psi\right]^* \\ &= -\langle\psi|\,\boldsymbol{L}\,|\psi\rangle^* = -\langle\psi|\,\boldsymbol{L}\,|\psi\rangle = -\langle \boldsymbol{L} \rangle \ .\end{aligned}$$

To show: $\langle \boldsymbol{J} \rangle_C = -\langle \boldsymbol{J} \rangle$, with $\boldsymbol{J} = \boldsymbol{L} + \boldsymbol{S}$.

$$\langle \boldsymbol{J} \rangle_C = \langle \boldsymbol{L} \rangle_C + \langle \boldsymbol{S} \rangle_C = -\left[\langle \boldsymbol{L} \rangle + \langle \boldsymbol{S} \rangle\right] = -\langle \boldsymbol{J} \rangle \ .$$

To show: $\langle H(-e) \rangle_C = \langle H(e) \rangle$, with $H(e) = c\boldsymbol{\alpha}\left(\boldsymbol{p} - \frac{e}{c}\boldsymbol{A}\right) + eA^0 + \beta m_0c^2$.
We have

$$\gamma^2\boldsymbol{\alpha}^*\gamma^2 = -\boldsymbol{\alpha} \Longrightarrow \langle \boldsymbol{\alpha}\boldsymbol{p} \rangle_C = -\langle \boldsymbol{\alpha}\boldsymbol{p} \rangle \ , \ \langle \boldsymbol{\alpha}\boldsymbol{A} \rangle_C = \langle \boldsymbol{\alpha}\boldsymbol{A} \rangle$$

and

$$\left\langle A^0\right\rangle_C = \left\langle A^0\right\rangle \ , \ \langle \beta \rangle_C = -\langle \beta \rangle \ .$$

$$\begin{aligned}\Longrightarrow \langle H(-e) \rangle_C &= c\left(\langle \boldsymbol{\alpha}\boldsymbol{p} \rangle_C + \frac{e}{c}\langle \boldsymbol{\alpha}\boldsymbol{A} \rangle_C\right) - e\left\langle A^0\right\rangle_C + m_0c^2\langle \beta \rangle_C \\ &= -\left[c\left(\langle \boldsymbol{\alpha}\boldsymbol{p} \rangle - \frac{e}{c}\langle \boldsymbol{\alpha}\boldsymbol{A} \rangle\right) + e\left\langle A^0\right\rangle + m_0c^2\langle \beta \rangle\right] \\ &= -\langle H(e) \rangle \ .\end{aligned}$$

To show: $\rho_C = \rho$, with $\rho = \psi^\dagger\psi$.

$$\begin{aligned}\rho_C &= \psi_C^\dagger\psi_C = \left(\mathrm{i}\gamma^2\psi^*\right)^\dagger\mathrm{i}\gamma^2\psi^* = \psi^T\gamma^{2\dagger}\gamma^2\psi^* = -\psi^T\gamma^2\gamma^2\psi^* \\ &= \psi^T\psi^* = \left(\psi^\dagger\psi\right)^* = \psi^\dagger\psi = \rho \ .\end{aligned}$$

To show: $\boldsymbol{j}_C = \boldsymbol{j}$, with $\boldsymbol{j} = \psi^\dagger c\boldsymbol{\alpha}\psi$.

$$\begin{aligned}\boldsymbol{j}_C &= \psi_C^\dagger c\boldsymbol{\alpha}\psi_C = c\left(\mathrm{i}\gamma^2\psi^*\right)^\dagger\boldsymbol{\alpha}\mathrm{i}\gamma^2\psi^* = c\psi^T\gamma^{2\dagger}\boldsymbol{\alpha}\gamma^2\psi^* \\ &= -c\psi^T\gamma^2\boldsymbol{\alpha}\gamma^2\psi^* = c\psi^T\boldsymbol{\alpha}^*\psi^* = \left(\psi^\dagger c\boldsymbol{\alpha}\psi\right)^* = \psi^\dagger c\boldsymbol{\alpha}\psi = \boldsymbol{j} \ .\end{aligned}$$

15. Dirac equation for structured particles. The Dirac equation of a structured particle (e.g. proton or neutron) in an external electromagnetic field exhibits an additional term which describes the interaction of the particle's anomalous magnetic moment with this field:

$$\left[\gamma^\mu\left(p_\mu - \frac{e}{c}A_\mu\right) - \frac{\hbar\delta}{4m_0c}\sigma^{\mu\nu}F_{\mu\nu} - m_0c\right]\psi(x) = 0 ,$$

with

$$\sigma^{\mu\nu} = \frac{\mathrm{i}}{2}[\gamma^\mu, \gamma^\nu] , \quad F_{\mu\nu} = \partial_\mu A_\nu - \partial_\nu A_\mu .$$

a) Show that, with this gauge and Lorentz-invariant term, the Dirac operator is Hermitean and the probability remains conserved.
b) Express $\sigma^{\mu\nu}F_{\mu\nu}$ in terms of the electromagnetic fields $\boldsymbol{E}$ and $\boldsymbol{B}$ in the Dirac representation.

Solution.

To a) First of all, we notice that, in the above equation, it is not the bracket term that must be Hermitean but the Dirac-Hamilton operator which is obtained by rewriting the equation in canonical form:

$$\mathrm{i}\hbar\frac{\partial\psi}{\partial t} = \left[c\boldsymbol{\alpha}\left(\boldsymbol{p} - \frac{e}{c}\boldsymbol{A}\right) + eA^0 + \frac{\hbar\delta}{4m_0c}\gamma^0\sigma^{\mu\nu}F_{\mu\nu} + m_0c^2\gamma^0\right]\psi . \tag{2.52}$$

From this the Hermitecity condition

$$\gamma^0\sigma^{\mu\nu}F_{\mu\nu} = \left(\gamma^0\sigma^{\mu\nu}F_{\mu\nu}\right)^\dagger \tag{2.53}$$

follows, and (since $F_{\mu\nu}$ is a real tensor field)

$$\sigma^{\mu\nu\dagger} = \gamma^0\sigma^{\mu\nu}\gamma^0 .$$

This condition can be immediately verified with the help of (2.19):

$$\begin{aligned}(\sigma^{\mu\nu})^\dagger &= -\frac{\mathrm{i}}{2}[\gamma^{\nu\dagger}, \gamma^{\mu\dagger}] = -\frac{\mathrm{i}}{2}[\gamma^0\gamma^\nu\gamma^0, \gamma^0\gamma^\mu\gamma^0] = \frac{\mathrm{i}}{2}\gamma^0[\gamma^\mu, \gamma^\nu]\gamma^0 \\ &= \gamma^0\sigma^{\mu\nu}\gamma^0 .\end{aligned}$$

To prove the conservation of probability, we argue similarly to what led to (2.11) and (2.12), now on the basis of the modified Dirac equation (2.52). With (2.53) taken into account, the same additional term appears on the right hand side of both equations (2.11) and (2.12):

$$\frac{\hbar\delta}{4m_0c}\psi^\dagger\gamma^0\sigma^{\mu\nu}F_{\mu\nu}\psi .$$

Subtraction of both equations leads again to the continuity equation (2.13) with the same probability density and probability current density.

To b) Exploiting the antisymmetry of $\sigma^{\mu\nu}$ and $F_{\mu\nu}$ as well as the explicit form of $F_{\mu\nu}$,

$$(F_{\mu\nu}) = \begin{pmatrix} 0 & E_1 & E_2 & E_3 \\ -E_1 & 0 & -B_3 & B_2 \\ -E_2 & B_3 & 0 & -B_1 \\ -E_3 & -B_2 & B_1 & 0 \end{pmatrix} \; , \; F_{\mu\nu} = g_{\mu\alpha} g_{\nu\beta} F^{\alpha\beta} \; ,$$

the term $\sigma^{\mu\nu} F_{\mu\nu}$ can be rewritten as

$$\sigma^{\mu\nu} F_{\mu\nu} = 2 \sum_{\mu<\nu} \sigma^{\mu\nu} F_{\mu\nu} = 2 \sum_i \sigma^{0i} E_i + 2(-\sigma^{12} B_3 + \sigma^{13} B_2 - \sigma^{23} B_1) \; .$$

Furthermore, we have in the Dirac and Weyl representation

$$\sigma^{0i} = \mathrm{i}\alpha_i \; , \; \sigma^{ij} = \epsilon_{ijk} \hat{\sigma}_k \; , \; \hat{\sigma}_k = \begin{pmatrix} \sigma_k & 0 \\ 0 & \sigma_k \end{pmatrix} \; ,$$

so that

$$\sigma^{\mu\nu} F_{\mu\nu} = 2(\mathrm{i}\boldsymbol{\alpha}\boldsymbol{E} - \hat{\boldsymbol{\sigma}}\boldsymbol{B}) \; . \tag{2.54}$$

As we show in the next section, the matrix $\sigma^{\mu\nu}$ is connected to the bispinor transformations of proper Lorentz transformations.

16. Quadratic form of the Dirac equation. Show that the Dirac equation (2.25) can be cast into the following representation-independent form:

$$\left[\left(p^\mu - \frac{e}{c} A^\mu\right)\left(p_\mu - \frac{e}{c} A_\mu\right) - \frac{\hbar e}{2c} \sigma^{\mu\nu} F_{\mu\nu} - m_0^2 c^2\right] \psi = 0 \; . \tag{2.55}$$

Solution. Multiplying (2.25) from the left by

$$\gamma^\nu \left(p_\nu - \frac{e}{c} A_\nu\right) + m_0 c$$

and respecting the anticommutator relation $\{\gamma^\mu, \gamma^\nu\} = 2g^{\mu\nu}$ leads to

$$\begin{aligned}
0 &= \left[\gamma^\nu \left(p_\nu - \frac{e}{c} A_\nu\right) + m_0 c\right] \left[\gamma^\mu \left(p_\mu - \frac{e}{c} A_\mu\right) - m_0 c\right] \psi \\
&= \left[\gamma^\nu \gamma^\mu \left(p_\nu p_\mu + \frac{e^2}{c^2} A_\nu A_\mu\right) - \frac{e}{c} \gamma^\nu \gamma^\mu (A_\nu p_\mu + p_\nu A_\mu) - m_0^2 c^2\right] \psi \\
&= \left[\frac{1}{2} \{\gamma^\nu, \gamma^\mu\} \left(p_\nu p_\mu + \frac{e^2}{c^2} A_\nu A_\mu\right)\right. \\
&\quad \left. - \frac{e}{2c} (\gamma^\nu \gamma^\mu - \gamma^\mu \gamma^\nu + \{\gamma^\nu, \gamma^\mu\})(A_\nu p_\mu + p_\nu A_\mu) - m_0^2 c^2\right] \psi \\
&= \left[p^\mu p_\mu + \frac{e^2}{c^2} A^\mu A_\mu - \frac{e}{c} (A^\mu p_\mu + p^\mu A_\mu)\right. \\
&\quad \left. - \frac{e}{2c} (\gamma^\nu \gamma^\mu - \gamma^\mu \gamma^\nu)(A_\nu p_\mu + p_\nu A_\mu) - m_0^2 c^2\right] \psi \\
&= \left[\left(p^\mu - \frac{e}{c} A^\mu\right)\left(p_\mu - \frac{e}{c} A_\mu\right) - \frac{e}{2c} [\gamma^\nu, \gamma^\mu](p_\nu A_\mu) - m_0^2 c^2\right] \psi \; .
\end{aligned}$$

In the last-but-one term $(p_\nu A_\mu)$ the operator p_ν only acts on A_μ and not on ψ. Using $p_\mu = \mathrm{i}\hbar\partial_\mu$ and (all derivatives only act on A_μ again)

$$\begin{aligned}\sigma^{\mu\nu}F_{\mu\nu} &= \frac{\mathrm{i}}{2}[\gamma^\mu,\gamma^\nu](\partial_\mu A_\nu - \partial_\nu A_\mu) = \mathrm{i}(\gamma^\mu\gamma^\nu\partial_\mu A_\nu - \gamma^\nu\gamma^\mu\partial_\mu A_\nu)\\ &= \mathrm{i}[\gamma^\mu,\gamma^\nu]\partial_\mu A_\nu\ ,\end{aligned}$$

we finally arrive at (2.55).

17. Lagrange density and energy-momentum tensor of the free Dirac field. Determine the Lagrange density of the free Dirac field similarly to Exercise 2. Subsequently, show with the help of the energy-momentum tensor that the energy is

$$E = \int \mathrm{d}^3x\psi^\dagger H^{(0)}\psi\ ,\ H^{(0)} = c\boldsymbol{\alpha}\boldsymbol{p} + \beta m_0c^2\ .$$

What is the interpretation of this result?

Solution. The Lagrange density of the free Dirac field is

$$\mathcal{L} = \bar{\psi}(\mathrm{i}\hbar c\gamma^\mu\partial_\mu - m_0c^2)\psi$$

since the variation of the corresponding action functional $I = \int \mathrm{d}^4x\mathcal{L}$ with respect to the components of $\bar{\psi}$ and ψ leads to

$$\frac{\partial I}{\partial\bar{\psi}} = 0 \Longrightarrow \frac{\partial\mathcal{L}}{\partial\bar{\psi}} - \partial_\mu\frac{\partial\mathcal{L}}{\partial(\partial_\mu\bar{\psi})} = 0 \quad \text{(Lagrange equation)}$$

$$\frac{\partial\mathcal{L}}{\partial\bar{\psi}} = (\mathrm{i}\hbar c\gamma^\mu\partial_\mu - m_0c^2)\psi\ ,\ \frac{\partial\mathcal{L}}{\partial(\partial_\mu\bar{\psi})} = 0$$

$$\Longrightarrow (\mathrm{i}\hbar c\gamma^\mu\partial_\mu - m_0c^2)\psi = 0 \quad \text{(free Dirac equation)}$$

and

$$\frac{\partial I}{\partial\psi} = 0 \Longrightarrow \frac{\partial\mathcal{L}}{\partial\psi} - \partial_\mu\frac{\partial\mathcal{L}}{\partial(\partial_\mu\psi)} = 0 \quad \text{(Lagrange equation)}$$

$$\frac{\partial\mathcal{L}}{\partial\psi} = -m_0c^2\bar{\psi}\ ,\ \frac{\partial\mathcal{L}}{\partial(\partial_\mu\psi)} = \mathrm{i}\hbar c\bar{\psi}\gamma^\mu$$

$$\Longrightarrow \mathrm{i}\hbar c\partial_\mu\bar{\psi}\gamma^\mu + m_0c^2\bar{\psi} = 0 \quad \text{(free adjoint Dirac equation)}\ .$$

The energy density T^{00} follows from the energy-momentum tensor

$$T^{\mu\nu} = \frac{\partial\mathcal{L}}{\partial(\partial_\mu\psi)}\partial^\nu\psi + \frac{\partial\mathcal{L}}{\partial(\partial_\mu\bar{\psi})}\partial^\nu\bar{\psi} - g^{\mu\nu}\mathcal{L}$$

as

$$\begin{aligned}T^{00} &= \frac{\partial\mathcal{L}}{\partial(\partial_0\psi)}\partial_0\psi + \frac{\partial\mathcal{L}}{\partial(\partial_0\bar{\psi})}\partial_0\bar{\psi} - \mathcal{L}\\ &= \mathrm{i}\hbar c\bar{\psi}\gamma^0\partial_0\psi - \bar{\psi}(\mathrm{i}\hbar c\gamma^\mu\partial_\mu - m_0c^2)\psi\\ &= -\mathrm{i}\hbar c\bar{\psi}\gamma^i\partial_i\psi + m_0c^2\bar{\psi}\psi\\ &= \psi^\dagger(c\boldsymbol{\alpha}\boldsymbol{p} + \beta m_0c^2)\psi = \psi^\dagger H^{(0)}\psi\ .\end{aligned}$$

Therefore, the energy is

$$E = \int \mathrm{d}^3 x T^{00} = \int \mathrm{d}^3 x \psi^\dagger H^{(0)} \psi = \left\langle \psi \middle| H^{(0)} \middle| \psi \right\rangle .$$

Comparing this result with the corresponding result of the Klein-Gordon case in Exercise 2, we find that in both cases the field energy coincides with the (G-)expectation value for positive as well as for negative solutions.

In the Klein-Gordon case we saw that for the description of spin-0 antiparticles we have to use negative solutions and not their charge conjugates (definition of the G-scalar product, Subsection 1.3.1). From this we concluded that, with respect to the G-expectation values, it is justified to consider the original negative solutions to be antiparticle wave functions. This is supported by the consideration of the field energy that yields positive (physical) values for both types of solutions. In the case in hand this is no longer true: for positive solutions we have positive (physical) values of the field energy and the energy expectation value, whereas for negative solutions we have negative (unphysical) ones. This implies that the negative Dirac solutions cannot be the wave functions of spin-1/2 antiparticles. In fact, we have already found that, for the description of antiparticles, we must insert the charge conjugates of the negative solutions into the corresponding expectation values (definition on page 108).

2.2 Symmetry Transformations

As in the symmetry considerations of the Klein-Gordon theory in Section 1.2, we now turn to the symmetry properties of the Dirac equation. Here we again resort to the differentiation between active and passive transformations. Our task consists in explicitly constructing the bispinor transformations $D(\Lambda)$ belonging to the homogeneous (i.e. proper and improper) Lorentz transformations Λ, thus completing the proof of the covariance of the Dirac equation under the full Poincaré group.[8] The basic prerequisites to be fulfilled are presented already in Subsection 2.1.2. Moreover, in connection with spatial rotations, we provide a transformation theoretical argument for Dirac's spin operator introduced in Subsection 2.1.4. Finally, we also consider non-Lorentz-like symmetry operations from which the important Feynman-Stückelberg interpretation is derived once again.

2.2.1 Proper Lorentz Transformations

Proper Lorentz transformations (with $\Lambda^0{}_0 > 0$, $\det \Lambda = +1$) are distinguished by the fact that they can be composed of a series of infinitesimal transformations; that is why they are also called *continuous*. We therefore initially

[8] In addition to the homogeneous Lorentz transformations, the Poincaré group also contains space-time translations for which the proof can be carried out easily. See footnote 2 on page 94.

consider an infinitesimal proper Lorentz transformation for which we can generally assume the form

$$\Lambda^{\mu}{}_{\nu} = g^{\mu}{}_{\nu} + \Delta\omega^{\mu}{}_{\nu} \ , \ \Delta_{\mu\nu} = -\Delta_{\nu\mu} \ . \tag{2.56}$$

The last relation follows from (A.2) in the appendix as, up to linear order of $\Delta\omega$, we have

$$\begin{aligned}
&\Lambda^{\mu}{}_{\alpha} g_{\mu\nu} \Lambda^{\nu}{}_{\beta} = g_{\alpha\beta} \\
&\Longleftrightarrow (g^{\mu}{}_{\alpha} + \Delta\omega^{\mu}{}_{\alpha}) g_{\mu\nu} (g^{\nu}{}_{\beta} + \Delta\omega^{\nu}{}_{\beta}) = g_{\alpha\beta} \\
&\Longleftrightarrow g^{\mu}{}_{\alpha} g_{\mu\nu} g^{\nu}{}_{\beta} + \Delta\omega^{\mu}{}_{\alpha} g_{\mu\nu} g^{\nu}{}_{\beta} + g^{\mu}{}_{\alpha} g_{\mu\nu} \Delta\omega^{\nu}{}_{\beta} = g_{\alpha\beta} \\
&\Longleftrightarrow g_{\alpha\beta} + \Delta\omega^{\mu}{}_{\alpha} g_{\mu\beta} + g_{\alpha\nu} \Delta\omega^{\nu}{}_{\beta} = g_{\alpha\beta} \\
&\Longleftrightarrow g_{\beta\mu} \Delta\omega^{\mu}{}_{\alpha} + g_{\alpha\nu} \Delta\omega^{\nu}{}_{\beta} = 0 \\
&\Longleftrightarrow \Delta\omega_{\beta\alpha} = -\Delta\omega_{\alpha\beta} \ .
\end{aligned}$$

Since with Λ the corresponding bispinor transformation $D(\Lambda)$ will also deviate only infinitesimally from the unity transformation, we make the following ansatz:

$$D = 1 - \frac{\mathrm{i}}{4}\sigma_{\mu\nu}\Delta\omega^{\mu\nu} \ , \ D^{-1} = 1 + \frac{\mathrm{i}}{4}\sigma_{\mu\nu}\Delta\omega^{\mu\nu} \ , \ \sigma_{\mu\nu} = -\sigma_{\nu\mu} \ . \tag{2.57}$$

If we now insert (2.56) and (2.57) into the conditional equation (2.26) from Theorem 2.2 which is valid for Lorentz-like bispinor transformations, we obtain to linear order of $\Delta\omega$

$$\Delta\omega^{\mu}{}_{\nu}\gamma^{\nu} = -\frac{\mathrm{i}}{4}\Delta\omega^{\alpha\beta}[\gamma^{\mu}, \sigma_{\alpha\beta}] \ .$$

From this and taking into account the antisymmetry of $\Delta\omega^{\mu}{}_{\nu}$,

$$\Delta\omega^{\mu}{}_{\nu} = \frac{1}{2}\Delta\omega^{\alpha\beta}(g^{\mu}{}_{\alpha} g_{\nu\beta} - g^{\mu}{}_{\beta} g_{\nu\alpha}) \ ,$$

it follows that

$$2\mathrm{i}(g^{\mu}{}_{\alpha}\gamma_{\beta} - g^{\mu}{}_{\beta}\gamma_{\alpha}) = [\gamma^{\mu}, \sigma_{\alpha\beta}] \ . \tag{2.58}$$

Thus, the construction of bispinor transformations for proper Lorentz transformations is reduced to finding an antisymmetric matrix $\sigma_{\alpha\beta}$ that obeys (2.58). We can easily see that this matrix is given by (compare to Exercises 15 and 16)

$$\sigma_{\alpha\beta} = \frac{\mathrm{i}}{2}[\gamma_{\alpha}, \gamma_{\beta}]$$

since, due to (2.18), we have

$$
\begin{aligned}
\frac{\mathrm{i}}{2}[\gamma^\mu,[\gamma_\alpha,\gamma_\beta]] &= \frac{\mathrm{i}}{2}\left([\gamma^\mu,\gamma_\alpha\gamma_\beta]-[\gamma^\mu,\gamma_\beta,\gamma_\alpha]\right)\\
&= \frac{\mathrm{i}}{2}\left([\gamma^\mu,\gamma_\alpha\gamma_\beta]-2[\gamma^\mu,g_{\alpha\beta}]+[\gamma^\mu,\gamma_\alpha,\gamma_\beta]\right)\\
&= \mathrm{i}[\gamma^\mu,\gamma_\alpha\gamma_\beta]\\
&= \mathrm{i}\left(\gamma^\mu\gamma_\alpha\gamma_\beta-2g^\mu{}_\beta\gamma_\alpha+\gamma_\alpha\gamma^\mu\gamma_\beta\right)\\
&= \mathrm{i}\left(\gamma^\mu\gamma_\alpha\gamma_\beta-2g^\mu{}_\beta\gamma_\alpha+2g^\mu{}_\alpha\gamma_\beta-\gamma^\mu\gamma_\alpha\gamma_\beta\right)\\
&= 2\mathrm{i}\left(g^\mu{}_\alpha\gamma_\beta-g^\mu{}_\beta\gamma_\alpha\right)\ .
\end{aligned}
$$

We are now in a position to construct finite bispinor transformations by a repeated application of infinitesimal transformations via

$$
D(\Lambda)=\lim_{N\to\infty}\left(1-\frac{\mathrm{i}}{4}\sigma_{\mu\nu}\Delta\omega^{\mu\nu}\right)^N=\exp\left(-\frac{\mathrm{i}}{4}\sigma_{\mu\nu}\Delta^{\mu\nu}\right)\ .
$$

To do this, it is advantageous to rewrite the infinitesimal quantity $\Delta\omega^\mu{}_\nu$ as

$$
\Delta\omega^\mu{}_\nu=\Delta\omega(I_n)^\mu{}_\nu\ ,\ \lim_{N\to\infty}N\Delta\omega=\omega\ ,
$$

where $\Delta\omega$ denotes the infinitesimal "rotation angle" around an axis in n-direction, ω the corresponding finite rotation angle, and I_n the 4×4-coefficient matrix (in space and time) of the "unity Lorentz rotation" around this axis.

Theorem 2.6: Bispinor transformations for proper Lorentz transformations

To the proper Lorentz transformation

$$
\Lambda:\quad x^\mu\longrightarrow x'^\mu=\Lambda^\mu{}_\nu x^\nu
$$

around the rotation axis n with the rotation angle ω there belongs the passive bispinor transformation

$$
D(\Lambda):\quad \psi(x)\longrightarrow\psi'(x')=D(\Lambda)\psi(x)\ ,
$$

with

$$
D(\Lambda)=\exp\left(-\frac{\mathrm{i}}{4}\omega\sigma_{\mu\nu}(I_n)^{\mu\nu}\right)\ ,\ \sigma_{\mu\nu}=\frac{\mathrm{i}}{2}[\gamma_\mu,\gamma_\nu]\ ,
$$

where I_n denotes the unity Lorentz rotation around the axis n.

This theorem completes the proof for the form invariance of the Dirac equation under proper Lorentz transformations. The proof for improper Lorentz transformations as well as for non-Lorentz-like transformations will be carried out in Subsection 2.2.3.

Lorentz boosts. As a first concrete application of Theorem 2.6, we calculate the bispinor transformation for a Lorentz boost describing the transition to an inertial system moving with velocity $\boldsymbol{v}$ relative to the original reference frame.

To do this, we initially determine the corresponding infinitesimal Lorentz boosts along the three spatial axes,

$$(\Lambda^{(1)\mu}{}_{\nu}) = \begin{pmatrix} \cosh\Delta\omega_1 & \sinh\Delta\omega_1 & 0 & 0 \\ \sinh\Delta\omega_1 & \cosh\Delta\omega_1 & 0 & 0 \\ 0 & 0 & 1 & 0 \\ 0 & 0 & 0 & 1 \end{pmatrix}$$

$$\stackrel{\Delta\omega_1 \ll 1}{=} (g^{\mu}{}_{\nu}) + \Delta\omega_1 \begin{pmatrix} 0&1&0&0\\1&0&0&0\\0&0&0&0\\0&0&0&0 \end{pmatrix}$$

$$(\Lambda^{(2)\mu}{}_{\nu}) = \begin{pmatrix} \cosh\Delta\omega_2 & 0 & \sinh\Delta\omega_2 & 0 \\ 0 & 1 & 0 & 0 \\ \sinh\Delta\omega_2 & 0 & \cosh\Delta\omega_2 & 0 \\ 0 & 0 & 0 & 1 \end{pmatrix}$$

$$\stackrel{\Delta\omega_2 \ll 1}{=} (g^{\mu}{}_{\nu}) + \Delta\omega_2 \begin{pmatrix} 0&0&1&0\\0&0&0&0\\1&0&0&0\\0&0&0&0 \end{pmatrix}$$

$$(\Lambda^{(3)\mu}{}_{\nu}) = \begin{pmatrix} \cosh\Delta\omega_3 & 0 & 0 & \sinh\Delta\omega_3 \\ 0 & 1 & 0 & 0 \\ 0 & 0 & 1 & 0 \\ \sinh\Delta\omega_3 & 0 & 0 & \cosh\Delta\omega_3 \end{pmatrix}$$

$$\stackrel{\Delta\omega_3 \ll 1}{=} (g^{\mu}{}_{\nu}) + \Delta\omega_3 \begin{pmatrix} 0&0&0&1\\0&0&0&0\\0&0&0&0\\1&0&0&0 \end{pmatrix} ,$$

to construct the infinitesimal Lorentz boost in the direction of $\boldsymbol{v} = v \begin{pmatrix} \cos\theta_1 \\ \cos\theta_2 \\ \cos\theta_3 \end{pmatrix}$:

$$\Lambda^{\mu}{}_{\nu} \stackrel{\Delta\omega \ll 1}{=} g^{\mu}{}_{\nu} + \Delta\omega I^{\mu}{}_{\nu} \; , \; (I^{\mu}{}_{\nu}) = \begin{pmatrix} 0 & \cos\theta_1 & \cos\theta_2 & \cos\theta_3 \\ \cos\theta_1 & 0 & 0 & 0 \\ \cos\theta_2 & 0 & 0 & 0 \\ \cos\theta_3 & 0 & 0 & 0 \end{pmatrix} .$$

For the bispinor transformation belonging to the finite Lorentz boost, we have according to Theorem 2.6

$$\sigma_{\mu\nu} I^{\mu\nu} = \sigma_{\mu\nu} g^{\nu\rho} I^{\mu}{}_{\rho} = -2 \sum_i \sigma_{0i} \cos\theta_i = 2\mathrm{i} \frac{\boldsymbol{\alpha v}}{v}$$

$$\Longrightarrow D(\Lambda_{\boldsymbol{v}}) = \exp\left(\frac{\omega}{2} \frac{\boldsymbol{\alpha v}}{v}\right) .$$

With the help of the representation-independent relations

$$(\boldsymbol{\alpha v}) = \boldsymbol{v}^{2n} \ , \ (\boldsymbol{\alpha v})^{2n+1} = v^{2n}\boldsymbol{\alpha v} \ ,$$

the exponential term can be carried out as follows:

$$\begin{aligned}
\exp\left(\frac{\omega}{2v}\boldsymbol{\alpha v}\right) &= \sum_{n=0}^{\infty}\frac{1}{n!}\left(\frac{\omega}{2v}\right)^n(\boldsymbol{\alpha v})^n \\
&= \sum_{n=0}^{\infty}\frac{1}{(2n)!}\left(\frac{\omega}{2v}\right)^{2n}(\boldsymbol{\alpha v})^{2n} \\
&\quad +\sum_{n=0}^{\infty}\frac{1}{(2n+1)!}\left(\frac{\omega}{2v}\right)^{2n+1}(\boldsymbol{\alpha v})^{2n+1} \\
&= \sum_{n=0}^{\infty}\frac{1}{(2n)!}\left(\frac{\omega}{2}\right)^{2n} + \frac{\boldsymbol{\alpha v}}{v}\sum_{n=0}^{\infty}\frac{1}{(2n+1)!}\left(\frac{\omega}{2}\right)^{2n+1} \\
&= \cosh\left(\frac{\omega}{2}\right) + \frac{\boldsymbol{\alpha v}}{v}\sinh\left(\frac{\omega}{2}\right) \ .
\end{aligned}$$

Taking into account

$$\cosh\omega = \frac{1}{\sqrt{1-\beta^2}} \ , \ \beta = \frac{v}{c}$$

as well as

$$\cosh\left(\frac{\omega}{2}\right) = \sqrt{\frac{1}{2}(\cosh\omega + 1)} \ , \ \sinh\left(\frac{\omega}{2}\right) = -\sqrt{\cosh^2\left(\frac{\omega}{2}\right) - 1} \ ,$$

we finally obtain the result

$$D(\Lambda_{\boldsymbol{v}}) = \sqrt{\frac{1}{2}\left(\frac{1}{\sqrt{1-\beta^2}}+1\right)} - \frac{\boldsymbol{\alpha v}}{v}\sqrt{\frac{1}{2}\left(\frac{1}{\sqrt{1-\beta^2}}-1\right)} \ . \tag{2.59}$$

If the original reference frame is equal to the rest system of the particle, the particle possesses the momentum $\boldsymbol{p} \sim -\boldsymbol{v}$ in the Lorentz transformed system. In this case, using $\sqrt{1-\beta^2} = m_0c/p_0$, $p_0 > 0$, we can rewrite (2.59) as[9]

$$D(\boldsymbol{p}) = \frac{p_0 + m_0c + \boldsymbol{\alpha p}}{\sqrt{2m_0c(p_0+m_0c)}} \ . \tag{2.60}$$

In return, $D(-\boldsymbol{p})$ transforms to the rest system of a particle with momentum $\boldsymbol{p}$ in the original system.

Spatial rotations. Next we consider the spatial Lorentz rotation around a unit axis $\boldsymbol{u}$, $|\boldsymbol{u}| = 1$ by the rotation angle φ and calculate the corresponding bispinor transformation. We can proceed, as in the boost case, by calculating the mathematically positive infinitesimal rotations around the three Cartesian axes which, in total, lead to the infinitesimal rotation

[9] Be aware of the notational difference in (2.59) and (2.60): $D(\Lambda_{\boldsymbol{v}})$ and $D(\boldsymbol{p})$ transform the original reference frame in opposite directions.

$$\Lambda^\mu{}_\nu \overset{\Delta\varphi\ll 1}{=} g^\mu{}_\nu + \Delta\varphi I^\mu{}_\nu \ , \ (I^\mu{}_\nu) = \begin{pmatrix} 0 & 0 & 0 & 0 \\ 0 & 0 & u_3 & -u_2 \\ 0 & -u_3 & 0 & u_1 \\ 0 & u_2 & -u_1 & 0 \end{pmatrix} . \tag{2.61}$$

For the bispinor transformation belonging to the finite Lorentz rotation there follows in the Dirac and Weyl representations (but not in every representation!)

$$\begin{aligned} \sigma_{\mu\nu} I^{\mu\nu} &= \sigma_{\mu\nu} I^\mu{}_\rho g^{\rho\nu} = -2(\sigma_{12}u_3 + \sigma_{31}u_2 + \sigma_{23}u_1) \\ &= -2\boldsymbol{u}\hat{\boldsymbol{\sigma}} \ , \ \hat{\boldsymbol{\sigma}} = \begin{pmatrix} \boldsymbol{\sigma} & 0 \\ 0 & \boldsymbol{\sigma} \end{pmatrix} \end{aligned}$$

$$\Longrightarrow D(\Lambda_\varphi) = \exp\left(\frac{\mathrm{i}}{2}\varphi\boldsymbol{u}\hat{\boldsymbol{\sigma}}\right) . \tag{2.62}$$

With the help of

$$(\boldsymbol{u}\hat{\boldsymbol{\sigma}})^{2n} = 1 \ , \ (\boldsymbol{u}\hat{\boldsymbol{\sigma}})^{2n+1} = \boldsymbol{u}\hat{\boldsymbol{\sigma}} \ ,$$

we can carry out the exponential term in (2.62) in a similar way as in the boost case. This results in

$$\begin{aligned} \exp\left(\frac{\mathrm{i}}{2}\varphi\boldsymbol{u}\hat{\boldsymbol{\sigma}}\right) &= \sum_{n=0}^{\infty} \frac{\mathrm{i}^n}{n!}\left(\frac{\varphi}{2}\right)^n (\boldsymbol{u}\hat{\boldsymbol{\sigma}})^n \\ &= \sum_{n=0}^{\infty} \frac{\mathrm{i}^{2n}}{(2n)!}\left(\frac{\varphi}{2}\right)^{2n} + \boldsymbol{u}\hat{\boldsymbol{\sigma}} \sum_{n=0}^{\infty} \frac{\mathrm{i}^{2n+1}}{(2n+1)!}\left(\frac{\varphi}{2}\right)^{2n+1} \\ &= \sum_{n=0}^{\infty} \frac{(-1)^n}{(2n)!}\left(\frac{\varphi}{2}\right)^{2n} + \mathrm{i}\boldsymbol{u}\hat{\boldsymbol{\sigma}} \sum_{n=0}^{\infty} \frac{(-1)^n}{(2n+1)!}\left(\frac{\varphi}{2}\right)^{2n+1} \end{aligned}$$

$$\Longrightarrow D(\Lambda_\varphi) = \cos\left(\frac{\varphi}{2}\right) + \mathrm{i}\boldsymbol{u}\hat{\boldsymbol{\sigma}} \sin\left(\frac{\varphi}{2}\right) .$$

Note that, after a rotation by 2π, the original state is not reached. Instead, we have

$$D(\Lambda_{2n\pi}) = (-1)^n \ ,$$

which is a general characteristic of half-integer spins. Therefore, physically observable quantities must always be bilinear in the fields $\psi(x)$ or must contain an even power of them. Only then, physical observations return to their original states at spatial rotations by 2π, in accordance with our experience.

2.2.2 Spin of Dirac Solutions

On the basis of (2.62) and the general relationship between passive and active transformations, we can now give a transformation theoretical argument for the spin operator (2.31) introduced in Subsection 2.1.4 and thus for the fact that Dirac solutions describe spin-1/2 particles. First we note that each active rotation of a physical system around a unit axis $\boldsymbol{u}$ by the angle $-\Delta\varphi$ can be expressed, with the help of the total angular momentum $\boldsymbol{J} = \boldsymbol{L} + \boldsymbol{S}$, as

$$\psi'(x) = \exp\left(\frac{\mathrm{i}\Delta\varphi\boldsymbol{u}\boldsymbol{J}}{\hbar}\right)\psi(x) \stackrel{\Delta\varphi\ll 1}{=} \left(1 + \frac{\mathrm{i}\Delta\varphi\boldsymbol{u}\boldsymbol{J}}{\hbar}\right)\psi(x) \tag{2.63}$$

or, equally, that $\boldsymbol{L}$, $\boldsymbol{S}$, and $\boldsymbol{J}$ are defined by this relation. On the other hand, we know that each active rotation can also be derived from the corresponding passive transformation of the reference system in the opposite direction [see (1.30) and (1.31) in Subsection 1.2.1]:

$$\left.\begin{array}{l}\text{passive rotation}\\ \text{around } \boldsymbol{u} \text{ by } +\Delta\varphi\end{array}\right\} : \psi'(x') = D(\Lambda_{\Delta\varphi})\psi(x)$$

$$\Updownarrow$$

$$\left.\begin{array}{l}\text{Active rotation}\\ \text{around } \boldsymbol{u} \text{ by } -\Delta\varphi\end{array}\right\} : \psi'(x) = D(\Lambda_{\Delta\varphi})\psi(\Lambda_{\Delta\varphi}^{-1}x) = D(\Lambda_{\Delta\varphi})\psi(\Lambda_{-\Delta\varphi}x) \ .$$

Restricting ourselves to infinitesimal rotation angles, we can rewrite the last relation with the help of (2.61) and (2.62) as

$$\begin{aligned}
\psi'(x) &\stackrel{\Delta\varphi\ll 1}{=} \left(1 + \frac{\mathrm{i}\Delta\varphi\boldsymbol{u}\hat{\boldsymbol{\sigma}}}{2}\right)\psi[x^0, x^1 + \Delta\varphi(-u_3x^2 + u_2x^3), \\
&\qquad\qquad x^2 + \Delta\varphi(u_3x^1 - u_1x^3), \\
&\qquad\qquad x^3 + \Delta\varphi(-u_2x^1 + u_1x^2)] \\
&= \left(1 + \frac{\mathrm{i}\Delta\varphi\boldsymbol{u}\hat{\boldsymbol{\sigma}}}{2}\right)\left\{\psi(x) + \Delta\varphi\left[\frac{\partial\psi}{\partial x^1}(-u_3x^2 + u_2x^3)\right.\right. \\
&\qquad\qquad + \frac{\partial\psi}{\partial x^2}(u_3x^1 - u_1x^3) \\
&\qquad\qquad \left.\left. + \frac{\partial\psi}{\partial x^3}(-u_2x^1 + u_1x^2)\right]\right\} \\
&= \left(1 + \frac{\mathrm{i}\Delta\varphi\boldsymbol{u}\hat{\boldsymbol{\sigma}}}{2}\right)\left(1 + \frac{\mathrm{i}\Delta\varphi\boldsymbol{u}\boldsymbol{L}}{\hbar}\right)\psi(x) \\
&= \left[1 + \frac{\mathrm{i}\Delta\varphi\boldsymbol{u}}{\hbar}\left(\boldsymbol{L} + \frac{\hbar\hat{\boldsymbol{\sigma}}}{2}\right)\right]\psi(x) \ .
\end{aligned}$$

Comparison with (2.63) finally leads to the already known expression for the Dirac spin operator:

$$\boldsymbol{S} = \frac{\hbar}{2}\hat{\boldsymbol{\sigma}} \ .$$

2.2.3 Discrete Transformations

We now come to the case of improper Lorentz transformations (with $\det\Lambda = -1$) as well as to other non-Lorentz-like transformations that cannot be constructed by repeated application of infinitesimal operations wherefore they are called *discrete*. Here we proceed similarly to the Klein-Gordon case (Subsection 1.2.3). The corresponding physical implications have already been discussed there and can be completely carried over.

Parity transformation P. The improper orthochronous Lorentz-like parity transformation or space reflection is defined by

$$(\Lambda^\nu{}_\mu) = \begin{pmatrix} 1 & 0 & 0 & 0 \\ 0 & -1 & 0 & 0 \\ 0 & 0 & -1 & 0 \\ 0 & 0 & 0 & -1 \end{pmatrix} = g^{\mu\nu} \ , \ \det(\Lambda) = -1 \ , \ \Lambda^0{}_0 > 0 \ .$$

In order to show that Dirac's equation is form invariant under this operation, we need to find a bispinor transformation $D(\Lambda)$ obeying the conditional equation (2.26) for Lorentz-like bispinor transformations. Introducing the notation P for $D(\Lambda)$, this means

$$P^{-1}\gamma^\mu P = \Lambda^\mu{}_\nu \gamma^\nu = g^{\mu\mu}\gamma^\mu \quad \text{(no summation over } \mu\text{)} \ . \tag{2.64}$$

Furthermore and as in spatial rotations by 4π, we demand that four space reflections transform a bispinor back into itself, i.e.

$$P^4 = 1 \ . \tag{2.65}$$

As can be seen easily, this is achieved by the representation-independent choice $P = \lambda_P \gamma^0$ with $\lambda_P^4 = 1$. In the passive case we then have

$$\left.\begin{array}{rl} \boldsymbol{x} \longrightarrow & \boldsymbol{x}' = -\boldsymbol{x} \ , \ t \longrightarrow t' = t \\ \psi(\boldsymbol{x}, t) \longrightarrow & \psi_P(\boldsymbol{x}', t') = P\psi(\boldsymbol{x}, t) \ , \ P = \lambda_P\gamma^0 \ , \ \lambda_P^4 = 1 \\ A^0(\boldsymbol{x}, t) \longrightarrow & A_P^0(\boldsymbol{x}', t') = A^0(\boldsymbol{x}, t) \\ \boldsymbol{A}(\boldsymbol{x}, t) \longrightarrow & \boldsymbol{A}_P(\boldsymbol{x}', t') = -\boldsymbol{A}(\boldsymbol{x}, t) \end{array}\right\} \begin{array}{c}\text{passive} \\ \text{space} \\ \text{reflection } P\end{array}$$

and in the active case

$$\left.\begin{array}{rl} \psi(\boldsymbol{x}, t) \longrightarrow & \psi_P(\boldsymbol{x}, t) = P\psi(-\boldsymbol{x}, t) \\ A^0(\boldsymbol{x}, t) \longrightarrow & A_P^0(\boldsymbol{x}, t) = A^0(-\boldsymbol{x}, t) \\ \boldsymbol{A}(\boldsymbol{x}, t) \longrightarrow & \boldsymbol{A}_P(\boldsymbol{x}, t) = -\boldsymbol{A}(-\boldsymbol{x}, t) \end{array}\right\} \begin{array}{l}\text{active space} \\ \text{reflection } P.\end{array} \tag{2.66}$$

Applying the active space reflection to a free plane Dirac solution yields (see Exercise 20)

$$\left\{\begin{array}{c} \psi_{\epsilon,p,n}(x) \\ p = \begin{pmatrix} p_0 \\ \boldsymbol{p} \end{pmatrix}, n = \begin{pmatrix} n_0 \\ \boldsymbol{n} \end{pmatrix} \end{array}\right\} \longrightarrow \left\{\begin{array}{c} \psi_{\epsilon,p',n'}(x) \\ p' = \begin{pmatrix} p_0 \\ -\boldsymbol{p} \end{pmatrix}, n = \begin{pmatrix} -n_0 \\ \boldsymbol{n} \end{pmatrix} \end{array}\right\} . \tag{2.67}$$

Hence, a free solution with opposite spatial momentum index is created, whereas the spatial spin index is unchanged. Since ϵ is unchanged, too, this relationship can be carried over to the particle level: the space reflection reverses the momentum of a spin-1/2 particle and leaves its spin orientation unchanged.

Time reversal transformation T. The time reversal or, rather, motion reversal transformation is a non-Lorentz-like symmetry operation which, from the passive point of view, is defined by

$$\left.\begin{array}{rcl} \boldsymbol{x} & \longrightarrow & \boldsymbol{x}' = \boldsymbol{x} \ , \ t \longrightarrow t' = -t \\ A^0(\boldsymbol{x},t) & \longrightarrow & A^0_T(\boldsymbol{x}',t') = A^0(\boldsymbol{x},t) \\ \boldsymbol{A}(\boldsymbol{x},t) & \longrightarrow & \boldsymbol{A}_T(\boldsymbol{x}',t') = -\boldsymbol{A}(\boldsymbol{x},t) \end{array}\right\} \begin{array}{c}\text{passive time}\\ \text{reversal } T\end{array} \qquad (2.68)$$

so that

$$\mathrm{i}\hbar\partial^0 \longrightarrow \mathrm{i}\hbar\partial'^0 = -\mathrm{i}\hbar\partial^0 \ , \ \mathrm{i}\hbar\partial^i \longrightarrow \mathrm{i}\hbar\partial'^i = \mathrm{i}\hbar\partial^i \ .$$

Due to the non-Lorentz-like character of this operation, it is clear that we cannot make use of the relation (2.26) in order to construct the corresponding bispinor transformation. We therefore start with the Dirac equation in the transformed (primed) system,

$$\left[\gamma^\mu\left(\mathrm{i}\hbar\partial'_\mu - \frac{e}{c}A_{T,\mu}(x')\right) - m_0c\right]\psi'(x') = 0 \ , \qquad (2.69)$$

make the (antilinear and reciprocal) ansatz

$$\psi'(x') = T\psi^*(x) \ , \ T^2 = 1 \ , \ T \text{ linear} \ ,$$

and express (2.69) by the original (unprimed) quantities:

$$\left[\gamma^0\left(-\mathrm{i}\hbar\partial_0 - \frac{e}{c}A_0(x)\right) + \gamma^i\left(\mathrm{i}\hbar\partial_i + \frac{e}{c}A_i(x)\right) - m_0c\right]T\psi^*(x) = 0 \ . \qquad (2.70)$$

If we now require that

$$T^{-1}\gamma^\mu T = g^{\mu\mu}\gamma^{\mu *} \quad \text{(no summation over } \mu) \ , \qquad (2.71)$$

then multiplying (2.70) from the left by T^{-1} and subsequently taking the complex conjugate leads to the equation

$$\left[\gamma^0\left(\mathrm{i}\hbar\partial_0 - \frac{e}{c}A_0(x)\right) + \gamma^i\left(\mathrm{i}\hbar\partial_i - \frac{e}{c}A_i(x)\right) - m_0c\right]\psi(x) = 0$$

in the original system. This is formally identical to (2.69). The solution to (2.71) is not hard to find and is given by $T = \mathrm{i}\lambda_T\gamma^1\gamma^3$ in the Dirac or Weyl representation. To complete (2.68), we can therefore write:

$$\psi(\boldsymbol{x},t) \longrightarrow \psi_T(\boldsymbol{x}',t') = T\psi^*(\boldsymbol{x},t) \ , \ \left.\begin{array}{c} T = \mathrm{i}\lambda_T\gamma^1\gamma^3 \\ |\lambda_T| = 1 \end{array}\right\} \begin{array}{c}\text{passive time}\\ \text{reversal } T.\end{array}$$

Accordingly, for the active time reversal, we have

$$\left.\begin{array}{rcl} \psi(\boldsymbol{x},t) & \longrightarrow & \psi_T(\boldsymbol{x},t) = T\psi^*(\boldsymbol{x},-t) \\ A^0(\boldsymbol{x},t) & \longrightarrow & A^0_T(\boldsymbol{x},t) = A^0(\boldsymbol{x},-t) \\ \boldsymbol{A}(\boldsymbol{x},t) & \longrightarrow & \boldsymbol{A}_T(\boldsymbol{x},t) = -\boldsymbol{A}(\boldsymbol{x},-t) \end{array}\right\} \begin{array}{c}\text{active time}\\ \text{reversal } T.\end{array} \qquad (2.72)$$

Applied to a free plane Dirac solution, the active time reversal yields (see Exercise 20)

$$\left\{\begin{array}{c}\psi_{\epsilon,p,n}(x)\\ p=\begin{pmatrix}p_0\\ \boldsymbol{p}\end{pmatrix},n=\begin{pmatrix}n_0\\ \boldsymbol{n}\end{pmatrix}\end{array}\right\}\longrightarrow\left\{\begin{array}{c}\psi_{\epsilon,p',n'}(x)\\ p'=\begin{pmatrix}p_0\\ -\boldsymbol{p}\end{pmatrix},n=\begin{pmatrix}n_0\\ -\boldsymbol{n}\end{pmatrix}\end{array}\right\}. \quad (2.73)$$

On the level of wave functions the spatial momentum and spin indices are reversed. On the particle level this implies, due to the unchanged ϵ, that the time reversal reverses the momentum and spin of a spin-1/2 particle.

***PCT*-transformation (no symmetry transformation).** Instead of the charge conjugation C from Subsection 2.1.6, we can, as in the Klein-Gordon case, apply a combination of the three transformations C, P, and T to negative Dirac solutions $\psi^{(-)}$ in order to generate wave functions for antiparticles. Taking into account Theorem 2.5 as well as (2.66) and (2.72), this yields for the active case in the Dirac or Weyl representation (possible phases ignored)

$$\psi^{(-)}(x)\longrightarrow\psi^{(-)}_{PCT}(x)=\mathrm{i}\gamma^5\psi^{(-)}(-x)\ .$$

This, in turn, leads again to the Feynman-Stückelberg interpretation to which we will refer particularly at the description of scattering processes in Chapter 3 (compare to Theorem 1.5).

Theorem 2.7: Feynman-Stückelberg interpretation in the Dirac theory

Due to the *PCT*-transformation, the wave function of a physical spin-1/2 antiparticle of charge $-e$ can be interpreted as a negative Dirac solution of charge $+e$ moving backward in space and time.

As in the Klein-Gordon case, this interpretation can be verified by applying the *PCT*-transformation to the eigenvalue equation of a negative Dirac state with charge $+e$ and then observing that the resulting equation corresponds to the eigenvalue equation for a positive Dirac state of charge $-e$ with opposite direction of motion in space and time.

Extended charge conjugation $\mathcal{C}$. We can again extend the mathematical equivalence operation C to a symmetry operation by also transforming the electromagnetic fields in an appropriate way:

$$\left.\begin{array}{rcl}\psi(\boldsymbol{x},t) &\longrightarrow& \psi_{\mathcal{C}}(\boldsymbol{x},t)=\mathrm{i}\lambda_{\mathcal{C}}\gamma^2\psi^*(\boldsymbol{x},t)\ ,\ |\lambda_{\mathcal{C}}|=1\\ A^0(\boldsymbol{x},t) &\longrightarrow& A^0_{\mathcal{C}}(\boldsymbol{x},t)=-A^0(\boldsymbol{x},t)\\ \boldsymbol{A}(\boldsymbol{x},t) &\longrightarrow& \boldsymbol{A}_{\mathcal{C}}(\boldsymbol{x},t)=-\boldsymbol{A}(\boldsymbol{x},t)\end{array}\right\}\begin{array}{c}\text{active charge}\\ \text{conjugation } \mathcal{C}.\end{array} \quad (2.74)$$

On the level of wave functions the action of this extended charge conjugation is, for example, that the Dirac equation for a positive solution $\psi^{(+)}$ with charge $+e$ within the potential $+A^\mu$,

$$\left[\gamma^\mu\left(\mathrm{i}\hbar\partial_\mu - \frac{e}{c}A^\mu(x)\right) - m_0 c\right]\psi^{(-)}(x) = 0\ ,$$

is transformed into the Dirac equation for a negative solution $\psi_C^{(+)} = \mathrm{i}\gamma^2\psi^{(+)*}$ with the same charge $+e$ within the potential $A_C^\mu = -A^\mu$, i.e.

$$\left[\gamma^\mu\left(\mathrm{i}\hbar\partial_\mu + \frac{e}{c}A^\mu(x)\right) - m_0 c\right]\psi_C^{(+)}(x) = 0\ .$$

However, due to the original C-transformation, the last equation can be interpreted as the Dirac equation for a positive solution $\psi_{CC}^{(+)}$ with charge $-e$ within the potential $-A^\mu$ possessing the same quantum numbers as the original solution $\psi^{(+)}$.

On the particle level this can be interpreted analogously to the Klein-Gordon case as follows: the charge conjugation $\mathcal{C}$ turns a fermion into an antifermion with an opposite charge but in all other respects identical quantum numbers. In other words, a fermion of charge $+e$ within the potential $+A^\mu$ behaves exactly as the corresponding antifermion of charge $-e$ within the potential $-A^\mu$, in accordance with our expectations.

Further symmetry considerations. Having discussed the most important symmetry properties of the Dirac theory, which are in principle the same as those of the Klein-Gordon theory, we supplement the final statements of Section 1.2 with the following remarks: generally, the theoretical and experimental investigation of symmetry principles is a very important instrument for finding or verifying theoretical descriptions of microscopic physical processes. For example, we know that all three discrete symmetries $\mathcal{C}$, $\mathcal{P}$, and $\mathcal{T}$ are conserved within electromagnetic interactions, such as the electron-electron scattering, so that the Dirac-Hamilton operator can only contain terms that do not disturb these symmetries. This is ensured by the minimal coupling (for the description of elementary particles) and by certain additional terms (for the description of nonelementary particles with anomalous magnetic moment, see Exercise 15), which give additional support for the correctness of the Dirac theory.

As another example, let us consider the weak process of β-decay (neutron decay),

$$\nu + n \longrightarrow p + e\ ,$$

where n, p, e, and ν stand for neutron, proton, electron, and neutrino, respectively. Here one initially assumed that the Lorentz-invariant amplitude of this process can be expressed by the product of two weak vector currents as

$$\left.\begin{array}{l} M \sim G j_{(p,n)}^{(\mathrm{weak}),\mu} \cdot j_{(e,\nu),\mu}^{(\mathrm{weak})} \quad (G\text{=Fermi constant}) \\ j_{(p,n)}^{(\mathrm{weak}),\mu} = \bar{\psi}_p\gamma^\mu\psi_n\ ,\ j_{(e,\nu)}^{(\mathrm{weak}),\mu} = \bar{\psi}_e\gamma^\mu\psi_\nu \end{array}\right\}\ , \tag{2.75}$$

in complete analogy to the electromagnetic process of electron-proton scattering which is described to first order perturbation theory by the Lorentz-invariant amplitude

$$M \sim \frac{e^2}{q^2} j^{(\mathrm{em}),\mu}_{(p,p)} \cdot j^{(\mathrm{em})}_{(e,e),\mu} \qquad \left(\begin{array}{c} q\text{=four-momentum transfer of the} \\ \text{virtual one-photon exchange} \end{array}\right) ,$$

with the electromagnetic vector currents

$$j^{(\mathrm{em}),\mu}_{(p,p)} = \bar{\psi}_p \gamma^\mu \psi_p \ , \ j^{(\mathrm{em}),\mu}_{(e,e)} = \bar{\psi}_e \gamma^\mu \psi_e$$

(see Subsection 3.3.2).[10] However, the choice of the vector operator γ^μ within the weak currents is very special, and there is a priori no reason for not taking another γ-matrix combination discussed in Subsection 2.1.3 in order to build covariant bilinear forms (currents). It is true that the amplitude in (2.75) is capable of describing some properties of the β-decays, but not others. Therefore, a vast number of β-decay experiments have been carried out in order to find the correct form of the weak interaction amplitude. This culminated in Lee and Yang in 1956 proposing certain experiments which revealed that the parity is not preserved within weak interaction processes. This implies, among other things, that only left-handed neutrinos (with negative helicity) and right-handed antineutrinos (with positive helicity) appear, but no right-handed neutrinos or left-handed antineutrinos. From this it follows that the $\mathcal{C}$-invariance must be violated, too, since the $\mathcal{C}$-transformation turns a left-handed neutrino into a left-handed antineutrino which is never observed. From all this can be concluded that vector and pseudo-vector interactions are present (but no scalar, pseudo-scalar or tensor interactions) that must appear in a certain combination to produce a mixture with no well-defined parity. Eventually, the amplitude

$$M \sim G[\bar{\psi}_p \gamma^\mu (1 - \lambda\gamma^5)\psi_n][\bar{\psi}_e \gamma_\mu (1 + \gamma^5)\psi_\nu] \ , \ \lambda \approx -1.25$$

proved to be the correct choice for the above β-decay where λ denotes the mixing ratio of the vector-pseudo-vector coupling (also called vector-axial vector coupling) in the hadronic current. This example demonstrates, once again, how helpful symmetry considerations are for finding the correct mathematical description of physical processes.

Summary

- The proof of the Lorentz covariance of the Dirac theory is completed by the explicit construction of bispinor transformations to given Lorentz transformations.
- The Dirac theory is covariant under the full Poincaré group. Discrete symmetry transformations of the theory are the improper Lorentz transformation of **space reflection** P as well as the non-Lorentz-like trans-

▷

[10] Here the proton and the neutron are considered approximately to be structureless (point-like).

formations of **time reversal** T and the extended **charge conjugation** $\mathcal{C}$.

- Exploiting the relationship between passive and active transformations and the definition of the total angular momentum via active rotations, one obtains a transformation theoretical argument for the fact that Dirac solutions describe spin-1/2 particles.
- The wave function of a spin-1/2 antiparticle of charge $-e$ can be seen as a negative Dirac solution of charge $+e$ moving backward in space and time (**Feynman-Stückelberg interpretation**).

Exercises

18. Completeness and orthogonality relations of free bispinors. Verify the completeness and orthogonality relations (2.15) and (2.16) with the help of the bispinor transformation (2.60) for Lorentz boosts.

Solution. For the proof of (2.15), we turn to the rest system and exploit $D^\dagger(\boldsymbol{p}) = D(\boldsymbol{p})$:

$$\begin{aligned}
\omega^{(r)\dagger}(\epsilon_r \boldsymbol{p})\omega^{(r')}(\epsilon_{r'}\boldsymbol{p}) &= \omega^{(r)\dagger}(\boldsymbol{0})D^\dagger(\epsilon_r\boldsymbol{p})D(\epsilon_{r'}\boldsymbol{p})\omega^{r'}(\boldsymbol{0}) \\
&= \omega^{(r)\dagger}(\boldsymbol{0})D(\epsilon_r\boldsymbol{p})D(\epsilon_{r'}\boldsymbol{p})\omega^{r'}(\boldsymbol{0}) \\
&= \omega^{(r)\dagger}(\boldsymbol{0})\frac{(p_0+m_0c)^2+\epsilon_r\epsilon_{r'}\boldsymbol{p}^2}{2m_0c(p_0+m_0c)}\omega^{(r')}(\boldsymbol{0}) \\
&\quad +\omega^{(r)\dagger}(\boldsymbol{0})\frac{(p_0+m_0c)(\epsilon_r+\epsilon_{r'})\boldsymbol{\alpha}\boldsymbol{p}}{2m_0c(p_0+m_0c)}\omega^{(r')}(\boldsymbol{0}) \ .
\end{aligned}$$

Since $\omega^{(r)\dagger}(\boldsymbol{0})\boldsymbol{\alpha}\boldsymbol{p}\omega^{(r')}(\boldsymbol{0})$ is nonzero only for $\epsilon_r \neq \epsilon_{r'}$, the second term does not contribute. Therefore, we find

$$\omega^{(r)\dagger}(\epsilon_r\boldsymbol{p})\omega^{(r')}(\epsilon_{r'}\boldsymbol{p}) = \frac{(p_0+m_0c)^2+\boldsymbol{p}^2}{2m_0c(p_0+m_0c)}\delta_{rr'} = \frac{p_0}{m_0c}\delta_{rr'} \ .$$

It is sufficient to prove the first equation of (2.16) in the rest system as the left hand side is a Lorentz scalar.

The second equation can also be easily verified by transforming it to the rest system where it is certainly true:

$$\begin{aligned}
\sum_r \epsilon_r\omega_\alpha^{(r)}(\boldsymbol{p})\bar{\omega}_\beta^{(r)}(\boldsymbol{p}) &= \sum_{r,\alpha',\beta'}\epsilon_r D_{\alpha\alpha'}(\boldsymbol{p})\omega_{\alpha'}^{(r)}(\boldsymbol{0})\bar{\omega}_{\beta'}^{(r)}(\boldsymbol{0})D^{-1}_{\beta'\beta}(\boldsymbol{p}) \\
&= \sum_{\alpha\alpha'}\delta_{\alpha'\beta'}D_{\alpha\alpha'}(\boldsymbol{p})D^{-1}_{\beta'\beta}(\boldsymbol{p}) = \delta_{\alpha\beta} \ .
\end{aligned}$$

For the third relation there follows similarly

$$\begin{aligned}\sum_r \omega_\alpha^{(r)}(\epsilon_r \boldsymbol{p})\omega_\beta^{(r)\dagger}(\epsilon_r \boldsymbol{p}) &= \sum_{r,\alpha',\beta'} D_{\alpha\alpha'}(\epsilon_r \boldsymbol{p})\omega_{\alpha'}^{(r)}(\mathbf{0})\omega_{\beta'}^{(r)\dagger}(\mathbf{0})D^\dagger_{\beta'\beta}(\epsilon_r \boldsymbol{p}) \\ &= \sum_{r,\alpha',\beta'} \delta_{\alpha'\beta'}\delta_{\alpha' r} D_{\alpha,\alpha'}(\epsilon_r \boldsymbol{p})D^\dagger_{\beta'\beta}(\epsilon_r \boldsymbol{p}) \\ &= \sum_r D_{\alpha r}(\epsilon_r \boldsymbol{p})D_{r\beta}(\epsilon_r \boldsymbol{p}) \\ &= \frac{p_0}{m_0 c}\sum_r U^\dagger_{\alpha r}U_{r\beta} = \frac{p_0}{m_0 c}\delta_{\alpha\beta} ,\end{aligned}$$

with

$$U = \frac{p_0 + m_0 c + \beta\boldsymbol{\alpha p}}{\sqrt{2p_0(p_0+m_0c)}} , \quad U^\dagger = U^{-1} = \frac{p_0 + m_0 c - \beta\boldsymbol{\alpha p}}{\sqrt{2p_0(p_0+m_0c)}} .$$

19. Nonunitarity of bispinor transformations (II). Show the validity of (2.23) for proper Lorentz transformations, i.e.

$$D^\dagger(\Lambda) = \gamma^0 D^{-1}(\Lambda)\gamma^0 ,$$

by using the explicit form of $D(\Lambda)$.

Solution. The most simple solution consists in starting from the infinitesimal representation of $D(\Lambda)$:

$$D = 1 - \frac{\mathrm{i}}{4}\sigma_{\mu\mu}\Delta\omega^{\mu\nu} , \quad \sigma_{\mu\nu} = \frac{\mathrm{i}}{2}[\gamma_\mu, \gamma_\nu] .$$

Then, with

$$\sigma^{\mu\nu\dagger} = \gamma^0\sigma^{\mu\nu}\gamma^0 ,$$

the corresponding adjoint transformation immediately follows as

$$\begin{aligned} D^\dagger &= 1 + \frac{\mathrm{i}}{4}\sigma^\dagger_{\mu\nu}\Delta\omega^{\mu\nu} = 1 + \frac{\mathrm{i}}{4}\gamma^0\sigma_{\mu\nu}\gamma^0\Delta\omega^{\mu\nu} = \gamma^0\left(1 + \frac{\mathrm{i}}{4}\sigma_{\mu\nu}\Delta\omega^{\mu\nu}\right)\gamma^0 \\ &= \gamma^0 D^{-1}\gamma^0 .\end{aligned}$$

20. Free Dirac states under space reflection and time reversal. Verify the relations (2.67) and (2.73).

Solution. For a plane free Dirac wave function with energy sign ϵ, four-momentum index p, and four-polarization index n,

$$\psi_{\epsilon,p,n}(x) = \left(\frac{\epsilon\gamma^\mu p_\mu + m_0 c}{2m_0 c}\right)\left(\frac{1+\gamma^5\gamma^\mu n_\mu}{2}\right)\psi_{\epsilon,p,n}(x) ,$$

and with (2.64) as well as $\{P, \gamma^5\} = 0$, we have

$$\begin{aligned}[\psi_{\epsilon,p,n}]_P(\boldsymbol{x}, t) = P&\left(\frac{\epsilon\gamma^\mu p_\mu + m_0 c}{2m_0 c}\right)P^{-1}P\left(\frac{1+\gamma^5\gamma^\mu n_\mu}{2}\right)P^{-1} \\ &\times P\psi_{\epsilon,p,n}(-\boldsymbol{x}, t)\end{aligned}$$

$$= \left(\frac{\epsilon g^{\mu\mu}\gamma^\mu p_\mu + m_0 c}{2m_0 c}\right)\left(\frac{1-\gamma^5 g^{\mu\mu}\gamma^\mu n_\mu}{2}\right) P\psi_{\epsilon,p,n}(-\boldsymbol{x},t)$$
$$= \left(\frac{\epsilon \gamma^\mu p'_\mu + m_0 c}{2m_0 c}\right)\left(\frac{1+\gamma^5 \gamma^\mu n'_\mu}{2}\right) [\psi_{\epsilon,p,n}]_P(\boldsymbol{x},t) ,$$

where

$$p'^\mu = g^{\mu\mu}p^\mu \ , \ n'^\mu = -g^{\mu\mu}n^\mu \ .$$

On the other hand, using (2.71) and $[T,\gamma^5]=0$, it follows that

$$[\psi_{\epsilon,p,n}]_T(\boldsymbol{x},t) = T\left(\frac{\epsilon\gamma^{*\mu}p_\mu + m_0 c}{2m_0 c}\right)T^{-1}T\left(\frac{1+\gamma^5\gamma^{*\mu}n_\mu}{2}\right)T^{-1}$$
$$\times T\psi^*_{\epsilon,p,n}(\boldsymbol{x},-t)$$
$$= \left(\frac{\epsilon g^{\mu\mu}\gamma^\mu p_\mu + m_0 c}{2m_0 c}\right)\left(\frac{1+\gamma^5 g^{\mu\mu}\gamma^\mu n_\mu}{2}\right) T\psi^*_{\epsilon,p,n}(\boldsymbol{x},-t)$$
$$= \left(\frac{\epsilon \gamma^\mu p'_\mu + m_0 c}{2m_0 c}\right)\left(\frac{1+\gamma^5 \gamma^\mu n'_\mu}{2}\right) [\psi_{\epsilon,p,n}]_T(\boldsymbol{x},t) ,$$

with

$$p'^\mu = g^{\mu\mu}p^\mu \ , \ n'^\mu = g^{\mu\mu}n^\mu \ .$$

21. Expectation values of time-reversed Dirac states. Show the following relations between time-reversed Dirac states:

$$j_T^\mu(x') = j_\mu(x) \ , \ \langle\boldsymbol{x}\rangle_T = \langle\boldsymbol{x}\rangle \ , \ \langle\boldsymbol{p}\rangle_T = -\langle\boldsymbol{p}\rangle \ .$$

Solution. First we note that, due to the properties of the γ-matrices, we have in the Dirac and Weyl representations

$$T = \mathrm{i}\gamma^1\gamma^3 = T^\dagger = T^{-1} \ , \ \psi_T(t') = T\psi^*(t) \ , \ \psi_T^\dagger(t') = \psi^T(t)T^{-1} \ ,$$

where the spatial argument is suppressed. From this and $[T,\gamma^0]=0$ as well as $T^{-1}\gamma^\mu T = \gamma^*_\mu$, the current density follows as

$$j_T^\mu(t') = \bar\psi_T(t')\gamma^\mu\psi_T(t') = \psi^T(t)T^{-1}\gamma^0\gamma^\mu T\psi^*(t)$$
$$= \psi^T(t)\gamma^0 T^{-1}\gamma^\mu T\psi^*(t) = \psi^T(t)\gamma^0\gamma^*_\mu\psi^*(t)$$
$$= \psi_\alpha(t)[\gamma^0\gamma^*_\mu]_{\alpha\beta}\psi^*_\beta(t) = \psi^*_\beta(t)[\gamma^0\gamma^*_\mu]^T_{\beta\alpha}\psi_\alpha(t) = \psi^*(t)\gamma^\dagger_\mu\gamma^0\psi(t)$$
$$= \psi^*(t)\gamma^0\gamma_\mu\gamma^0\gamma^0\psi(t) = \bar\psi(t)\gamma_\mu\psi(t) = j_\mu(t) \ .$$

For the position expectation value we obtain ($[T,\boldsymbol{x}]=0$)

$$\langle\boldsymbol{x}\rangle_T = \int \mathrm{d}^3x\psi_T^\dagger(t')\boldsymbol{x}\psi_T(t') = \int \mathrm{d}^3x\psi^T(t)T^{-1}\boldsymbol{x}T\psi^*(t)$$
$$= \int \mathrm{d}^3x \underbrace{\psi^T(t)\boldsymbol{x}\psi^*(t)}_{\text{real}} = \int \mathrm{d}^3x\psi^\dagger\boldsymbol{x}\psi(t) = \langle\boldsymbol{x}\rangle$$

and for the momentum expectation value ($[T, \boldsymbol{p}] = 0$)

$$\begin{aligned}
\langle \boldsymbol{p} \rangle_T &= \int \mathrm{d}^3x \psi^T(t) T^{-1} \boldsymbol{p} T \psi^*(t) = \int \mathrm{d}^3x \psi^T(t) \boldsymbol{p} \psi^*(t) \\
&= -\mathrm{i}\hbar \int \mathrm{d}^3x [\boldsymbol{\nabla}\psi^\dagger(t)]\psi(t) \\
&= -\mathrm{i}\hbar \int \mathrm{d}^3x \boldsymbol{\nabla}[\psi^\dagger(t)\psi(t)] + \mathrm{i}\hbar \int \mathrm{d}^3x \psi^\dagger(t) \boldsymbol{\nabla}\psi(t) \\
&= \underbrace{-\mathrm{i}\hbar \int_{\partial V} \mathrm{d}\boldsymbol{F} \psi^\dagger(t)\psi(t)}_{0} - \int \mathrm{d}^3x \psi^\dagger(t) \boldsymbol{p} \psi(t) = -\langle \boldsymbol{p} \rangle \ .
\end{aligned}$$

22. Lorentz behavior of the *PCT*-symmetry transformation (II). Similarly to Exercise 3, show the Lorentz-like behavior of the PCT-transformation in the Dirac case by considering the improper and nonorthochronous Lorentz transformation of the Racah time reflection.

Solution. To determine the bispinor transformation belonging to the Racah time reflection, we can use the relation (2.26), i.e.

$$R^{-1}\gamma^\mu R = \Lambda^\mu{}_\nu \gamma^\nu \ , \ (\Lambda^\mu{}_\nu) = \begin{pmatrix} -1 & 0 & 0 & 0 \\ 0 & 1 & 0 & 0 \\ 0 & 0 & 1 & 0 \\ 0 & 0 & 0 & 1 \end{pmatrix} .$$

As can easily be proven, its solution is

$$R = \gamma^1\gamma^2\gamma^3 \Longrightarrow R^{-1} = -\gamma^3\gamma^2\gamma^1 = -\gamma^0 R^\dagger \gamma^0 \ .$$

For the passive and active transformation laws of the Racah time reflection we therefore obtain

$$\left.\begin{aligned}
\boldsymbol{x} &\longrightarrow \boldsymbol{x}' = \boldsymbol{x} \ , \ t \longrightarrow t' = -t \\
\psi(\boldsymbol{x}, t) &\longrightarrow \psi_R(\boldsymbol{x}', t') = R\psi(\boldsymbol{x}, t) \ , \ \begin{array}{l} R = \lambda_R \gamma^1\gamma^2\gamma^3 \\ \lambda_R = \pm 1 \end{array} \\
A^0(\boldsymbol{x}, t) &\longrightarrow A^0_R(\boldsymbol{x}', t') = -A^0(\boldsymbol{x}, t) \\
\boldsymbol{A}(\boldsymbol{x}, t) &\longrightarrow \boldsymbol{A}_R(\boldsymbol{x}', t') = \boldsymbol{A}(\boldsymbol{x}, t)
\end{aligned}\right\} \text{passive time reflection } R$$

and

$$\left.\begin{aligned}
\psi(\boldsymbol{x}, t) &\longrightarrow \psi_R(\boldsymbol{x}, t) = R\psi(\boldsymbol{x}, -t) \\
A^0(\boldsymbol{x}, t) &\longrightarrow A^0_R(\boldsymbol{x}, t) = -A^0(\boldsymbol{x}, -t) \\
\boldsymbol{A}(\boldsymbol{x}, t) &\longrightarrow \boldsymbol{A}_R \boldsymbol{x}, t) = \boldsymbol{A}(\boldsymbol{x}, -t)
\end{aligned}\right\} \text{active time reflection } R.$$

As in Exercise 3, a comparison of the last relations with the active transformation laws of the time reversal T and the charge conjugation $\mathcal{C}$, i.e. (2.72) and (2.74), shows that the Racah time reflection is identical to the combination of $\mathcal{C}$ and T:

$$\mathcal{C}\mathcal{T} = \mathcal{R} \Longrightarrow \mathcal{P}\mathcal{C}\mathcal{T} = \mathcal{P}\mathcal{R} \ .$$

This in turn means that the $\mathcal{PCT}$-transformation is Lorentz-like.

Note that the adjoint bispinor $\bar{\psi}$ transforms under the Racah time reflection as

$$\bar{\psi} \longrightarrow \bar{\psi}_R = \psi^\dagger R^\dagger \gamma^0 = -\psi^\dagger \gamma^0 R^{-1} \gamma^0 \gamma^0 = -\bar{\psi} R^{-1}$$

in accordance with our general result (2.24) for nonorthochronous Lorentz transformations ($b = -1$). Therefore, the current density transforms as a pseudo-vector:

$$j_R^\mu = \bar{\psi}_R \gamma^\mu \psi_R = -\Lambda^\mu{}_\nu \bar{\psi} \gamma^\nu \psi \ .$$

However, since A^μ transforms as a four-vector, the field equation for the *radiation field*, $\partial_\nu \partial^\nu A^\mu = 4\pi e j^\mu$, is not invariant under the Racah time reflection (see Subsection 3.3.2).

2.3 One-Particle Interpretation of the Dirac Theory

As in the Klein-Gordon case in Section 1.3, we now turn back to the one-particle interpretation of the Dirac theory and deal with the clarification of the still open questions from Subsection 2.1.6, namely

[1] what prerequisites are required for the complete decoupling of the Dirac theory into two one-particle theories and

[2] how can physically sensible one-particle operators be constructed, i.e. operators that do not mix positive and negative solutions.

To this end, we first address the second point and then the first one. At the end we discuss the Klein paradox in order to highlight some contradictions – in principle the same as in the Klein-Gordon case – arising outside the validity range of the one-particle concept.

2.3.1 One-Particle Operators and Feshbach-Villars Representation

Based on our considerations in Subsection 1.3.2, we may assume that in the Dirac case, too, not every relativistic operator is physically sensible in relation to the one-particle picture. To see this, we go back to the Ehrenfest theorem (1.42), now, of course, without the index G:

$$\frac{\partial \mathcal{O}}{\partial t} = 0 \Longrightarrow \frac{\mathrm{d}\langle \mathcal{O} \rangle}{\mathrm{d}t} = \frac{1}{\mathrm{i}\hbar} \langle [\mathcal{O}, H] \rangle \ . \tag{2.76}$$

In the free case [$H = H^{(0)}$ from (2.8)] this leads to the "velocity operator"

$$\langle \boldsymbol{v} \rangle = \frac{\langle \mathrm{d}\boldsymbol{x} \rangle}{\mathrm{d}t} = \frac{1}{\mathrm{i}\hbar} \left\langle \left[\boldsymbol{x}, H^{(0)}\right] \right\rangle = \langle c\boldsymbol{\alpha} \rangle \Longrightarrow \boldsymbol{v} = c\boldsymbol{\alpha} \ .$$

Obviously, this operator has no formal similarity with the corresponding classical relation $\boldsymbol{v} = c\boldsymbol{p}/p_0$ as one would expect from the correspondence principle. Furthermore, we have $[\boldsymbol{\alpha}, H^{(0)}] \neq 0$ so that, for free particles, $\boldsymbol{v}$ is not constant. Finally, due to $[\alpha_i, \alpha_{j\neq i}] \neq 0$, the components v_i cannot be simultaneously measured, which seems to be unphysical, too.

The reason for these unsatisfactory findings is related, as in the Klein-Gordon case, to the fact that $\boldsymbol{\alpha}$ turns positive Dirac solutions into negative ones and vice versa, i.e. that $\boldsymbol{\alpha}$ is an odd operator.[11] However, it is clear that, with respect to the one-particle concept, only even operators can be accepted, i.e. one-particle operators that do not mix positive and negative states. From a relativistic operator

$$\mathcal{O} = [\mathcal{O}] + \{\mathcal{O}\} \;,\; [\mathcal{O}] = \text{even} \;,\; \{\mathcal{O}\} = \text{odd}$$

we therefore have to isolate its even part $[\mathcal{O}]$.

Feshbach-Villars representation. For the explicit construction of one-particle operators, we can completely resort to the corresponding discussion in Subsection 1.3.2. This means that we first diagonalize the Hamilton operator in the α-matrix space to move to a representation where $[\mathcal{O}]$ is the diagonal part of $\mathcal{O}$. As in the Klein-Gordon case, it also holds here that an exact diagonalization is possible only for the free case. As before, the corresponding representation is called Feshbach-Villars representation. To get to it we need an appropriate transformation that can be obtained by the following line of argument: in the Schrödinger momentum representation the eigenbasis of the free Dirac-Hamilton operator is $\{\omega^{(r)}(\epsilon_r \boldsymbol{p})\}$ (see Theorem 2.1) with the energy eigenvalues $\epsilon_r c p_0$. They compose an orthogonal system due to (2.15). The inverse of the unitary transformation U mediating between the lastly mentioned and the canonical basis is therefore given by

$$\begin{aligned} U^{-1} &= \sqrt{\frac{m_0 c}{p_0}}\left[\omega^{(1)}(\boldsymbol{p}), \omega^{(2)}(\boldsymbol{p}), \omega^{(3)}(-\boldsymbol{p}), \omega^{(4)}(-\boldsymbol{p})\right] \\ &= \frac{p_0 + m_0 c - \beta\boldsymbol{\alpha}\boldsymbol{p}}{\sqrt{2p_0(p_0 + m_0 c)}} \;. \end{aligned}$$

From this follows that

$$U = U^{-1\dagger} = \frac{p_0 + m_0 c + \beta\boldsymbol{\alpha}\boldsymbol{p}}{\sqrt{2p_0(p_0 + m_0 c)}} \;.$$

With the help of U, we can now perform the transition from the Schrödinger momentum representation to the corresponding Feshbach-Villars representation to find

---

[11] An even operator $\mathcal{O}$ is defined by the relation $\mathcal{O}\psi^{(\pm)} = \psi'^{(\pm)}$, where $\psi^{(\pm)}$ and $\psi'^{(\pm)}$ denote arbitrary positive (+) and negative (−) Dirac solutions respectively. The operator $\mathcal{O}$ is called odd, if $\mathcal{O}\psi^{(\pm)} = \psi'^{(\mp)}$.

$$\tilde{\omega}^{(1)} = U\omega^{(1)}(\boldsymbol{p}) = \begin{pmatrix} 1 \\ 0 \\ 0 \\ 0 \end{pmatrix} , \ \tilde{\omega}^{(2)} = U\omega^{(2)}(\boldsymbol{p}) = \begin{pmatrix} 0 \\ 1 \\ 0 \\ 0 \end{pmatrix}$$

$$\tilde{\omega}^{(3)} = U\omega^{(3)}(-\boldsymbol{p}) = \begin{pmatrix} 0 \\ 0 \\ 1 \\ 0 \end{pmatrix} , \ \tilde{\omega}^{(4)} = U\omega^{(4)}(-\boldsymbol{p}) = \begin{pmatrix} 0 \\ 0 \\ 0 \\ 1 \end{pmatrix}$$

and (see Exercise 23)

$$\tilde{H}^{(0)} = UH^{(0)}U^{-1} = cp_0\beta \ , \ \tilde{H}^{(0)}\tilde{\omega}^{(r)} = \epsilon_r cp_0\tilde{\omega}^{(r)} \tag{2.77}$$

as well as

$$\tilde{\boldsymbol{p}} = U\boldsymbol{p}U^{-1} = UU^{-1}\boldsymbol{p} = \boldsymbol{p} \ .$$

As in the Klein-Gordon case, the operators $H^{(0)}$ and $\boldsymbol{p}$ are even: $H^{(0)} = [H^{(0)}]$, $\boldsymbol{p} = [\boldsymbol{p}]$.

One-particle operators for position and velocity. We now determine the one-particle position operator $[\boldsymbol{x}]$ and the one-particle velocity operator $[\boldsymbol{v}]$ using the well-known transformation and separation scheme

coordinate repres. → momentum repres. → FV-momentum repres. →
→ isolation of the diag. part → momentum repres. → coordinate repres..

The corresponding detailed calculations are presented in Exercise 23.

Position operator in coordinate representation:

$$\boldsymbol{x} = \mathbb{C}\text{-number} \qquad (\boldsymbol{p} = -\mathrm{i}\hbar\boldsymbol{\nabla}) \ .$$

Position operator in momentum representation:

$$\boldsymbol{x} = \mathrm{i}\hbar\boldsymbol{\nabla}_{\boldsymbol{p}} \qquad (\boldsymbol{p} = \mathbb{C}\text{-number}) \ .$$

Position operator in FV-momentum representation:

$$\tilde{\boldsymbol{x}} = U\boldsymbol{x}U^{\dagger} = \mathrm{i}\hbar\boldsymbol{\nabla}_{\boldsymbol{p}} + \mathrm{i}\hbar\left(\frac{\mathrm{i}\hat{\boldsymbol{\sigma}} \times \boldsymbol{p}}{2p_0(p_0 + m_0c)} + \frac{\beta(\boldsymbol{\alpha p})\boldsymbol{p}}{2p_0^2(p_0 + m_0c)} - \frac{\beta\boldsymbol{\alpha}}{2p_0}\right) . \tag{2.78}$$

One-particle position operator in FV-momentum representation:

$$[\tilde{\boldsymbol{x}}] = \mathrm{i}\hbar\boldsymbol{\nabla}_{\boldsymbol{p}} + \mathrm{i}\hbar\frac{\mathrm{i}\hat{\boldsymbol{\sigma}} \times \boldsymbol{p}}{2p_0(p_0 + m_0c)} \ , \ [[\tilde{\boldsymbol{x}}]_i, [\tilde{\boldsymbol{p}}]_j] = \mathrm{i}\hbar\left[\frac{\partial}{\partial p_i}, p_j\right] = \mathrm{i}\hbar\delta_{ij} \ . \tag{2.79}$$

One-particle position operator in momentum representation:

$$[\boldsymbol{x}] = U^{\dagger}[\tilde{\boldsymbol{x}}]U = \mathrm{i}\hbar\boldsymbol{\nabla}_{\boldsymbol{p}} + \mathrm{i}\hbar\left(\frac{\mathrm{i}\hat{\boldsymbol{\sigma}} \times \boldsymbol{p}}{2p_0^2} + \frac{m_0c\beta\boldsymbol{\alpha}}{2p_0^2}\right) \ . \tag{2.80}$$

One-particle position operator in coordinate representation:

$$[\boldsymbol{x}] = \boldsymbol{x} + \mathrm{i}\hbar \left(\frac{\mathrm{i}\hat{\boldsymbol{\sigma}} \times \boldsymbol{p}}{2p_0^2} + \frac{m_0 c\beta \boldsymbol{\alpha}}{2p_0^2} \right) . \tag{2.81}$$

Velocity operator in coordinate or momentum representation:

$$\boldsymbol{v} = c\boldsymbol{\alpha} .$$

Velocity operator in FV-momentum representation:

$$\tilde{\boldsymbol{v}} = U\boldsymbol{v}U^\dagger = c\boldsymbol{\alpha} + \frac{c\beta\boldsymbol{p}}{p_0} - \frac{c\boldsymbol{p}(\boldsymbol{\alpha}\boldsymbol{p})}{p_0(p_0 + m_0 c)} . \tag{2.82}$$

One-particle velocity operator in FV-momentum representation:

$$[\tilde{\boldsymbol{v}}] = \frac{c\beta\boldsymbol{p}}{p_0}$$

or, via Ehrenfest's theorem (2.76):

$$\langle [\tilde{\boldsymbol{v}}] \rangle = \frac{1}{\mathrm{i}\hbar} \left\langle \left[[\tilde{\boldsymbol{x}}], \tilde{H}^{(0)} \right] \right\rangle = \left\langle \frac{c\beta\boldsymbol{p}}{p_0} \right\rangle \Longrightarrow [\tilde{\boldsymbol{v}}] = \frac{c\beta\boldsymbol{p}}{p_0} ,$$

with $[\tilde{\boldsymbol{x}}]$ from (2.79) and $\tilde{H}^{(0)} = [\tilde{H}^{(0)}] = cp_0\beta$.

One-particle velocity operator in momentum or coordinate representation:

$$[\boldsymbol{v}] = U^\dagger [\tilde{\boldsymbol{v}}] U = \frac{m_0 c^2 \beta \boldsymbol{p}}{p_0^2} + \frac{c\boldsymbol{p}(\boldsymbol{\alpha}\boldsymbol{p})}{p_0^2} . \tag{2.83}$$

As in the Klein-Gordon case, the following items must be kept in mind:

- The Feshbach-Villars transformation U is a nonlocal transformation where the transformed wave function $\tilde{\psi}(x)$ results from the original wave function $\psi(x)$ by smearing the position argument $\boldsymbol{x}$ over a region comparable to the Compton wave length of the particle. This can be seen, for example, in the eigenfunctions of the one-particle position operator $[\boldsymbol{x}]$ from (2.81) that are not pure δ-functions any more but possess an extent of size $\sim \hbar/m_0 c$.
- In terms of the one-particle concept, the one-particle velocity operator now seems to be acceptable as the "true velocity operator". In the FV-momentum representation we find for positive Dirac solutions the same relation between $[\tilde{\boldsymbol{v}}]$ and $[\tilde{\boldsymbol{p}}]$ as in relativistic mechanics. For negative solutions this is true only for the absolute values.
- Together with $H^{(0)}$ and $\boldsymbol{p}$, the one-particle velocity operator possesses a common eigenbasis which, in coordinate representation, is given by $\psi_{\boldsymbol{p}}^{(r)}(x)$, with the energy eigenvalues $\epsilon_r c p_0$, momentum eigenvalues $\epsilon_r \boldsymbol{p}$, and the one-particle velocity eigenvalue $c\boldsymbol{p}/p_0$. Accordingly, for negative eigensolutions the eigenvalue (or expectation value) of $\boldsymbol{v}$ is opposite to the eigenvalue (or expectation value) of $\boldsymbol{p}$. This seemingly contradictory behavior is again a consequence of the propagation of negative solutions backward in time (See Theorem 2.7).

Theorem 2.8: One-particle operators and FV-representation in the Dirac theory

With respect to the one-particle interpretation of the Dirac theory, only even Hermitean operators are sensible for the description of physical quantities that do not mix positive and negative Dirac states. In the case of free spin-1/2 particles the corresponding Hamilton operator $H^{(0)}$ can be diagonalized by applying the unitary Feshbach-Villars transformation

$$U = \frac{p_0 + m_0c + \beta\boldsymbol{\alpha}\boldsymbol{p}}{\sqrt{2p_0(p_0 + m_0c)}}$$

leading to the Feshbach-Villars representation. In this representation the even part of an operator can be determined most easily since it is simply given by its diagonal part.
Contrary to $H^{(0)}$ and $\boldsymbol{p}$, the position operator $\boldsymbol{x}$ and the velocity operator $\boldsymbol{v}$ are not even operators. Transforming them into the Feshbach-Villars representation, separating the diagonal parts, and subsequently transforming the latter back to the original representation, one finds the one-particle position operator $[\boldsymbol{x}]$ and the one-particle velocity operator $[\boldsymbol{v}]$ in the usual coordinate or momentum representation to be

$$[\boldsymbol{x}] = \boldsymbol{x} + \mathrm{i}\hbar\left(\frac{\mathrm{i}\boldsymbol{\sigma}\times\boldsymbol{p}}{2p_0^2} + \frac{m_0c\beta\boldsymbol{\alpha}}{2p_0^2}\right) \; , \; [\boldsymbol{v}] = \frac{m_0c^2\beta\boldsymbol{p}}{p_0^2} + \frac{c\boldsymbol{p}(\boldsymbol{\alpha}\boldsymbol{p})}{p_0^2} \; .$$

2.3.2 Validity Range of the One-Particle Concept

So far we have progressed well on our way to a consistent one-particle interpretation of the Dirac theory by giving the positive and the (charge-conjugated) negative Dirac solutions a physically meaningful interpretation and by setting up a formal framework which, with respect to expectation values, is aligned even more closely to the nonrelativistic quantum mechanical formalism as in in the Klein-Gordon case. However, point [1] remains to be clarified, i.e. under which circumstances a complete decoupling of the Dirac theory is possible and, therefore, a meaningful separation of particles and antiparticles in terms of the one-particle concept is feasible.

First of all, the general plausibility arguments in the introduction to Chapter 1 hold, namely that, within the considered physical process, the involved energies must be sufficiently small so that particle changing processes can be ignored. In order to see what additional constraints arise for a complete decoupling of the Dirac theory into two one-particle theories with pure positive or pure negative solutions, we proceed analogously to the Klein-Gordon case in Subsection 1.3.3. This means that we ask again, under which conditions a Dirac wave packet contains (almost) pure positive or negative solutions.

It is clear that a free wave packet which was originally built by solely positive solutions will not develop components with negative energies in the absence of external forces. On the other hand, a wave packet originally localized to a finite area generally encompasses solutions of both energy signs, whereas their ratio will presumably depend on the wave packet's initial localization. For this we consider a resting spin-1/2 particle whose wave packet has the following Gaussian distribution at time $t = 0$:

$$\psi(\boldsymbol{x}, t = 0) = (\pi\Delta^2)^{-3/4} \mathrm{e}^{-\boldsymbol{x}^2/(2\Delta^2)} \omega^{(1)}(\boldsymbol{0}) \; , \; \langle\psi|\,\psi\rangle = 1 \; .$$

A Fourier decomposition of this expression yields

$$\psi(\boldsymbol{x}, t = 0) = \left(\frac{\Delta^2}{\pi\hbar^2}\right)^{3/4} \int \frac{\mathrm{d}^3p}{(2\pi\hbar^2)^{3/2}} \mathrm{e}^{-\boldsymbol{p}^2\Delta^2/(2\hbar^2)} \mathrm{e}^{\mathrm{i}\boldsymbol{p}\boldsymbol{x}/\hbar} \omega^{(1)}(\boldsymbol{0}) \; . \quad (2.84)$$

Comparing this with the general solution for $t = 0$,

$$\psi(\boldsymbol{x}, t = 0) = \int \mathrm{d}^3p \sum_{r=1}^{4} a^{(r)}(\boldsymbol{p}) \psi_{\boldsymbol{p}}^{(r)}(\boldsymbol{x}, t = 0) \; ,$$

it follows that

$$\sqrt{\frac{m_0c}{p_0}} \sum_{r=1}^{4} a^{(r)}(\epsilon_r\boldsymbol{p}) \omega^{(r)}(\epsilon_r\boldsymbol{p}) = \left(\frac{\Delta^2}{\pi\hbar^2}\right)^{3/4} \mathrm{e}^{-\boldsymbol{p}^2\Delta^2/(2\hbar^2)} \omega^{(1)}(\boldsymbol{0}) \; .$$

With the help of (2.15), this becomes

$$a^{(r)}(\epsilon_r\boldsymbol{p}) = \sqrt{\frac{m_0c}{p_0}} \left(\frac{\Delta^2}{\pi\hbar^2}\right)^{3/4} \mathrm{e}^{-\boldsymbol{p}^2\Delta^2/(2\hbar^2)} \omega^{(r)\dagger}(\epsilon_r\boldsymbol{p}) \omega^{(1)}(\boldsymbol{0})$$

so that

$$\left|\frac{a^{(3,4)}(-\boldsymbol{p})}{a^{(1,2)}(\boldsymbol{p})}\right| = \left|\frac{\omega^{(3,4)\dagger}(-\boldsymbol{p})\omega^{(1)}(\boldsymbol{0})}{\omega^{(1,2)\dagger}(\boldsymbol{p})\omega^{(1)}(\boldsymbol{0})}\right| = \frac{|\boldsymbol{p}|}{p_0 + m_0c} \; .$$

Similarly to the Klein-Gordon case, we find that negative solutions contribute significantly to the wave packet for Fourier momenta $|\boldsymbol{p}| \gtrsim m_0c$. According to (2.84), they are suppressed only if

$$\Delta \gg \frac{\hbar}{|\boldsymbol{p}|} \Longrightarrow \Delta \gg \frac{\hbar}{m_0c} = \lambda_{\mathrm{c}} \; .$$

Hence we see that the requirement of a complete decoupling of the Dirac theory leads again to the well-known constraint, namely to wave packets with a large extent compared to the corresponding Compton wave length.

All in all we find that the statements concerning the validity range of the one-particle concept within the Klein-Gordon and Dirac theories (up to the necessity of a positive or a negative definite charge density in the Klein-Gordon case) are identical so that we can dispense here with a theorem corresponding to Theorem 1.7.

2.3.3 Klein Paradox

In the Dirac case, as in the Klein-Gordon case, the Klein paradox is a prime example for highlighting interpretational difficulties of the one-particle concept beyond its range of validity. To illustrate this we consider, analogously to the discussion in Subsection 1.3.4, a onedimensional electron (more generally: spin-1/2 particle) which, coming from the left along the z-axis, is scattered against a potential step of the form

$$eA^0(z) = V(z) = \left\{\begin{array}{l} 0 \text{ for } z < 0 \text{ (area I)} \\ V_0 \text{ for } z > 0 \text{ (area II)} \end{array}\right\} \;,\; V_0 > 0 \;,\; \boldsymbol{A} = \boldsymbol{0}$$

(see Figure 1.3). The stationary energy solutions in area I ($z < 0$) are composed of free incoming and reflected waves for which we choose the ansatz (see Theorem 2.1)

$$\psi_{\mathrm{I}}(z,t) = \mathrm{e}^{-\mathrm{i}Et\hbar}\Psi(z) \;,\; \Psi(z) = \Psi_{\mathrm{ein}}(z) + \Psi_{\mathrm{ref}}(z)$$

$$\Psi_{\mathrm{ein}}(z) = A\mathrm{e}^{\mathrm{i}k_1 z}\begin{pmatrix} 1 \\ 0 \\ \frac{c\hbar k_1}{E+m_0c^2} \\ 0 \end{pmatrix} \;,\; k_1 = \sqrt{\frac{E^2 - m_0^2c^4}{c^2\hbar^2}} \quad \left(\begin{array}{c}\text{rest spin} \\ \text{in } z\text{-direction}\end{array}\right)$$

$$\Psi_{\mathrm{ref}}(z) = B\mathrm{e}^{-\mathrm{i}k_1 z}\begin{pmatrix} 1 \\ 0 \\ \frac{-c\hbar k_1}{E+m_0c^2} \\ 0 \end{pmatrix} + C\mathrm{e}^{-\mathrm{i}k_1 z}\begin{pmatrix} 0 \\ 1 \\ 0 \\ \frac{-c\hbar k_1}{E+m_0c^2} \end{pmatrix} \;,$$

where the time-independent expressions fulfill the free time-independent Dirac equation

$$H\Psi = E\Psi \;,\; H = -\mathrm{i}\hbar c\alpha_3\frac{\mathrm{d}}{\mathrm{d}z} + V(z) + \beta m_0c^2 \tag{2.85}$$

with $V(z) = 0$. Note that for Ψ_{ref} a potentially contributing term with opposite spin is taken into account.

For the transmitted wave we need the solutions to (2.85) in the presence of a constant potential $V(z) = V_0$. They differ from the free solutions only by the substitution $E \to E - V_0$ so that we can write for area II ($z > 0$)

$$\psi_{\mathrm{II}}(z,t) = \mathrm{e}^{-\mathrm{i}Et/\hbar}\Psi_{\mathrm{trans}}(z)$$

$$\Psi_{\mathrm{trans}}(z) = D\mathrm{e}^{\mathrm{i}k_2 z}\begin{pmatrix} 1 \\ 0 \\ \frac{c\hbar k_2}{E-V_0+m_0c^2} \\ 0 \end{pmatrix} + E\mathrm{e}^{\mathrm{i}k_2 z}\begin{pmatrix} 0 \\ 1 \\ 0 \\ \frac{-c\hbar k_2}{E-V_0+m_0c^2} \end{pmatrix} \;,$$

with

$$k_2 = \sqrt{\frac{(E-V_0)^2 - m_0^2c^4}{c^2\hbar^2}} \;.$$

The amplitudes follow from the continuity constraint of the solutions at $z = 0$ due to current conservation. We obtain for them

$$C = E = 0$$

(no spin flip on the level of wave functions) and

$$B = \frac{(1-r)A}{1+r} \, , \; D = \frac{2A}{1+r} \, , \; r = \frac{k_2(E + m_0c^2)}{k_1(E - V_0 + m_0c^2)} \, .$$

Depending on the choice of V_0 or E, we consider the following cases (compare to the cases in Subsection 1.3.4 and to Figure 1.4):

1. case: $E > V_0 + m_0c^2$. The wave number k_2 is real so that in area II the transmitted wave oscillates and $r > 0$ holds. Furthermore, for the current densities of the incoming, reflected, and transmitted parts in z-direction, we have

$$T = \frac{j_{\rm trans}}{j_{\rm in}} = \frac{4r}{(1+r)^2} \, , \; R = -\frac{j_{\rm ref}}{j_{\rm in}} = \frac{(1-r)^2}{(1+r)^2} = 1 - T$$

and thus, in accordance with our expectation, $r > 0 \Longrightarrow 0 < R, T < 1$.

2. case: $V_0 - m_0c^2 < E < V_0 + m_0c^2$, $E > m_0c^2$. Since k_2 is imaginary, the transmitted wave is exponentially damped.

3. case: $m_0c^2 < E < V_0 - m_0c^2 \Longrightarrow V_0 > 2m_0c^2$. As in the 1. case, k_2 is real. On the other hand, we now have $r < 0$, i.e. we obtain a negative transmission current as well as a reflection current whose absolute value is even larger than that of the incoming current.

Similarly to the Klein-Gordon case, the first two cases can be sensibly interpreted within the framework of the one-particle picture as scattering of a particle with charge $+e$ against (from its point of view) repulsive potential barrier. In contrast, the 3. case again seems to be incomprehensible due to the oscillating transmission wave since, for the considered energy interval, the potential step should be inpenetrable. As before, this is caused essentially by too-strong a localization of the particle since a potential step of height $V_0 \approx E$ delimits the penetration depth in area II to an extent of the magnitude $1/k_2 \approx \hbar/m_0c$.

However, even outside the strict one-particle view, the 3. case holds some puzzles. Here we first note that the transmitted wave has a negative energy relative to the potential V_0. Therefore, as in the Klein-Gordon case, it is reasonable to replace k_2 by $-k_2$ so that $\psi_{\rm trans}$ corresponds to an antiparticle of charge $-e$ moving to the right with momentum $+\hbar|k_2|$ which, for its part, is described by $\mathrm{i}\gamma^2\psi^*_{\rm trans}$. This leads to the following consequences: for the reflection and transmission coefficients we now have $0 < R, T < 1$. This means that, contrary to the Klein-Gordon case, the conception of pair production at the border $z = 0$ where particles move to the left and antiparticles move to the right cannot be sustained here. In fact, some of the particles coming from the left are transformed into antiparticles moving to the right. Obviously, this particle transformation leads to a violation of charge conservation, whereas, in the Klein-Gordon case, the total charge is explicitly conserved. Furthermore, the transformation implies a spin flip on the particle level. Due to these facts, we often speak of the Klein super paradox.

As in Subsection 1.3.4, we now give the remaining energy intervals, including their interpretation:

4. case: $-m_0c^2 < E < m_0c^2$. A solution for a particle movement from the left to the right does not exist here.

5. case: $E < -m_0c^2$. If we choose $k_1 = -|k_1|$ and $k_2 = -|k_2|$, this case can again be interpreted within the one-particle picture as an antiparticle with charge $-e$ moving to the right and scattering against the (from its point of view) attractive potential barrier, with $r > 0 \Longrightarrow 0 < R, T < 1$.

Summary

- In terms of the one-particle interpretation, only those Hermitean operators that are **even operators** can be regarded as observables, i.e. those that do not mix positive and negative Dirac solutions (**one-particle operators**). The even part of an operator is determined most easily within a representation where the Hamilton operator is diagonal. In the free case, this is given by the **Feshbach-Villars representation**.
- The **Feshbach-Villars transformation** is a nonlocal transformation. Here the position argument $\boldsymbol{x}$ of a wave function $\psi(x)$ is averaged over a region whose extent is comparable to the Compton wave length of the particle.
- The validity range of the one-particle probabilistic interpretation is limited, on the one hand, to small energies where particle creation processes can be ignored and, on the other hand, to Dirac wave packets whose extent is large compared to the corresponding wave length.
- The **Klein paradox** is a simple example of interpretational difficulties of the one-particle concept stemming from too-strong a localization of Dirac wave packets. Even outside the one-particle picture one encounters contradictions, for example the nonconservation of the total charge or the spin flip on the particle level.

Exercises

23. Feshbach-Villars transformation (II). Verify the relationships (2.77), (2.78), (2.80), (2.82), and (2.83).

Solution. For the subsequent calculations we need:

$$(\boldsymbol{\alpha p})\boldsymbol{\alpha} = \mathrm{i}\hat{\boldsymbol{\sigma}} \times \boldsymbol{p} + \boldsymbol{p}\ ,\ \{\boldsymbol{\alpha}, \boldsymbol{\alpha p}\} = 2\boldsymbol{p}\ ,\ (\boldsymbol{\alpha p})(\boldsymbol{\alpha p}) = \boldsymbol{p}^2$$

$$(\boldsymbol{\alpha p})\boldsymbol{\alpha}(\boldsymbol{\alpha p}) = 2\boldsymbol{p}(\boldsymbol{\alpha p}) - \boldsymbol{\alpha p}^2$$

$$[\hat{\boldsymbol{\sigma}}, \beta] = 0\ ,\ [\hat{\sigma}_i, \alpha_j] = 2\mathrm{i}\epsilon_{ijk}\alpha_k\ ,\ [\hat{\boldsymbol{\sigma}} \times \boldsymbol{p}, \boldsymbol{\alpha p}] = 2\mathrm{i}[\boldsymbol{\alpha p}^2 - (\boldsymbol{\alpha p})\boldsymbol{p}]$$

$$(\boldsymbol{\alpha p})\hat{\boldsymbol{\sigma}} \times \boldsymbol{p}(\boldsymbol{\alpha p}) = -\boldsymbol{p}^2\hat{\boldsymbol{\sigma}} \times \boldsymbol{p}\ .$$

To (2.77).

$$\begin{aligned}
\tilde{H}^{(0)} &= UH^{(0)}U^\dagger \\
&= \frac{(p_0 + m_0c + \beta\boldsymbol{\alpha p})(c\boldsymbol{\alpha p} + \beta m_0c^2)(p_0 + m_0c - \beta\boldsymbol{\alpha p})}{2p_0(p_0 + m_0c)} \\
&= \frac{[cp_0\boldsymbol{\alpha p} + \beta(p_0m_0c^2 + m_0^2c^3 + c\boldsymbol{p}^2)](p_0 + m_0c - \beta\boldsymbol{\alpha p})}{2p_0(p_0 + m_0c)} \\
&= \frac{[c\boldsymbol{\alpha p} + c\beta(p_0 + m_0c)](p_0 + m_0c - \beta\boldsymbol{\alpha p})}{2(p_0 + m_0c)} \\
&= c\beta\frac{(p_0 + m_0c)^2 + \boldsymbol{p}^2}{2(p_0 + m_0c)} = cp_0\beta = \beta\left(m_0^2c^4 + c^2\boldsymbol{p}^2\right)^{1/2}\ .
\end{aligned}$$

To (2.78).

$$\begin{aligned}
(\boldsymbol{\nabla}_{\boldsymbol{p}}U^\dagger) &= \boldsymbol{\nabla}_{\boldsymbol{p}}\frac{p_0 + m_0c - \beta\boldsymbol{\alpha p}}{\sqrt{2p_0(p_0 + m_0c)}} \\
&= \frac{\boldsymbol{p}/p_0 - \beta\boldsymbol{\alpha}}{\sqrt{2p_0(p_0 + m_0c)}} - U^\dagger\frac{(2p_0 + m_0c)\boldsymbol{p}}{2p_0^2(p_0 + m_0c)} \\
U(\boldsymbol{\nabla}_{\boldsymbol{p}}U^\dagger) &= \frac{(p_0 + m_0c + \beta\boldsymbol{\alpha p})(\boldsymbol{p}/p_0 - \beta\boldsymbol{\alpha})}{2p_0(p_0 + m_0c)} - \frac{(2p_0 + m_0c)\boldsymbol{p}}{2p_0^2(p_0 + m_0c)} \\
&= \frac{(\boldsymbol{\alpha p})\boldsymbol{\alpha}}{2p_0(p_0 + m_0c)} + \frac{\beta(\boldsymbol{\alpha p})\boldsymbol{p}}{2p_0^2(p_0 + m_0c)} - \frac{\beta\boldsymbol{\alpha}}{2p_0} - \frac{\boldsymbol{p}}{2p_0(p_0 + m_0c)} \\
&= \frac{\mathrm{i}\hat{\boldsymbol{\sigma}} \times \boldsymbol{p}}{2p_0(p_0 + m_0c)} + \frac{\beta(\boldsymbol{\alpha p})\boldsymbol{p}}{2p_0^2(p_0 + m_0c)} - \frac{\beta\boldsymbol{\alpha}}{2p_0}
\end{aligned}$$

$$\begin{aligned}
\Longrightarrow \tilde{\boldsymbol{x}} &= U\boldsymbol{x}U^\dagger = \mathrm{i}\hbar\boldsymbol{\nabla}_{\boldsymbol{p}} + \mathrm{i}\hbar U(\boldsymbol{\nabla}_{\boldsymbol{p}}U^\dagger) \\
&= \mathrm{i}\hbar\boldsymbol{\nabla}_{\boldsymbol{p}} + \mathrm{i}\hbar\left(\frac{\mathrm{i}\hat{\boldsymbol{\sigma}} \times \boldsymbol{p}}{2p_0(p_0 + m_0c)} + \frac{\beta(\boldsymbol{\alpha p})\boldsymbol{p}}{2p_0^2(p_0 + m_0c)} - \frac{\beta\boldsymbol{\alpha}}{2p_0}\right)\ .
\end{aligned}$$

To (2.80).

$$\begin{aligned}
[\boldsymbol{x}] &= U^\dagger[\tilde{\boldsymbol{x}}]U = U^\dagger\boldsymbol{x}U + \mathrm{i}\hbar U^\dagger\frac{\mathrm{i}\hat{\boldsymbol{\sigma}} \times \boldsymbol{p}}{2p_0(p_0 + m_0c)}U \\
U^\dagger\boldsymbol{x}U &= \mathrm{i}\hbar\boldsymbol{\nabla}_{\boldsymbol{p}} + \mathrm{i}\hbar\left(\frac{\mathrm{i}\hat{\boldsymbol{\sigma}} \times \boldsymbol{p}}{2p_0(p_0 + m_0c)} - \frac{\beta(\boldsymbol{\alpha p})\boldsymbol{p}}{2p_0^2(p_0 + m_0c)} + \frac{\beta\boldsymbol{\alpha}}{2p_0}\right)
\end{aligned}$$

$$
\begin{aligned}
U^\dagger \frac{\mathrm{i}\hat{\boldsymbol{\sigma}} \times \boldsymbol{p}}{2p_0(p_0+m_0c)} U &= \frac{(p_0+m_0c-\beta\boldsymbol{\alpha p})\mathrm{i}\hat{\boldsymbol{\sigma}} \times \boldsymbol{p}(p_0+m_0c+\beta\boldsymbol{\alpha p})}{4p_0^2(p_0+m_0c)^2} \\
&= \mathrm{i}\frac{(p_0+m_0c)^2\hat{\boldsymbol{\sigma}} \times \boldsymbol{p} + (p_0+m_0c)\beta[\hat{\boldsymbol{\sigma}} \times \boldsymbol{p}, \boldsymbol{\alpha p}]}{4p_0^2(p_0+m_0c)^2} \\
&\quad + \mathrm{i}\frac{(\boldsymbol{\alpha p})\hat{\boldsymbol{\sigma}} \times \boldsymbol{p}(\boldsymbol{\alpha p})}{4p_0^2(p_0+m_0c)^2} \\
&= \frac{\mathrm{i}m_0c\hat{\boldsymbol{\sigma}} \times \boldsymbol{p}}{2p_0^2(p_0+m_0c)} - \frac{\boldsymbol{p}^2\beta\boldsymbol{\alpha}}{2p_0^2(p_0+m_0c)} + \frac{\beta(\boldsymbol{\alpha p})\boldsymbol{p}}{2p_0^2(p_0+m_0c)}
\end{aligned}
$$

$$
\Longrightarrow [\boldsymbol{x}] = \mathrm{i}\hbar\boldsymbol{\nabla}_{\boldsymbol{p}} + \mathrm{i}\hbar\left(\frac{\mathrm{i}\hat{\boldsymbol{\sigma}} \times \boldsymbol{p}}{2p_0^2} + \frac{m_0c\beta\boldsymbol{\alpha}}{2p_0^2}\right) .
$$

To (2.82).

$$
\begin{aligned}
\tilde{\boldsymbol{v}} = U\boldsymbol{v}U^\dagger &= \frac{(p_0+m_0c+\beta\boldsymbol{\alpha p})c\boldsymbol{\alpha}(p_0+m_0c-\beta\boldsymbol{\alpha p})}{2p_0(p_0+m_0c)} \\
&= \frac{c\boldsymbol{\alpha}(p_0+m_0c)^2 - c(\boldsymbol{\alpha p})\boldsymbol{\alpha}(\boldsymbol{\alpha p}) + c\beta\{\boldsymbol{\alpha}, \boldsymbol{\alpha p}\}(p_0+m_0c)}{2p_0(p_0+m_0c)} \\
&= c\boldsymbol{\alpha} + \frac{c\beta\boldsymbol{p}}{p_0} - \frac{c\boldsymbol{p}(\boldsymbol{\alpha p})}{p_0(p_0+m_0c)} .
\end{aligned}
$$

To (2.83).

$$
\begin{aligned}
[\boldsymbol{v}] = U^\dagger[\tilde{\boldsymbol{v}}]U &= \frac{(p_0+m_0c-\beta\boldsymbol{\alpha p})c\beta\boldsymbol{p}(p_0+m_0c+\beta\boldsymbol{\alpha p})}{2p_0^2(p_0+m_0c)} \\
&= \frac{m_0c^2\beta\boldsymbol{p}}{p_0^2} + \frac{c\boldsymbol{p}(\boldsymbol{\alpha p})}{p_0^2} .
\end{aligned}
$$

24. Construction of one-particle operators using the sign operator (II). Instead of using the Feshbach-Villars representation, Dirac one-particle operators can be constructed with less effort by taking into account that the Hermitean sign operator

$$
\Lambda = \frac{H^{(0)}}{\sqrt{H^{(0)2}}} = \frac{\boldsymbol{\alpha p} + m_0c\beta}{p_0}
$$

possesses the eigenfunctions $\psi_{\boldsymbol{p}}^{(r)}(x)$ with the eigenvalues (energy signs) ϵ_r (compare to Exercise 6). Construct the one-particle operators for position and velocity, $[\boldsymbol{x}]$ and $[\boldsymbol{v}]$, by exploiting this fact.

Solution. Based on the same argument as in Exercise 6, the even part $[\mathcal{O}]$ and odd part $\{\mathcal{O}\}$ of an operator $\mathcal{O}$ are given by

$$
[\mathcal{O}] = \frac{1}{2}(\mathcal{O} + \Lambda\mathcal{O}\Lambda) , \ \{\mathcal{O}\} = \frac{1}{2}(\mathcal{O} - \Lambda\mathcal{O}\Lambda) ,
$$

where $\Lambda\mathcal{O}\Lambda$ and, therefore, $[\mathcal{O}]$ and $\{\mathcal{O}\}$ are Hermitean, if $\mathcal{O}$ is Hermitean. Now we calculate in the momentum representation ($\boldsymbol{x} = \mathrm{i}\hbar\boldsymbol{\nabla}_{\boldsymbol{p}}$, $\boldsymbol{p} = \mathbb{C}$-number) as follows:

$$\begin{aligned}
(\nabla_{\boldsymbol{p}}\Lambda) &= \nabla_{\boldsymbol{p}}\frac{\boldsymbol{\alpha p}+m_0c\beta}{p_0} = \frac{\boldsymbol{\alpha}}{p_0} - \Lambda\frac{\boldsymbol{p}}{p_0^2} \\
\Lambda(\nabla_{\boldsymbol{p}}\Lambda) &= \frac{(\boldsymbol{\alpha p}+m_0c\beta)\boldsymbol{\alpha}}{p_0^2} - \frac{\boldsymbol{p}}{p_0^2} = \frac{\mathrm{i}\hat{\boldsymbol{\sigma}}\times\boldsymbol{p}}{p_0^2} + \frac{m_0c\beta\boldsymbol{\alpha}}{p_0^2} \\
\Longrightarrow \Lambda\boldsymbol{x}\Lambda &= \mathrm{i}\hbar\nabla_{\boldsymbol{p}} + \mathrm{i}\hbar\Lambda(\nabla_{\boldsymbol{p}}\Lambda) = \mathrm{i}\hbar\nabla_{\boldsymbol{p}} + \frac{\mathrm{i}\hat{\boldsymbol{\sigma}}\times\boldsymbol{p}}{p_0^2} + \frac{m_0c\beta\boldsymbol{\alpha}}{p_0^2}\ .
\end{aligned}$$

In momentum or position representation the one-particle position operator follows as (compare to Theorem 2.8)

$$[\boldsymbol{x}] = \frac{1}{2}(\boldsymbol{x}+\Lambda\boldsymbol{x}\Lambda) = \boldsymbol{x} + \mathrm{i}\hbar\left(\frac{\mathrm{i}\hat{\boldsymbol{\sigma}}\times\boldsymbol{p}}{2p_0^2} + \frac{m_0c\beta\boldsymbol{\alpha}}{2p_0^2}\right)\ .$$

For the one-particle velocity operator we obtain from a similar calculation (compare to Theorem 2.8)

$$\begin{aligned}
\Lambda\boldsymbol{v}\Lambda &= \frac{(\boldsymbol{\alpha p}+m_0c\beta)c\boldsymbol{\alpha}(\boldsymbol{\alpha p}+m_0c\beta)}{p_0^2} \\
&= \frac{c(\boldsymbol{\alpha p})\boldsymbol{\alpha}(\boldsymbol{\alpha p}) + m_0c^2\beta\{\boldsymbol{\alpha},\boldsymbol{\alpha p}\} - m_0^2c^3\boldsymbol{\alpha}}{p_0^2} \\
&= \frac{2m_0c^2\beta\boldsymbol{p}}{p_0^2} + \frac{2c\boldsymbol{p}(\boldsymbol{\alpha p})}{p_0^2} - c\boldsymbol{\alpha} \\
\Longrightarrow [\boldsymbol{v}] &= \frac{1}{2}(\boldsymbol{v}+\Lambda\boldsymbol{v}\Lambda) = \frac{m_0c^2\beta\boldsymbol{p}}{p_0^2} + \frac{c\boldsymbol{p}(\boldsymbol{\alpha p})}{p_0^2}\ .
\end{aligned}$$

25. Gordon decomposition. Show that for two arbitrary solutions ψ_1 and ψ_2 to the free Dirac equation the following holds:

$$\bar{\psi}_2\gamma^\mu\psi_1 = \frac{1}{2m_0c}\left[\bar{\psi}_2p^\mu\psi_1 - (p^\mu\bar{\psi}_2)\psi_1\right] - \frac{\mathrm{i}}{2m_0c}p_\nu(\bar{\psi}_2\sigma^{\mu\nu}\psi_1)\ . \tag{2.86}$$

Solution. For two arbitrary four-vectors a^μ and b^μ we have

$$\begin{aligned}
\gamma^\mu a_\mu\gamma^\nu b_\nu &= a_\mu b_\nu\left[\frac{1}{2}(\gamma^\mu\gamma^\nu+\gamma^\nu\gamma^\mu) + \frac{1}{2}(\gamma^\mu\gamma^\nu-\gamma^\nu\gamma^\mu)\right] \\
&= a_\mu b_\nu\left(\frac{1}{2}\{\gamma^\mu,\gamma^\nu\} + \frac{1}{2}[\gamma^\mu,\gamma^\nu]\right) \\
&= a^\mu b_\mu - \mathrm{i}a_\mu b_\nu\sigma^{\mu\nu}\ .
\end{aligned}$$

This and using the free Dirac equation as well as its adjoint ($\overleftarrow{p_\mu}$ acts to the left) yield

$$0 = \bar{\psi}_2(-\gamma^\mu\overleftarrow{p_\mu} - m_0c)\gamma^\nu a_\nu\psi_1 + \bar{\psi}_2\gamma^\nu a_\nu(\gamma^\mu p_\mu - m_0c)\psi_1$$

and

$$\begin{aligned}
2m_0c\bar{\psi}_2\gamma^\nu a_\nu\psi_1 &= -\bar{\psi}_2\gamma^\mu\overleftarrow{p_\mu}\gamma^\nu a_\nu\psi_1 + \bar{\psi}_2\gamma^\nu a_\nu\gamma^\mu p_\mu\psi_1 \\
&= -\bar{\psi}_2\left(\overleftarrow{p^\mu}a_\mu + \mathrm{i}\overleftarrow{p_\nu}a_\mu\sigma^{\mu\nu}\right)\psi_1 + \bar{\psi}_2\left(p^\mu a_\mu - \mathrm{i}p_\nu a_\mu\sigma^{\mu\nu}\right)\ .
\end{aligned}$$

With $a_\mu = \delta_{\mu\rho}$ there follows (2.86). Physically, the Gordon decomposition separates the Dirac probability current density $j^\mu = c\bar{\psi}\gamma^\mu\psi$ into a *convection current density*

$$j^\mu_{\mathrm{K}} = \frac{1}{2m_0}\left[\bar{\psi}p^\mu\psi - (p^\mu\bar{\psi})\psi\right] \tag{2.87}$$

(similar to the nonrelativistic probability current density or the Klein-Gordon charge current density) and a *spin current density*

$$j^\mu_{\mathrm{S}} = -\frac{\mathrm{i}}{2m_0}p_\nu(\bar{\psi}\sigma^{\mu\nu}\psi) \ . \tag{2.88}$$

26. Shaky movement (II). Calculate the mean current of an arbitrary free Dirac wave packet similarly to Exercise 7. Show that the interference terms of positive and negative solutions contain a temporally oscillating movement.

Solution. Our starting point is the wave packet

$$\begin{aligned}
\psi(x) &= \psi^{(+)}(x) + \psi^{(-)}(x) \\
\psi^{(+)}(x) &= \int \mathrm{d}^3p \sum_{r=1}^{2} a^{(r)}(\boldsymbol{p})\psi^{(r)}_{\boldsymbol{p}}(x) \\
\psi^{(-)}(x) &= \int \mathrm{d}^3p \sum_{r=3}^{4} a^{(r)}(\boldsymbol{p})\psi^{(r)}_{\boldsymbol{p}}(x) \ .
\end{aligned}$$

To calculate the mean spatial convection current (2.87) we use the identity

$$(\boldsymbol{p}\bar{\psi})\psi = -(\boldsymbol{p}\psi)^\dagger\gamma^0\psi = -\psi^T\gamma^{0,T}(\boldsymbol{p}\psi)^* = -(\psi^\dagger\gamma^0\boldsymbol{p}\psi)^* = -(\bar{\psi}\boldsymbol{p}\psi)^*$$

from which follows that

$$\boldsymbol{j}_{\mathrm{K}} = \frac{1}{m_0}\mathrm{Re}\left(\bar{\psi}\boldsymbol{p}\psi\right) \ .$$

Exploiting the adjunction relation $\langle\phi|\,\boldsymbol{A}\,|\psi\rangle = \langle\psi|\,\boldsymbol{A}^\dagger\,|\phi\rangle^*$, we further have

$$\begin{aligned}
\langle\boldsymbol{j}\rangle_{\mathrm{K}} &= \frac{1}{m_0}\left\langle\psi\right|\gamma^0\boldsymbol{p}\left|\psi\right\rangle \\
&= \frac{1}{m_0}\left\langle\psi^{(+)}+\psi^{(-)}\right|\gamma^0\boldsymbol{p}\left|\psi^{(+)}+\psi^{(-)}\right\rangle \\
&= \frac{1}{m_0}\left[\left\langle\psi^{(+)}\right|\gamma^0\boldsymbol{p}\left|\psi^{(+)}\right\rangle + \left\langle\psi^{(-)}\right|\gamma^0\boldsymbol{p}\left|\psi^{(-)}\right\rangle\right. \\
&\quad \left. +\left\langle\psi^{(+)}\right|\gamma^0\boldsymbol{p}\left|\psi^{(-)}\right\rangle + \left\langle\psi^{(-)}\right|\gamma^0\boldsymbol{p}\left|\psi^{(+)}\right\rangle\right] \\
&= \frac{1}{m_0}\left[\left\langle\psi^{(+)}\right|\gamma^0\boldsymbol{p}\left|\psi^{(+)}\right\rangle + \left\langle\psi^{(-)}\right|\gamma^0\boldsymbol{p}\left|\psi^{(-)}\right\rangle\right] \\
&\quad +\frac{2}{m_0}\mathrm{Re}\left(\left\langle\psi^{(+)}\right|\gamma^0\boldsymbol{p}\left|\psi^{(-)}\right\rangle\right)
\end{aligned}$$

$$= \underbrace{\int d^3p \frac{c\boldsymbol{p}}{p_0} \sum_{r=1}^{2} \left|a^{(r)}(\boldsymbol{p})\right|^2}_{\langle \boldsymbol{j}\rangle_{\mathrm{K}}^{(+)}} + \underbrace{\int d^3p \frac{c\boldsymbol{p}}{p_0} \sum_{r=3}^{4} \left|a^{(r)}(\boldsymbol{p})\right|^2}_{\langle \boldsymbol{j}\rangle_{\mathrm{K}}^{(-)}}$$

$$+2\mathrm{Re}\left[\int d^3p \frac{c\boldsymbol{p}}{p_0} \mathrm{e}^{2\mathrm{i}p_0x^0/\hbar} \times \sum_{\substack{r=1,2\\ r'=3,4}} a^{(r)*}(\boldsymbol{p})a^{(r')}(-\boldsymbol{p})\bar{\omega}^{(r)}(\boldsymbol{p})\omega^{(r')}(-\boldsymbol{p})\right] . \tag{2.89}$$

For the evaluation of the mean spatial spin current

$$\langle j^k\rangle_{\mathrm{S}} = -\frac{\mathrm{i}}{2m_0}\int d^3x p_\nu \left(\bar{\psi}^{(+)} + \bar{\psi}^{(-)}\right)\sigma^{k\nu}\left(\psi^{(+)} + \psi^{(-)}\right)$$

[see (2.88)], we have to take into account that the individual terms $\int d^3x p_\nu \left(\bar{\psi}^{(\pm)}\sigma^{k\nu}\psi^{(\pm)}\right)$ lead to integrals of the form

$$\int d^3p \int d^3p'(p'_\nu - p_\nu)\delta(\boldsymbol{p}' - \boldsymbol{p})\ldots$$

and hence do not contribute. On the other hand, the interference terms $\int d^3x p_\nu \left(\bar{\psi}^{(\pm)}\sigma^{k\nu}\psi^{(\mp)}\right)$ yield the integrals

$$\int d^3p \int d^3p'(p'_\nu + p_\nu)\delta(\boldsymbol{p}' + \boldsymbol{p})\ldots \begin{cases} \neq 0 & \text{for } \nu = 0 \\ = 0 & \text{else} . \end{cases}$$

Therefore, it remains that

$$\langle j^k\rangle_{\mathrm{S}} = -\frac{\mathrm{i}}{2m_0}\int d^3x \left[p_0\left(\bar{\psi}^{(+)}\sigma^{k0}\psi^{(-)}\right) + p_0\left(\bar{\psi}^{(-)}\sigma^{k0}\psi^{(+)}\right)\right] .$$

Finally, this expression can be further simplified using

$$\begin{aligned} p_0\left(\bar{\psi}^{(-)}\sigma^{k0}\psi^{(+)}\right) &= p_0\left(\psi^{(-)\dagger}\gamma^0\sigma^{k0}\psi^{(+)}\right) = p_0\left(\psi^{(+)T}\sigma^{k0,T}\gamma^{0,T}\psi^{(-)*}\right) \\ &= p_0\left(\psi^{(+)\dagger}\sigma^{k0\dagger}\gamma^0\psi^{(-)}\right)^* = p_0\left(\bar{\psi}^{(+)}\gamma^0\sigma^{k0\dagger}\gamma^0\psi^{(-)}\right)^* \\ &= p_0\left(\bar{\psi}^{(+)}\sigma^{k0}\psi^{(-)}\right)^* = -\left[p_0\left(\bar{\psi}^{(+)}\sigma^{k0}\psi^{(-)}\right)\right]^* \end{aligned}$$

to become

$$\begin{aligned} \langle j^k\rangle_{\mathrm{S}} &= \frac{1}{m_0}\mathrm{Im}\left[\int d^3x p_0\left(\bar{\psi}^{(+)}\sigma^{k0}\psi^{(-)}\right)\right] \\ &= 2\mathrm{Im}\left[c\int d^3p \mathrm{e}^{2\mathrm{i}p_0x^0/\hbar} \times \sum_{\substack{r=1,2\\ r'=3,4}} a^{(r)*}(\boldsymbol{p})a^{(r')}(-\boldsymbol{p})\bar{\omega}^{(r)}(\boldsymbol{p})\sigma^{k0}\omega^{(r')}(-\boldsymbol{p})\right] . \end{aligned} \tag{2.90}$$

With (2.89) and (2.90), one realizes that the movement of a Dirac wave packet contains a temporally oscillating movement (shaky movement) if and only if it encompasses positive and negative components. This fact is already known from the corresponding considerations of the Klein-Gordon case (Exercise 7).

2.4 Nonrelativistic Approximation of the Dirac Theory

In our previous discussion of the Dirac theory we have not yet addressed one important question, namely whether it is possible to derive the correct equations of nonrelativistic quantum mechanics from it. Of course, this should be the case for the Dirac theory to be an acceptable relativistic enhancement. In this section we focus on this issue closely following the discussion of the Klein-Gordon case in Section 1.4. First we consider the nonrelativistic limit to leading order of v/c which will lead us to the nonrelativistic *Pauli equation* for spin-1/2 particles. Subsequently, we make use of the Fouldy-Wouthuysen transformation to include higher relativistic corrections, i.e. to diagonalize the Hamilton operator in higher orders of v/c. All considerations in this section refer to the Dirac representation.

2.4.1 Nonrelativistic Limit

The starting point for our discussion is the Dirac equation (2.14) in canonical form,

$$\mathrm{i}\hbar\frac{\partial\psi(x)}{\partial t} = \left[c\boldsymbol{\alpha}\left(\boldsymbol{p}-\frac{e}{c}\boldsymbol{A}\right)+eA^0+\beta m_0c^2\right]\psi(x)\ . \tag{2.91}$$

In order to study its nonrelativistic limit, it is advantageous to introduce the two-component spinors

$$\psi_{\mathrm{u}} = \begin{pmatrix}\psi_1\\ \psi_2\end{pmatrix}\ ,\ \psi_{\mathrm{d}} = \begin{pmatrix}\psi_3\\ \psi_4\end{pmatrix}\ ,\ \psi = \begin{pmatrix}\psi_{\mathrm{u}}\\ \psi_{\mathrm{d}}\end{pmatrix}$$

so that (2.91) can be transformed into the equation system

$$\left.\begin{aligned}\mathrm{i}\hbar\frac{\partial\psi_{\mathrm{u}}}{\partial t} &= c\boldsymbol{\sigma}\left(\boldsymbol{p}-\frac{e}{c}\boldsymbol{A}\right)\psi_{\mathrm{d}}+(eA^0+m_0c^2)\psi_{\mathrm{u}}\\ \mathrm{i}\hbar\frac{\partial\psi_{\mathrm{d}}}{\partial t} &= c\boldsymbol{\sigma}\left(\boldsymbol{p}-\frac{e}{c}\boldsymbol{A}\right)\psi_{\mathrm{u}}+(eA^0-m_0c^2)\psi_{\mathrm{d}}\end{aligned}\right\}\ , \tag{2.92}$$

where the index u stands for "up" (the two upper components) and d for "down" (the two lower components). Taking into account

$$\left(\mathrm{i}\hbar\frac{\partial}{\partial t}-eA^0\right)\psi^{(\pm)}_{\mathrm{u,d}} = m_0c^2\left[\pm 1+\mathcal{O}\left(\frac{v^2}{c^2}\right)\right]\psi^{(\pm)}_{\mathrm{u,d}}\ ,$$

the second equation of (2.92) yields for positive solutions (+)

$$\psi_{\mathrm{d}}^{(+)} = \frac{\boldsymbol{\sigma}}{2m_0c}\left(\boldsymbol{p} - \frac{e}{c}\boldsymbol{A}\right)\psi_{\mathrm{u}}^{(+)} + \mathcal{O}\left(\frac{v^2}{c^2}\right) , \tag{2.93}$$

whereas for negative solutions (−), the first equation leads to

$$\psi_{\mathrm{u}}^{(-)} = -\frac{\boldsymbol{\sigma}}{2m_0c}\left(\boldsymbol{p} - \frac{e}{c}\boldsymbol{A}\right)\psi_{\mathrm{d}}^{(-)} + \mathcal{O}\left(\frac{v^2}{c^2}\right) . \tag{2.94}$$

This means that in the case of positive solutions the spinor ψ_{d} relative to ψ_{u}, and in the case of negative solutions, the spinor ψ_{u} relative to ψ_{d} is suppressed by a factor of v/c. Inserting (2.93) or (2.94) into the remaining equation of (2.92), we have for positive solutions

$$\begin{aligned} \psi &= \begin{pmatrix} 1 \\ \mathcal{O}(v/c) \end{pmatrix}\psi_{\mathrm{u}} \\ \mathrm{i}\hbar\frac{\partial\psi_{\mathrm{u}}}{\partial t} &= \left\{\frac{1}{2m_0}\left[\boldsymbol{\sigma}\left(\boldsymbol{p} - \frac{e}{c}\boldsymbol{A}\right)\right]^2 + m_0c^2 + eA^0 + \mathcal{O}\left(\frac{v^3}{c^3}\right)\right\}\psi_{\mathrm{u}} \end{aligned}$$

and for negative solutions

$$\begin{aligned} \psi &= \begin{pmatrix} \mathcal{O}(v/c) \\ 1 \end{pmatrix}\psi_{\mathrm{d}} \\ \mathrm{i}\hbar\frac{\partial\psi_{\mathrm{d}}}{\partial t} &= \left\{-\frac{1}{2m_0}\left[\boldsymbol{\sigma}\left(\boldsymbol{p} - \frac{e}{c}\boldsymbol{A}\right)\right]^2 - m_0c^2 + eA^0 + \mathcal{O}\left(\frac{v^3}{c^3}\right)\right\}\psi_{\mathrm{d}} . \end{aligned}$$

With the help of the identities

$$(\boldsymbol{\sigma A})(\boldsymbol{\sigma B}) = (\boldsymbol{AB}) + \mathrm{i}\boldsymbol{\sigma}(\boldsymbol{A}\times\boldsymbol{B}) , \ \left(\boldsymbol{p} - \frac{e}{c}\boldsymbol{A}\right)\times\left(\boldsymbol{p} - \frac{e}{c}\boldsymbol{A}\right) = -\frac{e\hbar}{\mathrm{i}c}\boldsymbol{B} ,$$

these relations can finally be summarized in the Dirac equation

$$\left.\begin{aligned} \mathrm{i}\hbar\frac{\partial\psi}{\partial t} &= H^{\mathrm{nr}}\psi \\ H^{\mathrm{nr}} &= \beta\left[m_0c^2 + \frac{1}{2m_0}\left(\boldsymbol{p} - \frac{e}{c}\boldsymbol{A}\right)^2 - \frac{e\hbar}{2m_0c}\hat{\boldsymbol{\sigma}}\boldsymbol{B}\right] + eA^0 \\ &\quad + \mathcal{O}\left(\frac{v^3}{c^3}\right) , \ \hat{\boldsymbol{\sigma}} = \begin{pmatrix} \boldsymbol{\sigma} & 0 \\ 0 & \boldsymbol{\sigma} \end{pmatrix} , \end{aligned}\right\} \tag{2.95}$$

correct up to order $\mathcal{O}\left(v^2/c^2\right)$, with the diagonal and Hermitean Hamilton operator H^{nr} [compare to the nonrelativistic Klein-Gordon equation (1.57)]. If we restrict ourselves to positive solutions, i.e. to the two upper components, this equation is – up to the rest energy m_0c^2 – identical to the nonrelativistic *Pauli equation* for spin-1/2 particles within an electromagnetic field. Especially noteworthy is the fact that the nonrelativistic limiting process of the Dirac equation automatically leads to an interaction term $-\boldsymbol{MB}$ between the magnetic moment (or spin) of the particle and the external magnetic field which, in the case of the elementary electron, has the correct magnetic moment or, rather, the correct *gyromagnetic ratio*

$$M^{(e)} = \frac{e\hbar}{2m_0c}\boldsymbol{\sigma} = \frac{eg}{2m_0c}\boldsymbol{S} \;,\; g = 2 \quad (\textit{Landé factor}) \;.$$

By contrast, in the Pauli equation this term has to be introduced by hand. Historically it was this feature that gave an important impetus for the confidence in the Dirac theory.

For nonelementary particles, such as protons or neutrons, the above limiting process leads to the wrong results $\boldsymbol{M}^{(p)} = -e\boldsymbol{S}/(m_pc)$ (proton) and $\boldsymbol{M}^{(n)} = \boldsymbol{0}$ (neutron). Obviously, in those cases the minimal coupling is not sufficient for taking external electromagnetic fields into account. However, even for those particles, we can obtain the corresponding nonrelativistic equations with the correct magnetic moments by adding phenomenologically motivated terms (see Exercises 15 and 27).

For the sake of completeness, we also note the expressions of the probability density and the probability current density belonging to (2.95) and correct up to order $\mathcal{O}\left(v^2/c^2\right)$:

$$\rho = \psi^\dagger\psi \;,\; \boldsymbol{j} = \frac{\hbar}{2\mathrm{i}m}\left[\psi^\dagger\beta\boldsymbol{\nabla}\psi - (\boldsymbol{\nabla}\psi^\dagger)\beta\psi - \frac{2\mathrm{i}e}{\hbar c}\boldsymbol{A}\psi^\dagger\beta\psi\right] \;.$$

They are connected via the continuity equation $\partial\rho/\partial t + \boldsymbol{\nabla}\boldsymbol{j} = 0$ and, in the case of positive solutions, coincide with the corresponding formulae of the nonrelativistic theory.

2.4.2 Relativistic Corrections

In the previous subsection the reduction of the Dirac theory to the nonrelativistic Pauli theory is correct up to order $\mathcal{O}\left(v^2/c^2\right)$, and the error of the Hamilton operator in (2.95) is of order $\mathcal{O}\left(v^3/c^3\right)$. In this limit H^{nr} is diagonal, and the positive and negative solutions are completely decoupled. In order to diagonalize the Hamilton operator in higher orders systematically, i.e. to take higher relativistic corrections into account, we can, as in the Klein-Gordon case, utilize the Fouldy-Wouthuysen method which we now apply to the general Dirac equation (2.91). Here the same considerations hold as in Subsection 1.4.2, especially that an exact diagonalization of the Dirac theory is not possible due to vacuum polarization effects.

In order to simplify the treatment of orders in v/c, we start our discussion by rewriting the Dirac equation (2.91) in the form

$$m_0c^2K\psi = 0 \;,\; K = \beta + \epsilon + \omega \;,$$

with

$$\epsilon = -\frac{1}{m_0c^2}\left(\mathrm{i}\hbar\frac{\partial}{\partial t} - eA^0\right) = \mathcal{O}(1) + \mathcal{O}\left(\frac{v^2}{c^2}\right) \;,\; \beta + \epsilon = \mathcal{O}\left(\frac{v^2}{c^2}\right)$$

and

$$\omega = \frac{c\boldsymbol{\alpha}}{m_0c^2}\left(\boldsymbol{p} - \frac{e}{c}\boldsymbol{A}\right) = \mathcal{O}\left(\frac{v}{c}\right) \;,$$

where ϵ and $\beta+\epsilon$ are even (diagonal) and ω odd (antidiagonal) operators. With the help of appropriately chosen Fouldy-Wouthuysen transformations $U = e^{iS}$, $U' = e^{iS'}$, ... we aim to change to new representations where ω is of higher and higher order of v/c so that disregarding it leads to a diagonal K operator correct up to the respective order of v/c. So, after the first transformation, we should have

$$m_0c^2K'\psi' = 0\ ,\ \psi' = U\psi\ ,\ K' = UKU^{-1}$$

$$K' = \beta+\epsilon'+\omega'\ ,\ \beta+\epsilon' = \mathcal{O}\left(\frac{v^2}{c^2}\right)\ ,\ \omega' = \mathcal{O}\left(\frac{v^3}{c^3}\right)\ \text{(or higher)}\ ,$$

and after the second one

$$m_0c^2K''\psi'' = 0\ ,\ \psi'' = U'\psi\ ,\ K'' = U'K'U'^{-1}$$

$$K'' = \beta+\epsilon''+\omega''\ ,\ \beta+\epsilon'' = \mathcal{O}\left(\frac{v^2}{c^2}\right)\ ,\ \omega'' = \mathcal{O}\left(\frac{v^5}{c^5}\right)\ \text{(or higher)}\ ,$$

and so forth. As in (1.61), a good choice for the first transformation is

$$U = e^{iS}\ ,\ S = -\frac{i\beta\omega}{2}\ . \tag{2.96}$$

We can again make use of the Baker-Hausdorff expansion (1.60) as well as of the formulae (1.62) with the replacement $\tau_3 \to \beta$ to calculate the resulting K'. This yields

$$K' = \beta+\epsilon'+\omega'\ ,$$

with

$$\epsilon' = \underbrace{\epsilon}_{\mathcal{O}\left(\frac{v^2}{c^2}\right)} \underbrace{+\frac{\beta\omega^2}{2}}_{\mathcal{O}\left(\frac{v^2}{c^2}\right)} \underbrace{-\frac{\beta\omega^4}{8}}_{\mathcal{O}\left(\frac{v^4}{c^4}\right)} \underbrace{-\frac{1}{8}[\omega,[\omega,\epsilon]]}_{\mathcal{O}\left(\frac{v^4}{c^4}\right)} + \ldots = \mathcal{O}\left(\frac{v^2}{c^2}\right)$$

and

$$\omega' = -\frac{\omega^3}{3}+\frac{\beta}{2}[\omega,\epsilon]+\frac{\beta}{48}[\omega,[\omega,[\omega,\epsilon]]]+\ldots = \mathcal{O}\left(\frac{v^3}{c^3}\right)\ .$$

As one can see, ω' is now raised by two orders of v/c. From this we obtain the even operator $K' = \beta+\epsilon$ correct up to order $\mathcal{O}\left(v^2/c^2\right)$ resulting in the Pauli equation (2.95).

In order to suppress the odd part of the K-operators even further, we perform a second Fouldy-Wouthuysen transformation on K' with

$$U' = e^{iS'}\ ,\ S' = -\frac{i\beta\omega'}{2}\ .$$

From this follows that

$$K'' = \beta+\epsilon''+\omega''\ ,$$

with

$$\begin{array}{ccccccccc} & \mathcal{O}\left(\frac{v^2}{c^2}\right) & \mathcal{O}\left(\frac{v^6}{c^6}\right) & \mathcal{O}\left(\frac{v^{12}}{c^{12}}\right) & \mathcal{O}\left(\frac{v^8}{c^8}\right) & \\ & \downarrow & \downarrow & \downarrow & \downarrow & \\ \epsilon'' = & \epsilon' & +\frac{\beta\omega'^2}{2} & -\frac{\beta\omega'^4}{8} & -\frac{1}{8}[\omega',[\omega',\epsilon']] & +\ldots = \mathcal{O}\left(\frac{v^2}{c^2}\right) \end{array}$$

and

$$\omega'' = -\frac{\omega'^3}{3} + \frac{\beta}{2}[\omega',\epsilon'] + \frac{\beta}{48}[\omega',[\omega',[\omega',\epsilon']]] + \ldots = \mathcal{O}\left(\frac{v^5}{c^5}\right) .$$

Disregarding all terms of order $\mathcal{O}\left(v^5/c^5\right)$ (and higher), we obtain the even operator

$$K'' = \beta + \epsilon + \frac{\beta\omega^2}{2} - \frac{\beta\omega^4}{8} - \frac{1}{8}[\omega,[\omega,\epsilon]] + \mathcal{O}\left(\frac{v^5}{c^5}\right) \tag{2.97}$$

finally leading to the Dirac equation

$$\mathrm{i}\hbar\frac{\partial\psi''}{\partial t} = H''\psi''$$

correct up to order $\mathcal{O}\left(v^4/c^4\right)$, with the diagonal and Hermitean Hamilton operator (see Exercise 28)

$$\begin{aligned} H'' = \beta & \left[m_0c^2 + \frac{1}{2m_0}\left(\boldsymbol{p} - \frac{e}{c}\boldsymbol{A}\right)^2 - \frac{e\hbar}{2m_0c}\hat{\boldsymbol{\sigma}}\boldsymbol{B}\right] + eA^0 \\ & -\beta\left[\frac{1}{8m_0^3c^2}\left(\boldsymbol{p} - \frac{e}{c}\boldsymbol{A}\right)^4 + \frac{e^2\hbar^2}{8m_0^3c^4}\boldsymbol{B}^2\right. \\ & \left. - \frac{e\hbar}{8m_0^3c^3}\left\{\hat{\boldsymbol{\sigma}}\boldsymbol{B}, \left(\boldsymbol{p} - \frac{e}{c}\boldsymbol{A}\right)^2\right\}\right] \\ & -\frac{e\hbar^2}{8m_0^2c^2}\boldsymbol{\nabla}\boldsymbol{E} - \frac{\mathrm{i}e\hbar^2}{8m_0^2c^2}\hat{\boldsymbol{\sigma}}(\boldsymbol{\nabla}\times\boldsymbol{E}) - \frac{e\hbar}{4m_0^2c^2}\hat{\boldsymbol{\sigma}}(\boldsymbol{E}\times\boldsymbol{p}) \\ & +\mathcal{O}\left(\frac{v^5}{c^5}\right) \end{aligned} \tag{2.98}$$

and the wave function

$$\psi''(x) = \mathrm{e}^{-\mathrm{i}\beta\omega'/2}\mathrm{e}^{-\mathrm{i}\beta\omega/2}\psi(x) .$$

All in all, we see that the successive diagonalization of the Dirac-Hamilton operator to higher orders of v/c can be performed quite similarly to the Klein-Gordon case in Subsection 1.4.2. However, as before, we should bear in mind the following:

- Since S, S', ... are Hermitean, all Fouldy-Wouthuysen transformations U, U', ... of the form (2.96) are unitary transformations. This implies the invariance of expectation values that transform as $U[\cdot]U^{-1}$.

- For the Dirac-Hamilton operator, this is only true if $\partial \boldsymbol{A}/\partial t = \boldsymbol{0}$, since the transition
$$K\psi = 0 \longrightarrow K'\psi' = 0 \ , \ K' = UKU^{-1} = UKU^{\dagger} \ , \ \psi' = U\psi$$
is equivalent to
$$\mathrm{i}\hbar\frac{\partial\psi}{\partial t} = H\psi \longrightarrow \mathrm{i}\hbar\frac{\partial\psi'}{\partial t} = H'\psi' \ , \ H' = U\left(H - \mathrm{i}\hbar\frac{\partial}{\partial t}\right)U^{\dagger} \ .$$

- One-particle operators are obtained within a Fouldy-Wouthuysen representation by accordingly transforming the original (relativistic) operators and subsequently separating their diagonal parts. As in the Klein-Gordon case, $[\boldsymbol{x}, U] \neq 0$, i.e. the Fouldy-Wouthuysen method is nonlocal and causes a smearing of the coordinate wave function with an extent comparable to the Compton wave length of the particle.

- The Fouldy-Wouthuysen method is only applicable to physical problems within the validity range of the one-particle picture where the Fouldy-Wouthuysen expansion converges.

Theorem 2.9: Fouldy-Wouthuysen transformation in the Dirac Theory

The Fouldy-Wouthuysen transformation provides a systematic procedure for diagonalizing the Dirac-Hamilton operator up to any (finite) order of v/c. Writing the Dirac equation (2.91) in the form
$$m_0c^2K^{(0)}\psi^{(0)} = 0 \ , \ K^{(0)} = \beta + \epsilon^{(0)} + \omega^{(0)} \ ,$$
with the dimensionless even operators $\epsilon^{(0)}$, $\beta + \epsilon^{(0)} = \mathcal{O}\left(v^2/c^2\right)$, the odd operator $\omega^{(0)} = \mathcal{O}\left(v/c\right)$, and iterating the relations according to
$$\begin{aligned} K^{(n)} &= \beta + \epsilon^{(n)} + \omega^{(n)} = U^{(n-1)}K^{(n-1)}U^{(n-1)\dagger} \\ \psi^{(n)}(x) &= U^{(n-1)}\psi^{(n-1)}(x) \\ U^{(n)} &= \exp\left(-\frac{\mathrm{i}\beta\omega^{(n)}}{2}\right) \qquad \text{(unitary)} \ , \end{aligned}$$
we obtain new representations of the Dirac theory where
$$\beta + \epsilon^{(n)} = \mathcal{O}\left(\frac{v^2}{c^2}\right) \ , \ \omega^{(n)} = \mathcal{O}\left(\frac{v^{2n+1}}{c^{2n+1}}\right) \ .$$
Disregarding the odd operator, the even part of $K^{(n)}$ leads to two explicitly decoupled one-particle theories for particle and antiparticle, correct up to order $\mathcal{O}\left(v^{2n-1}/c^{2n-1}\right)$.

Electron in an electrostatic central potential. Let us, at the end of this section, turn back to (2.98). This equation can be cast into a well-established form by considering the case of an electron within a centrally symmetric potential:

$$eA^0 = V(|\boldsymbol{x}|) = V(r) \; , \; \boldsymbol{A} = \boldsymbol{0} \; .$$

In this case we have

$$\boldsymbol{B} = \boldsymbol{0} \; , \; \boldsymbol{E} = -\boldsymbol{\nabla} A^0 = -\frac{1}{e}\frac{\boldsymbol{x}}{r}\frac{\partial V}{\partial r} \; , \; \boldsymbol{\nabla} \times \boldsymbol{E} = \boldsymbol{0} \; .$$

Restricting ourselves to the upper two components, the corresponding Hamilton operator is

$$H''_{\mathrm{u}} = m_0 c^2 + \frac{\boldsymbol{p}^2}{2m_0} + V(r) - \frac{\boldsymbol{p}^4}{8m_0^3 c^2} + \frac{\hbar^2}{8m_0^2 c^2}\boldsymbol{\nabla}^2 V + \frac{\hbar}{4m_0^2 c^2}\frac{1}{r}\frac{\partial V}{\partial r}\boldsymbol{\sigma}\boldsymbol{L} \; .$$

The fourth term on the right hand side is a relativistic correction to the kinetic energy. The fifth term is a relativistic correction to the central potential known as *Darwin term* and can be accredited to the shaky movement of the electron. The last term contains the interaction energy between the electron's spin (or magnetic moment) and its orbital angular momentum (spin-orbit coupling). Note that, in this term, the *Thomas precession* is taken into account correctly by a factor of 4 in the denominator.[12] In the case of a Coulomb potential $V(r) = -Ze^2/r$, the last two terms are

$$\frac{\pi Z e^2 \hbar^2}{2m_0^2 c^2}\delta(r) \quad \text{and} \quad \frac{Ze^2\hbar}{4m_0^2 c^2 r^3}\boldsymbol{\sigma}\boldsymbol{L} \; .$$

Here the Darwin term only influences the s-states.

Summary

- To lowest order (**nonrelativistic limit**) the nonrelativistic approximation of the Dirac theory leads to a diagonal and Hermitean Hamilton operator. From this follow two explicitly decoupled one-particle theories for particle and antiparticle, the former being identical to the nonrelativistic **Pauli equation** for spin-1/2 particles.
- Generally and contrary to the field-free case, the Dirac-Hamilton operator is diagonalizable only approximately. This can be achieved by using the **Fouldy-Wouthuysen method** where the Hamilton operator is dia- ▷

[12] In nonrelativistic quantum mechanics this term is explained classically as follows: in the rest system of the electron the force center produces a magnetic field at the electron's position that interacts with its spin. However, since this argument disregards (among other things) the nonuniform motion of the electron, the term turns out to be too large by a factor of 2.

gonalized successively to higher and higher orders of v/c. The respective even part yields a diagonal and Hermitean Hamilton operator correct up to the considered order of v/c, from which two explicitly decoupled one-particle theories for particle and antiparticle can be derived.

- The **Fouldy-Wouthuysen transformation** is, like the Feshbach-Villars transformation, a nonlocal transformation and leads to a smearing of the position argument comparable to the Compton wave length.
- The Fouldy-Wouthuysen method is reasonable only for those cases where, on the one hand, the v/c-expansion converges and, on the other hand, the one-particle interpretation is applicable.

Exercises

27. Anomalous magnetic moment of structured particles. Show that adding the term

$$\frac{\hbar\delta\beta}{4m_0c}\sigma_{\mu\nu}F^{\mu\nu} \ , \ F^{\mu\nu} = \partial^\mu A^\nu - \partial^\nu A^\mu$$

to (2.91) yields an equation in the nonrelativistic limit that describes a particle with the magnetic moment

$$\boldsymbol{M}^{(i)} = \frac{\hbar(e_i + \delta)}{2m_ic}\boldsymbol{\sigma} \ ,$$

where e_i and m_i denote the particle's charge and rest mass respectively.

Solution. Starting with (2.91), incorporating the above term, and repeating the line of argument resulting in (2.95), one obtains the modified Pauli equation

$$\mathrm{i}\hbar\frac{\partial\psi}{\partial t} = \left\{\beta\left[m_ic^2 + \frac{1}{2m_i}\left(\boldsymbol{p} - \frac{e_i}{c}\boldsymbol{A}\right)^2 - \frac{e_i\hbar}{2m_ic}\hat{\boldsymbol{\sigma}}\boldsymbol{B} + \frac{\hbar\delta}{4m_ic}\sigma_{\mu\nu}F^{\mu\nu}\right] + e_iA^0\right\}\psi \ . \tag{2.99}$$

The second last term of this equation was already calculated in Exercise 15 and is

$$\sigma_{\mu\nu}F^{\mu\nu} = 2(\mathrm{i}\boldsymbol{\alpha}\boldsymbol{E} - \hat{\boldsymbol{\sigma}}\boldsymbol{B}) \ .$$

Thus, the third and second last terms in (2.99) lead to the above mentioned magnetic moment (plus electric terms suppressed by a factor of v/c).

For the elementary electron as well as for the composed proton and neutron experiments determine the following values of δ (e =electron charge):

$$\begin{aligned}
\text{electron}: \ \delta = 0 \quad & \Longrightarrow \boldsymbol{M}^{(e)} = \frac{e\hbar}{2m_e c}\boldsymbol{\sigma} \\
\text{proton}: \ \delta \approx 3.79e \quad & \Longrightarrow \boldsymbol{M}^{(p)} \approx \frac{\hbar(-e+3.79e)}{2m_p c}\boldsymbol{\sigma} = \frac{2.79e\hbar}{2m_p c}\boldsymbol{\sigma} \\
\text{neutron}: \ \delta \approx -1.91e & \Longrightarrow \boldsymbol{M}^{(n)} \approx \frac{\hbar(0-1.91e)}{2m_n c}\boldsymbol{\sigma} = \frac{-1.91e\hbar}{2m_n c}\boldsymbol{\sigma}\ .
\end{aligned}$$

28. Fouldy-Wouthuysen transformation. Show the transition from (2.97) to (2.98).

Solution.

$$\begin{aligned}
\omega^2 &= \frac{1}{m_0^2c^2}\left[\boldsymbol{\alpha}\left(\boldsymbol{p}-\frac{e}{c}\boldsymbol{A}\right)\right]\left[\boldsymbol{\alpha}\left(\boldsymbol{p}-\frac{e}{c}\boldsymbol{A}\right)\right] \\
&= \frac{1}{m_0^2c^2}\sum_{i,j}\alpha_i\alpha_j\left(p_i-\frac{e}{c}A_i\right)\left(p_j-\frac{e}{c}A_j\right) \\
&= \frac{\mathrm{i}}{m_0^2c^2}\sum_{i,j,k}\epsilon_{ijk}\hat{\sigma}_k\left(p_i-\frac{e}{c}A_i\right)\left(p_j-\frac{e}{c}A_j\right)+\frac{1}{m_0^2c^2}\left(\boldsymbol{p}-\frac{e}{c}\boldsymbol{A}\right)^2 \\
&= -\frac{\mathrm{i}e}{m_0c^3}\hat{\boldsymbol{\sigma}}(\boldsymbol{p}\times\boldsymbol{A})+\frac{1}{m_0^2c^2}\left(\boldsymbol{p}-\frac{e}{c}\boldsymbol{A}\right)^2 \\
&= -\frac{e\hbar}{m_0^2c^3}\hat{\boldsymbol{\sigma}}\boldsymbol{B}+\frac{1}{m_0^2c^2}\left(\boldsymbol{p}-\frac{e}{c}\boldsymbol{A}\right)^2 \\
[\omega,\epsilon] &= -\frac{1}{m_0^2c^3}\left[\boldsymbol{\alpha}\left(\boldsymbol{p}-\frac{e}{c}\boldsymbol{A}\right), \mathrm{i}\hbar\frac{\partial}{\partial t}-eA^0\right] \\
&= \frac{1}{m_0^2c^3}\left\{e[\boldsymbol{\alpha p},A^0]+\frac{\mathrm{i}e\hbar}{c}\left[\boldsymbol{A},\frac{\partial}{\partial t}\right]\right\} \\
&= -\frac{\mathrm{i}e\hbar}{m_0^2c^3}\boldsymbol{\alpha}\left(\boldsymbol{\nabla}A^0+\frac{1}{c}\dot{\boldsymbol{A}}\right)=\frac{\mathrm{i}e\hbar}{m_0^2c^3}\boldsymbol{\alpha E} \\
[\omega,[\omega,\epsilon]] &= \frac{\mathrm{i}e\hbar}{m_0^3c^4}\left[\boldsymbol{\alpha}\left(\boldsymbol{p}-\frac{e}{c}\boldsymbol{A}\right),\boldsymbol{\alpha E}\right] \\
&= \frac{\mathrm{i}e\hbar}{m_0^3c^4}[\boldsymbol{\alpha p},\boldsymbol{\alpha E}] \\
&= \frac{\mathrm{i}e\hbar}{m_0^3c^4}\sum_{i,j}\alpha_i\alpha_j(p_iE_j-E_ip_j) \\
&= \frac{\mathrm{i}e\hbar}{m_0^3c^4}\sum_{i,j}\{\alpha_i\alpha_j(p_iE_j)+[\alpha_i,\alpha_j]E_jp_i\} \\
&= \frac{\mathrm{i}e\hbar}{m_0^3c^4}\left\{\sum_{i,j,k}(\mathrm{i}\epsilon_{ijk}\hat{\sigma}_k+\delta_{ij})(p_iE_j)+\sum_{i,j}2\mathrm{i}\epsilon_{ijk}\hat{\sigma}_kE_jp_i\right\} \\
&= \frac{\mathrm{i}e\hbar^2}{m_0^3c^4}\hat{\boldsymbol{\sigma}}(\boldsymbol{\nabla}\times\boldsymbol{E})+\frac{e\hbar^2}{m_0^3c^4}\boldsymbol{\nabla E}+\frac{2e\hbar}{m_0^3c^4}\hat{\boldsymbol{\sigma}}(\boldsymbol{E}\times\boldsymbol{p})\ .
\end{aligned}$$

Here it was made use of the identities

$$\alpha_i \alpha_j = \mathrm{i}\epsilon_{ijk}\hat{\sigma}_k + \delta_{ij} \ , \ [\alpha_i, \alpha_j] = 2\mathrm{i}\epsilon_{ijk}\hat{\sigma}_k \ .$$

2.5 Simple One-Particle Systems

In order to complete the parallelism of the discussions of the Klein-Gordon and Dirac cases, we finish this chapter with some simple examples of Dirac one-particle systems proceeding in analogy to the Klein-Gordon case, Section 1.5. First we extend our considerations of the Klein paradox from Subsection 2.3.3 to the case of a onedimensional potential well. Subsequently we turn to the problem of centrally symmetric potentials that, as in the Klein-Gordon theory, can be transformed into a radial problem by separating the angular-dependent part. Concrete examples are the free particle, the centrally symmetric potential well, and, finally, the Coulomb potential. As before, all considerations refer to the Dirac representation.

2.5.1 Potential Well

To start, we consider a spin-1/2 particle in the presence of a onedimensional potential well of the form

$$eA^{(0)}(z) = V(z) = \left\{ \begin{array}{lll} 0 \ \text{for} \ -a < z < a & \text{(area II)} \\ V_0 \quad \text{else} & \text{(area I,III)} \end{array} \right\}, \ V_0 > 0. \quad (2.100)$$

With respect to the qualitative discussion of the possible solutions, we can completely carry over the statements from Subsection 1.5.1 as well as Figure 1.5. For a more detailed analysis, we first separate the time-dependent part of the Dirac wave function via

$$\psi(z,t) = \Psi(z)\mathrm{e}^{-\mathrm{i}Et/\hbar}$$

so that $\Psi(z)$ fulfills the stationary equation (2.85) with $V(z)$ from (2.100). Next we concentrate on the scattering cases (1., 3., and 5. case), then the tunneling case (4. case), and, finally, the bound case (2. case).

1., 3., and 5. case in detail. For these scattering cases we assume a (anti)particle coming from the left along the z axis and scattering against the potential well. Therefore, our ansatz for the solution of (2.85) within the areas I ($z < -a$), II ($-a < z < a$), and III ($z > a$) is (rest spin in z-direction)

$$\Psi_{\mathrm{I}}(z) = \Psi_{\mathrm{in}}(z) + \Psi_{\mathrm{ref}}(z)$$

$$\Psi_{\mathrm{in}}(z) = A\mathrm{e}^{\mathrm{i}k_1 z}\begin{pmatrix} 1 \\ 0 \\ \lambda_1 \\ 0 \end{pmatrix} \ , \ \Psi_{\mathrm{ref}}(z) = B\mathrm{e}^{-\mathrm{i}k_1 z}\begin{pmatrix} 1 \\ 0 \\ -\lambda_1 \\ 0 \end{pmatrix}$$

$$\Psi_{\mathrm{II}}(z) = C\mathrm{e}^{\mathrm{i}k_2 z}\begin{pmatrix}1\\0\\\lambda_2\\0\end{pmatrix} + D\mathrm{e}^{-\mathrm{i}k_2 z}\begin{pmatrix}1\\0\\-\lambda_2\\0\end{pmatrix}$$

$$\Psi_{\mathrm{III}}(z) = \Psi_{\mathrm{trans}}(z) = E\mathrm{e}^{\mathrm{i}k_1 z}\begin{pmatrix}1\\0\\\lambda_1\\0\end{pmatrix} ,$$

with

$$k_1 = \pm\sqrt{\frac{(E-V_0)^2 - m_0^2c^4}{\hbar^2c^2}} \ , \ k_2 = \pm\sqrt{\frac{E^2 - m_0^2c^4}{\hbar^2c^2}}$$

$$\lambda_1 = \frac{c\hbar k_1}{E - V_0 + m_0c^2} \ , \ \lambda_2 = \frac{c\hbar k_2}{E + m_0c^2} ,$$

and $k_1 = +|k_1|, k_2 = +|k_2|$ in the 1. case, $k_1 = -|k_1|, k_2 = +|k_2|$ in the 3. case, and $k_1 = -|k_1|, k_2 = -|k_2|$ in the 5. case. This approach takes into account our findings in Subsection 2.3.3, namely that, at the area borders $z = \pm a$, there is no spin flip on the level of wave functions. The continuity conditions $\Psi_{\mathrm{I}}(-a) = \Psi_{\mathrm{II}}(-a)$ and $\Psi_{\mathrm{II}}(a) = \Psi_{\mathrm{III}}(a)$ lead to the following conditional equations for the integration constants A to E:

$$\begin{aligned}
A\mathrm{e}^{-\mathrm{i}k_1 a} + B\mathrm{e}^{\mathrm{i}k_1 a} &= C\mathrm{e}^{-\mathrm{i}k_1 a} + D\mathrm{e}^{\mathrm{i}k_1 a} \\
\lambda_1\left(A\mathrm{e}^{-\mathrm{i}k_1 a} - B^{\mathrm{i}k_1 a}\right) &= \lambda_2\left(C\mathrm{e}^{-\mathrm{i}k_2 a} - D\mathrm{e}^{\mathrm{i}k_2 a}\right) \\
C\mathrm{e}^{\mathrm{i}k_2 a} + D\mathrm{e}^{-\mathrm{i}k_2 a} &= E\mathrm{e}^{\mathrm{i}k_1 a} \\
\lambda_2\left(C\mathrm{e}^{\mathrm{i}k_2 a} - D\mathrm{e}^{-\mathrm{i}k_2 a}\right) &= \lambda_1 E\mathrm{e}^{\mathrm{i}k_1 a} .
\end{aligned}$$

They are formally identical to the corresponding equations (1.68) of the Klein-Gordon case with the prefactor replacement $k_i \to \lambda_i$. Therefore, the same replacement in (1.69) immediately leads to the reflection and transmission coefficients

$$\left.\begin{aligned}
R &= -\frac{j_{\mathrm{ref}}}{j_{\mathrm{in}}} = \frac{(\lambda_1^2-\lambda_2^2)^2\sin^2 2k_2a}{4\lambda_1^2\lambda_2^2 + (\lambda_1^2-\lambda_2^2)^2\sin^2 2k_2a} \\
T &= \frac{j_{\mathrm{trans}}}{j_{\mathrm{in}}} = \frac{4\lambda_1^2\lambda_2^2}{4\lambda_1^2\lambda_2^2 + (\lambda_1^2-\lambda_2^2)^2\sin^2 2k_2a} = 1 - R .
\end{aligned}\right\} \quad (2.101)$$

As in the Klein-Gordon case, both coefficients oscillate between 0 and 1 depending on k_2 or E. For $\sin 2k_2a = 0$, i.e. for

$$E^2 = n^2\frac{c^2\hbar^2\pi^2}{4a^2} + m_0^2c^4 \ , \ n = 1, 2, \ldots ,$$

the reflection coefficient vanishes exactly.

4. case in detail. For this tunneling case the above ansatz can be adopted, whereas k_1, k_2, λ_1, and λ_2 should be chosen as

$$\begin{aligned} k_1 &= -\sqrt{\frac{(E-V_0)^2 - m_0^2c^4}{\hbar^2c^2}} \ , \ k_2 = \mathrm{i}\kappa_2 \ , \ \kappa_2 = \sqrt{\frac{m_0^2c^4 - E^2}{\hbar^2c^2}} \\ \lambda_1 &= \frac{c\hbar k_1}{E - V_0 + m_0c^2} \ , \ \lambda_2 = \mathrm{i}\xi_2 \ , \ \xi_2 = \frac{c\hbar\kappa_2}{E + m_0c^2} \ . \end{aligned}$$

The corresponding reflection and transmission coefficients follow from (2.101) as

$$\begin{aligned} R &= \frac{(\kappa_1^2 + \xi_2^2)^2 \sinh^2 2\kappa_2 a}{4\kappa_1^2\xi_2^2 + (\kappa_1^2 + \xi_2^2)^2 \sinh^2 2\kappa_2 a} \\ T &= \frac{4\kappa_1^2\xi_2^2}{4\kappa_1^2\xi_2^2 + (\kappa_1^2 + \xi_2^2)^2 \sinh^2 2\kappa_2 a} = 1 - R \ . \end{aligned}$$

As before, both coefficients lie between 0 and 1, whereas the transmission coefficient is now decreasing exponentially with a and increasing exponentially with E.

2. case in detail. As in the corresponding 2. Klein-Gordon case, we choose the bound wave functions to be

$$\begin{aligned} \Psi_{\mathrm{I}}(z) &= A\mathrm{e}^{\kappa_1 z}\begin{pmatrix} 1 \\ 0 \\ -\mathrm{i}\xi_1 \\ 0 \end{pmatrix} \\ \Psi_{\mathrm{II}}(z) &= \begin{pmatrix} B\cos k_2 z + C\sin k_2 z \\ 0 \\ \mathrm{i}\lambda_2(B\sin k_2 z - C\cos k_2 z) \\ 0 \end{pmatrix} \\ \Psi_{\mathrm{III}}(z) &= D\mathrm{e}^{-\kappa_1 z}\begin{pmatrix} 1 \\ 0 \\ \mathrm{i}\xi_1 \\ 0 \end{pmatrix} \ , \end{aligned}$$

with

$$\begin{aligned} \kappa_1 &= \sqrt{\frac{m_0^2c^4 - (E-V_0)^2}{\hbar^2c^2}} \ , \ k_2 = \sqrt{\frac{E^2 - m_0^2c^4}{\hbar^2c^2}} \\ \xi_1 &= \frac{c\hbar\kappa_1}{E - V_0 + m_0c^2} \ , \ \lambda_2 = \frac{c\hbar k_2}{E + m_0c^2} \ . \end{aligned}$$

In this case the continuity conditions at the area borders lead to

$$\begin{aligned} A\mathrm{e}^{-\kappa_1 a} &= B\cos k_2 a - C\sin k_2 a \\ \xi_1 A\mathrm{e}^{-\kappa_1 a} &= \lambda_2 B\sin k_2 a + \lambda_2 C\cos k_2 a \\ D\mathrm{e}^{-\kappa_1 a} &= B\cos k_2 a + C\sin k_2 a \\ \xi_1 D\mathrm{e}^{-\kappa_1 a} &= \lambda_2 B\sin k_2 a - \lambda_2 C\cos k_2 a \ . \end{aligned}$$

Combining the first two and the last two equations, one obtains

$$\xi_1 = \lambda_2 \frac{B \sin k_2 a + C \cos k_2 a}{B \cos k_2 a - C \sin k_2 a} = \lambda_2 \frac{B \sin k_2 a - C \cos k_2 a}{B \cos k_2 a + C \sin k_2 a} ,$$

from which the condition $BC = 0$ follows. Thus, we have to distinguish again the following two cases leading to different quantization conditions for the energy E:

2.a: $C = 0 \Longrightarrow A = D$.

$$\tan k_2 a = \frac{\xi_1}{\lambda_2} .$$

2.b: $B = 0 \Longrightarrow A = -D$.

$$-\cot k_2 a = \tan\left(k_2 a + \frac{\pi}{2}\right) = \frac{\xi_1}{\lambda_2} .$$

Analogously to footnote 26 on page 65, we point out that, due to the form of $V(z)$, the Hamilton operator in (2.85) commutes with the parity transformation of Subsection 2.2.3. Up to an irrelevant constant, we have in the active case

$$[H(z)\Psi(z)]_P = \gamma^0 H(-z)\Psi(-z) = H(z)\gamma^0\Psi(-z) = H(z)\Psi_P(z) .$$

Thus, the wave function $\Psi(z)$ as well as its parity transformed $\Psi_P(z)$ are both solutions to (2.85) with the same energy eigenvalue:

$$E\Psi(z) = H(z)\Psi(z) \Longrightarrow E\Psi_P(z) = [H(z)\Psi(z)]_P = H(z)\Psi_P(z) .$$

Due to the linearity of the Dirac equation, both solutions can be combined to give new solutions with defined parity,

$$\Psi^{(\pm)}(z) = \Psi(z) \pm \Psi_P(z) \; , \; \Psi_P^{(\pm)}(z) = \pm\Psi^{(\pm)}(z) .$$

Case 2.a corresponds to even solutions (+) and case 2.b to odd solutions (−).

2.5.2 Radial Form of the Dirac Equation

Now we turn to the case of a spin-1/2 particle in a centrally symmetric potential of the form $eA^0(x) = V(\boldsymbol{x}) = V(|\boldsymbol{x}|)$, $\boldsymbol{A} = \boldsymbol{0}$. Due to the rotational symmetry of the corresponding Dirac-Hamilton operator, it is useful to pass over to spherical coordinates,

$$x = r\cos\varphi\sin\theta \; , \; y = r\sin\varphi\sin\theta \; , \; z = r\cos\theta \; ,$$

in order to isolate the angular-dependent and radial parts. To this end, we start with the Dirac equation

$$\mathrm{i}\hbar\frac{\partial\psi(x)}{\partial t} = \left[c\boldsymbol{\alpha}\boldsymbol{p} + \beta m_0 c^2 + V(r)\right]\psi(x) \; , \; r = |\boldsymbol{x}|$$

and rewrite it as the time-independent (stationary) equation

$$H\Psi(\boldsymbol{x}) = E\Psi(\boldsymbol{x}) \ , \ H = c\boldsymbol{\alpha p} + \beta m_0 c^2 + V(r) \ , \ \Psi(\boldsymbol{x}) = \psi(x)\mathrm{e}^{-\mathrm{i}Et/\hbar} \ . \tag{2.102}$$

To separate the radial and angular-dependent parts in the momentum term $\boldsymbol{\alpha p}$, we introduce the *radial momentum*

$$p_r = -\mathrm{i}\hbar \frac{1}{r}\frac{\partial}{\partial r} r = -\mathrm{i}\hbar\left(\frac{\partial}{\partial r} + \frac{1}{r}\right)$$

and the *radial velocity*

$$\alpha_r = \frac{\boldsymbol{\alpha x}}{r} \ .$$

From this and taking into account

$$(\boldsymbol{\sigma A})(\boldsymbol{\sigma B}) = \boldsymbol{AB} + \mathrm{i}\boldsymbol{\sigma}(\boldsymbol{A}\times\boldsymbol{B}) \ , \ \boldsymbol{x\nabla} = r\partial/\partial r \ ,$$

it follows that

$$(\boldsymbol{\alpha x})(\boldsymbol{\alpha p}) = \boldsymbol{xp} + \mathrm{i}\hat{\boldsymbol{\sigma}}\boldsymbol{L} = rp_r + \mathrm{i}\left(\hbar + \frac{2\boldsymbol{SL}}{\hbar}\right) \ . \tag{2.103}$$

Multiplying this expression from the left by α_r/r finally yields

$$\boldsymbol{\alpha p} = \alpha_r\left[p_r + \frac{\mathrm{i}}{r}\left(\hbar + \frac{\boldsymbol{J}^2 - \boldsymbol{S}^2 - \boldsymbol{L}^2}{\hbar}\right)\right] \ , \tag{2.104}$$

where $\boldsymbol{J} = \boldsymbol{L} + \boldsymbol{S}$ denotes the total angular momentum of the particle. As one can easily show, the operators $\{H, \boldsymbol{J}^2, J_z\}$ together with the parity transformation P form a complete set of commuting observables. Therefore, we construct the solutions to (2.102) in such a way that they are eigenvectors of these four operators. If we restrict ourselves to two solutional components, we can immediately quote the eigenfunctions of $\boldsymbol{J}^2$ and J_z (and $\boldsymbol{L}^2$, $\boldsymbol{S}^2$) by resorting to the corresponding results of nonrelativistic quantum mechanics:

$$\mathcal{Y}_{J,M}^{(l)}(\theta,\varphi) = \sum_{m+m_s=M}\left\langle l, m_l; \frac{1}{2}, m_s \middle| J, M\right\rangle Y_{l,m}(\theta,\varphi)\chi(m_s) \ , \ l = J \mp \frac{1}{2} \ ,$$

with

$$\boldsymbol{J}^2\mathcal{Y}_{J,M}^{(l)} = \hbar^2 J(J+1)\mathcal{Y}_{J,M}^{(l)} \ , \ J_z\mathcal{Y}_{J,M}^{(l)} = \hbar M\mathcal{Y}_{J,M}^{(l)} \ , \ M = -J, \ldots, J \ .$$

These *spinor spherical harmonics* are composed of the spherical harmonics $Y_{l,m}(\theta,\varphi)$, with

$$\boldsymbol{L}^2 Y_{l,m} = \hbar^2 l(l+1)Y_{l,m} \ , \ L_z Y_{l,m} = \hbar m Y_{l,m} \ , \ m = -l, \ldots, l \ ,$$

and the spinors $\chi(m_s)$, with

$$\boldsymbol{S}^2\chi(m_s) = \frac{3\hbar^2}{4}\chi(m_s) \ , \ S_z\chi(m_s) = \hbar m_s\chi(m_s) \ , \ \boldsymbol{S} = \frac{\hbar}{2}\boldsymbol{\sigma} \ , \ m_s = \pm\frac{1}{2} \ .$$

$\langle\ldots|\ldots\rangle$ denote the usual Clebsch-Gordan coefficients. Next we have to combine two spinor spherical harmonics with the same J and M so that the resulting bispinor has a defined parity. Note that the behavior of the spherical harmonics under space reflection,

$$Y_{l,m}(\pi-\theta,\varphi+\pi)=(-1)^l Y_{l,m}(\theta,\varphi)\ ,$$

imply for the spinor spherical harmonics that

$$\mathcal{Y}^{(l)}_{J,M}(\pi-\theta,\varphi+\pi)=(-1)^l Y^{(l)}_{J,M}(\theta,\varphi)\ .$$

On the other hand, applying the active parity transformation to a bispinor yields up to an irrelevant phase

$$\begin{pmatrix}\Psi_{\mathrm{u}}(\boldsymbol{x})\\ \Psi_{\mathrm{d}}(\boldsymbol{x})\end{pmatrix}\longrightarrow\begin{pmatrix}\Psi_{\mathrm{u}}(\boldsymbol{x})\\ \Psi_{\mathrm{d}}(\boldsymbol{x})\end{pmatrix}_P=\begin{pmatrix}\Psi_{\mathrm{u}}(-\boldsymbol{x})\\ -\Psi_{\mathrm{d}}(-\boldsymbol{x})\end{pmatrix}\ .$$

From this it follows that we have to combine two $\mathcal{Y}^{(l)}_{J,M}$ whose l-values differ by 1 in order to get states with defined parity. Including the radial dependence, the two possible combinations can be written as

$$\Psi^{(\omega)}_{J,M}(r,\theta,\varphi)=\frac{1}{r}\begin{pmatrix}F_{J+\omega/2}(r)\mathcal{Y}^{(J+\omega/2)}_{J,M}(\theta,\varphi)\\ \mathrm{i}G_{J-\omega/2}(r)\mathcal{Y}^{(J-\omega/2)}_{J,M}(\theta,\varphi)\end{pmatrix}\ ,\ \omega=\pm 1\ ,$$

with

$$\left[\Psi^{(\omega)}_{J,M}\right]_P(\theta,\varphi)=(-1)^{J+\omega/2}\Psi^{(\omega)}_{J,M}(\theta,\varphi)\ .$$

Here $F_l(r)$ and $G_l(r)$ denote two yet unspecified, solely r-dependent scalar functions and ω the parity quantum number. Due to

$$\boldsymbol{L}^2\Psi^{(\omega)}_{J,M}=\hbar^2 l(l+1)\Psi^{(\omega)}_{J,M}=\hbar^2\left[J(J+1)+\frac{1}{4}+\frac{\omega}{2}\beta(2J+1)\right]\Psi^{(\omega)}_{J,M}\ ,$$

the bracket term in (2.104) can be cast as

$$\left(\hbar+\frac{\boldsymbol{J}^2-\boldsymbol{S}^2-\boldsymbol{L}^2}{\hbar}\right)\Psi^{(\omega)}_{J,M}=-\frac{\hbar\omega}{2}(2J+1)\beta\Psi^{(\omega)}_{J,M}$$

so that, in total, (2.102) becomes

$$\left[c\alpha_r\left(p_r-\frac{\mathrm{i}\hbar\omega\left(J+\frac{1}{2}\right)}{r}\beta\right)+\beta m_0c^2+V(r)\right]\Psi^{(\omega)}_{J,M}=E\Psi^{(\omega)}_{J,M}\ .$$

Using the identities (see Exercise 29)

$$\frac{\boldsymbol{\sigma}\boldsymbol{x}}{r}\mathcal{Y}^{(J\pm\omega/2)}_{J,M}=-\mathcal{Y}^{(J\mp\omega/2)}_{J,M} \tag{2.105}$$

and

$$p_r\left(\frac{f(r)}{r}\right)=-\mathrm{i}\hbar\frac{1}{r}\frac{\mathrm{d}f}{\mathrm{d}r}\ ,$$

we finally arrive at

Theorem 2.10: Radial Dirac equations for centrally symmetric potentials

The solutions to the time-independent Dirac equation with a centrally symmetric potential

$$H\Psi(\boldsymbol{x}) = E\Psi(\boldsymbol{x}) \ , \ H = c\boldsymbol{\alpha p} + \beta m_0 c^2 + V(r) \ ,$$

can be written in spherical coordinates as

$$\Psi_{J,M}^{(\omega)}(r,\theta,\varphi) = \frac{1}{r}\begin{pmatrix} F_{J+\omega/2}(r)\mathcal{Y}_{J,M}^{(J+\omega/2)}(\theta,\varphi) \\ \mathrm{i}G_{J-\omega/2}(r)\mathcal{Y}_{J,M}^{(J-\omega/2)}(\theta,\varphi)\end{pmatrix} ,$$

where the functions F_l and G_l fulfill the *radial Dirac equations*

$$\left.\begin{aligned} \left[-\frac{\mathrm{d}}{\mathrm{d}r} + \frac{\omega\left(J+\frac{1}{2}\right)}{r}\right] G_{J-\omega/2}(r) &= \frac{E - m_0c^2 - V}{c\hbar}F_{J+\omega/2}(r) \\ \left[\frac{\mathrm{d}}{\mathrm{d}r} + \frac{\omega\left(J+\frac{1}{2}\right)}{r}\right] F_{J+\omega/2}(r) &= \frac{E + m_0c^2 - V}{c\hbar}G_{J-\omega/2}(r). \end{aligned}\right\} \tag{2.106}$$

Furthermore, these solutions obey the relationships

$$\begin{aligned} \boldsymbol{J}^2\Psi_{J,M}^{(\omega)}(r,\theta,\varphi) &= \hbar^2 J(J+1)\Psi_{J,M}^{(\omega)}(r,\theta,\varphi) \ , \ J = \frac{1}{2},\frac{3}{2},\ldots \\ J_z\Psi_{J,M}^{(\omega)}(r,\theta,\varphi) &= \hbar M\Psi_{J,M}^{(\omega)}(r,\theta,\varphi) \ , \ M = -J,\ldots,J \\ \left[\Psi_{J,M}^{(\omega)}\right]_P(r,\theta,\varphi) &= (-1)^{J+\omega/2}\Psi_{J,M}^{(\omega)}(r,\theta,\varphi) \ , \ \omega = \pm 1 \ . \end{aligned}$$

2.5.3 Free Particle and Centrally Symmetric Potential Well

The simplest application of Theorem 2.10 is a free spin-1/2 particle ($V = 0$). If, in this case, the second radial equation is solved for G,

$$G_{J-\omega/2}(r) = \frac{c\hbar}{E + m_0c^2}\left[\frac{\mathrm{d}}{\mathrm{d}r} + \frac{\omega\left(J+\frac{1}{2}\right)}{r}\right] F_{J+\omega/2}(r) \ , \tag{2.107}$$

and subsequently inserted into the first, one obtains

$$\left[-\frac{\mathrm{d}^2}{\mathrm{d}r^2} + \frac{\left(J+\frac{1}{2}\right)\left(J+\frac{1}{2}+\omega\right)}{r^2}\right] F_{J+\omega/2}(r) = \frac{E^2 - m_0^2c^4}{c^2\hbar^2}F_{J+\omega/2}(r) \ .$$

Taking into account

$$l = J + \frac{\omega}{2} \Longrightarrow \left(J+\frac{1}{2}\right)\left(J+\frac{1}{2}+\omega\right) = l(l+1) \ ,$$

it follows that

$$\left[\frac{\mathrm{d}^2}{\mathrm{d}r^2} - \frac{l(l+1)}{r^2} + k^2\right] F_l(r) = 0 \;,\; k^2 = \frac{E^2 - m_0^2 c^4}{c^2\hbar^2} \;. \tag{2.108}$$

Obviously, this equation is formally identical to the radial Klein-Gordon equation (1.72) in Theorem 1.9. For further evaluation, we can therefore resort to the corresponding calculations of the Klein-Gordon case. Making the substitutions

$$\rho = kr \;,\; F_l(r) = \rho \hat{F}_l(\rho) \;,\; G_l(r) = \rho \hat{G}_l(\rho) \;,$$

(2.107) and (2.108) pass over to the equation system

$$\left.\begin{aligned} &\left[\frac{\mathrm{d}^2}{\mathrm{d}\rho^2} + \frac{2}{\rho}\frac{\mathrm{d}}{\mathrm{d}\rho} - \frac{l(l+1)}{\rho^2} + 1\right] \hat{F}_l(\rho) = 0 \\ &\hat{G}_{J-\omega/2}(\rho) = \frac{c\hbar k}{E + m_0 c^2}\left[\frac{\mathrm{d}}{\mathrm{d}\rho} + \frac{\omega\left(J+\frac{1}{2}\right)}{\rho}\right] \hat{F}_{J+\omega/2}(\rho) \;. \end{aligned}\right\} \tag{2.109}$$

The first equation is the spherical Bessel differential equation which was already discussed in Subsection 1.5.3 and whose solutions are given by the spherical Bessel functions j_l, n_l, $h_l^{(\pm)}$. The following recursion formulae hold for them ($\hat{F}_l = j_l, n_l, h_l^{(\pm)}$):

$$\hat{F}_{l-1}(\rho) = \frac{d}{d\rho}\hat{F}_l(\rho) + \frac{l+1}{\rho}\hat{F}_l(\rho) \;,\; \hat{F}_{l+1}(\rho) = -\frac{d}{d\rho}\hat{F}_l(\rho) + \frac{l}{\rho}\hat{F}_l(\rho) \;.$$

Thus, the second equation of (2.109) yields for G_l:

$$\hat{G}_{J-\omega/2}(\rho) = \frac{c\hbar k\omega}{E + m_0 c^2}\hat{F}_{J-\omega/2}(\rho) \;.$$

With respect to the physical content of these solutions, we again have to distinguish the following cases:

- $|E| < m_0 c^2$: in this energy range, there is no solution to the spherical Bessel differential equation that is bounded at infinity and regular at the origin. As in the Klein-Gordon case, a free (anti)particle with energy E within the "forbidden range" $-m_0 c^2 < E < m_0 c^2$ is not possible.
- $|E| > m_0 c^2$: here we have exactly one solution to the Bessel equation bounded everywhere, namely $\hat{F}_l(\rho) = j_l(\rho)$. Returning to the original quantities, the physical solution to (2.106) is therefore

$$\left.\begin{aligned} F_{J+\omega/2}(r) &= A_{J+\omega/2}\, r j_{J+\omega/2}(kr) \;,\; k = \sqrt{\frac{E^2 - m_0^2 c^4}{c^2\hbar^2}} \\ G_{J-\omega/2}(r) &= A_{J+\omega/2}\frac{c\hbar k\omega}{E + m_0 c^2}\, r j_{J-\omega/2}(kr) \;, \end{aligned}\right\} \tag{2.110}$$

  with the normalization constant $A_{J+\omega/2}$. Thus, to each energy value $|E| > m_0 c^2$, we obtain a free spherical wave with angular momentum J, M, and parity $(-1)^{J+\omega/2}$.

Spherically symmetric potential well. It is easy to extend our considerations to the case of a spherically symmetric potential well given by

$$eA^0(r) = V(r) = \left\{ \begin{array}{c} -V_0 \text{ for } r < a \text{ (area I)} \\ 0 \quad \text{for } r > a \text{ (area II)}\ . \end{array} \right\} \ , \ V_0 > 0$$

(see Figure 1.7). Concerning the upper radial function F_l, the same line of argument hold as for the Klein-Gordon case in Subsection 1.5.3. Thus, using the abbreviations $l = J + \omega/2$, $l' = J - \omega/2$, the regular solution to (2.106) in area I is

- $|E + V_0| > m_0c^2$: $$\left\{ \begin{array}{l} F_l^{(\mathrm{I})}(r) = A_l r j_l(k_1 r)\ , \ k_1 = \sqrt{\dfrac{(E+V_0)^2 - m_0^2c^4}{c^2\hbar^2}} \\ G_{l'}^{(\mathrm{I})}(r) = A_l \dfrac{c\hbar k_1 \omega}{E + V_0 + m_0c^2} r j_{l'}(k_1 r) \end{array} \right.$$

or

- $|E + V_0| < m_0c^2$: $$\left\{ \begin{array}{l} F_l^{(\mathrm{I})}(r) = A_l r j_l(\mathrm{i}\kappa_1 r)\ , \ \kappa_1 = \sqrt{\dfrac{m_0^2c^4 - (E+V_0)^2}{c^2\hbar^2}} \\ G_{l'}^{(\mathrm{I})}(r) = A_l \dfrac{\mathrm{i}c\hbar \kappa_1 \omega}{E + V_0 + m_0c^2} r j_{l'}(\mathrm{i}\kappa_1 r) \end{array} \right.$$

which is obtained from the free solution (2.110) with the replacement $E \to E + V_0$. The following two cases must be distinguished in the outer area II:

- $|E| < m_0c^2$ (bound states): in this case the only solution to (2.106) bounded at infinity is given by

  $$F_l^{(\mathrm{II})}(r) = B_l r h_l^{(+)}(\mathrm{i}\kappa_2 r)\ , \ \kappa_2 = \sqrt{\frac{m_0^2c^4 - E^2}{c^2\hbar^2}}$$
  $$G_{l'}^{(\mathrm{II})}(r) = B_l \frac{\mathrm{i}c\hbar\kappa_2\omega}{E + m_0c^2} r h_{l'}^{(+)}(\mathrm{i}\kappa_2 r)\ .$$

  The continuity conditions at the area border $r = a$,

  $$\frac{F_l^{(\mathrm{I})}(a)}{a} = \frac{F_l^{(\mathrm{II})}(a)}{a}\ , \ \left.\frac{\mathrm{d}}{\mathrm{d}r}\frac{F_l^{(\mathrm{I})}(r)}{r}\right|_{r=a} = \left.\frac{\mathrm{d}}{\mathrm{d}r}\frac{F_l^{(\mathrm{II})}(r)}{r}\right|_{r=a}\ ,$$

  can be simultaneously satisfied only for discrete energy values E, the energy levels of the bound states. For l=0-states (with respect to the F-function, i.e. $J = 1/2$, $\omega = -1$) and $E + V_0 > m_0c^2$, one obtains the quantization condition

  $$\tan k_1 a = -\frac{k_1}{\kappa_2}\ ,$$

  which is formally identical to the Klein-Gordon condition [see (1.74)]

- $|E| > m_0c^2$ (unbound states): here a linear combination of the spherical Bessel functions seems to be feasible. We therefore write for the solution of (2.106):

$$F_l^{(\mathrm{II})}(r) = B_l r[j_l(k_2r)\cos\delta_l + n_l(k_2r)\sin\delta_l] \ , \ k_2 = \sqrt{\frac{E^2 - m_0^2c^4}{c^2\hbar^2}}$$

$$G_{l'}^{(\mathrm{II})}(r) = B_l \frac{c\hbar k_2\omega}{E + m_0c^2} r[j_{l'}(k_2r)\cos\delta_l + n_{l'}(k_2r)\sin\delta_l] \ .$$

If we focus on l=0-states (with respect to the F-function), the corresponding continuity conditions determine the phase δ_0 to be

$$\tan(k_2a + \delta_0) = \frac{k_2}{k_1}\tan k_1a \quad \text{for } |E + V_0| > m_0c^2$$

or

$$\tan(k_2a + \delta_0) = \frac{k_2}{\kappa_1}\tanh \kappa_1a \quad \text{for } |E + V_0| < m_0c^2 \ .$$

These relations are also formally identical to the Klein-Gordon relations [compare to (1.75), (1.76)].

At this stage, we leave out a more detailed division and interpretation of these solutions since they result in the five cases discussed in Subsection 1.5.3.

2.5.4 Coulomb Potential

As the last example of centrally symmetric problems, we now turn to a bound spin-1/2 particle in a Coulomb potential of the form (*hydrogen-like atom*)

$$eA^0(r) = V(r) = -\frac{Ze^2}{r} = -\frac{Z\hbar c\alpha_e}{r} \ , \ \alpha_e = \frac{e^2}{\hbar c} = 1/137.03602 \ .$$

As in the corresponding Klein-Gordon problem (Subsection 1.5.4), we first investigate the asymptotic regions of the radial Dirac equations for small and large distances in order to find an appropriate power series expansion to solve these equations.

$r \to \infty$: in the limit of large r the radial Dirac equations (2.106) become

$$-\frac{\mathrm{d}G}{\mathrm{d}r} = \frac{E - m_0c^2}{c\hbar}F \ , \ \frac{\mathrm{d}F}{\mathrm{d}r} = \frac{E + m_0c^2}{c\hbar}G \ ,$$

whereas here and in the following the indices $J \pm \omega/2$ are suppressed. Combining these two equations yields the relation

$$\frac{\mathrm{d}^2F}{\mathrm{d}r^2} = -\frac{E^2 - m_0^2c^4}{c^2\hbar^2}F \ .$$

Its normalizable solution (descending at infinity) is given by

$$F(r \to \infty) \sim \mathrm{e}^{-kr} \ , \ k = \sqrt{\frac{m_0^2c^4 - E^2}{c^2\hbar^2}} \ .$$

$\boldsymbol{r \longrightarrow 0}$: in this case (2.106) passes over to the equations

$$\left[-\frac{\mathrm{d}}{\mathrm{d}r}+\frac{\omega\left(J+\frac{1}{2}\right)}{r}\right]G=\frac{Z\alpha_e}{r}F$$

$$\left[\frac{\mathrm{d}}{\mathrm{d}r}+\frac{\omega\left(J+\frac{1}{2}\right)}{r}\right]F=\frac{Z\alpha_e}{r}G$$

that can be combined to give

$$\left\{r\frac{\mathrm{d}^2}{\mathrm{d}r^2}+\frac{\mathrm{d}}{\mathrm{d}r}+\frac{1}{r}\left[(Z\alpha_e)^2-\left(J+\frac{1}{2}\right)^2\right]\right\}F=0\ .$$

Its regular solution[13] is

$$F(r\to 0)\sim r^s\ ,\ s=+\sqrt{\left(J+\frac{1}{2}\right)^2-(Z\alpha_e)^2}\ .$$

In order to facilitate the following calculations, we introduce the substitutions

$$\left.\begin{array}{l}\rho=kr\ ,\ F(r)=\hat{F}(\rho)\ ,\ G(r)=\hat{G}(\rho)\ ,\ k=\sqrt{\dfrac{m_0^2c^4-E^2}{c^2\hbar^2}}\\ \tau=\omega\left(J+\dfrac{1}{2}\right)\ ,\ \nu=\sqrt{\dfrac{m_0c^2-E}{m_0c^2+E}}\end{array}\right\}\quad(2.111)$$

so that the original radial equations (2.106) become

$$\left(-\frac{\mathrm{d}}{\mathrm{d}\rho}+\frac{\tau}{\rho}\right)\hat{G}=\left(-\nu+\frac{Z\alpha_e}{\rho}\right)\hat{F}$$

$$\left(\frac{\mathrm{d}}{\mathrm{d}\rho}+\frac{\tau}{\rho}\right)\hat{F}=\left(\frac{1}{\nu}+\frac{Z\alpha_e}{\rho}\right)\hat{G}\ .$$

To solve them we choose, due to the above considerations, the ansatz

$$\left.\begin{array}{l}\hat{F}(\rho)=\rho^s\mathrm{e}^{-\rho}\sum\limits_i a_i\rho^i\ ,\ s=\sqrt{\tau^2-(Z\alpha_e)^2}\\ \hat{G}(\rho)=\rho^s\mathrm{e}^{-\rho}\sum\limits_i b_i\rho^i\ ,\end{array}\right\}\quad(2.112)$$

which leads to the following recursive conditional equations for the expansion coefficients a_i and b_i:

$$\begin{aligned}(\tau-s)b_0&=Z\alpha_e a_0\\ (\tau+s)a_0&=Z\alpha_e b_0\\ b_{i-1}+(\tau-s-i)b_i&=-\nu a_{i-1}+Z\alpha_e a_i\ ,\ i\geq 1\\ -a_{i-1}+(\tau+s+i)a_i&=\frac{b_{i-1}}{\nu}+Z\alpha_e b_i\ ,\ i\geq 1\ .\end{aligned}$$

[13] Here we assume that $Z\alpha_e < J+1/2$. Fortunately, this is the case for $Z < 137$, i.e. for all nuclei that can be found in nature (otherwise, the discussion of the regularity conditions at the origin would be more extensive).

From the first two equations we obtain again $s = \sqrt{\tau^2 - (Z\alpha_e)^2}$, while the last two equations yield

$$b_i = \frac{Z\alpha_e - \nu(\tau + s + i)}{\tau - s - i - \nu Z\alpha_e} a_i$$

and

$$a_{i+1} \frac{(Z\alpha_e)^2 + (\tau + s + i + 1)(-\tau + s + i + 1)}{\tau - s - i - 1 - \nu Z\alpha_e}$$
$$= a_i \frac{Z\alpha_e \nu^2 + 2\nu(s + i) - Z\alpha_e}{\nu(\tau - s - i - \nu Z\alpha_e)} . \quad (2.113)$$

For $\hat{F}(\rho)$ and $\hat{G}(\rho)$ to have the desired (descending) behavior at infinity, the power series in (2.112) needs to terminate at some $i = n'$. This means that

$$Z\alpha_e - 2\nu(n' + s) - \nu^2 Z\alpha_e = 0 .$$

From this we get the quantization condition

$$\nu = -\frac{n' + s}{Z\alpha_e} + \sqrt{\left(\frac{n' + s}{Z\alpha_e}\right)^2 + 1}$$

and, together with (2.111), the possible energy levels of the hydrogen atom for bound states,

$$E_{n',J} = \frac{m_0 c^2}{\sqrt{1 + \frac{(Z\alpha_e)^2}{\left(n' + \sqrt{\left(J + \frac{1}{2}\right)^2 - (Z\alpha_e)^2}\right)^2}}} .$$

Introducing the principal quantum number

$$n = n' + J + \frac{1}{2} ,$$

the final result is

$$E_{n,J} = \frac{m_0 c^2}{\sqrt{1 + \frac{(Z\alpha_e)^2}{\left(n - \left(J + \frac{1}{2}\right) + \sqrt{\left(J + \frac{1}{2}\right)^2 - (Z\alpha_e)^2}\right)^2}}} , \quad (2.114)$$

where n and J can take on the values

$$n = 1, 2, \ldots, \infty , \quad J = \frac{1}{2}, \frac{3}{2}, \ldots, n - \frac{1}{2} .$$

Equation (2.114) is formally identical to the corresponding relation (1.81) of the Klein-Gordon case, if, in the latter, l is replaced by J. Consequently, the series expansion of (2.114) in powers of $Z\alpha_e$ can be immediately read off from (1.82):

$$E_{n,J} = m_0 c^2 \left[1 - \frac{(Z\alpha_e)^2}{2n^2} - \frac{(Z\alpha_e)^4}{2n^4}\left(\frac{n}{J + \frac{1}{2}} - \frac{3}{4}\right) + \ldots\right] . \quad (2.115)$$

The first term is the mass term. The second one complies with the quantity predicted from nonrelativistic quantum mechanics. All subsequent terms are relativistic corrections. Their effect is that the nonrelativistic degeneracy of all levels with the same n is removed. Instead, we now encounter a degeneracy of levels with the same n and J. At fixed n the energy of each level is slightly raised depending on J.

For an easier distinction between the various solutions, one usually adopts the nonrelativistic spectroscopic notation nl_J, whereas the l-value conventionally refers to the two upper components of the Dirac bispinor, i.e. $l = J + \omega/2$. Table 2.2 provides an overview of the electron energies of hydrogen-like atoms for the ground state as well as for the first excited states together with their spectroscopic labels and the corresponding quantum numbers. To each J

n	J	ω	l	nl_J	E_{nJ}/m_0c^2
1	1/2	−1	0	$1s_{1/2}$	$\sqrt{1-(Z\alpha_e)^2}$
2	1/2	−1	0	$2s_{1/2}$	$\sqrt{\frac{1+\sqrt{1-(Z\alpha_e)^2}}{2}}$
2	1/2	+1	1	$2p_{1/2}$	"
2	3/2	−1	1	$2p_{3/2}$	$\frac{\sqrt{4-(Z\alpha_e)^2}}{2}$
3	1/2	−1	0	$3s_{1/2}$	$\frac{2+\sqrt{1-(Z\alpha_e)^2}}{\sqrt{5+4\sqrt{1-(Z\alpha_e)^2}}}$
3	1/2	+1	1	$3p_{1/2}$	"
3	3/2	−1	1	$3p_{3/2}$	$\frac{1+\sqrt{4-(Z\alpha_e)^2}}{\sqrt{5+2\sqrt{4-(Z\alpha_e)^2}}}$
3	3/2	+1	2	$3d_{3/2}$	"
3	5/2	−1	2	$3d_{5/2}$	$\frac{\sqrt{9-(Z\alpha_e)^2}}{3}$

Tab. 2.2. Energy levels of hydrogen-like atoms.

there belong two series of $2J + 1$ solutional functions with opposite parities, except for $J = n - 1/2$ for which only one series of $2J + 1$ solutions with parity $(-1)^{n-1}$ exists. The latter is a consequence of the fact that the right hand side of (2.113) becomes singular for $n' = 0$ and $\omega = +1$ (and only for this combination).

Until 1947 the spectroscopic observations of the hydrogen atom (and hydrogen-like atoms, particularly He^+) agreed well with the above results after they had been complemented by the interaction effects between the spins of the electron and the nucleus (*hyperfine structure splitting*). However, in 1947 Lamb and Retherford observed a small shift of the hydrogen level

$2s_{1/2}$ upward, which amounts to roughly one tenth of the distance between the $2p_{3/2}$- and $2p_{1/2}$-levels (see Figure 2.2). This effect, known as *Lamb shift*,

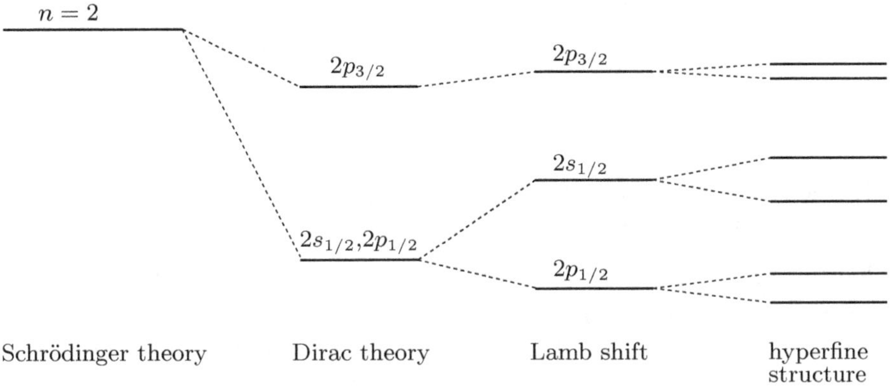

Fig. 2.2. Line splittings of the nonrelativistic n=2-hydrogen level including relativistic effects (Dirac theory, main part: fine structure splitting), the Lamb shift, as well as the hyperfine structure splitting.

is today seen as a consequence of the interaction between the electron and the fluctuations of the quantized radiation field and can fully be understood only within the framework of *quantum electrodynamics* (see Section 3.4, particularly Subsection 3.4.4). The Dirac theory deals with the major part of this interaction, i.e. the Coulomb potential, and the Lamb shift represents the *radiation corrections* within this approximation.

We now quote the Dirac wave functions for the ground state of the hydrogen atom. In this case, we have

$$n = 1 \ , \ J = \frac{1}{2} \ , \ M = \pm\frac{1}{2} \ , \ \omega = -1$$

and

$$k = \frac{m_0 c Z\alpha_e}{\hbar} \ , \ s = \sqrt{1 - (Z\alpha_e)^2} \ , \ \nu = \frac{1-s}{Z\alpha_e}$$

so that

$$\Psi^{(\omega=-1)}_{J=1/2,M=+1/2}(r,\theta,\varphi) = N(2kr)^{s-1}\mathrm{e}^{-kr}\begin{pmatrix} 1 \\ 0 \\ \frac{\mathrm{i}(1-s)}{Z\alpha_e}\cos\theta \\ \frac{\mathrm{i}(1-s)}{Z\alpha_e}\sin\theta\mathrm{e}^{\mathrm{i}\varphi} \end{pmatrix}$$

$$\Psi^{(\omega=-1)}_{J=1/2,M=-1/2}(r,\theta,\varphi) = N(2kr)^{s-1}\mathrm{e}^{-kr}\begin{pmatrix} 0 \\ 1 \\ \frac{\mathrm{i}(1-s)}{Z\alpha_e}\sin\theta\mathrm{e}^{-\mathrm{i}\varphi} \\ -\frac{\mathrm{i}(1-s)}{Z\alpha_e}\cos\theta \end{pmatrix} ,$$

with

$$N = \frac{(2k)^{3/2}}{\sqrt{4\pi}} \sqrt{\frac{1+s}{2\Gamma(1+2s)}}$$

and the Gamma-function $\Gamma(x)$. The normalization constant N is chosen so that

$$\int \mathrm{d}^3 x \Psi^{(\omega=-1)\dagger}_{J=1/2,M=\pm 1/2}(r,\theta,\varphi) \Psi^{(\omega=-1)}_{J=1/2,M=\pm 1/2}(r,\theta,\varphi) = 1 \ .$$

In the nonrelativistic limit $s \longrightarrow 1$ and $(s-1)/Z\alpha_e \longrightarrow 0$, the two upper components pass over to the Schrödinger wave functions multiplied by the Pauli spinors $\begin{pmatrix} 1 \\ 0 \end{pmatrix}$ and $\begin{pmatrix} 0 \\ 1 \end{pmatrix}$ respectively. Contrary to the nonrelativistic expressions, the relativistic wave functions exhibit a weak but square-integrable singularity r^{s-1}. For $Z\alpha_e > 1$, s becomes imaginary and the solutions start to oscillate. At $Z\alpha_e = 1$ the "slope" of the energy $E_{1,1/2}$ is

$$\left.\frac{\mathrm{d}E_{1,1/2}}{\mathrm{d}Z}\right|_{Z\alpha_e=1} = -\left.\frac{m_0 c^2 Z\alpha_e^2}{\sqrt{1-(Z\alpha_e)^2}}\right|_{Z\alpha_e=1} \longrightarrow -\infty \ .$$

At the end, we point out that the Dirac equation with an external Coulomb potential is just a rough approximation for the description of hydrogen-like bound states. With respect to the accompanying exclusions, similar statements hold as in the Klein-Gordon case, Subsection 1.5.4.

Summary

- Depending on the particle energy, the discussion of the onedimensional potential well leads to different Dirac solutions which can be interpreted within the one-particle picture more or less consistently as **scattering** or **binding** of particles or antiparticles.
- Within the Dirac equation with a centrally symmetric potential the radial and angular-dependent parts can be separated, the latter being solved by the **spinor spherical harmonics**, i.e. the eigenfunctions of $\boldsymbol{J}^2$, J_z, and of the parity transformation P. For the solution of the radial part there remain two **radial Dirac equations**.
- For the free case (as well as that of constant potential sections), the radial Dirac equations lead to the **spherical Bessel differential equation** whose solutions are given by the **spherical Bessel functions**.
- Similar to pion atoms in the Klein-Gordon case, hydrogen-like atoms can be described approximately using the Coulomb potential. Here bound energy levels with the same principal and total angular momentum quantum numbers turn out be degenerate.

Exercises

29. Properties of spinor spherical harmonics. Prove the identity [see (2.105)]

$$\frac{\boldsymbol{\sigma x}}{r}\mathcal{Y}_{J,M}^{(l=J\pm 1/2)} = -\mathcal{Y}_{J,M}^{(l=J\mp 1/2)} .$$

Tip: exploit the fact that $\mathcal{Y}_{J,M}^{(l)}$ are eigenfunctions of $\boldsymbol{\sigma L}$ and calculate the commutator $[\boldsymbol{\sigma x}/r, \boldsymbol{\sigma L}]$.

Solution. To calculate the first half of the commutator, we use (2.103) to obtain

$$(\boldsymbol{\sigma x})(\boldsymbol{\sigma p}) = (\boldsymbol{xp}) + \mathrm{i}\boldsymbol{\sigma L} .$$

Multiplying this equation by $(\boldsymbol{\sigma x})$ yields

$$r^2(\boldsymbol{\sigma p}) = (\boldsymbol{\sigma x})(\boldsymbol{xp}) + \mathrm{i}(\boldsymbol{\sigma x})(\boldsymbol{\sigma L})$$

and finally

$$\frac{\boldsymbol{\sigma x}}{r}\boldsymbol{\sigma L} = -\mathrm{i}r(\boldsymbol{\sigma p}) + \mathrm{i}\frac{(\boldsymbol{\sigma x})(\boldsymbol{xp})}{r} . \tag{2.116}$$

Written down explicitly, the other half of the commutator is

$$\begin{aligned}\boldsymbol{\sigma L}\frac{\boldsymbol{\sigma x}}{r} &= -\mathrm{i}\hbar\sum_{i,j,k,l}\epsilon_{ijk}\sigma_i x_j\frac{\partial}{\partial x_k}\sigma_l\frac{x_l}{r} \\ &= -\mathrm{i}\hbar\sum_{i,j,k,l}\sigma_i\sigma_l\epsilon_{ijk}x_j\left(\frac{\delta_{kl}}{r} - \frac{x_l x_k}{r^3} + \frac{x_l}{r}\frac{\partial}{\partial x_k}\right) .\end{aligned}$$

Here the first term yields

$$-\mathrm{i}\hbar\sum_{i,j,k,l}\sigma_i\sigma_l\epsilon_{ijk}x_j\frac{\delta_{kl}}{r} = -2\hbar\frac{\boldsymbol{\sigma x}}{r} .$$

The second term does not contribute whereas the third term is calculated as

$$\begin{aligned}-\mathrm{i}\hbar\sum_{i,j,k,l}\sigma_i\sigma_l\epsilon_{ijk}\frac{x_j x_l}{r}\frac{\partial}{\partial x_k} &= \frac{\hbar}{r}(\boldsymbol{x}\times\boldsymbol{\sigma})(\boldsymbol{x}\times\boldsymbol{\nabla}) \\ &= \frac{\mathrm{i}}{r}(\boldsymbol{x}\times\boldsymbol{\sigma})(\boldsymbol{x}\times\boldsymbol{p}) \\ &= \mathrm{i}r(\boldsymbol{\sigma p}) - \mathrm{i}\frac{(\boldsymbol{\sigma x})(\boldsymbol{xp})}{r} .\end{aligned}$$

Thus, we have

$$\boldsymbol{\sigma L}\frac{\boldsymbol{\sigma x}}{r} = -2\hbar\frac{\boldsymbol{\sigma x}}{r} + \mathrm{i}r(\boldsymbol{\sigma p}) - \mathrm{i}\frac{(\boldsymbol{\sigma x})(\boldsymbol{xp})}{r}$$

and, together with (2.116), it follows that

$$\boldsymbol{\sigma L}\frac{\boldsymbol{\sigma x}}{r} = -\frac{\boldsymbol{\sigma x}}{r}\boldsymbol{\sigma L} - 2\hbar\frac{\boldsymbol{\sigma x}}{r} .$$

Now we make use of the fact that $\mathcal{Y}_{J,M}^{(l)}$ are eigenfunctions of $\boldsymbol{\sigma L}$ and write

$$\begin{aligned}\boldsymbol{\sigma L} &= \hbar\left[J(J+1) - l(l+1) - \frac{3}{4}\right] \\ \boldsymbol{\sigma L}\frac{\boldsymbol{\sigma x}}{r} &= -\hbar\left[J(J+1) - l(l+1) + \frac{5}{4}\right]\frac{\boldsymbol{\sigma x}}{r} .\end{aligned}$$

Applying this to $\mathcal{Y}_{J,M}^{(l=J\pm\omega/2)}$ yields

$$\begin{aligned}\boldsymbol{\sigma L}\mathcal{Y}_{J,M}^{(l=J+\omega/2)} &= -\hbar\left(\omega J + \frac{\omega}{2} + 1\right)\mathcal{Y}_{J,M}^{(l=J+\omega/2)} \\ \boldsymbol{\sigma L}\frac{\boldsymbol{\sigma x}}{r}\mathcal{Y}_{J,M}^{(l=J-\omega/2)} &= -\hbar\left(\omega J + \frac{\omega}{2} + 1\right)\frac{\boldsymbol{\sigma x}}{r}\mathcal{Y}_{J,M}^{(l=J-\omega/2)} .\end{aligned}$$

The function $\frac{\boldsymbol{\sigma x}}{r}\mathcal{Y}_{J,M}^{(l=J-\omega/2)}$ is an eigenfunction of $\boldsymbol{\sigma L}$ with the same eigenvalue as $\mathcal{Y}_{J,M}^{(l=J+\omega/2)}$, meaning that

$$\frac{\boldsymbol{\sigma x}}{r}\mathcal{Y}_{J,M}^{(l=J-\omega/2)} = c_l\mathcal{Y}_{J,M}^{(l=J+\omega/2)} . \tag{2.117}$$

Due to $(\boldsymbol{\sigma x}/r)^2 = 1$, the constant of proportionality c_l is restricted to $|c_l| = 1$. Furthermore, if we take into account the parities of both sides of (2.117) as well as the fact that $\boldsymbol{\sigma x}/r$ is a pseudo-vector (and does not change under space reflection), we finally arrive at $c_l = c = -1$.

3. Relativistic Scattering Theory

Having laid down the framework for the relativistic quantum mechanical description of spin-0 and spin-1/2 particles with a special regard for the one-particle interpretation, we will now deal with the scattering of those particles.

The study of scattering processes is an important instrument, particularly for the exploration of microscopic interaction effects as, due to their narrowness, they are not directly accessible to the human senses and therefore need to be magnified appropriately. In practice, this is achieved by, for example, directing a collimated particle beam toward a fixed target and comparing the measured angular distribution of the scattering products with theoretical calculations. In this way, a large number of new particles have been discovered. Many of them are not only subject to the electromagnetic force but also to two other very short-ranged forces, namely the *strong* and the *weak interaction*. Today it is assumed that all three types of interactions can be described by *quantum field theories*: the electromagnetic interaction by *quantum electrodynamics*, the strong one by *quantum chromodynamics*, and the weak one by *quantum flavordynamics*.

Of these three elementary types of interactions, as in the preceding two chapters, we restrict ourselves to the electromagnetic interaction. Since the theoretical description of relativistic scattering processes at not too small energies must inevitably take into account particle creation and annihilation processes, one may suspect that, at this stage, we have to give up our formalism with focus on the one-particle interpretation developed so far and, instead, have to consider the above mentioned quantum electrodynamics, as a many-particle theory (with infinitely many degrees of freedom). However, as Feynman and Stückelberg have shown, relativistic scattering processes can, similarly to nonrelativistic scattering processes, also be described using propagator techniques that directly tie in with the level of knowledge in the preceding chapters, particularly with the interpretation of solutions with negative energy (Feynman-Stückelberg interpretation). Apart from the fact that propagator techniques seem to be more appropriate for our intention to suppress quantum field theoretical aspects as much as possible (relativistic quantum mechanics "in the narrow sense"), they provide the following advantages:

A. Wachter, Relativistic Scattering Theory. In: A. Wachter, Relativistic Quantum Mechanics, Theoretical and Mathematical Physics, pp. 177–348 (2011)
DOI 10.1007/978-90-481-3645-2_3

- Contrary to quantum electrodynamics, propagator techniques provide a less deductive but intuitively more comprehensible approach to the description of relativistic processes, which seems to be particularly suited for the non-expert.
- Via propagator methods one arrives relatively fast to the mathematical techniques necessary for the calculation of concrete scattering processes.

For these reasons, the following pages will center on propagator methods, while we emphasize that the deeper grounds for the formalisms developed here, particularly the *radiation corrections* at higher orders, can only be fully understood within quantum electrodynamics.

The first section of this chapter restates the formalism of nonrelativistic scattering theory and introduces the necessary mathematical concepts of *propagator*, *scattering matrix*, and *cross section*. In the second section these concepts are transferred and adequately extended to the relativistic case of Dirac's theory. The third section deals with the concrete calculation of relativistic spin-1/2 scattering processes to the lowest orders of the scattering theory as well as with the development of the *Feynman rules*. With their help, arbitrarily complex scattering processes can be formulated mathematically. The fourth section discusses quantum field theoretical corrections in higher orders. As we will see, certain divergences stemming from radiation corrections appear, which, however, can be eliminated using the program of *renormalization*. At the end we transfer the developed formalisms to the scattering processes of spin-0 particles and extend the Feynman rules in such a way that they also cover the spin-0 case.

Note. In this chapter we diverge from the order used so far, in that we address the scattering of spin-0 particles at the end and a little more succinctly. This is because all known spin-0 particles are not elementary but composed of quarks. They, in turn, are subject to strong and weak interactions that usually overlap electromagnetic effects. By contrast, electrons and positrons that we discuss predominantly in the context of spin-1/2 particle scatterings are, indeed, to be regarded as elementary (structureless) so that, in this case, electromagnetic interactions can be studied theoretically and experimentally in the purest form. It is exactly in this domain where all predictions of quantum electrodynamics turn out to be very precise. That is why quantum electrodynamics is regarded as one of the most successful theories in physics.

3.1 Review: Nonrelativistic Scattering Theory

In this section we recall the nonrelativistic treatment of scattering processes using propagator techniques and prepare the accompanying concepts with a view to a subsequent relativistic generalization. First we turn our attention to the solution of the general Schrödinger equation using the *Green function calculus*. From this we derive the *retarded* and *advanced propagators*. They can be used to describe and calculate, at least approximately, the propagation of Schrödinger solutions in temporally forward and backward directions. After some intermediate considerations on propagators, we come to the actual scattering of particles and utilize the propagators to cast the *scattering amplitudes* into a calculable form. In this context, we also derive the relationship between scattering amplitudes and the experimentally interesting quantity, the *differential cross section*. At the end, the developed formalisms are applied to the concrete problem of Coulomb scattering.

3.1.1 Solution of the General Schrödinger Equation

The starting point of our considerations is the time-dependent Schrödinger equation[1]

$$\left(\mathrm{i}\hbar\frac{\partial}{\partial t'} - H'\right)\psi(\boldsymbol{x}', t') = 0 \ , \ H' = H'^{(0)} + V(\boldsymbol{x}', t') \ , \ H'^{(0)} = \frac{\boldsymbol{p}'^2}{2m_0} , \quad (3.1)$$

for whose solution we wish to develop a general formalism using the Green function calculus.

First of all, we can assert that (3.1) is a differential equation of first order in time. This means that the temporal evolution of a wave function $\psi(\boldsymbol{x}, t)$ known at a specific time t is uniquely determined for all future times $t' > t$ as well as for all past times $t' < t$. Furthermore, (3.1) is linear so that, firstly, solutions are superposable, and, secondly, the relation between solutions at different times must be linear. From this follows that the wave function ψ must obey a linear integral equation of the form

$$\psi(\boldsymbol{x}', t') = \mathrm{i}\int \mathrm{d}^3x G(\boldsymbol{x}', t', \boldsymbol{x}, t)\psi(\boldsymbol{x}, t)$$

or

$$\psi(x') = \mathrm{i}\int \mathrm{d}^3x G(x', x)\psi(x) \ , \ x = (\boldsymbol{x}, t) \ , \quad (3.2)$$

where the integration is carried out over the whole space. $G(x', x)$ denotes the *Green function* containing the entire information with respect to the evolution from $\psi(x)$ to $\psi(x')$ during the time from t to t'. Obviously, the relation (3.2) does not distinguish between the two temporal directions of evolution, $t' > t$ and $t' < t$. However, with a view to our actual intention, namely the description of quantum mechanical scattering processes, such a distinction would be preferable. This can be achieved by splitting up (3.2) according to both temporal directions of evolution as follows:[2]

[1] Note that, in the following, the free position and time arguments are often denoted by $\boldsymbol{x}'$ und t' (instead of $\boldsymbol{x}$ and t).

[2] The positive sign holds for the upper equation, the negative sign for the lower one.

$$\left\{\begin{array}{l}\Theta(t'-t)\\ \Theta(t-t')\end{array}\right\}\psi(x') = \pm\mathrm{i}\int \mathrm{d}^3x G^{(\pm)}(x',x)\psi(x) \;, \tag{3.3}$$

where $\Theta(t)$ denotes the *step function* defined by

$$\Theta(t) = \left\{\begin{array}{l}0 \text{ for } t<0\\ \\ 1 \text{ for } t>0 \;.\end{array}\right.$$

As can be easily perceived from (3.3), now the Green functions $G^{(\pm)}(x',x)$ allow a propagation from $\psi(x)$ to $\psi(x')$ only in a positive or a negative temporal direction. Due to this causal behavior, one calls $G^{(+)}$ the *retarded propagator* and $G^{(-)}$ the *advanced propagator.*

Differential equation for $G^{(\pm)}$. In order to obtain a conditional equation for the propagators $G^{(\pm)}$, we apply the operator in (3.1) to the upper equation of (3.3) to get

$$\begin{aligned}\left(\mathrm{i}\hbar\frac{\partial}{\partial t'} - H'\right)\Theta(t'-t)\psi(x') &= \mathrm{i}\hbar\delta(t'-t)\psi(x')\\ &= \mathrm{i}\int \mathrm{d}^3x \left(\mathrm{i}\hbar\frac{\partial}{\partial t'} - H'\right) G^{(+)}(x',x)\psi(x)\end{aligned}$$

$$\Longrightarrow 0 = \int \mathrm{d}^3x \left[\left(\mathrm{i}\hbar\frac{\partial}{\partial t'} - H'\right) G^{(+)}(x',x) - \hbar\delta(t'-t)\delta(\boldsymbol{x}'-\boldsymbol{x})\right]\psi(x)$$

$$\Longrightarrow \left(\mathrm{i}\hbar\frac{\partial}{\partial t'} - H'\right) G^{(+)}(x',x) = \hbar\delta(x'-x) \;,$$

with

$$\delta(x'-x) = \delta(t'-t)\delta(\boldsymbol{x}'-\boldsymbol{x}) \;.$$

Applying $(\mathrm{i}\hbar\partial/\partial t' - H')$ to the lower equation of (3.3) leads to the exactly same differential equation for $G^{(-)}$ so that both cases can finally be combined in[3]

$$\left(\mathrm{i}\hbar\frac{\partial}{\partial t'} - H'\right) G^{(\pm)}(x',x) = \hbar\delta(x'-x) \;. \tag{3.4}$$

This equation along with the boundary conditions of the temporal direction of evolution,

$$G^{(+)}(x',x) = 0 \text{ for } t'<t \qquad \text{and} \qquad G^{(-)}(x',x) = 0 \text{ for } t'>t \;, \tag{3.5}$$

determines the propagators $G^{(\pm)}$ uniquely.

[3] This is the reason why the relative sign between the two equations was introduced in (3.3).

Integral equation for $G^{(\pm)}$ and ψ. At first sight, the introduction of the propagators $G^{(\pm)}$ does not seem to be advantageous for solving the Schrödinger equation (3.1), since, for the propagators, we also have to solve a Schrödinger-like differential equation (3.4) [with a space-time point source of "unity strength" $\hbar\delta(x'-x)$]. However, as we see shortly, the advantage of using propagators is that they lead to integral equations that, in general, can also not be solved exactly but at least approximately.

Writing the differential equation (3.4) in the form

$$\left(\mathrm{i}\hbar\frac{\partial}{\partial t'}-H'^{(0)}\right)G^{(\pm)}(x',x)=\hbar\delta(x'-x)+V(x')G^{(\pm)}(x',x)\;, \tag{3.6}$$

one can show that it is most generally solved by

$$\begin{aligned}G^{(\pm)}(x',x) &= G^{(0,\pm)}(x',x)\\ &\quad+\frac{1}{\hbar}\int \mathrm{d}^4x_1 G^{(0,\pm)}(x',x_1)V(x_1)G^{(\pm)}(x_1,x)\;,\end{aligned} \tag{3.7}$$

where $G^{(0,\pm)}$ denote the free retarded and advanced propagators that, in turn, fulfill the differential equation

$$\left(\mathrm{i}\hbar\frac{\partial}{\partial t'}-H'^{(0)}\right)G^{(0,\pm)}(x',x)=\hbar\delta(x'-x)\;. \tag{3.8}$$

This is because inserting (3.7) into (3.6) together with (3.8) yields

$$\begin{aligned}&\left(\mathrm{i}\hbar\frac{\partial}{\partial t'}-H'^{(0)}\right)G^{(\pm)}(x',x)\\ &\quad=\hbar\delta(x'-x)\\ &\qquad+\frac{1}{\hbar}\int \mathrm{d}^4x_1\left(\mathrm{i}\hbar\frac{\partial}{\partial t'}-H'^{(0)}\right)G^{(0,\pm)}(x',x_1)V(x_1)G^{(\pm)}(x_1,x)\\ &\quad=\hbar\delta(x'-x)+\int \mathrm{d}^4x_1\delta(x'-x_1)V(x_1)G^{(\pm)}(x_1,x)\\ &\quad=\hbar\delta(x'-x)+V(x')G^{(\pm)}(x',x)\;.\end{aligned}$$

Combining (3.7) with (3.3), we obtain an integral equation for ψ,

$$\begin{aligned}\psi(x') &= \pm\mathrm{i}\lim_{t\to\mp\infty}\int \mathrm{d}^3x G^{(\pm)}(x',x)\psi(x)\\ &= \pm\mathrm{i}\lim_{t\to\mp\infty}\left[\int \mathrm{d}^3x G^{(0,\pm)}(x'x)\psi(x)\right.\\ &\quad\left.+\frac{1}{\hbar}\int \mathrm{d}^3x\int \mathrm{d}^4x_1 G^{(0,\pm)}(x',x_1)V(x_1)G^{(\pm)}(x_1,x)\psi(x)\right]\\ &= \psi_{\mathrm{free}}(x')+\frac{1}{\hbar}\int \mathrm{d}^4x_1 G^{(0,\pm)}(x',x_1)V(x_1)\psi(x_1)\;,\end{aligned} \tag{3.9}$$

where

$$\psi_{\text{free}}(x') = \pm \mathrm{i} \lim_{t\to\mp\infty} \int \mathrm{d}^3x G^{(0,\pm)}(x',x)\psi(x)$$

solves the free Schrödinger equation. Thus, once the explicit form of the free propagators $G^{(0,\pm)}$ is known, the full propagators $G^{(\pm)}$ as well as ψ can be approximately determined for all times by iterating the relation (3.7) or (3.9)[4], whereas the causality principle (3.5) is automatically carried over from $G^{(0,\pm)}$ to $G^{(\pm)}$.

Theorem 3.1: Solution of the general Schrödinger equation in the propagator formalism

The solution to the general Schrödinger equation

$$\left(\mathrm{i}\hbar\frac{\partial}{\partial t'} - H'\right)\psi(x') = 0\ ,\ H' = H'^{(0)} + V(x')\ ,\ H'^{(0)} = \frac{\boldsymbol{p}'^2}{2m_0}$$

with the constraint $\psi(x)$ at time t is given for a later or sooner time t' by

$$\left\{\begin{matrix}\Theta(t'-t)\\ \Theta(t-t')\end{matrix}\right\}\psi(x') = \pm\mathrm{i}\int \mathrm{d}^3x G^{(\pm)}(x',x)\psi(x)\ . \tag{3.10}$$

$G^{(+)}$ and $G^{(-)}$ are called retarded and advanced propagators. They contain the entire temporally forward or backward directed dynamics of the problem and obey the differential equations

$$\left(\mathrm{i}\hbar\frac{\partial}{\partial t'} - H'\right)G^{(\pm)}(x',x) = \hbar\delta(x'-x) \tag{3.11}$$

as well as the integral equations

$$\begin{aligned} G^{(\pm)}(x',x) &= G^{(0,\pm)}(x',x) \\ &\quad + \frac{1}{\hbar}\int \mathrm{d}^4x_1 G^{(0,\pm)}(x',x_1)V(x_1)G^{(\pm)}(x_1,x)\ . \end{aligned} \tag{3.12}$$

$G^{(0,\pm)}$ denote the free retarded and free advanced propagators whose Fourier decomposition is given by

$$\left.\begin{aligned} G^{(0,\pm)}(x',x) &= G^{(0,\pm)}(x'-x) \\ &= \int\frac{\mathrm{d}^3p}{(2\pi\hbar)^3}\int\frac{\mathrm{d}E}{2\pi\hbar}\mathrm{e}^{\mathrm{i}\boldsymbol{p}(\boldsymbol{x}'-\boldsymbol{x})/\hbar}\mathrm{e}^{-\mathrm{i}E(t'-t)/\hbar}\tilde{G}^{(0,\pm)}(E,\boldsymbol{p}) \\ \tilde{G}^{(0,\pm)}(E,\boldsymbol{p}) &= \frac{\hbar}{E-\frac{\boldsymbol{p}^2}{2m_0}\pm\mathrm{i}\epsilon} \end{aligned}\right\} \tag{3.13}$$

(see Exercise 31). For ψ we have the integral equation ▷

[4] The correctness of (3.9) can be proven immediately by inserting it into the Schrödinger equation $\left(\mathrm{i}\hbar\partial/\partial t' - H^{(0)}\right)\psi(x') = V(x')\psi(x')$.

$$\psi(x') = \psi_{\text{free}}(x') + \frac{1}{\hbar}\int d^4x_1 G^{(0,\pm)}(x', x_1)V(x_1)\psi(x_1) \; . \tag{3.14}$$

In Subsection 3.1.2 we show that, besides (3.10), the relations

$$\left\{\begin{matrix} \Theta(t-t') \\ \Theta(t'-t) \end{matrix}\right\} \psi^*(x') = \pm \mathrm{i}\int d^3x \psi^*(x) G^{(\pm)}(x, x') \tag{3.15}$$

also hold. They describe the temporally backward and forward propagations of the complex conjugated wave function ψ^*, respectively.

3.1.2 Propagator Decomposition by Schrödinger Solutions

Before we combine Theorem 3.1 with quantum mechanical scattering processes, we first consider more concretely the form of the propagators and then show the validity of (3.15). We assume that a complete set of orthonormal solutions $\{\psi_n\}$ to the general Schrödinger equation is known, i.e.

$$\left(\mathrm{i}\hbar\frac{\partial}{\partial t'} - H'\right)\psi_n(x') = 0 \; , \; \sum_n \psi_n(\boldsymbol{x}', t')\psi_n^*(\boldsymbol{x}, t') = \delta(\boldsymbol{x}' - \boldsymbol{x}) \; ,$$

where the sum $\sum\limits_n$ is meant to be the generalized sum or integral over the spectrum of the quantum numbers n. It then follows that

$$\left.\begin{aligned} G^{(+)}(x', x) &= -\mathrm{i}\Theta(t'-t)\sum_n \psi_n(x')\psi_n^*(x) \\ G^{(-)}(x', x) &= +\mathrm{i}\Theta(t-t')\sum_n \psi_n(x')\psi_n^*(x) \end{aligned}\right\} \tag{3.16}$$

solve the differential equation

$$\left(\mathrm{i}\hbar\frac{\partial}{\partial t'} - H'\right) G^{(\pm)}(x', x) = \hbar\delta(x' - x) \; ,$$

because

$$\begin{aligned} &\left(\mathrm{i}\hbar\frac{\partial}{\partial t'} - H'\right) G^{(\pm)}(x', x) \\ &\quad = \hbar\delta(t'-t)\sum_n \psi_n(x')\psi_n^*(x) \\ &\quad\quad \mp \mathrm{i}\left\{\begin{matrix} \Theta(t'-t) \\ \Theta(t-t') \end{matrix}\right\}\sum_n \left[\left(\mathrm{i}\hbar\frac{\partial}{\partial t'} - H'\right)\psi_n(x')\right]\psi_n^*(x) \\ &\quad = \hbar\delta(t'-t)\delta(\boldsymbol{x}' - \boldsymbol{x}) = \hbar\delta(x' - x) \; . \end{aligned}$$

Since, in addition, the expressions (3.16) obey explicitly the causality principle, they are indeed the retarded and advanced propagators. With the help of (3.16) and the well-known solutions to the free Schrödinger equation,

$$\psi_{\boldsymbol{p}}(x) = \frac{1}{(2\pi\hbar)^{3/2}} \mathrm{e}^{\mathrm{i}\boldsymbol{p}\boldsymbol{x}/\hbar}\mathrm{e}^{-\mathrm{i}Et/\hbar} \;,\quad \int \mathrm{d}^3x\psi^*_{\boldsymbol{p}'}(x)\psi_{\boldsymbol{p}}(x) = \delta(\boldsymbol{p}'-\boldsymbol{p}) \;,$$

we are now in a position to derive explicit expressions for the free propagators $G^{(0,\pm)}$. Using $E = \boldsymbol{p}^2/2m_0$ and quadratic completion, we have

$$\begin{aligned}
\sum_n \psi_n(x')\psi_n^*(x) &= \int \mathrm{d}^3p\psi_{\boldsymbol{p}}(x')\psi^*_{\boldsymbol{p}}(x) \\
&= \frac{1}{(2\pi\hbar)^3}\int \mathrm{d}^3p \mathrm{e}^{-\mathrm{i}E(t'-t)/\hbar}\mathrm{e}^{\mathrm{i}\boldsymbol{p}(\boldsymbol{x}'-\boldsymbol{x})/\hbar} \\
&= \frac{1}{(2\pi\hbar)^3}\int \mathrm{d}^3p \exp\left(-\frac{\mathrm{i}\boldsymbol{p}^2(t'-t)}{2\hbar m_0} + \frac{\mathrm{i}\boldsymbol{p}(\boldsymbol{x}'-\boldsymbol{x})}{\hbar}\right) \\
&= \frac{1}{(2\pi\hbar)^3}\exp\left(\frac{\mathrm{i}m_0(\boldsymbol{x}'-\boldsymbol{x})^2}{2\hbar(t'-t)}\right) \\
&\quad \times \int \mathrm{d}^3p \exp\left[-\frac{\mathrm{i}(t'-t)}{2\hbar m_0}\left(\boldsymbol{p} - \frac{m_0(\boldsymbol{x}'-\boldsymbol{x})^2}{t'-t}\right)^2\right] \\
&= \left(\frac{m_0}{2\pi\mathrm{i}\hbar(t'-t)}\right)^{3/2} \exp\left(\frac{\mathrm{i}m_0(\boldsymbol{x}'-\boldsymbol{x})^2}{2\hbar(t'-t)}\right) \;.
\end{aligned}$$

Thus, in total, it follows that

$$\left.\begin{aligned}
G^{(0,+)}(x',x) &= -\mathrm{i}\Theta(t'-t) \\
&\quad \times \left(\frac{m_0}{2\pi\mathrm{i}\hbar(t'-t)}\right)^{3/2} \exp\left(\frac{\mathrm{i}m_0(\boldsymbol{x}'-\boldsymbol{x})^2}{2\hbar(t'-t)}\right) \\
G^{(0,-)}(x',x) &= +\mathrm{i}\Theta(t-t') \\
&\quad \times \left(\frac{m_0}{2\pi\mathrm{i}\hbar(t'-t)}\right)^{3/2} \exp\left(\frac{\mathrm{i}m_0(\boldsymbol{x}'-\boldsymbol{x})^2}{2\hbar(t'-t)}\right)
\end{aligned}\right\} \tag{3.17}$$

and

$$G^{(0,\pm)}(x',x) = G^{(0,\pm)}(x'-x) \;.$$

The latter is a consequence of the homogeneity of space and time and is generally true only for the free propagators. The relations (3.17) can be derived alternatively by passing over from the time-space to the energy-momentum representation and solving the differential equation (3.11) directly. In doing so, one obtains a complex energy integral for $G^{(0,\pm)}$ with one singularity at $E = \boldsymbol{p}^2/2m_0$, whose shift into the lower or upper half plane is equivalent to incorporating the causality principle belonging to $G^{(0,+)}$ or $G^{(0,-)}$ [see (3.13)]. In Exercise 31 we show the corresponding calculation, and we encounter some more of the same type in the following.

Relationship between temporally forward and backward directed evolutions. In order to show the validity of relation (3.15) stated at the end of Theorem 3.1, we first focus on the retarded case, multiply the first equation of (3.16) by $\psi_m^*(x')$ and then perform an integration over $\boldsymbol{x}'$. This, in conjunction with the orthonormality of the ψ_n, leads to the relation

$$\begin{aligned}\mathrm{i}\int \mathrm{d}^3x'\psi_m^*(x')G^{(+)}(x',x) &= \Theta(t'-t)\sum_n \underbrace{\int \mathrm{d}^3x'\psi_m^*(x')\psi_n(x')}_{\delta_{mn}}\,\psi_n^*(x)\\ &= \Theta(t'-t)\psi_m^*(x)\ .\end{aligned}$$

A similar relation holds for the advanced case so that, finally, both cases can be combined to give (3.15). Hence, the same propagators $G^{(\pm)}$ that describe the temporally forward and backward directed evolutions of a Schrödinger wave function according to (3.10) also determine the temporally backward and forward directed evolutions of the complex conjugated wave function ψ^*.

3.1.3 Scattering Formalism

We are now able to connect the propagator formalism developed in the preceding two subsections with nonrelativistic quantum mechanical scattering processes. To do this, we concentrate on the scattering of particles against a fixed scattering center.

A typical scattering experiment is depicted in Figure 3.1. A homoenergetic collimated particle beam is directed toward a fixed target. At a sufficiently large distance, the particles scattered by a certain angle θ relative to the incoming beam are registered by a detector.[5] This means that the detector essentially counts all scattered particles with a momentum directed toward the spherical surface element $\mathrm{d}\Omega$. Usually the setting of such experiments is chosen in a way so that the following prerequisites are fulfilled:

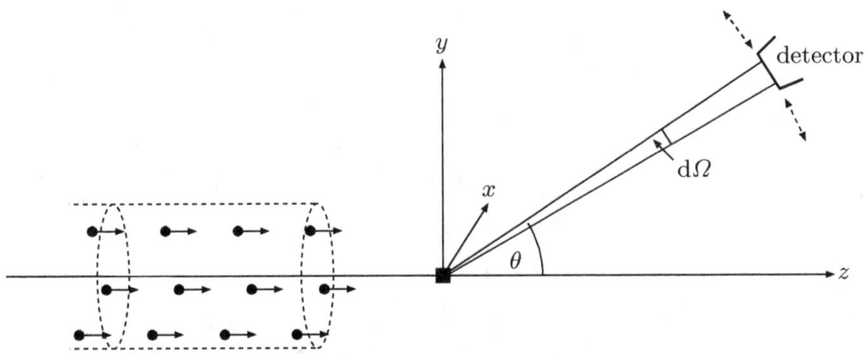

Fig. 3.1. Experimental arrangement of particle scattering against a fixed target.

[5] For rotationally symmetric potentials the measurements are independent from the azimuth angle φ.

1. The interaction potential of the target is spatially confined:

$$\lim_{|\boldsymbol{x}|\to\infty} V(\boldsymbol{x},t) = 0 \ .$$

2. The particle beam is created at a sufficiently large distance from the target so that the incoming particles can be regarded as initially free.

3. The detector is located at a sufficiently large distance from the target so that the registered scattered particles can also be treated as free particles.

With a view to a quantum mechanical description of such scattering processes, we need to tackle the following question: *given a free wave packet with mean momentum* $\boldsymbol{p}_i$ *(i=incident) moving toward the target. What shape does the wave packet possess "long after" the scattering where it can be regarded as free or, putting it differently, what is the probability for the wave packet to have the sharp momentum* $\boldsymbol{p}_f$ *(f=final) long after the scattering?* Once we have answered this question, we can sum over the individual probabilities for scattering momenta directed toward $\mathrm{d}\Omega$ and relate them to experimentally detected particle numbers.

For the mathematical realization of this program, it is much easier to describe the initially created particles propagating toward the target by plane waves instead of wave packets, which we do in the following. However, this implies that the actual scattering process is no longer confined in space and time since incoming and scattered waves now coexist (see Figure 3.2). Consequently, the above 2. and 3. prerequisites are violated. A way out of this is provided by the *adiabatic approximation.* Here it is assumed that the interaction potential V is temporally localized according to

$$\lim_{t\to\pm\infty} V(\boldsymbol{x},t) \to 0$$

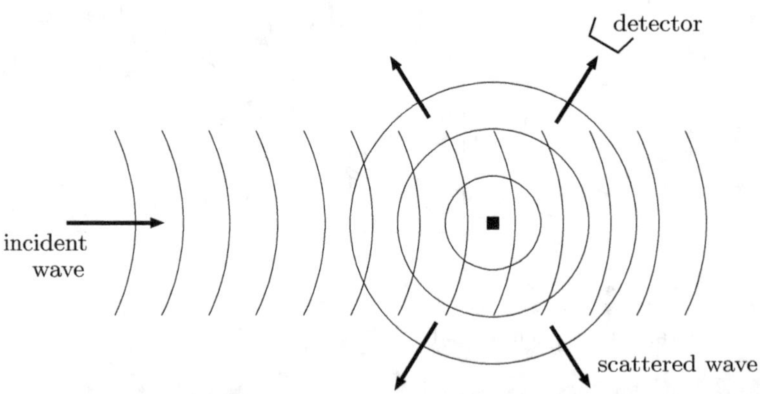

Fig. 3.2. Idealized description of a quantum mechanical scattering process where incoming particles are represented as initially free plane waves. Due to the infinite extent of plane waves, the space-time confinement of the actual scattering is lost.

in such a way that it is *adiabatically switched on* in the far past $t \to -\infty$ and *adiabatically switched off* in the far future $t \to +\infty$.[6] To some extent, this approximation turns the space-time confinement of scattering processes given automatically for wave packets into a pure temporal confinement for plane waves. Similarly to the above 2. and 3. prerequisites, this allows us to regard incoming and scattering waves as being free in the far past and the far future and, particularly,

- to represent the incoming wave function as a plane wave with momentum $\boldsymbol{p}_i$ in the limit $t \to -\infty$,
- to study the projection of the scattered wave onto plane waves with scattering momenta $\boldsymbol{p}_f$ in the direction of $\mathrm{d}\Omega$ in the limit $t \to +\infty$.

Clearly, interference effects between the incoming and scattering waves have to be excluded when doing so.

Scattering amplitude, scattering matrix. Let us now assume that[7]

$$\Psi_i(x) = \frac{1}{(2\pi\hbar)^{3/2}} \mathrm{e}^{\mathrm{i}\boldsymbol{p}_i\boldsymbol{x}/\hbar} \mathrm{e}^{-\mathrm{i}E_i t/\hbar}$$

is the incoming plane wave with momentum $\boldsymbol{p}_i$, ψ_i is the temporally forward moving scattered wave,

$$\lim_{t\to-\infty} \psi_i(x) = \Psi_i(x) \ ,$$

and

$$\Psi_f(x) = \frac{1}{(2\pi\hbar)^{3/2}} \mathrm{e}^{\mathrm{i}\boldsymbol{p}_f\boldsymbol{x}/\hbar} \mathrm{e}^{-\mathrm{i}E_f t/\hbar}$$

is a plane wave with momentum $\boldsymbol{p}_f$. According to the above, we are interested in the projection of ψ_i onto Ψ_f in the far future, i.e.

$$S_{fi} = \lim_{t'\to+\infty} \int \mathrm{d}^3x' \Psi_f^*(x')\psi_i(x') \ . \tag{3.18}$$

This expression is called *scattering amplitude, transition amplitude* or *probability amplitude* for the transition $\Psi_i \to \Psi_f$. The totality of all the scattering

6 In this context, adiabatic means that the solutions to the Schrödinger equation are approximated by the stationary eigensolutions to the instantaneous Hamilton operator so that, at a given time, a certain eigenfunction passes over continuously to the corresponding eigenfunction at a later time. The justification of the adiabatic approximation becomes apparent in calculations of concrete scattering processes where the wave function is initially normalized to a finite volume V so that the actual scattering process is confined to the time range $[-T/2 : T/2]$. It is then easy to imagine that the potential is adiabatically switched on in the time interval $[-\infty : -T/2]$ and adiabatically switched off in the interval $[T/2 : \infty]$; not until the end of the calculations is the limiting process $V, T \to \infty$ performed.

7 Here and in the following the plane solutions to the free Schrödinger equation are denoted by the capital symbol Ψ.

amplitudes are the elements of the *Heisenberg scattering matrix*, also called the *S-matrix*. For further evaluation of (3.18), we can make use of Theorem 3.1. Taking into account [see (3.14)][8]

$$\psi_i(x') = \Psi_i(x') + \frac{1}{\hbar}\int \mathrm{d}^4x_1 G^{(0,+)}(x'-x_1)V(x_1)\psi_i(x_1)$$

and

$$\Psi_f^*(x_1) = \lim_{t'\to+\infty} \mathrm{i}\int \mathrm{d}^3x' \Psi_f^*(x')G^{(0,+)}(x'-x_1)$$

$$\int \mathrm{d}^3x' \Psi_f^*(x')\Psi_i(x') = \delta(\boldsymbol{p}_f - \boldsymbol{p}_i)\ ,$$

it follows from (3.18) that[9]

$$\begin{aligned} S_{fi} &= \lim_{t'\to+\infty}\left[\int \mathrm{d}^3x' \Psi_f^*(x')\Psi_i(x')\right. \\ &\quad \left. + \frac{1}{\hbar}\int \mathrm{d}^4x_1 \int \mathrm{d}^3x' \Psi_f^*(x')G^{(0,+)}(x'-x_1)V(x_1)\psi_i(x_1)\right] \\ &= \delta(\boldsymbol{p}_f - \boldsymbol{p}_i) - \frac{\mathrm{i}}{\hbar}\int \mathrm{d}^4x_1 \Psi_f^*(x_1)V(x_1)\psi_i(x_1)\ . \end{aligned}$$

Iterating ψ_i in the way

$$\begin{aligned} \psi_i(x_1) &= \Psi_i(x_1) \\ &\quad + \frac{1}{\hbar}\int \mathrm{d}^4x_2 G^{(0,+)}(x_1-x_2)V(x_2)\Psi(x_2) \\ &\quad + \frac{1}{\hbar^2}\int \mathrm{d}^4x_2 \int \mathrm{d}^4x_3 G^{(0,+)}(x_1-x_2)V(x_2) \\ &\quad \times G^{(0,+)}(x_2-x_3)V(x_3)\Psi(x_2) \\ &\quad + \ldots\ , \end{aligned}$$

we finally obtain

Theorem 3.2: Scattering matrix in the Schrödinger theory

With respect to the quantum mechanical scattering of a particle against a target, the scattering matrix element S_{fi} is defined as the probability amplitude for the transition $\Psi_i \longrightarrow \Psi_f$, where Ψ_i denotes the free initial state long before and Ψ_f the free final state long after the scattering. In other terms, S_{fi} is the projection of ψ_i onto Ψ_f, where ψ_i is the evolutionary state during the scattering process originating from Ψ_i: ▷

[8] Since we wish to study the scattering process in temporally forward direction, the retarded case has to be chosen.

[9] In concrete calculations of scattering amplitudes we will use wave functions normalized to a box volume V instead of the continuum normalization. This implies the replacement $\delta(\boldsymbol{p}_f - \boldsymbol{p}_i) \to \delta_{fi}$.

$$S_{fi} = \lim_{t' \to +\infty} \int \mathrm{d}^3x' \Psi_f^*(x')\psi_i(x') \;, \quad \lim_{t \to -\infty} \psi_i(x) = \Psi_i(x) \;. \tag{3.19}$$

Using the propagator formalism, S_{fi} can be expanded in the following way:

$$\begin{aligned} S_{fi} &= \delta(\boldsymbol{p}_f - \boldsymbol{p}_i) - \frac{\mathrm{i}}{\hbar} \int \mathrm{d}^4x_1 \Psi_f^*(x_1) V(x_1) \psi_i(x_1) \\ &= \delta(\boldsymbol{p}_f - \boldsymbol{p}_i) \\ &\quad - \frac{\mathrm{i}}{\hbar} \int \mathrm{d}^4x_1 \Psi_f^*(x_1) V(x_1) \Psi_i(x_1) \\ &\quad - \frac{\mathrm{i}}{\hbar^2} \int \mathrm{d}^4x_1 \int \mathrm{d}^4x_2 \Psi_f^*(x_2) V(x_2) G^{(0,+)}(x_2 - x_1) V(x_1) \Psi_i(x_1) \\ &\quad - \frac{\mathrm{i}}{\hbar^3} \int \mathrm{d}^4x_1 \int \mathrm{d}^4x_2 \int \mathrm{d}^4x_3 \Psi_f^*(x_3) V(x_3) G^{(0,+)}(x_3 - x_2) \\ &\quad \times V(x_2) G^{(0,+)}(x_2 - x_1) V(x_1) \Psi_i(x_1) \\ &\quad - \ldots \;. \end{aligned} \tag{3.20}$$

V denotes the interaction potential of the target and $G^{(0,+)}$ the free retarded propagator.
Due to the adiabatic approximation, it is justified to consider Ψ_i to be a free plane wave in the limit $t \to -\infty$ and to study the projection of ψ_i onto plane waves in the limit $t \to +\infty$.

As one can see, at the transition from (3.19) to (3.20), the unknown wave function ψ_i has completely disappeared. Instead, the known free propagator $G^{(0,+)}$ appears in a series of *multiple scatterings* which encompasses the entire dynamics of the scattering process. For a better understanding of this theorem, the following items have to be kept in mind:

- A general property of the scattering matrix S, following from the Hermitecity of the Schrödinger-Hamilton operator, is its unitarity (see Exercise 33) expressing, once again, the conservation of the total probability. Furthermore, it is easy to realize that the S matrix shares all the properties of the Hamilton operator.
- From the Hermitecity of the Hamilton operator follows that the S matrix can be defined equivalently via

  $$S_{fi} = \lim_{t \to -\infty} \int \mathrm{d}^3x \psi_f^*(x) \Psi_i(x) \;, \quad \lim_{t' \to +\infty} \psi_f(x') = \Psi_f(x') \;,$$

  where ψ_f denotes the wave that, originating from Ψ_f in the far future, is moving backward in time and is scattered against V. This is because, using

  $$\begin{aligned} \psi_i(x') &= \lim_{t \to -\infty} \mathrm{i} \int \mathrm{d}^3x G^{(+)}(x', x) \Psi_i(x) \\ \psi_f^*(x') &= \lim_{t' \to +\infty} \mathrm{i} \int \mathrm{d}^3x' \Psi_f^*(x') G^{(+)}(x', x) \;, \end{aligned}$$

we have

$$\begin{aligned} S_{fi} &= \lim_{t\to-\infty} \int \mathrm{d}^3x \psi_f^*(x)\Psi_i(x) \\ &= \lim_{\substack{t\to-\infty \\ t'\to+\infty}} \mathrm{i}\int \mathrm{d}^3x \int \mathrm{d}^3x' \Psi_f^*(x') G^{(+)}(x',x)\Psi_i(x) \\ &= \lim_{t'\to+\infty} \int \mathrm{d}^3x' \Psi_f^*(x')\psi_i(x') \ . \end{aligned}$$

Consequently, it makes no difference whether we let the initial state Ψ_i propagate in temporally forward direction or the final state Ψ_f in temporally backward direction.

- Equation (3.20) essentially represents an expansion in powers of the interaction potential. In practice, only the first few terms need to be calculated depending on how fast the expansion converges.
- Taking into account

$$\begin{aligned} \Psi_i(x_1) &= \lim_{t\to-\infty} \mathrm{i}\int \mathrm{d}^3x G^{(0,+)}(x_1 - x)\Psi_i(x) \\ \Psi_f^*(x_n) &= \lim_{t'\to+\infty} \mathrm{i}\int \mathrm{d}^3x' \Psi_f^*(x') G^{(0,+)}(x' - x_n) \ , \end{aligned}$$

the individual terms in (3.20) can be interpreted as follows: during its temporal evolution between the space-time points $x = (\boldsymbol{x}, t)$ and $x' = (\boldsymbol{x}', t')$, the particle is scattered at different intermediate points x_i with the probability amplitude $V(x_i)$ per space-time volume unit into a particle moving undisturbed with the probability amplitude $G^{(0,+)}(x_{i+1} - x_i)$ toward the next interaction point x_{i+1}. Thereby, the integration is carried out over all possible intermediate points. Another way of putting it is: at the *vortex* x_i, the potential $V(x_i)$ destroys a particle that has moved freely to x_i and creates a particle that will move freely to x_{i+1}. Here it is of vital importance that the propagators $G^{(0,+)}$ allow only chronological scatterings (in the temporally forward direction). This implies that there cannot exist any terms leading to $\vee$- or $\wedge$-like diagrams (see Figure 3.3). As we will see later, this is different in relativistic scattering theories.

Total cross section, differential cross section. Having found a reliable formalism for the description of nonrelativistic quantum mechanical transitions in terms of the scattering amplitudes S_{fi}, we now have to relate it to the particle numbers detected in scattering experiments. To this end, we initially ask for the *transition rate* R, i.e. the number of particles scattered in any direction (except for the direction of incidence) within the time T:

$$R = \frac{\text{number of scattered particles}}{T} \ .$$

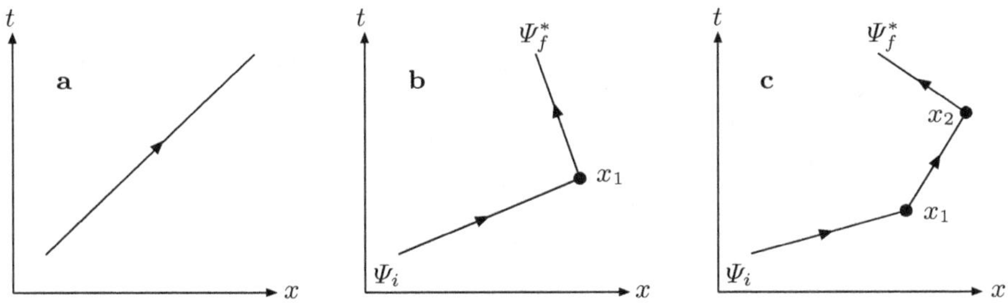

Fig. 3.3. Pictorial representations (*Feynman diagrams*) of the series expansion given by (3.20). **a** represents the first δ-term (zeroth order), i.e. the free propagation of a particle. **b** symbolizes the second term (first order). At the intermediate point $x_1 = (\boldsymbol{x}_1, t_1)$ the particle suffers one scattering at the potential V. **c** depicts the third term (second order) where the particle is scattered at two intermediate points $x_1 = (\boldsymbol{x}_1, t_1)$ and $x_2 = (\boldsymbol{x}_2, t_2)$. In principle, only those space-time paths are possible that are directed temporally forward.

If we know the number N of particles moving through the cross sectional area A of the incident particle beam per time interval T, as well as the *total cross section* σ of the target, i.e. the area which has to be crossed perpendicularly by the incident particles to be deflected at all, we have for R:

$$R = \frac{N}{T}\frac{\sigma}{A} = \rho v \sigma = |\boldsymbol{j}_i|\sigma \ , \ \rho = \frac{N}{vTA} \ . \tag{3.21}$$

ρ denotes the particle density, v the particle velocity and $|\boldsymbol{j}_i|$ the particle current density of the incident particle beam (see Figure 3.4).[10] On the other

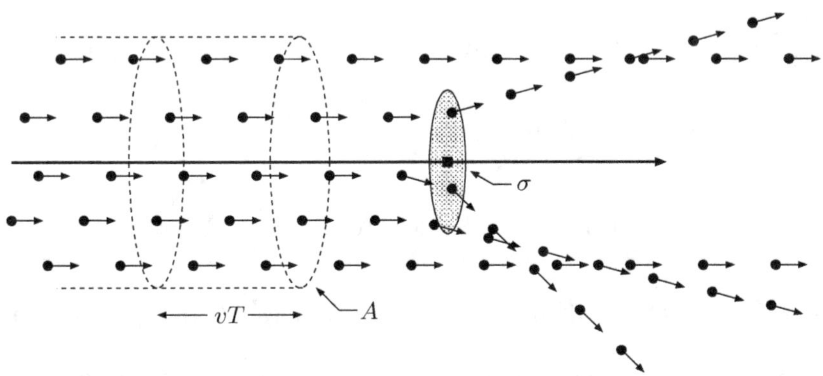

Fig. 3.4. Scattering of a particle beam with particle current density $|\boldsymbol{j}_i| = \rho v$ and cross sectional area A against a fixed scattering center with total cross section σ.

[10] Strictly speaking, this is the transition rate per target particle since, in real scattering experiments, the target consists of many particles. However, there mutual distances can be assumed to be large compared to σ.

hand, the transition rate R can also be expressed using the scattering amplitude S_{fi} since, in the wave picture, it denotes the probability for the transition $\Psi_i \longrightarrow \Psi_f$. Consequently, in the experimentally relevant particle picture, it gives the statistical fraction of the incident particles with momentum $\boldsymbol{p}_i$ scattered into states with momentum $\boldsymbol{p}_f$ so that

$$R = \frac{N}{T}\sum_f |S_{fi}|^2 = \frac{N}{T}\int |S_{fi}|^2 \mathrm{d}N_f \ , \tag{3.22}$$

where $\mathrm{d}N_f$ denotes the state density or the number of all possible states in the momentum interval $[\boldsymbol{p}_f : \boldsymbol{p}_f + \mathrm{d}^3 p_f]$. Combining the equations (3.21) and (3.22) leads to the total cross section

$$\sigma = \frac{N}{T|\boldsymbol{j}_i|}\int |S_{fi}|^2 \mathrm{d}N_f \ , \tag{3.23}$$

with the expected dimension of an area. For further evaluation of this expression, we have to keep in mind that we aim to describe the entire quantum mechanical scattering process with the help of plane waves, instead of localized wave packets. This means that, due to the plane waves' normalization to the whole space, the state density $\mathrm{d}N_f$ becomes arbitrarily large and the particle number in any considered volume vTA arbitrarily small. However, we can circumvent these ill-defined mathematical infinities by initially normalizing the plane waves to a finite volume V and performing the limiting process $V, T \to \infty$ at the end. As a result we have $N = 1$ (there is exactly one particle in the volume V). Furthermore, the state density becomes finite due to the requirement of standing waves within the volume V (i.e. periodical boundary conditions at the volume's border). Therefore, we can write

$$V = \prod_k L_k \ , \ \mathrm{e}^{\mathrm{i} p_{f\,k} L_k/\hbar} = 1 \Longrightarrow p_{f\,k} L_k/\hbar = 2\pi n_k \Longrightarrow \mathrm{d}n_k = \frac{L_k}{2\pi\hbar}\mathrm{d}p_{f\,k}$$

$$\Longrightarrow \mathrm{d}N_f = \prod_k \mathrm{d}n_k = \frac{V\mathrm{d}^3 p_f}{(2\pi\hbar)^3} \ \textit{(phase space factor)} \ .$$

Thus, taking into account $\mathrm{d}^3 p_f = \boldsymbol{p}_f^2 \mathrm{d}|\boldsymbol{p}_f|\mathrm{d}\Omega$ (compare to Figure 3.1), (3.23) turns into

$$\sigma = \frac{V}{(2\pi\hbar)^3 T|\boldsymbol{j}_i|}\int |S_{fi}|^2 \mathrm{d}^3 p_f = \frac{V}{(2\pi\hbar)^3 T|\boldsymbol{j}_i|}\int \mathrm{d}|\boldsymbol{p}_f|\boldsymbol{p}_f^2 \int \mathrm{d}\Omega |S_{fi}|^2 \ . \tag{3.24}$$

With regard to the experimental situation, this expression can be interpreted as

$$\sigma = \frac{\text{number of scattered particles}/T}{\text{incident particle current density}} \ .$$

Differentiating (3.24) with respect to Ω and suppressing the remaining $\boldsymbol{p}_f$-integral sign, we obtain the dimensionless *differential cross section*

$$\frac{\mathrm{d}\sigma}{\mathrm{d}\Omega} = \frac{|S_{fi}|^2}{T|\boldsymbol{j}_i|} \frac{V\boldsymbol{p}_f^2 \mathrm{d}|\boldsymbol{p}_f|}{(2\pi\hbar)^3} ,$$

with the experimental interpretation

$$\frac{\mathrm{d}\sigma}{\mathrm{d}\Omega} = \frac{\text{number of particles scattered toward } \mathrm{d}\Omega/T}{\text{incident particle current density}\cdot\mathrm{d}\Omega} .$$

This is exactly the quantity measured in scattering experiments if $\mathrm{d}\Omega$ is replaced by the small but finite detector surface.

Theorem 3.3: Cross section

The differential cross section of a quantum mechanical scattering is given by

$$\mathrm{d}\sigma = \frac{|S_{fi}|^2}{T|\boldsymbol{j}_i|} \frac{V\mathrm{d}^3 p_f}{(2\pi\hbar)^3} \left(\frac{\text{number of particles scattered toward } [\boldsymbol{p}_f : \boldsymbol{p}_f + \mathrm{d}^3 \boldsymbol{p}_f] \Big/ T}{\text{incident particle current density}} \right)$$

or ($\boldsymbol{p}_f$-integral sign suppressed)

$$\frac{\mathrm{d}\sigma}{\mathrm{d}\Omega} = \frac{|S_{fi}|^2}{T|\boldsymbol{j}_i|} \frac{V\boldsymbol{p}_f^2 \mathrm{d}|\boldsymbol{p}_f|}{(2\pi\hbar)^3} \left(\frac{\text{number of particles scattered toward } \mathrm{d}\Omega/T}{\text{incident particle current density}\cdot\mathrm{d}\Omega} \right) .$$

S_{fi} denotes the scattering amplitude for the transition $\Psi_i \to \Psi_f$, V the normalization volume of $\Psi_{i,f}$, $\boldsymbol{j}_i$ the particle current density of Ψ_i (toward the target), T the duration of the scattering process, and $\mathrm{d}\Omega$ the considered spherical surface element. Within concrete calculations and in the limit $V, T \to \infty$, the quantities V and T are removed.
Integrating the last expression over Ω yields the total cross section σ. In the particle picture it represents the effective area of the target that incident particles have to cross to be deflected at all.

Note that this theorem does not rely on specifically nonrelativistic relationships. Therefore, it is also valid for relativistic scattering theories. All in all, it can be ascertained that with the last three Theorems 3.1 to 3.3 we have found a viable formalism for the description of nonrelativistic quantum mechanical scattering processes that particularly allows a direct comparison with experimental measurements. In the following our task will be an adequate extension of this formalism to relativistic scattering processes within the scope of the Dirac and Klein-Gordon theories.

3.1.4 Coulomb Scattering

At the end of our review of nonrelativistic scattering theory we demonstrate its application by considering the concrete problem of Coulomb scattering (Rutherford scattering). We assume an interaction potential of the form

$$V(x) = \frac{\alpha}{|\boldsymbol{x}|} \; , \; \alpha = -Ze^2$$

that is hit by a particle beam directed along the z-axis, and we restrict ourselves to the calculation of the scattering amplitude as well as the differential cross section to leading order. First we need the plane solutions Ψ_i and Ψ_f to the free Schrödinger equation normalized to a volume V:

$$\Psi_i(x) = \frac{1}{\sqrt{V}} \mathrm{e}^{\mathrm{i}\boldsymbol{p}_i \boldsymbol{x}/\hbar} \mathrm{e}^{-\mathrm{i}E_i t/\hbar} \; , \; \Psi_f(x) = \frac{1}{\sqrt{V}} \mathrm{e}^{\mathrm{i}\boldsymbol{p}_f \boldsymbol{x}/\hbar} \mathrm{e}^{-\mathrm{i}E_f t/\hbar} \; .$$

Following Theorem 3.2, the scattering amplitude can then be written as ($f \neq i$):

$$\begin{aligned} S_{fi} &= -\frac{\mathrm{i}}{\hbar} \int \mathrm{d}^4 x \Psi_f^*(x) V(x) \Psi_i(x) \\ &= -\frac{\mathrm{i}}{\hbar V} \int\limits_{-T/2}^{T/2} \mathrm{d}t \mathrm{e}^{\mathrm{i}(E_f - E_i)t/\hbar} \int\limits_V \mathrm{d}^3 x \mathrm{e}^{-\mathrm{i}\boldsymbol{q}\boldsymbol{x}/\hbar} \frac{\alpha}{|\boldsymbol{x}|} \; , \; \boldsymbol{q} = \boldsymbol{p}_f - \boldsymbol{p}_i \; , \end{aligned}$$

with the momentum transfer $\boldsymbol{q}$. Note that the temporal confinement of the scattering in the interval $[-T/2 : T/2]$ is accomplished by the spatial confinement of $\Psi_{i,f}$ to the volume V. For the time and space integrals follows that

$$\int\limits_{-T/2}^{T/2} \mathrm{d}t \mathrm{e}^{\mathrm{i}(E_f - E_i)t/\hbar} \stackrel{T \to \infty}{=} 2\pi\hbar\delta(E_f - E_i) \tag{3.25}$$

$$\begin{aligned} \int\limits_V \mathrm{d}^3 x \frac{\mathrm{e}^{-\mathrm{i}\boldsymbol{q}\boldsymbol{x}/\hbar}}{|\boldsymbol{x}|} &= -\frac{\hbar^2}{\boldsymbol{q}^2} \int\limits_V \mathrm{d}^3 x \frac{1}{|\boldsymbol{x}|} \boldsymbol{\nabla}^2 \mathrm{e}^{-\mathrm{i}\boldsymbol{q}\boldsymbol{x}/\hbar} \\ &= -\frac{\hbar^2}{\boldsymbol{q}^2} \int\limits_V \mathrm{d}^3 x \left(\boldsymbol{\nabla}^2 \frac{1}{|\boldsymbol{x}|} \right) \mathrm{e}^{-\mathrm{i}\boldsymbol{q}\boldsymbol{x}/\hbar} \\ &= -\frac{\hbar^2}{\boldsymbol{q}^2} \int\limits_V \mathrm{d}^3 x [-4\pi\delta(\boldsymbol{x})] \mathrm{e}^{-\mathrm{i}\boldsymbol{q}\boldsymbol{x}/\hbar} = \frac{4\pi\hbar^2}{\boldsymbol{q}^2} \; , \end{aligned} \tag{3.26}$$

so that the scattering amplitude finally turns into

$$S_{fi} = -\mathrm{i} \frac{4\pi\hbar\alpha[2\pi\hbar\delta(E_f - E_i)]}{V\boldsymbol{q}^2} \; .$$

Now we need to supply the particle current density in z-direction to calculate the differential cross section:

$$|\boldsymbol{j}_i| = \frac{\hbar}{2\mathrm{i}m_0} |\Psi_i^* \boldsymbol{\nabla}\Psi_i - \Psi_i \boldsymbol{\nabla}\Psi_i^*| = \frac{|\boldsymbol{p}_i|}{m_0 V} \; .$$

Hence, according to Theorem 3.3, we obtain

$$\frac{\mathrm{d}\sigma}{\mathrm{d}\Omega} = \frac{2\alpha^2 m_0 [2\pi\hbar\delta(E_f - E_i)]^2 \boldsymbol{p}_f^2 \mathrm{d}|\boldsymbol{p}_f|}{\pi\hbar T |\boldsymbol{p}_i| \boldsymbol{q}^4} , \tag{3.27}$$

where the integration has to be carried out over all scattering momenta $\boldsymbol{p}_f$. Obviously, this formula contains the mathematically ill-defined expression $\delta^2(E_f - E_i)$ resulting from the fact that we have performed the limiting process $T \to \infty$ in (3.25) too early. In Exercise 34 we show that for a finite scattering time T, it is justified to make the replacement

$$[2\pi\hbar\delta(E_f - E_i)]^2 \longrightarrow 2\pi T\hbar\delta(E_f - E_i) \tag{3.28}$$

in (3.27) so that the scattering amplitude becomes

$$\frac{\mathrm{d}\sigma}{\mathrm{d}\Omega} = \int_0^\infty \mathrm{d}|\boldsymbol{p}_f| \frac{4\alpha^2 m_0 \delta(E_f - E_i) \boldsymbol{p}_f^2}{|\boldsymbol{p}_i| \boldsymbol{q}^4} .$$

This equation is well-defined and the normalization volume V as well as the scattering time T have disappeared. Furthermore, the remaining δ-function expresses explicitly the energy conservation of the scattering process. Using the identities

$$E_{i,f} = \frac{\boldsymbol{p}_{i,f}^2}{2m_0} , \quad \mathrm{d}|\boldsymbol{p}_f| = \frac{m_0 \mathrm{d}E_f}{|\boldsymbol{p}_f|} , \quad \boldsymbol{p}_i\boldsymbol{p}_f\big|_{|\boldsymbol{p}_f|=|\boldsymbol{p}_i|} = \boldsymbol{p}_i^2 \cos\theta$$

$$\boldsymbol{q}^2_{|\boldsymbol{p}_f|=|\boldsymbol{p}_i|} = 4\boldsymbol{p}_i^2 \sin^2\frac{\theta}{2} = 8m_0 E_i \sin^2\frac{\theta}{2} ,$$

$\mathrm{d}\sigma/\mathrm{d}\Omega$ can be further simplified to give the famous Rutherford scattering formula

$$\begin{aligned}\frac{\mathrm{d}\sigma}{\mathrm{d}\Omega} = \left(\frac{\mathrm{d}\sigma}{\mathrm{d}\Omega}\right)_{\mathrm{Ruth}} &= \int_0^\infty \mathrm{d}E_f \frac{4\alpha^2 m_0^2 \delta(E_f - E_i) \boldsymbol{p}_f^2}{|\boldsymbol{p}_i||\boldsymbol{p}_f| \boldsymbol{q}^4} \\ &= \frac{4\alpha^2 m_0^2}{\boldsymbol{q}^4}\bigg|_{|\boldsymbol{p}_f|=|\boldsymbol{p}_i|} = \left(\frac{\alpha}{4E_i \sin^2\frac{\theta}{2}}\right)^2 .\end{aligned} \tag{3.29}$$

Note that this equation is true exactly and not only in the demonstrated approximation.

Summary

- The solution of the general Schrödinger equation can be related back to an integral equation for the **retarded** and **advanced propagators** as well as for the wave function itself which is solvable approximately through iteration.
- The retarded propagator describes the temporally forward evolution of a wave function known at a fixed time, while the advanced propagator describes its temporally backward evolution.

▷

- Quantum mechanical scattering processes are described by the **scattering matrix**. With the help of the propagator formalism, the **scattering amplitudes** can be expanded in powers of the interaction potential (series of **multiple scatterings**).
- Due to the **adiabatic approximation**, incident wave functions can be regarded as free plane waves in the limit $t \to -\infty$. In the limit $t \to +\infty$ the scattered wave can also be regarded as free (but in general not as plane).
- In scattering experiments one usually directs a collimated particle beam toward a target and measures the **differential cross section**, i.e. the number of particles scattered toward different spherical surface elements. The scattering amplitudes are of vital importance for a theoretical description of these measures.
- Integrating the differential cross section over all spherical surface elements, one obtains the **total cross section**. It is equal to the (fictitious) area of the target that incident particles have to cross to be deflected at all.

Exercises

30. Integral representation of the Θ-function. Show that the step function

$$\Theta(\tau) = \begin{cases} 1 \text{ for } \tau > 0 \\ 0 \text{ for } \tau < 0 \end{cases}$$

can be expressed by

$$\Theta(\tau) = -\frac{1}{2\pi \mathrm{i}} \lim_{\epsilon \to 0} \int_{-\infty}^{\infty} \mathrm{d}\omega \frac{\mathrm{e}^{-\mathrm{i}\omega\tau}}{\omega + \mathrm{i}\epsilon} \;,\; \epsilon > 0 \;. \tag{3.30}$$

Solution. To calculate this expression we perform the integration in the complex ω-plane along a closed integration path K depending on τ.

$\tau < 0$: in this case the upper semicircle $\mathrm{H}(0, R, +)$ shown in Figure 3.5 seems to be appropriate in the limit $R \to \infty$ since then the integration along the circular arc $\mathrm{B}(0, R, +)$ vanishes so that we are left with the integration along the real ω-axis. This is because, using the parametrization

$$\mathrm{B}(0, R, +) : \; t \to \omega(t) = \mathrm{e}^{\mathrm{i}Rt} \;,\; t \in [0 : \pi] \;,$$

we can make the following estimation:

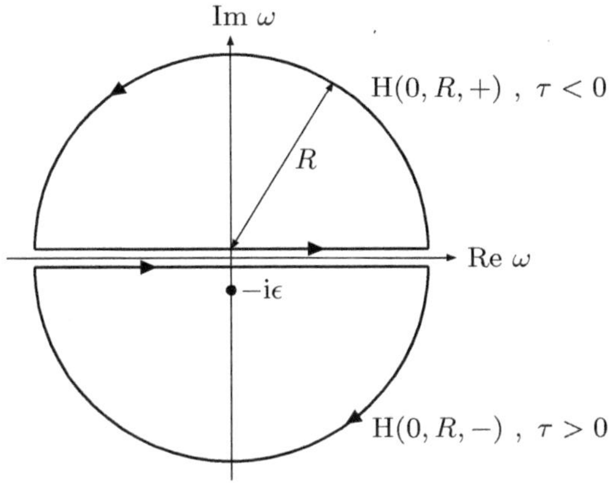

Fig. 3.5. Integration paths in the complex ω-plane. The upper closed semicircle $\mathrm{H}(0,R,+)$ in the mathematically positive direction is to be chosen for $\tau < 0$ and the lower closed semicircle $\mathrm{H}(0,R,-)$ in the mathematically negative direction for $\tau > 0$.

$$\begin{aligned}\left|\int_{\mathrm{B}(0,R,+)} \mathrm{d}\omega \frac{\mathrm{e}^{-\mathrm{i}\omega\tau}}{\omega}\right| &= \left|\int_0^\pi \mathrm{d}t \frac{\mathrm{e}^{-\mathrm{i}\tau R(\cos t + \mathrm{i}\sin t)}}{R\mathrm{e}^{\mathrm{i}t}} \mathrm{i}R\mathrm{e}^{\mathrm{i}t}\right| \\ &= \left|\mathrm{i}\int_0^\pi \mathrm{d}t \mathrm{e}^{-\mathrm{i}\tau R\cos t}\mathrm{e}^{\tau R\sin t}\right| \\ &< \left|\mathrm{i}\int_0^\pi \mathrm{d}t \mathrm{e}^{-|\tau| R\sin t}\right| \stackrel{R\to\infty}{\longrightarrow} 0 \,.\end{aligned}$$

Now the only singularity $\omega_{\mathrm{s}} = -\mathrm{i}\epsilon$ of the integrand lies outside the region enclosed by $\mathrm{H}(0,R,+)$ so that, according to Cauchy's integral theorem, we have[11]

$$\Theta(\tau) = -\lim_{\substack{\epsilon\to 0\\ R\to\infty}} \oint_{\mathrm{H}(0,R,+)} \mathrm{d}\omega \frac{\mathrm{e}^{-\mathrm{i}\omega\tau}}{\omega + \mathrm{i}\epsilon} = -\lim_{\epsilon\to 0} \oint_{\mathrm{H}(0,R,+)} \mathrm{d}\omega \frac{\mathrm{e}^{-\mathrm{i}\omega\tau}}{\omega + \mathrm{i}\epsilon} = 0 \,.$$

$\boldsymbol{\tau > 0}$**:** here we choose the lower semicircle $H(0,R,-)$ shown in Figure 3.5 in the limit $R \to \infty$ so that, due to the same reasoning as above, the integration along the circular arc $\mathrm{B}(0,R,-)$ vanishes. Hence, it follows that

$$\Theta(\tau) = -\frac{-1}{2\pi\mathrm{i}} \lim_{\epsilon\to 0} \oint_{\mathrm{H}(0,R,-)} \mathrm{d}\omega \frac{\mathrm{e}^{-\mathrm{i}\omega\tau}}{\omega + \mathrm{i}\epsilon} \,.$$

---

[11] Note that in the following the limiting process $R \to \infty$ will not be considered any more. As long as the integrand is regular up to its singularities, we can shrink the integration path to an arbitrary contour enclosing all singularities.

The additional sign on the right hand side results from the mathematically negative direction of H(0, R, −). For further evaluation of this expression, we have to take into account that now the singularity $\omega_{\mathrm{s}} = -\mathrm{i}\epsilon$ lies inside the area enclosed by H(0, R, −) so that the residue theorem must be applied:

$$f(\omega) = \frac{\mathrm{e}^{-\mathrm{i}\omega\tau}}{\omega + \mathrm{i}\epsilon} \Longrightarrow \oint\limits_{\mathrm{H}(0,R,-)} f(\omega)\mathrm{d}\omega = 2\pi\mathrm{i} \operatorname*{Res}_{\omega_{\mathrm{s}} = -\mathrm{i}\epsilon} f = 2\pi\mathrm{i}\mathrm{e}^{-\epsilon\tau} .$$

We therefore obtain

$$\Theta(\tau) = \frac{1}{2\pi\mathrm{i}} \lim_{\epsilon\to 0} \oint\limits_{\mathrm{H}(0,R,-)} f(\omega)\mathrm{d}\omega = 1 .$$

This completes the proof of the above proposition.

With a view to Exercise 31, we point out that the complex conjugate of (3.30) leads to the representation

$$\Theta(-\tau) = \frac{1}{2\pi\mathrm{i}} \lim_{\epsilon\to 0} \int\limits_{-\infty}^{\infty} \mathrm{d}\omega \frac{\mathrm{e}^{-\mathrm{i}\omega\tau}}{\omega - \mathrm{i}\epsilon} , \ \epsilon > 0 . \tag{3.31}$$

Here the singularity lies above the real axis.

31. Fourier decomposition of $G^{(0,\pm)}$. Show the validity of (3.13).

Solution. Our starting point is the differential equation

$$\left(\mathrm{i}\hbar\frac{\partial}{\partial t'} - H'^{(0)}\right) G^{(0,\pm)}(x' - x) = \hbar\delta(x' - x) \tag{3.32}$$

for the free retarded and advanced propagators. We use the fourdimensional Fourier representation of $G^{(0,\pm)}(x' - x)$ and $\delta(x' - x)$ for its solution:

$$\begin{aligned} G^{(0,\pm)}(x' - x) &= \int \frac{\mathrm{d}^3p}{(2\pi\hbar)^3} \int \frac{\mathrm{d}E}{2\pi\hbar} \\ &\quad \times \mathrm{e}^{\mathrm{i}\boldsymbol{p}(\boldsymbol{x}'-\boldsymbol{x})/\hbar} \mathrm{e}^{-\mathrm{i}E(t'-t)/\hbar} \tilde{G}^{(0,\pm)}(\boldsymbol{p}, E) \\ \delta(x' - x) &= \int \frac{\mathrm{d}^3p}{(2\pi\hbar)^3} \int \frac{\mathrm{d}E}{2\pi\hbar} \mathrm{e}^{\mathrm{i}\boldsymbol{p}(\boldsymbol{x}'-\boldsymbol{x})/\hbar} \mathrm{e}^{-\mathrm{i}E(t'-t)/\hbar} . \end{aligned} \tag{3.33}$$

Inserting these expressions into (3.32) yields a corresponding equation for $\tilde{G}^{(0,\pm)}(\boldsymbol{p}, E)$:

$$\begin{aligned} &\int \frac{\mathrm{d}p}{(2\pi\hbar)^3} \int \frac{\mathrm{d}E}{2\pi\hbar} \left(E - \frac{\boldsymbol{p}^2}{2m_0}\right) \mathrm{e}^{\mathrm{i}\boldsymbol{p}(\boldsymbol{x}'-\boldsymbol{x})/\hbar} \mathrm{e}^{-\mathrm{i}E(t'-t)/\hbar} \tilde{G}^{(0,\pm)}(\boldsymbol{p}, E) \\ &\qquad = \hbar \int \frac{\mathrm{d}^3p}{(2\pi)\hbar^3} \int \frac{\mathrm{d}E}{2\pi\hbar} \mathrm{e}^{\mathrm{i}\boldsymbol{p}(\boldsymbol{x}'-\boldsymbol{x})/\hbar} \mathrm{e}^{-\mathrm{i}E(t'-t)/\hbar} . \end{aligned}$$

The solution to this equation can be easily obtained. For $E \neq \boldsymbol{p}^2/2m_0$ it is

$$\tilde{G}^{(0,\pm)}(\boldsymbol{p}, E) = \frac{\hbar}{E - \frac{\boldsymbol{p}^2}{2m_0}} \; .$$

Thus, (3.33) turns into

$$G^{(0,\pm)}(x' - x) = \int \frac{\mathrm{d}^3 p}{(2\pi\hbar)^3} \mathrm{e}^{\mathrm{i}\boldsymbol{p}(\boldsymbol{x}'-\boldsymbol{x})/\hbar} \int \frac{\mathrm{d}E}{2\pi} \frac{\mathrm{e}^{-\mathrm{i}E(t'-t)/\hbar}}{E - \frac{\boldsymbol{p}^2}{2m_0}} \; . \tag{3.34}$$

Next we have to carry out the E- and $\boldsymbol{p}$-integrations bearing in mind the singularity at $E_\mathrm{s} = \boldsymbol{p}^2/2m_0$. We first consider the integration over E,

$$I = \int \frac{\mathrm{d}E}{2\pi} \frac{\mathrm{e}^{-\mathrm{i}E(t'-t)/\hbar}}{E - \frac{\boldsymbol{p}^2}{2m_0}} = \exp\left(-\frac{\mathrm{i}\boldsymbol{p}^2(t'-t)}{2\hbar m_0}\right) \int \frac{\mathrm{d}\omega}{2\pi} \frac{\mathrm{e}^{-\mathrm{i}\omega(t'-t)}}{\omega} \; , \tag{3.35}$$

where, on the right hand side, the substitution $\omega = (E - \boldsymbol{p}^2/2m_0)/\hbar$ has been made. Clearly, the ω-integral can be related back to the integrals (3.30) and (3.31) discussed in Exercise 30, if the singularity in the denominator is shifted by adding or subtracting an imaginary part $\mathrm{i}\epsilon$. If we decide in favor of the addition, comparison with (3.30) yields

$$\int \frac{\mathrm{d}\omega}{2\pi} \frac{\mathrm{e}^{-\mathrm{i}\omega(t'-t)}}{\omega} \longrightarrow \lim_{\epsilon \to 0} \int \frac{\mathrm{d}\omega}{2\pi} \frac{\mathrm{e}^{-\mathrm{i}\omega(t'-t)}}{\omega + \mathrm{i}\epsilon} = -\mathrm{i}\Theta(t'-t) \; . \tag{3.36}$$

This clearly corresponds to a temporally forward directed movement (retarded case). If, on the other hand, we choose to subtract $\mathrm{i}\epsilon$ in (3.35), comparison with (3.31) leads to

$$\int \frac{\mathrm{d}\omega}{2\pi} \frac{\mathrm{e}^{-\mathrm{i}\omega(t'-t)}}{\omega} \longrightarrow \lim_{\epsilon \to 0} \int \frac{\mathrm{d}\omega}{2\pi} \frac{\mathrm{e}^{-\mathrm{i}\omega(t'-t)}}{\omega - \mathrm{i}\epsilon} = +\mathrm{i}\Theta(t-t') \; , \tag{3.37}$$

expressing a temporally backward movement (advanced case). Hence, the correct Fourier decomposition of $G^{(0,\pm)}$ is indeed given by (3.13).

Combining the equations (3.34), (3.35), (3.36), and (3.37), we obtain for the retarded case the explicit form

$$G^{(0,+)}(x' - x) = -\mathrm{i}\Theta(t'-t) \int \frac{\mathrm{d}^3 p}{(2\pi\hbar)^3} \exp\left(-\frac{\mathrm{i}\boldsymbol{p}^2(t'-t)}{2\hbar m_0} + \frac{\mathrm{i}\boldsymbol{p}(\boldsymbol{x}'-x)}{\hbar}\right)$$

and, for the advanced case

$$G^{(0,-)}(x' - x) = +\mathrm{i}\Theta(t-t') \int \frac{\mathrm{d}^3 p}{(2\pi\hbar)^3} \exp\left(-\frac{\mathrm{i}\boldsymbol{p}^2(t'-t)}{2\hbar m_0} + \frac{\mathrm{i}\boldsymbol{p}(\boldsymbol{x}'-x)}{\hbar}\right) .$$

As expected, these results coincide with the equations that were derived differently in Subsection 3.1.2; the further evaluation of the $\boldsymbol{p}$-integral is the same as there.

32. General properties of $G^{(\pm)}$. Show that the following relations hold:

$$G^{(+)}(x',x) = G^{(-)*}(x,x') \tag{3.38}$$

$$G^{(+)}(x',x) = \mathrm{i}\int \mathrm{d}^3x_1 G^{(+)}(x',x_1)G^{(+)}(x_1,x) \quad \text{if } t' > t_1 > t \tag{3.39}$$

$$G^{(-)}(x',x) = -\mathrm{i}\int \mathrm{d}^3x_1 G^{(-)}(x',x_1)G^{(-)}(x_1,x) \quad \text{if } t' < t_1 < t \tag{3.40}$$

$$\delta(\boldsymbol{x}'-\boldsymbol{x}) = \int \mathrm{d}^3x_1 G^{(+)}(\boldsymbol{x}',t,\boldsymbol{x}_1,t_1)G^{(-)}(\boldsymbol{x}_1,t_1,\boldsymbol{x},t) \quad \text{if } t > t_1 \tag{3.41}$$

$$\delta(\boldsymbol{x}'-\boldsymbol{x}) = \int \mathrm{d}^3x_1 G^{(-)}(\boldsymbol{x}',t,\boldsymbol{x}_1,t_1)G^{(+)}(\boldsymbol{x}_1,t_1,\boldsymbol{x},t) \quad \text{if } t < t_1. \tag{3.42}$$

Solution.

To (3.38). This can be proven quickly by complex conjugation of (3.15) and subsequently comparing it with (3.10).

To (3.39) and (3.40). Here we start from the defining equation (3.10) of the retarded propagator,

$$\psi(x') = \mathrm{i}\int \mathrm{d}^3x G^{(+)}(x',x)\psi(x) \quad \text{if } t' > t\ , \tag{3.43}$$

where we are free to consider also $\psi(x)$ to be the result of a propagation from the earlier time t_1 to t, i.e.

$$\psi(x') = -\int \mathrm{d}^3x \int \mathrm{d}^3x_1 G^{(+)}(x',x)G^{(+)}(x,x_1)\psi(x_1) \quad \text{if } t' > t > t_1$$

or, after exchanging the variables $x \leftrightarrow x_1$,

$$\psi(x') = -\int \mathrm{d}^3x \int \mathrm{d}^3x_1 G^{(+)}(x',x_1)G^{(+)}(x_1,x)\psi(x) \quad \text{if } t' > t_1 > t\ .$$

Comparing this relation with (3.43) finally leads to (3.39). The proof of (3.40) proceeds analogously.

To (3.41) and (3.42). At a constant time t ψ can be represented as

$$\psi(\boldsymbol{x}',t) = \int \mathrm{d}^3x \delta(\boldsymbol{x}'-\boldsymbol{x})\psi(\boldsymbol{x},t)\ . \tag{3.44}$$

On the other hand, we have

$$\psi(\boldsymbol{x}',t) = \mathrm{i}\int \mathrm{d}^3x_1 G^{(+)}(\boldsymbol{x}',t,\boldsymbol{x}_1,t_1)\psi(\boldsymbol{x}_1,t_1) \quad \text{if } t > t_1$$

and

$$\psi(\boldsymbol{x}_1,t_1) = -\mathrm{i}\int \mathrm{d}^3x G^{(-)}(\boldsymbol{x}_1,t_1,\boldsymbol{x},t)\psi(\boldsymbol{x},t) \quad \text{if } t > t_1$$

as well as their combination

$$\psi(\boldsymbol{x}',t) = \int \mathrm{d}^3x \int \mathrm{d}^3x_1 G^{(+)}(\boldsymbol{x}',t,\boldsymbol{x}_1,t_1)G^{(-)}(\boldsymbol{x}_1,t_1,\boldsymbol{x},t)\psi(\boldsymbol{x},t) \quad \text{if } t > t_1 \ .$$

Comparing this relation with (3.44) yields (3.41). The proof of (3.42) can be carried out similarly.

33. Unitarity of the scattering matrix. Show that the scattering matrix S is unitary.

Solution. As long as we focus on wave functions normalized to a finite volume, we need to show that

a) $\sum\limits_k S_{fk}S^\dagger_{ki} = \sum\limits_k S_{fk}S^*_{ik} = \delta_{fi}$

b) $\sum\limits_k S^\dagger_{fk}S_{ki} = \sum\limits_k S^*_{kf}S_{ki} = \delta_{fi}$

[in the case of normalization to the whole space, the replacements $\sum\limits_k \to \int \mathrm{d}^3p_k$ and $\delta_{fi} \to \delta(\boldsymbol{p}_f - \boldsymbol{p}_i)$ have to be made].

To a) It holds that

$$S_{fk} = \mathrm{i} \lim_{\substack{t' \to +\infty \\ t \to -\infty}} \int \mathrm{d}^3x' \int \mathrm{d}^3x \Psi^*_f(\boldsymbol{x}',t')G^{(+)}(\boldsymbol{x}',t',\boldsymbol{x},t)\Psi_k(\boldsymbol{x},t)$$

and, due to (3.38),

$$\begin{aligned} S^*_{ik} &= -\mathrm{i} \lim_{\substack{t' \to +\infty \\ t \to -\infty}} \int \mathrm{d}^3y' \int \mathrm{d}^3y \Psi_i(\boldsymbol{y}',t')G^{(+)*}(\boldsymbol{y}',t',\boldsymbol{y},t)\Psi^*_k(\boldsymbol{y},t) \\ &= -\mathrm{i} \lim_{\substack{t' \to +\infty \\ t \to -\infty}} \int \mathrm{d}^3y' \int \mathrm{d}^3y \Psi^*_k(\boldsymbol{y},t)G^{(-)}(\boldsymbol{y},t,\boldsymbol{y}',t')\Psi_i(\boldsymbol{y}',t') \ . \end{aligned}$$

$\{\Psi_k(\boldsymbol{x},t)\}$ is a complete orthonormal solution system to the free Schrödinger equation with

$$\int \mathrm{d}^3x \Psi^*_k(\boldsymbol{x},t)\Psi_j(\boldsymbol{x},t) = \delta_{kj} \ , \ \sum_k \Psi_k(\boldsymbol{x},t)\Psi^*_k(\boldsymbol{y},t) = \delta(\boldsymbol{x}-\boldsymbol{y}) \ \ \forall\, t \ .$$

Therefore, along with (3.41), it follows that

$$\begin{aligned} \sum_k S_{fk}S^*_{ik} &= \lim_{\substack{t' \to +\infty \\ t \to -\infty}} \sum_k \int \mathrm{d}^3x' \int \mathrm{d}^3x \int \mathrm{d}^3y' \int \mathrm{d}^3y \\ &\quad \times \Psi^*_f(\boldsymbol{x}',t')G^{(+)}(\boldsymbol{x}',t',\boldsymbol{x},t)\Psi_k(\boldsymbol{x},t) \\ &\quad \times \Psi^*_k(\boldsymbol{y},t)G^{(-)}(\boldsymbol{y},t,\boldsymbol{y}',t')\Psi_i(\boldsymbol{y}',t') \\ &= \lim_{\substack{t' \to +\infty \\ t \to -\infty}} \int \mathrm{d}^3x' \int \mathrm{d}^3x \int \mathrm{d}^3y' \end{aligned}$$

$$\times \Psi_f^*(\boldsymbol{x}',t')G^{(+)}(\boldsymbol{x}',t',\boldsymbol{x},t)G^{(-)}(\boldsymbol{x},t,\boldsymbol{y}',t')\Psi_i(\boldsymbol{y}',t')$$

$$= \lim_{t'\to\infty} \int \mathrm{d}^3x' \Psi_f^*(\boldsymbol{x}',t')\Psi_i(\boldsymbol{x}',t') = \delta_{fi} \,.$$

The proof of b) proceeds analogously.

34. Square of the δ-function. Show by considering (3.25) for large but finite T that in (3.27) the expression $[2\pi\hbar\delta(E_f - E_i)]^2$ can be replaced by $2\pi T\hbar\delta(E_f - E_i)$.

Solution. In (3.25) we have equated

$$2\pi\hbar\delta(E_f - E_i) = \int_{-\infty}^{\infty} \mathrm{d}t \mathrm{e}^{\mathrm{i}(E_f - E_i)t/\hbar}$$

which is surely justified for an infinitely long scattering period T. Unfortunately, this leads to the badly defined expression $[2\pi\hbar\delta(E_f - E_i)]^2$. However, according to our scattering formalism, we have to bear in mind that we initially presume a finite scattering period T and do not perform the limiting process $T \to \infty$ until the end. Thus, we have

$$2\pi\hbar\delta(E_f - E_i) \to \int_{-T/2}^{T/2} \mathrm{d}t \mathrm{e}^{\mathrm{i}(E_f - E_i)t/\hbar} = \frac{\hbar}{\mathrm{i}(E_f - E_i)} \mathrm{e}^{(E_f - E_i)t/\hbar}\bigg|_{-T/2}^{T/2}$$

$$= \frac{2\hbar \sin[(E_f - E_i)T/2\hbar]}{E_f - E_i}$$

and

$$[2\pi\hbar\delta(E_f - E_i)]^2 \to \frac{4\hbar^2 \sin^2[(E_f - E_i)T/2\hbar]}{(E_f - E_i)^2} \,.$$

On the other hand,

$$\int_{-\infty}^{\infty} \mathrm{d}E_f \frac{4\hbar^2 \sin^2[(E_f - E_i)T/2\hbar]}{(E_f - E_i)^2} = 2T\hbar \int_{-\infty}^{\infty} \mathrm{d}x \frac{\sin^2 x}{x^2} = 2\pi\hbar T \qquad (3.45)$$

is valid. Therefore, we can conclude that

$$[2\pi\hbar\delta(E_f - E_i)]^2 = 2\pi\hbar\delta(0) 2\pi\hbar\delta(E_f - E_i) \to 2\pi\hbar T\delta(E_f - E_i) \,,$$

and hence $2\pi\hbar\delta(0) \to T$ so that, according to (3.45), the integration over E_f adds up to $2\pi\hbar T$.

3.2 Scattering of Spin-1/2 Particles

Building on the results of the preceding section, we now turn to the description of relativistic scattering processes of spin-1/2 particles. Analogously to the preceding section, we first develop an appropriate propagator formalism for solving the general Dirac equation in which, however, we need to implement a modified causality principle for the so-called *Feynman propagator* from the start in order to meet the Feynman-Stückelberg interpretation of positive and negative Dirac solutions. After the derivation of the free Feynman propagator in momentum space, we turn to the actual scattering formalism and look particularly at its specific relativistic aspects. At the end we discuss some identities related to the traces of γ-matrices. This will turn out to be very useful for the concrete calculation of scattering processes.

Note. Since in the following we will often encounter equations with numerous γ-matrix operations, we introduce the convenient "slash notation"

$$\not{a} = a_\mu \gamma^\mu$$

at this stage. Likewise, for convenience, we will often denote (elementary) spin-1/2 particles by "electron" and spin-1/2 antiparticles by "positron".

3.2.1 Solution of the General Dirac Equation

As in Subsection 3.1.1, we start with the time-dependent Dirac equation

$$\left[\not{p}' - \frac{e}{c}\not{A}(x') - m_0 c\right]\psi(x') = 0 . \tag{3.46}$$

For the same reasons as in the Schrödinger case, the solutions to this equation should be expressible by a linear homogeneous integral equation of the form

$$\psi(x') = \mathrm{i}\int \mathrm{d}^3x S(x', x)\gamma^0\psi(x) \ , \ x = (x^0, \boldsymbol{x}) = (ct, \boldsymbol{x}) \ .$$

Here and further on we denote the corresponding Green function by S. The factor γ^0 results from the covariant Dirac equation which is derived from the canonical equation by multiplying the latter by γ^0/c. Apparently, it is again appropriate to introduce a causality principle via

$$\left\{\begin{array}{c}\Theta(x'^0 - x^0)\\ \Theta(x^0 - x'^0)\end{array}\right\}\psi(x') = \pm\mathrm{i}\int \mathrm{d}^3x S^{(\pm)}(x', x)\gamma^0\psi(x) \ , \tag{3.47}$$

guaranteeing the temporally forward or backward evolution of $\psi(x)$ to $\psi(x')$ independently of the positive and negative energy contributions built up in the course of time. However, with respect to the description of relativistic scattering processes of electrons and positrons, it is more sensible to adjust the temporal evolutionary direction of ψ according to its positive and negative parts in such a way that the positive parts $\psi^{(+)}$ can propagate only in the temporally forward direction and the negative parts $\psi^{(-)}$ only in the temporally negative direction. In this way, the proportionate temporally forward evolution of ψ can be interpreted as forward directed particle propagation (electron) and, likewise, due to the Feynman-Stückelberg interpretation (Theorem 2.7), the proportionate temporally backward evolution of

ψ as forward directed antiparticle propagation (positron). We return to this point later on. Based on the above reasoning, we now introduce the following modified causal relationship[12] instead of (3.47):

$$\left\{\begin{matrix}\Theta(x'^0 - x^0)\\ \Theta(x^0 - x'^0)\end{matrix}\right\}\psi^{(\pm)}(x') = \pm\mathrm{i}\int \mathrm{d}^3x S_\mathrm{F}(x',x)\gamma^0\psi^{(\pm)}(x) . \tag{3.48}$$

This is the defining equation of the *Feynman fermion propagator.*

Differential equation for S_F. The following procedure is similar to that in Subsection 3.1.1 in that we first determine the differential equation for S_F and then derive iteratively solvable integral equations for S_F and ψ. Applying the operator in (3.46) to (3.48) yields

$$\begin{aligned}
&\left[\not{p}' - \frac{e}{c}\not{A}(x') - m_0c\right]\left\{\begin{matrix}\Theta(x'^0 - x^0)\\ \Theta(x^0 - x'^0)\end{matrix}\right\}\psi^{(\pm)}(x')\\
&\qquad = \pm\mathrm{i}\hbar\gamma^0\delta(x'^0 - x^0)\psi^{(\pm)}(x')\\
&\qquad = \pm\mathrm{i}\int \mathrm{d}^3x\left[\not{p}' - \frac{e}{c}\not{A}(x') - m_0c\right]S_\mathrm{F}(x',x)\gamma^0\psi^{(\pm)}(x)\\
&\Longrightarrow \int \mathrm{d}^3x\left\{\left[\not{p}' - \frac{e}{c}\not{A}(x') - m_0c\right]S_\mathrm{F}(x',x)\right.\\
&\qquad\qquad \left. - \hbar\delta(x'^0 - x^0)\delta(\boldsymbol{x}' - \boldsymbol{x})\right\}\gamma^0\psi^{(\pm)}(x) = 0\\
&\Longrightarrow \left[\not{p}' - \frac{e}{c}\not{A}(x') - m_0c\right]S_\mathrm{F}(x',x) = \hbar\delta(x' - x) .
\end{aligned}$$

Integral equation for S_F and ψ. Rewriting the last equation as

$$(\not{p}' - m_0c)\,S_\mathrm{F}(x',x) = \hbar\delta(x' - x) + \frac{e}{c}\not{A}(x')S_\mathrm{F}(x',x) , \tag{3.49}$$

we see that its solution is given by

$$S_\mathrm{F}(x',x) = S_\mathrm{F}^{(0)}(x',x) + \frac{e}{\hbar c}\int \mathrm{d}^4x_1 S_\mathrm{F}^{(0)}(x',x_1)\not{A}(x_1)S_\mathrm{F}(x_1,x) , \tag{3.50}$$

with the free fermion propagator $S_\mathrm{F}^{(0)}$ that, in turn, must obey the differential equation

$$(\not{p}' - m_0c)\,S_\mathrm{F}^{(0)}(x',x) = \hbar\delta(x' - x) .$$

The corresponding integral equation for $\psi = \psi^{(+)} + \psi^{(-)}$ follows from (3.48) and (3.50) as

[12] As before, the plus sign holds for the upper equation and the minus sign for the lower one.

$$\begin{aligned}\psi^{(\pm)}(x') &= \pm\mathrm{i}\lim_{t\to\mp\infty}\int \mathrm{d}^3x S_\mathrm{F}(x',x)\gamma^0\psi^{(\pm)}(x)\\ &= \pm\mathrm{i}\lim_{t\to\mp\infty}\left[\int \mathrm{d}^3x S_\mathrm{F}^{(0)}(x',x)\gamma^0\psi^{(\pm)}(x)\right.\\ &\quad\left.+\frac{e}{\hbar c}\int \mathrm{d}^3x\int \mathrm{d}^4x_1 S_\mathrm{F}^{(0)}(x',x_1)\not{A}(x_1)S_\mathrm{F}(x_1,x)\gamma^0\psi^{(\pm)}(x)\right]\\ &= \psi_\text{free}^{(\pm)}(x')+\frac{e}{\hbar c}\int \mathrm{d}^4x_1 S_\mathrm{F}^{(0)}(x',x_1)\not{A}(x_1)\psi^{(\pm)}(x_1)\end{aligned}$$

$$\Longrightarrow \psi(x') = \psi_\text{free}(x')+\frac{e}{\hbar c}\int \mathrm{d}^4x_1 S_\mathrm{F}^{(0)}(x',x_1)\not{A}(x_1)\psi(x_1)\ .$$

Analogously to the equations (3.12) and (3.14) of the nonrelativistic case, we arrive at integral equations for the fermion propagator S_F and the wave function ψ that can be solved approximately once the free propagator $S_\mathrm{F}^{(0)}$ is known. However, the fundamental difference to the nonrelativistic case is that there the integrands contribute either only for past times ($x'^0 > x_1^0 > x^0$, retarded case) or only for future times ($x'^0 < x_1^0 < x^0$, advanced case), whereas here both temporal directions, past and future, play a role.

Theorem 3.4: Solution of the general Dirac equation in the propagator formalism in consideration of the Feynman-Stückelberg interpretation

Given that the solution to the general Dirac equation

$$\left[\not{p}' - \frac{e}{c}\not{A}(x') - m_0c\right]\psi(x') = 0\ ,\ \psi(x') = \psi^{(+)}(x') + \psi^{(-)}(x')$$

is subject to the boundary condition $\psi(x)$ at time t, the Feynman fermion propagator S_F describes the temporal evolution of the positive parts $\psi^{(+)}$ in forward direction and of the negative parts $\psi^{(-)}$ in backward direction according to

$$\left\{\begin{matrix}\Theta(x'^0 - x^0)\\ \Theta(x^0 - x'^0)\end{matrix}\right\}\psi^{(\pm)}(x') = \pm\mathrm{i}\int \mathrm{d}^3x S_\mathrm{F}(x',x)\gamma^0\psi^{(\pm)}(x)\ . \qquad (3.51)$$

For S_F we have the differential equation

$$\left[\not{p}' - \frac{e}{c}\not{A}(x') - m_0c\right] S_\mathrm{F}(x',x) = \hbar\delta(x'-x)$$

as well as the integral equation

$$S_\mathrm{F}(x',x) = S_\mathrm{F}^{(0)}(x',x) + \frac{e}{\hbar c}\int \mathrm{d}^4x_1 S_\mathrm{F}^{(0)}(x',x_1)\not{A}(x_1)S_\mathrm{F}(x_1,x)\ .$$

$S_\mathrm{F}^{(0)}$ denotes the free fermion propagator given by the Fourier decomposition ▷

$$\left.\begin{aligned} S_{\mathrm{F}}^{(0)}(x',x) &= S_{\mathrm{F}}^{(0)}(x'-x) = \int \frac{\mathrm{d}^4p}{(2\pi\hbar)^4}\mathrm{e}^{-\mathrm{i}p_\mu(x'^\mu-x^\mu)/\hbar}\tilde{S}_{\mathrm{F}}^{(0)}(p) \\ \tilde{S}_{\mathrm{F}}^{(0)}(p) &= \frac{\hbar(\not{p}+m_0c)}{p_\mu p^\mu - m_0^2c^2+\mathrm{i}\epsilon} \end{aligned}\right\} \quad (3.52)$$

(see next subsection). For ψ the integral equation follows as

$$\psi(x') = \psi_{\mathrm{free}}(x') + \frac{e}{\hbar c}\int \mathrm{d}^4x_1 S_{\mathrm{F}}^{(0)}(x',x_1)\not{A}(x_1)\psi(x_1) \ .$$

In Exercise 36 we show (at least for the free case) that, besides (3.51), the causality relations

$$\left\{\begin{matrix}\Theta(x^0-x'^0)\\ \Theta(x'^0-x^0)\end{matrix}\right\}\bar{\psi}^{(\pm)}(x') = \pm\mathrm{i}\int \mathrm{d}^3x\bar{\psi}^{(\pm)}(x)\gamma^0 S_{\mathrm{F}}(x,x') \quad (3.53)$$

also hold. They describe the temporally backward and forward evolutions of $\bar{\psi}^{(+)}$ and $\bar{\psi}^{(-)}$, respectively.

3.2.2 Fourier Decomposition of the Free Fermion Propagator

In order to show the validity of (3.51) and (3.52), we proceed as in Exercise 31 and consider the differential equation

$$(\not{p}' - m_0c)\, S_{\mathrm{F}}^{(0)}(x',x) = \hbar\delta(x'-x) \quad (3.54)$$

for the free fermion propagator by passing from the time-space representation to the energy-momentum representation. Using

$$\begin{aligned} S_{\mathrm{F}}^{(0)}(x',x) &= S_{\mathrm{F}}^{(0)}(x'-x) \quad \text{(homogeneity of space and time)} \\ S_{\mathrm{F}}^{(0)}(x'-x) &= \int \frac{\mathrm{d}^4p}{(2\pi\hbar)^4}\mathrm{e}^{-\mathrm{i}p_\mu(x'^\mu-x^\mu)/\hbar}\tilde{S}_{\mathrm{F}}^{(0)}(p) \\ \delta(x'-x) &= \int \frac{\mathrm{d}^4p}{(2\pi\hbar)^4}\mathrm{e}^{-\mathrm{i}p_\mu(x'^\mu-x^\mu)/\hbar} \ , \end{aligned}$$

(3.54) becomes

$$\int \frac{\mathrm{d}^4p}{(2\pi\hbar)^4}(\not{p}'-m_0c)\,\mathrm{e}^{-\mathrm{i}p_\mu(x'^\mu-x^\mu)\hbar}\tilde{S}_{\mathrm{F}}^{(0)}(p) = \hbar\int\frac{\mathrm{d}^4p}{(2\pi\hbar)^4}\mathrm{e}^{-\mathrm{i}p_\mu(x'^\mu-x^\mu)/\hbar}.$$

From this follows that

$$\begin{aligned} (\not{p}-m_0c)\,\tilde{S}_{\mathrm{F}}^{(0)}(p) = \hbar &\Longrightarrow (\not{p}+m_0c)\,(\not{p}-m_0c)\,\tilde{S}_{\mathrm{F}}^{(0)}(p) = \hbar\,(\not{p}+m_0c) \\ &\Longrightarrow \left(p_\mu p^\mu - m_0^2c^2\right)\tilde{S}_{\mathrm{F}}^{(0)}(p) = \hbar\,(\not{p}+m_0c) \\ &\Longrightarrow \tilde{S}_{\mathrm{F}}^{(0)}(p) = \hbar\frac{\not{p}+m_0c}{p_\mu p^\mu - m_0^2c^2} \ , \ p_\mu p^\mu \neq m_0^2c^2 \end{aligned}$$

and finally

$$\begin{aligned} S_{\mathrm{F}}^{(0)}(x'-x) &= \hbar \int \frac{\mathrm{d}^4p}{(2\pi\hbar)^4} \frac{\not{p}+m_0c}{p_\mu p^\mu - m_0^2c^2} \mathrm{e}^{-\mathrm{i}p_\mu(x'^\mu - x^\mu)/\hbar} \\ &= \int \frac{\mathrm{d}^3p}{(2\pi\hbar)^3} \mathrm{e}^{\mathrm{i}\boldsymbol{p}(\boldsymbol{x}'-\boldsymbol{x})/\hbar} \int \frac{\mathrm{d}p_0}{2\pi} \frac{\not{p}+m_0c}{p_\mu p^\mu - m_0^2c^2} \mathrm{e}^{-\mathrm{i}p_0(x'^0 - x^0)/\hbar} . \end{aligned} \quad (3.55)$$

The evaluation of the energy integral

$$I = \int \frac{\mathrm{d}p_0}{2\pi} \frac{\not{p}+m_0c}{p_\mu p^\mu - m_0^2c^2} \mathrm{e}^{-\mathrm{i}p_0(x'^0 - x^0)/\hbar} \quad (3.56)$$

can again be best carried out by choosing an appropriate closed path within the complex p_0-plane with special regard for the singularities at $\pm\sqrt{\boldsymbol{p}^2 + m_0^2c^2}$:

- For $x'^0 > x^0$: a semicircle in the lower complex half plane and
- For $x'^0 < x^0$: a semicircle in the upper complex half plane.

The effect of this choice is that in the limit $R \to \infty$ the integration along the respective angular arc vanishes so that we are left with the integration along the real p_0-axis.

Next we need to clarify how the integration path along the real p_0-axis is to be passed around the singularities or, put differently, how the singularities are to be shifted into the complex half planes by adding or subtracting an imaginary part $\mathrm{i}\epsilon$. Obviously, we have more options here than in the energy integral (3.35) of the nonrelativistic propagator due to the quadratic energy-momentum dependence. They are all depicted in Figure 3.6 and lead to different causal relationships. As we see shortly, the case **d** of Figure 3.6 turns out to be the right choice and leads to the desired causal behavior (3.51). Clearly, this case is equivalent to shifting the pole $-\sqrt{\boldsymbol{p}^2 + m_0^2c^2}$ into

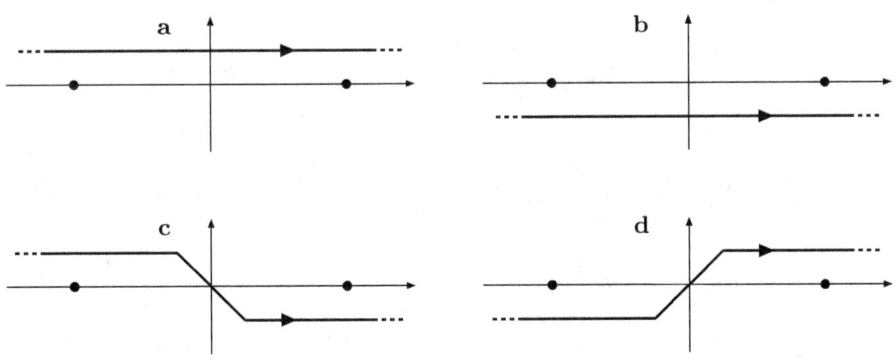

Fig. 3.6. Possible integration paths around the two singularities at $\pm\sqrt{\boldsymbol{p}^2 + m_0^2c^2}$ along the real p_0-axis. They all enforce different causal relationships. **a** corresponds to the first equation of (3.47)and **b** to the second equation (retarded or advanced propagator). **c** reflects (3.51) but with reversed temporal order. This means that here positive parts would propagate temporally backward and negative parts temporally forward. **d** finally yields the desired causality principle as defined in (3.51).

the upper and the pole $+\sqrt{\boldsymbol{p}^2+m_0^2c^2}$ into the lower complex half plane so that (3.56) can be rewritten as

$$I = \lim_{\epsilon\to 0} \oint\limits_{H(0,R,\pm)} \frac{\mathrm{d}p_0}{2\pi} \frac{(\gamma^0 p_0 - \boldsymbol{\gamma p} + m_0 c)\mathrm{e}^{-\mathrm{i}p_0(x'^0-x^0)/\hbar}}{\left(p_0+\sqrt{\boldsymbol{p}^2+m_0^2c^2}-\mathrm{i}\epsilon\right)\left(p_0-\sqrt{\boldsymbol{p}^2+m_0^2c^2}+\mathrm{i}\epsilon\right)} ,$$

where $H(0,R,\pm)$ denotes the semicircles with radius R in the upper and lower half planes, respectively. This equation can now be evaluated easily using the residue theorem.

$\boldsymbol{x'^0 > x^0}$: in this case only the pole at $+\sqrt{\boldsymbol{p}^2+m_0^2c^2}$ lies inside the region enclosed by $H(0,R,-)$. Taking into account the mathematically negative direction of circulation of $H(0,R,-)$, we therefore obtain

$$I = -\mathrm{i}\frac{(\gamma^0 p_0 - \boldsymbol{\gamma p} + m_0 c)\mathrm{e}^{-\mathrm{i}p_0(x'^0-x^0)/\hbar}}{2p_0} , \quad p_0 = +\sqrt{\boldsymbol{p}^2+m_0^2c^2} ,$$

with p_0 as the positive free energy (instead of the integration variable as before). Inserting this expression into (3.55) leads to

$$\begin{aligned} S_{\mathrm{F}}^{(0)}(x'-x) &= -\mathrm{i}\int \frac{\mathrm{d}^3p}{(2\pi\hbar)^3}\mathrm{e}^{\mathrm{i}\boldsymbol{p}(\boldsymbol{x}'-\boldsymbol{x})/\hbar}\mathrm{e}^{-\mathrm{i}p_0(x'^0-x^0)/\hbar}\frac{\gamma^0 p_0 - \boldsymbol{\gamma p} + m_0 c}{2p_0} \\ &= -\mathrm{i}\int \frac{\mathrm{d}^3p}{(2\pi\hbar)^3}\mathrm{e}^{\mathrm{i}\boldsymbol{p}(\boldsymbol{x}'-\boldsymbol{x})/\hbar}\mathrm{e}^{-\mathrm{i}p_0(x'^0-x^0)/\hbar}\frac{\not{p} + m_0 c}{2p_0} \\ &= -\mathrm{i}\int \frac{\mathrm{d}^3p}{(2\pi\hbar)^3}\mathrm{e}^{-\mathrm{i}p_\mu(x'^\mu-x^\mu)/\hbar}\frac{m_0c}{p_0}\Lambda_+(p) , \end{aligned}$$

with the energy projectors $\Lambda_\pm(p) = (\pm\not{p} + m_0 c)/2m_0 c$ from Theorem 2.4.

$\boldsymbol{x'^0 < x^0}$: here we have to consider the pole at $-\sqrt{\boldsymbol{p}^2+m_0^2c^2}$ enclosed by $H(0,R,+)$ so that

$$I = -\mathrm{i}\frac{(-\gamma^0 p_0 - \boldsymbol{\gamma p} + m_0 c)\mathrm{e}^{\mathrm{i}p_0(x'^0-x^0)/\hbar}}{2p_0} , \quad p_0 = +\sqrt{\boldsymbol{p}^2+m_0^2c^2} .$$

With this (3.55) turns into

$$\begin{aligned} S_{\mathrm{F}}^{(0)}(x'-x) &= -\mathrm{i}\int \frac{\mathrm{d}^3p}{(2\pi\hbar)^3}\mathrm{e}^{\mathrm{i}\boldsymbol{p}(\boldsymbol{x}'-\boldsymbol{x})/\hbar}\mathrm{e}^{\mathrm{i}p_0(x'^0-x^0)/\hbar}\frac{-\gamma^0 p_0 - \boldsymbol{\gamma p} + m_0 c}{2p_0} \\ &= -\mathrm{i}\int \frac{\mathrm{d}^3p}{(2\pi\hbar)^3}\mathrm{e}^{-\mathrm{i}\boldsymbol{p}(\boldsymbol{x}'-\boldsymbol{x})/\hbar}\mathrm{e}^{\mathrm{i}p_0(x'^0-x^0)/\hbar}\frac{-\gamma^0 p_0 + \boldsymbol{\gamma p} + m_0 c}{2p_0} \\ &= -\mathrm{i}\int \frac{\mathrm{d}^3p}{(2\pi\hbar)^3}\mathrm{e}^{-\mathrm{i}\boldsymbol{p}(\boldsymbol{x}'-\boldsymbol{x})/\hbar}\mathrm{e}^{\mathrm{i}p_0(x'^0-x^0)/\hbar}\frac{-\not{p} + m_0 c}{2p_0} \\ &= -\mathrm{i}\int \frac{\mathrm{d}^3p}{(2\pi\hbar)^3}\mathrm{e}^{\mathrm{i}p_\mu(x'^\mu-x^\mu)/\hbar}\frac{m_0c}{p_0}\Lambda_-(p) , \end{aligned}$$

where the replacement $\boldsymbol{p} \to -\boldsymbol{p}$ was made in the second row. Both cases can finally by summarized in[13]

$$
\begin{aligned}
S_{\mathrm{F}}^{(0)}(x'-x) = & -\mathrm{i}\Theta(x'^0 - x^0)\int \frac{\mathrm{d}^3p}{(2\pi\hbar)^3}\mathrm{e}^{-\mathrm{i}p_\mu(x'^\mu - x^\mu)/\hbar}\frac{m_0c}{p_0}\Lambda_+(p) \\
& -\mathrm{i}\Theta(x^0 - x'^0)\int \frac{\mathrm{d}^3p}{(2\pi\hbar)^3}\mathrm{e}^{\mathrm{i}p_\mu(x'^\mu - x^\mu)/\hbar}\frac{m_0c}{p_0}\Lambda_-(p)\ . \quad (3.57)
\end{aligned}
$$

As shown in Exercise 35, this expression can be rewritten, analogously to the nonrelativistic case, as

$$
\begin{aligned}
S_{\mathrm{F}}^{(0)}(x'-x) = & -\mathrm{i}\Theta(x'^0 - x^0)\int \mathrm{d}^3p \sum_{r=1}^{2} \psi_{\boldsymbol{p}}^{(r)}(x')\bar{\psi}_{\boldsymbol{p}}^{(r)}(x) \\
& +\mathrm{i}\Theta(x^0 - x'^0)\int \mathrm{d}^3p \sum_{r=3}^{4} \psi_{\boldsymbol{p}}^{(r)}(x')\bar{\psi}_{\boldsymbol{p}}^{(r)}(x)\ . \quad (3.58)
\end{aligned}
$$

With it the validity of the second causal relationship (3.53) for the free case is proven easily (Exercise 36).

All in all, we arrive at the result that the correct free fermion propagator in momentum space is given by

$$
\begin{aligned}
\tilde{S}_{\mathrm{F}}^{(0)}(p) &= \frac{\hbar(\gamma^\mu p_\mu + m_0c)}{\left(p_0 + \sqrt{\boldsymbol{p}^2 + m_0^2c^2} - \mathrm{i}\epsilon\right)\left(p_0 - \sqrt{\boldsymbol{p}^2 + m_0^2c^2} + \mathrm{i}\epsilon\right)} \\
&= \frac{\hbar(\gamma^\mu p_\mu + m_0c)}{p_\mu p^\mu - m_0^2c^2 + 2\mathrm{i}\epsilon\sqrt{\boldsymbol{p}^2 + m_0^2c^2} + \epsilon^2} \\
&\approx \frac{\hbar(\gamma^\mu p_\mu + m_0c)}{p_\mu p^\mu - m_0^2c^2 + \mathrm{i}\epsilon'}
\end{aligned}
$$

in accordance with (3.52).

At the end, we point out that the choice of the propagator depends decisively on the vacuum or on the states that are interpreted as electrons or positrons. In our preceding calculations we have tacitly assumed that the states of the negative energy continuum are occupied and that holes in it are to be regarded as positrons. However, one can also imagine situations (e.g. in the presence of strong fields) where certain states of the negative energy continuum need to be interpreted as electrons with a temporally forward direction of evolution. In such cases the integration path for $S_{\mathrm{F}}^{(0)}$ in the complex p_0-plane must be adjusted appropriately.

---

[13] It is easy to see from this that in the nonrelativistic limit $p_0 \approx m_0c + E/c$, $E \approx \boldsymbol{p}^2/2m_0$ the free fermion propagator $S_{\mathrm{F}}^{(0)}$ turns into the free retarded propagator $G^{(0,+)}$ from (3.16).

3.2.3 Scattering Formalism

With Theorem 3.4 we have a relativistic propagator formalism for the propagation of electrons and positrons on which we now build the corresponding scattering theory. We can carry over many of the prerequisites from Subsection 3.1.3, namely

- that we assume particle scatterings against a fixed target with a locally confined interaction potential,
$$\lim_{|\boldsymbol{x}|\to\infty} V(x) = 0 \ ,$$
- that, for the sake of simplicity, we describe the initially free incoming particles by plane waves instead of localized wave packets,
- which is justified because of the adiabatic approximation according to which the asymptotic freedom of localized wave packets in the limit $|\boldsymbol{x}| \to \infty$ can be replaced by the asymptotic freedom of an incident plane wave as well as of the resulting scattered wave in the limit $t \to \pm\infty$:
$$\lim_{t\to\pm\infty} V(x) = 0 \ .$$

Scattering amplitude, scattering matrix. As in the nonrelativistic case, we define the scattering amplitude S_{fi} by the projection of ψ_i onto the free plane wave Ψ_f long after the scattering, where ψ_i denotes the scattered wave evolving from the free plane wave Ψ_i:

$$S_{fi} = \lim_{t'\to\pm\infty} \int \mathrm{d}^3x' \Psi_f^\dagger(x')\psi_i(x') \ . \qquad (3.59)$$

The main difference to the nonrelativistic case is that we now have to consider one of two limits, $t' \to +\infty$ or $t' \to -\infty$, depending on the particle types involved in the scattering process. If we are interested in electronic scattering states Ψ_f is an electron wave with positive energy propagating forward in time so that the limit $t' \to +\infty$ is to be chosen. If, on the other hand, we wish to study scatterings into positron states then, according to the Feynman-Stückelberg interpretation, Ψ_f is an electron wave with negative energy propagating backward in time. In this case the limit $t' \to -\infty$ is relevant. Corresponding considerations hold for the incident particles: if they are electrons, we have $\lim\limits_{t\to-\infty} \psi_i(x) = \Psi_i(x)$, where Ψ_i is an electron wave with positive energy. In the case of positrons $\lim\limits_{t\to+\infty} \psi_i(x) = \Psi_i(x)$ follows with Ψ_i as an electron wave with negative energy.

With the help of Theorem 3.4, we can now conclusively derive further expressions for the scattering amplitude (3.59) with respect to the four scattering scenarios

electron or positron $\longrightarrow$ electron or positron

in the following way, where the upper limit is to be chosen for electrons ($r_f = 1, 2$; $\epsilon_f = +1$) and the lower limit for positrons ($r_f = 3, 4$; $\epsilon_f = -1$) in their final states: due to

$$\psi_i(x') = \Psi_i(x') + \frac{e}{\hbar c}\int \mathrm{d}^4x_1 S_{\mathrm{F}}^{(0)}(x'-x_1)\not{A}(x_1)\psi_i(x_1)$$

and[14]

$$\bar{\Psi}_f(x_1) = \lim_{t'\to\pm\infty} \mathrm{i}\epsilon_f \int \mathrm{d}^3x'\bar{\Psi}_f(x')\gamma^0 S_{\mathrm{F}}^{(0)}(x'-x_1)$$

$$\int \mathrm{d}^3x \Psi_f^{(r_f)\dagger}(x)\Psi_i^{(r_i)}(x) = \delta(\boldsymbol{p}_f - \boldsymbol{p}_i)\delta_{r_f r_i} \ ,$$

(3.59) turns into

$$\begin{aligned} S_{fi} &= \lim_{t'\to\pm\infty}\left[\int \mathrm{d}^3x'\Psi_f^\dagger(x')\psi_i(x')\right. \\ &\quad \left. + \frac{e}{\hbar c}\int \mathrm{d}^3x' \int \mathrm{d}^4x_1\bar{\Psi}_f(x')\gamma^0 S_{\mathrm{F}}^{(0)}(x'-x_1)\not{A}(x_1)\psi_i(x_1)\right] \\ &= \delta(\boldsymbol{p}_f - \boldsymbol{p}_i)\delta_{r_f r_i} - \frac{\mathrm{i}e\epsilon_f}{\hbar c}\int \mathrm{d}^4x_1\bar{\Psi}_f(x_1)\not{A}(x_1)\psi_i(x_1) \ . \end{aligned}$$

Iterating ψ_i in the way

$$\begin{aligned} \psi_i(x_1) &= \Psi_i(x_1) \\ &\quad + \frac{e}{\hbar c}\int \mathrm{d}^4x_2 S_{\mathrm{F}}^{(0)}(x_1-x_2)\not{A}(x_2)\Psi_i(x_2) \\ &\quad + \left(\frac{e}{\hbar c}\right)^2 \int \mathrm{d}^4x_2 \int \mathrm{d}^4x_3 S_{\mathrm{F}}^{(0)}(x_1-x_2)\not{A}(x_2) \\ &\quad \times S_{\mathrm{F}}^{(0)}(x_2-x_3)\not{A}(x_3)\Psi_i(x_3) \\ &\quad + \ldots \ , \end{aligned}$$

we finally obtain, analogously to Theorem 3.2:

Theorem 3.5: Scattering matrix in the Dirac theory

The scattering amplitude S_{fi} is defined by the projection of ψ_i onto Ψ_f long after the scattering against a target, where ψ_i is the scattered wave evolving from the free plane wave Ψ_i:

$$S_{fi} = \lim_{t'\to\pm\infty}\int \mathrm{d}^3x'\Psi_f^\dagger(x')\psi_i(x') \ , \ \lim_{t\to\mp\infty}\psi_i(x) = \Psi_i(x) \ . \qquad (3.60)$$

In the case of scattered electron [positron] states, Ψ_f is a plane electron wave with positive [negative] energy propagating forward [backward] in time so ▷

[14] See footnote 9 on page 188.

that, on the left hand side of (3.60), the limit $t' \to +\infty$ [$t' \to -\infty$] has to be considered. If the incident particles are given by electrons [positrons] then Ψ_i is a plane electron wave with positive [negative] energy and, on the right hand side of (3.60), the limit $t \to -\infty$ [$t \to +\infty$] is relevant.
With the help of the Feynman propagator formalism, the scattering amplitude S_{fi} can be expanded in a series of multiple scatterings:

$$\begin{aligned}
S_{fi} &= \delta(\boldsymbol{p}_f - \boldsymbol{p}_i)\delta_{r_f r_i} - \frac{\mathrm{i}e\epsilon_f}{\hbar c}\int \mathrm{d}^4x_1 \bar{\Psi}_f(x_1) A\!\!\!/(x_1)\psi_i(x_1) \\
&= \delta(\boldsymbol{p}_f - \boldsymbol{p}_i)\delta_{r_f r_i} \\
&\quad -\mathrm{i}\epsilon_f \frac{e}{\hbar c}\int \mathrm{d}^4x_1 \bar{\Psi}_f(x_1) A\!\!\!/(x_1)\Psi_i(x_1) \\
&\quad -\mathrm{i}\epsilon_f \left(\frac{e}{\hbar c}\right)^2 \int \mathrm{d}^4x_1 \int \mathrm{d}^4x_2 \bar{\Psi}_f(x_2) A\!\!\!/(x_2) S_\mathrm{F}^{(0)}(x_2 - x_1) A\!\!\!/(x_1)\Psi_i(x_1) \\
&\quad -\mathrm{i}\epsilon_f \left(\frac{e}{\hbar c}\right)^3 \int \mathrm{d}^4x_1 \int \mathrm{d}^4x_2 \int \mathrm{d}^4x_3 \bar{\Psi}_f(x_3) A\!\!\!/(x_3) S_\mathrm{F}^{(0)}(x_3 - x_2) \\
&\quad\quad \times A\!\!\!/(x_2) S_\mathrm{F}^{(0)}(x_2 - x_1) A\!\!\!/(x_1)\Psi_i(x_1) \\
&\quad - \ldots \ .
\end{aligned} \tag{3.61}$$

A^μ denotes the four-potential of the target, $S_\mathrm{F}^{(0)}$ the free fermion propagator, and ϵ_f the energy sign of Ψ_f.
This theorem is based on the adiabatic approximation and the Feynman-Stückelberg interpretation.

As in the remarks after Theorem 3.2, we note the following:

- The unitarity of the scattering matrix S follows again from the Hermitecity of the Dirac-Hamilton operator along with the resulting conservation of the total probability. However, the direct proof is a little more intricate than in the nonrelativistic case as here the sum of the unitarity condition $\sum\limits_k S^*_{kf}S_{ki} = \delta_{fi}$ needs to be taken over all states into which a given initial state can scatter. Therefore, the proof is best carried out within the framework of quantum electrodynamics.
- Similarly to the nonrelativistic case, we may consider the evolution of ψ_f starting from Ψ_f (instead of ψ_i starting from Ψ_i) in the opposite temporal direction. In this case the scattering matrix is defined via

$$S_{fi} = \lim_{t \to \mp\infty} \int \mathrm{d}^3x \psi_f^\dagger(x)\Psi_i(x) \ , \ \lim_{t' \to \pm\infty} \psi_f(x') = \Psi_f(x') \ ,$$

which, up to a phase, is identical to (3.60). This is because from

$$\begin{aligned}
\psi_i(x') &= \lim_{t \to \mp\infty} \mathrm{i}\epsilon_i \int \mathrm{d}^3x S_\mathrm{F}(x', x)\gamma^0 \Psi_i(x) \\
\bar{\psi}_f(x) &= \lim_{t' \to \pm\infty} \mathrm{i}\epsilon_f \int \mathrm{d}^3x' \bar{\psi}_f(x')\gamma^0 S_\mathrm{F}(x', x)
\end{aligned}$$

follows that

$$\begin{aligned} S_{fi} &= \lim_{t\to\mp\infty} \int \mathrm{d}^3x \psi_f^\dagger(x)\Psi_i(x) = \lim_{t\to\mp\infty} \int \mathrm{d}^3x \bar{\psi}_f(x)\gamma^0\Psi_i(x) \\ &= \lim_{\substack{t\,\to\,\mp\infty \\ t'\,\to\,\pm\infty}} \mathrm{i}\epsilon_f \int \mathrm{d}^3x \int \mathrm{d}^3x' \bar{\psi}_f(x')\gamma^0 S_\mathrm{F}(x',x)\gamma^0\Psi_i(x) \\ &= \lim_{t'\to\pm\infty} \epsilon_i\epsilon_f \int \mathrm{d}^3x' \Psi_f^\dagger(x')\psi_i(x') \ . \end{aligned}$$

- As one can see, the scattering series (3.61) is essentially an expansion in the fine structure constant in $\alpha_e = e^2/\hbar c \approx 1/137$ (in practice, $\not{A}$ itself contains the electric charge e). It is exactly the smallness of α_e that guarantees a fast convergence of (3.61) so that, in most cases, only the first few terms need to be taken into account.

Let us now consider the physical implications of Theorem 3.5 in more detail. First, along with

$$\begin{aligned} \Psi_i(x_1) &= \lim_{t\to\mp\infty} \mathrm{i}\epsilon_i \int \mathrm{d}^3x S_\mathrm{F}^{(0)}(x_1-x)\gamma^0\Psi_i(x) \\ \bar{\Psi}_f(x_n) &= \lim_{t'\to\pm\infty} \mathrm{i}\epsilon_f \int \mathrm{d}^3x' \bar{\psi}_f(x_n)\gamma^0 S_\mathrm{F}^{(0)}(x'-x_n) \ , \end{aligned}$$

the individual expansion terms of (3.61) can be interpreted, analogously to the nonrelativistic case, in such a way that a particle moves freely between the space-time points x and x' passing various intermediate points or vortices x_i with the probability amplitude $S_\mathrm{F}^{(0)}(x_{i+1}-x_i)$ where it suffers a scattering through the interaction potential A^μ. The total amplitude results from integration over all possible vortices. Contrary to the nonrelativistic case, the scatterings do not necessarily proceed in chronological order, which is a consequence of the construction of the fermion propagator or, rather, the causality principle (3.51) connected to it.

Electron scattering. If we are interested in the process of electron scattering

$$\left.\begin{aligned} \epsilon_i &= +1 \ , \ \Psi_i \text{ in the limit } t \to -\infty \\ \epsilon_f &= +1 \ , \ \Psi_f \text{ in the limit } t' \to +\infty \ , \end{aligned}\right\} \tag{3.62}$$

the relevant terms of S_{fi} to zeroth, first, and second order can again be represented by the Feynman diagrams **a**, **b**, and **c** in Figure 3.3 (with the replacement $\Psi^* \to \bar{\Psi}$). Additionally, in second order the "achronological" zigzag diagram in Figure 3.7 contributes, which, following the Feynman-Stückelberg interpretation, we can interpret on the particle level as follows: an electron-positron pair is created at x_2. While the electron is moving out of the region

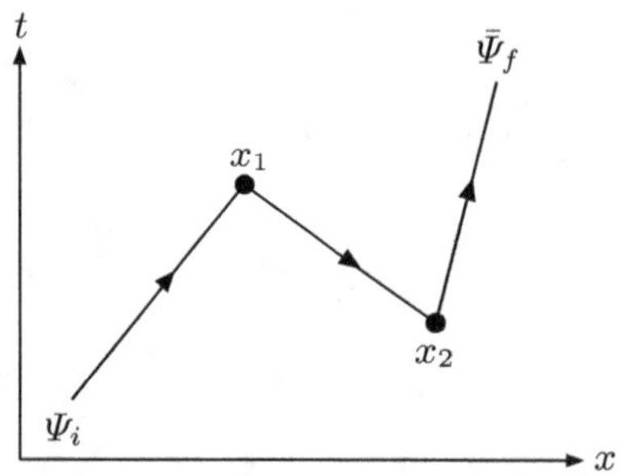

Fig. 3.7. Besides Figure 3.3**c**, this jagged Feynman diagram also contributes to the scattering amplitude for electron scattering to second order.

of interaction, the positron is moving toward x_1, where it is annihilated together with the incident electron.[15] Note that all particle movements proceed in temporal forward direction as physically required.

Within the hole theory, Figure 3.7 can also be interpreted in a way where, at x_2, an electron with negative energy is scattered into an electron with positive energy, leaving a hole in the Dirac sea. While the latter electron is moving out, the hole is propagating toward x_1 where it is finally filled (or destroyed) by the scattering of the incident electron.

Positron scattering. As one can easily see, for the positron scattering

$$\left.\begin{array}{l}\epsilon_i = -1 \ , \ \Psi_i \text{ in the limit } t \to +\infty \\ \\ \epsilon_f = -1 \ , \ \bar{\Psi}_f \text{ in the limit } t' \to -\infty \ ,\end{array}\right\} \quad (3.63)$$

the corresponding terms of S_{fi} lead to the same Feynman diagrams as those of the electron scattering rotated by 180°. They can also be interpreted consistently with the help of the Feynman-Stückelberg interpretation as well as with respect to the hole theory.

Pair creation. Apart from (3.62) and (3.63), we may also consider the combination

$$\begin{array}{l}\epsilon_i = -1 \ , \ \Psi_i \text{ in the limit } t \to +\infty \\ \\ \epsilon_f = +1 \ , \ \bar{\Psi}_f \text{ in the limit } t' \to +\infty \ ,\end{array}$$

which clearly corresponds to a positron and an electron both moving out of the region of interaction, i.e. the process of real electron-positron creation. Here, to first and second order, S_{fi} is represented by the Feynman diagrams of Figure 3.8**a**-**c**, where in **b** and **c** the second order is split up according to the different temporal orders of the two scatterings: in **b** the positron, and in **c** the electron suffers an additional scattering after its creation.

[15] Since the positron exists only for a very short time, it is also called *virtual*. In general, one distinguishes between *real particles* whose Feynman diagrams are open on one side (open or external lines) and *virtual particles* whose Feynman diagrams have a start and an end point (closed or internal lines).

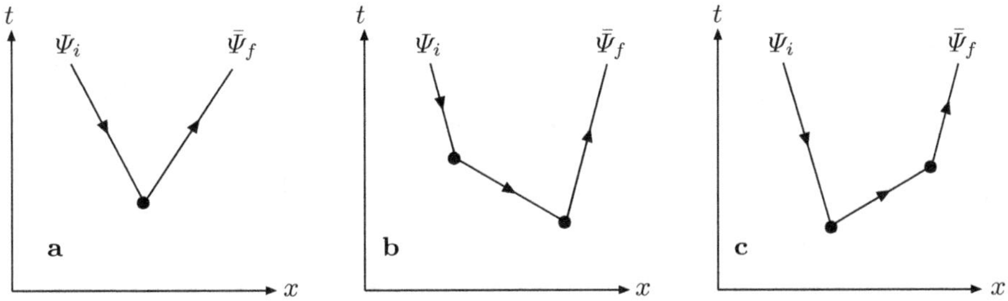

Fig. 3.8. Feynman diagrams of the scattering amplitude for the electron-positron creation to first and second order. In **b** and **c** the contribution to second order is split up according to the different temporal orders of both scatterings.

Pair annihilation. Finally, if one rotates the diagrams of Figure 3.8 by 180° one obtains the diagrams of S_{fi} to first and second order for the remaining combination

$$\epsilon_i = +1 \;,\; \Psi_i \text{ in the limit } t \to -\infty$$

$$\epsilon_f = -1 \;,\; \Psi_f \text{ in the limit } t' \to -\infty \;.$$

This corresponds to a situation where both electron and positron are entering the region of interaction thus destroying one another.

By these simple examples the physical meaning and necessity of the modified causality principle (3.51) that we have incorporated into the fermion propagator become very clear. Only using this, together with the Feynman-Stückelberg interpretation, is it possible to interpret the various expressions in S_{fi} as electron and positron scattering, pair creation and annihilation, i.e. as those phenomena that are indeed observed in nature.

To summarize, we can ascertain that Theorems 3.4 and 3.5 in conjunction with Theorem 3.3 provide a meaningful description of relativistic scattering processes of Dirac particles. In the following we will apply this formalism to several concrete problems and extend it to more complicated scattering situations. In so doing, we will encounter many complex but structurally similar bispinor-γ-matrix combinations. Therefore, we first investigate some properties of such combinations in more detail.

3.2.4 Trace Evaluations with γ-Matrices

For concrete scattering problems the calculation of $|S_{fi}|^2$ often leads to double spin sums of the form

$$\sum_{s_f, s_i} [\bar{u}(p_f, s_f)\Gamma_1 u(p_i, s_i)][\bar{u}(p_i, s_i)\Gamma_2 u(p_f, s_f)] \;,$$

with $u(p,s)$ or $v(p,s)$ from Theorem 2.4 and $\Gamma_{1,2}$ as operators containing certain γ-matrix combinations. Such expressions can generally be put down to traces and subsequently calculated further using some γ-matrix trace rules so that the explicit form of the bispinors does not need to be considered any more. In the following we summarize some relevant relationships concerning this matter and prove them directly afterward.

Theorem 3.6: Trace evaluations with γ-matrices

The following relations hold:

$$\left.\begin{aligned}\sum_{s_f,s_i}[\bar{u}(f)\Gamma_1 u(i)][\bar{u}(i)\Gamma_2 u(f)] &= \mathrm{tr}\left[\Lambda_+(p_f)\Gamma_1\Lambda_+(p_i)\Gamma_2\right]\\ \sum_{s_f,s_i}[\bar{v}(f)\Gamma_1 v(i)][\bar{v}(i)\Gamma_2 v(f)] &= \mathrm{tr}\left[\Lambda_-(p_f)\Gamma_1\Lambda_-(p_i)\Gamma_2\right]\\ \sum_{s_f,s_i}[\bar{u}(f)\Gamma_1 v(i)][\bar{v}(i)\Gamma_2 u(f)] &= -\mathrm{tr}\left[\Lambda_+(p_f)\Gamma_1\Lambda_-(p_i)\Gamma_2\right]\\ \sum_{s_f,s_i}[\bar{v}(f)\Gamma_1 u(i)][\bar{u}(i)\Gamma_2 v(f)] &= -\mathrm{tr}\left[\Lambda_-(p_f)\Gamma_1\Lambda_+(p_i)\Gamma_2\right]\ ,\end{aligned}\right\}\qquad(3.64)$$

with the abbreviatory notation $u(i) = u(p_i, s_i)$ and so on. Depending on the concrete form of the operators $\Gamma_{1,2}$, the traces can be calculated further with the help of the following identities:

$$\left.\begin{aligned}\mathrm{tr}(\not{a}\not{b}) &= 4a\cdot b\\ \mathrm{tr}(\not{a}_1\cdots\not{a}_n) &= 0 \text{ if } n \text{ is odd}\\ \mathrm{tr}(\not{a}_1\not{a}_2\cdots\not{a}_{2n}) &= \mathrm{tr}(\not{a}_{2n}\cdots\not{a}_1)\\ \mathrm{tr}(\not{a}_1\cdots\not{a}_n) &= a_1\cdot a_2\mathrm{tr}(\not{a}_3\cdots\not{a}_n)\\ &\quad -a_1\cdot a_3\mathrm{tr}(\not{a}_2\not{a}_4\cdots\not{a}_n)\\ &\quad +\ldots+(-1)^n a_1\cdot a_n\mathrm{tr}(\not{a}_2\cdots\not{a}_{n-1})\\ \mathrm{tr}(\gamma^5) &= 0\\ \mathrm{tr}(\gamma^5\not{a}_1\cdots\not{a}_n) &= 0 \text{ if } n \text{ is odd}\\ \mathrm{tr}(\gamma^5\not{a}\not{b}) &= 0\\ \mathrm{tr}(\gamma^5\not{a}\not{b}\not{c}\not{d}) &= -4\mathrm{i}\epsilon^{\alpha\beta\gamma\delta}a_\alpha b_\beta c_\gamma d_\delta\end{aligned}\right\}\qquad(3.65)$$

$$\left.\begin{aligned}\gamma_\mu\gamma^\mu &= 4\\ \gamma_\mu\not{a}\gamma^\mu &= -2\not{a}\\ \gamma_\mu\not{a}\not{b}\gamma^\mu &= 4a\cdot b\\ \gamma_\mu\not{a}\not{b}\not{c}\gamma^\mu &= -2\not{c}\not{b}\not{a}\\ \gamma_\mu\not{a}\not{b}\not{c}\not{d}\gamma^\mu &= 2\not{d}\not{a}\not{b}\not{c}+2\not{c}\not{b}\not{a}\not{d}\ .\end{aligned}\right\}\qquad(3.66)$$

Note the index saving notation $a\cdot b$ for $a^\mu b_\mu$ which will be used often in the following.

To (3.64). Due to the projection relations (see Theorem 2.4)

$$\bar{\omega}^{(1,2)}(\boldsymbol{p})\Lambda_+(p) = \bar{\omega}^{(1,2)}(\boldsymbol{p}) \quad , \quad \bar{\omega}^{(3,4)}(\boldsymbol{p})\Lambda_+(p) = 0$$
$$\bar{\omega}^{(3,4)}(\boldsymbol{p})\Lambda_-(p) = \bar{\omega}^{(3,4)}(\boldsymbol{p}) \quad , \quad \bar{\omega}^{(1,2)}(\boldsymbol{p})\Lambda_-(p) = 0$$

and the second equation of (2.16), we have

$$\begin{aligned}\sum_{s_i} u_\alpha(i)\bar{u}_\beta(i) &= \sum_{r=1}^{2} \omega_\alpha^{(r)}(\boldsymbol{p}_i)\bar{\omega}_\beta^{(r)}(\boldsymbol{p}_i) \\ &= \sum_{\epsilon,r=1}^{4} \epsilon_r \omega_\alpha^{(r)}(\boldsymbol{p}_i)\bar{\omega}_\delta^{(r)}(\boldsymbol{p}_i)\left[\Lambda_+(p_i)\right]_{\delta\beta} \\ &= \left[\Lambda_+(p_i)\right]_{\alpha\beta}\end{aligned} \tag{3.67}$$

$$\begin{aligned}\sum_{s_i} v_\alpha(i)\bar{v}_\beta(i) &= \sum_{r=3}^{4} \omega_\alpha^{(r)}(\boldsymbol{p}_i)\bar{\omega}_\beta^{(r)}(\boldsymbol{p}_i) \\ &= -\sum_{\epsilon,r=1}^{4} \epsilon_r \omega_\alpha^{(r)}(\boldsymbol{p}_i)\bar{\omega}_\delta^{(r)}(\boldsymbol{p}_i)\left[\Lambda_-(p_i)\right]_{\delta\beta} \\ &= -\left[\Lambda_+(p_i)\right]_{\alpha\beta} \,.\end{aligned} \tag{3.68}$$

Concerning the first equation of (3.64), it follows that

$$\begin{aligned}\sum_{s_f,s_i}[\bar{u}(f)\Gamma_1 u(i)][\bar{u}(i)\Gamma_2 u(f)] &= \sum_{\substack{\alpha,\beta,\delta,\epsilon \\ s_i,s_f}} \bar{u}_\alpha(f)[\Gamma_1]_{\alpha\beta}u_\beta(i)\bar{u}_\delta(i)[\Gamma_2]_{\delta\epsilon}u_\epsilon(f) \\ &= \sum_{\substack{\alpha,\beta,\delta,\epsilon \\ s_f}} \bar{u}_\alpha(f)[\Gamma_1]_{\alpha\beta}[\Lambda_+(p_i)]_{\beta\delta}[\Gamma_2]_{\delta\epsilon}u_\epsilon(f) \\ &= \sum_{\substack{\alpha,\beta,\delta,\epsilon \\ s_f}} u_\epsilon(f)\bar{u}_\alpha(f)[\Gamma_1]_{\alpha\beta}[\Lambda_+(p_i)]_{\beta\delta}[\Gamma_2]_{\delta\epsilon} \\ &= \sum_{\alpha,\beta,\delta,\epsilon} [\Lambda_+(p_f)]_{\epsilon\alpha}[\Gamma_1]_{\alpha\beta}[\Lambda_+(p_i)]_{\beta\delta}[\Gamma_2]_{\delta\epsilon} \,.\end{aligned}$$

Putting the matrix multiplications into one single matrix finally yields the proposition. The proof of the remaining equations of (3.64) proceeds analogously.

To prove (3.65) we particularly need the anticommutator relations $\{\gamma^\mu, \gamma^\nu\} = 2g^{\mu\nu}$ and $\{\gamma^5, \gamma^\mu\} = 0$ as well as the cyclic permutability of the trace: $\mathrm{tr}(AB) = \mathrm{tr}(BA)$. In the following the use of these relations is marked by the symbols a and c over the corresponding equal sign.

To 1. equation of (3.65).

$$\begin{aligned}\mathrm{tr}(\not{a}\not{b}) &\stackrel{\mathrm{c}}{=} \mathrm{tr}(\not{b}\not{a}) = \frac{1}{2}\mathrm{tr}(\not{a}\not{b}+\not{b}\not{a}) = \frac{1}{2}a_\mu b_\nu \mathrm{tr}(\gamma^\mu\gamma^\nu+\gamma^\nu\gamma^\mu)\\ &\stackrel{\mathrm{a}}{=} a_\mu b_\nu \mathrm{tr}(g^{\mu\nu}) = a_\mu b_\nu g^{\mu\nu}\mathrm{tr}(1) = 4a\cdot b\;.\end{aligned}$$

To 2. equation of (3.65).

$$\begin{aligned}\mathrm{tr}(\not{a}_1\cdots\not{a}_n) &= \mathrm{tr}(\not{a}_1\cdots\not{a}_n\gamma^5\gamma^5) \stackrel{\mathrm{c}}{=} \mathrm{tr}(\gamma^5\not{a}_1\cdots\not{a}_n\gamma^5)\\ &\stackrel{\mathrm{a}}{=} (-1)^n\mathrm{tr}(\not{a}_1\cdots\not{a}_n\gamma^5\gamma^5) = (-1)^n\mathrm{tr}(\not{a}_1\cdots\not{a}_n)\;.\end{aligned}$$

To 3. equation of (3.65). Here we make use of the charge conjugation transformation C from Subsection 2.1.6 with the property $C^{-1}\gamma^\mu C = -\gamma^{*\mu} = -\gamma^0\gamma^{\mu,T}\gamma^0$ [see (2.40)] and calculate as follows:

$$\begin{aligned}\mathrm{tr}(\not{a}_1\cdots\not{a}_{2n}) &= \mathrm{tr}(CC^{-1}\not{a}_1CC^{-1}\not{a}_2\cdots CC^{-1}\not{a}_{2n})\\ &\stackrel{\mathrm{c}}{=} \mathrm{tr}(C^{-1}\not{a}_1CC^{-1}\not{a}_2\cdots CC^{-1}\not{a}_{2n}C)\\ &= (-1)^{2n}\mathrm{tr}\left(\gamma^0\not{a}_1^T\gamma^0\gamma^0\not{a}_2^T\gamma^0\cdots\gamma^0\not{a}_{2n}^T\gamma^0\right)\\ &\stackrel{\mathrm{c}}{=} \mathrm{tr}\left(\not{a}_1^T\cdots\not{a}_{2n}^T\right) = \mathrm{tr}(\not{a}_{2n}\cdots\not{a}_1)^T = \mathrm{tr}(\not{a}_{2n}\cdots\not{a}_1)\;.\end{aligned}$$

To 4. equation of (3.65).

$$\begin{aligned}\mathrm{tr}(\not{a}_1\not{a}_2\cdots\not{a}_n) &\stackrel{\mathrm{a}}{=} \mathrm{tr}\left[(-\not{a}_2\not{a}_1+2a_1\cdot a_2)\not{a}_3\cdots\not{a}_n\right]\\ &= 2a_1\cdot a_2\mathrm{tr}(\not{a}_3\cdots\not{a}_n) - \mathrm{tr}(\not{a}_2\not{a}_1\not{a}_3\cdots\not{a}_n)\;.\end{aligned}$$

Further continuation of this procedure leads to

$$\begin{aligned}\mathrm{tr}(\not{a}_1\cdots\not{a}_n) &= 2a_1\cdot a_2\mathrm{tr}(\not{a}_3\cdots\not{a}_n) - 2a_1\cdot a_3\mathrm{tr}(\not{a}_2\not{a}_4\cdots\not{a}_n)+\ldots\\ &\quad +(-1)^n\mathrm{tr}(\not{a}_2\cdots\not{a}_n\not{a}_1)\;.\end{aligned}$$

From this, along with $\mathrm{tr}(\not{a}_2\cdots\not{a}_n\not{a}_1) \stackrel{\mathrm{c}}{=} \mathrm{tr}(\not{a}_1\cdots\not{a}_n)$, follows the asserted equation. This relation is exceedingly useful for calculating complex traces of γ-matrices. However, even for moderate n, the number of the arising individual terms can be considerably large. For example, in the case of $n=4$, the 4. and 1. equations of (3.65) yield

$$\mathrm{tr}(\not{a}\not{b}\not{c}\not{d}) = 4\left[(a_1\cdot a_2)(a_3\cdot a_4)+(a_1\cdot a_4)(a_2\cdot a_3)-(a_1\cdot a_3)(a_2\cdot a_4)\right]\;.$$

To 5. equation of (3.65).

$$\mathrm{tr}\left(\gamma^5\right) = \mathrm{tr}\left(\gamma^5\gamma^0\gamma^0\right) \stackrel{\mathrm{a}}{=} -\mathrm{tr}\left(\gamma^0\gamma^5\gamma^0\right) \stackrel{\mathrm{c}}{=} -\mathrm{tr}\left(\gamma^5\gamma^0\gamma^0\right) = 0\;.$$

To 6. equation of (3.65).

$$\mathrm{tr}(\gamma^5\not{a}_1\cdots\not{a}_n) \stackrel{\mathrm{c}}{=} \mathrm{tr}(\not{a}_1\cdots\not{a}_n\gamma^5) \stackrel{\mathrm{a}}{=} (-1)^n\mathrm{tr}(\gamma^5\not{a}_1\cdots\not{a}_n)\;.$$

To 7. equation of (3.65). The proof is adduced if we can show that $\mathrm{tr}\left(\gamma^5\gamma^\mu\gamma^\nu\right)=0$. We only need to consider the case $\mu \neq \nu$ due to the 5. equation of (3.65). For $\lambda \neq \mu, \nu$ follows that

$$\begin{aligned}\mathrm{tr}\left(\gamma^5\gamma^\mu\gamma^\nu\right) &= \mathrm{tr}\left[\gamma^5\gamma^\mu\gamma^\nu\left(\gamma^\lambda\right)^{-1}\gamma^\lambda\right] \stackrel{\mathrm{c}}{=} \mathrm{tr}\left[\gamma^\lambda\gamma^5\gamma^\mu\gamma^\nu\left(\gamma^\lambda\right)^{-1}\right] \\ &\stackrel{\mathrm{a}}{=} (-1)^3\mathrm{tr}\left[\gamma^5\gamma^\mu\gamma^\nu\gamma^\lambda\left(\gamma^\lambda\right)^{-1}\right] = (-1)^3\mathrm{tr}\left(\gamma^5\gamma^\mu\gamma^\nu\right) = 0\ .\end{aligned}$$

To 8. equation of (3.65). Here we consider the expression $\mathrm{tr}\left(\gamma^5\gamma^\alpha\gamma^\beta\gamma^\gamma\gamma^\delta\right)$. This trace vanishes if two indices take on the same value. This is because if, for example, the first and the third indices are equal, we have

$$\begin{aligned}\mathrm{tr}\left(\gamma^5\gamma^\alpha\gamma^\beta\gamma^\alpha\gamma^\delta\right) &\stackrel{\mathrm{a}}{=} \mathrm{tr}\left[\gamma^5\gamma^\alpha\left(2g^{\alpha\beta}-\gamma^\alpha\gamma^\beta\right)\gamma^\delta\right] \\ &= 2g^{\alpha\beta}\mathrm{tr}\left(\gamma^5\gamma^\alpha\gamma^\delta\right) - g^{\alpha\alpha}\mathrm{tr}\left(\gamma^5\gamma^\beta\gamma^\delta\right) \stackrel{\text{6.eq.}}{=} 0\ .\end{aligned}$$

Thus, only the trace

$$\mathrm{tr}\left(\gamma^5\gamma^0\gamma^1\gamma^2\gamma^3\right) = \mathrm{tr}\left(-\mathrm{i}\gamma^5\gamma^5\right) = -4\mathrm{i} = -4\mathrm{i}\epsilon^{0123}$$

contributes, as well as those with permuted indices. However, due to the above anticommutator relations, the result remains unchanged for even permutations, while for odd permutations there appears an additional sign, in accordance with $\epsilon^{\alpha\beta\gamma\delta}$.

To 1. equation of (3.66).

$$\gamma_\mu\gamma^\mu = g_{\mu\nu}\gamma^\nu\gamma^\mu \stackrel{\mathrm{a}}{=} g_{\mu\nu}\left(2g^{\mu\nu}-\gamma^\mu\gamma^\nu\right) = 2g_{\mu\nu}g^{\mu\nu} - \gamma_\mu\gamma^\mu = 8 - \gamma_\mu\gamma^\mu .$$

To 2. equation of (3.66).

$$\gamma_\mu \not{a}\gamma^\mu = \gamma_\mu a_\nu\gamma^\nu\gamma^\mu \stackrel{\mathrm{a}}{=} \gamma_\mu a_\nu\left(2g^{\mu\nu}-\gamma^\mu\gamma^\nu\right) \stackrel{\text{1.eq.}}{=} 2\not{a} - 4\not{a} = -2\not{a}\ .$$

To 3. equation of (3.66).

$$\begin{aligned}\gamma_\mu\not{a}\not{b}\gamma^\mu &= \gamma_\mu\not{a}b_\nu\gamma^\nu\gamma^\mu \stackrel{\mathrm{a}}{=} \gamma_\mu\not{a}b_\nu\left(2g^{\mu\nu}-\gamma^\mu\gamma^\nu\right) \stackrel{\text{2.eq.}}{=} 2\not{b}\not{a} + 2\not{a}\not{b} \\ &\stackrel{\mathrm{a}}{=} 4a\cdot b - 2\not{a}\not{b} + 2\not{a}\not{b} = 4a\cdot b\ .\end{aligned}$$

To 4. equation of (3.66).

$$\begin{aligned}\gamma_\mu\not{a}\not{b}\not{c}\gamma^\mu &= \gamma_\mu\not{a}\not{b}c_\nu\gamma^\nu\gamma^\mu \stackrel{\mathrm{a}}{=} \gamma_\mu\not{a}\not{b}c_\nu\left(2g^{\mu\nu}-\gamma^\mu\gamma^\nu\right) \stackrel{\text{3.eq.}}{=} 2\not{c}\not{a}\not{b} - 4a\cdot b\not{c} \\ &\stackrel{\mathrm{a}}{=} 4\not{c}a\cdot b - 2\not{c}\not{b}\not{a} - 4a\cdot b\not{c} = -2\not{c}\not{b}\not{a}\ .\end{aligned}$$

To 5. equation of (3.66).

$$\gamma_\mu\not{a}\not{b}\not{c}\not{d}\gamma^\mu = \gamma_\mu\not{a}\not{b}\not{c}d_\nu\gamma^\nu\gamma^\mu \stackrel{\mathrm{a}}{=} \gamma_\mu\not{a}\not{b}\not{c}d_\nu\,)(2g^{\mu\nu}-\gamma^\mu\gamma^\nu) \stackrel{\text{4.eq.}}{=} 2\not{d}\not{a}\not{b}\not{c} + 2\not{c}\not{b}\not{a}\not{d}.$$

Summary

- Similarly to the nonrelativistic case, the Dirac equation can be rewritten as an integral equation for the propagator and for the wave function itself that is solvable approximately.
- Due to the quadratic energy-momentum relation, there exist four possible boundary conditions (instead of two as in the nonrelativistic case) to determine the propagator uniquely. They all lead to different causal relationships.
- The **Feynman fermion propagator** describes the temporally forward [backward] evolution of a positive [negative] wave function. Through this and the Feynman-Stückelberg interpretation, the positive [negative] parts can be interpreted as temporally forward directed particle [antiparticle] propagation.
- Relativistic scattering processes are described by scattering amplitudes. Using the Feynman propagator formalism, they can be expanded in a series of multiple scatterings. Compared to the nonrelativistic case, the graphical representations of the individual expansion terms (**Feynman diagrams**) are more complicated reflecting the more versatile scattering constellations, particularly the possibility of particle creation and annihilation processes.
- The concrete calculation of scattering processes is based on the same considerations as in the nonrelativistic case (plane waves, adiabatic approximation and cross section).

Exercises

35. Decomposition of $S_{\mathrm{F}}^{(0)}$ by plane waves. Show the transition from (3.57) to (3.58).

Solution. For plane Dirac wave functions (see Theorem 2.4)

$$\begin{aligned}\psi_{\boldsymbol{p}}^{(1,2)}(x) &= \frac{1}{(2\pi\hbar)^{3/2}}\sqrt{\frac{m_0c}{p_0}}\mathrm{e}^{-\mathrm{i}p_\mu x^\mu/\hbar}u(p,\pm s)\\ \psi_{\boldsymbol{p}}^{(3,4)}(x) &= \frac{1}{(2\pi\hbar)^{3/2}}\sqrt{\frac{m_0c}{p_0}}\mathrm{e}^{+\mathrm{i}p_\mu x^\mu/\hbar}v(p,\mp s)\end{aligned}$$

we have the relations [see (3.67)and (3.68)]

$$\sum_s u(p,s)\bar{u}(p,s) = \Lambda_+(p)\ ,\ \sum_s v(p,s)\bar{v}(p,s) = -\Lambda_-(p)\ ,$$

from which follows that

$$(2\pi\hbar)^3 \sum_{r=1}^{2} \psi_{\boldsymbol{p}}^{(r)}(x')\bar{\psi}_{\boldsymbol{p}}^{(r)}(x) = \mathrm{e}^{-\mathrm{i}p_\mu(x'^\mu - x^\mu)/\hbar} \frac{m_0 c}{p_0} \sum_s u(p,s)\bar{u}(p,s)$$
$$= \mathrm{e}^{-\mathrm{i}p_\mu(x'^\mu - x^\mu)/\hbar} \frac{m_0 c}{p_0} \Lambda_+(p)$$
$$(2\pi\hbar)^3 \sum_{r=3}^{4} \psi_{\boldsymbol{p}}^{(r)}(x')\bar{\psi}_{\boldsymbol{p}}^{(r)}(x) = \mathrm{e}^{\mathrm{i}p_\mu(x'^\mu - x^\mu)/\hbar} \frac{m_0 c}{p_0} \sum_s v(p,s)\bar{v}(p,s)$$
$$= -\mathrm{e}^{\mathrm{i}p_\mu(x'^\mu - x^\mu)/\hbar} \frac{m_0 c}{p_0} \Lambda_-(p) \; .$$

Comparison with (3.57) yields (3.58).

36. Causality principle of $S_{\mathrm{F}}^{(0)}$. Show the validity of both causal relationships (3.51) and (3.53) for the free case by exploiting (3.58).

Solution. Let

$$\psi(x) = \psi^{(+)}(x) + \psi^{(-)}(x) = \int \mathrm{d}^3p' \sum_{r'=1}^{4} a^{(r')}(\boldsymbol{p}')\Psi_{\boldsymbol{p}'}^{(r')}(x)$$

be an arbitrary free Dirac wave packet. Then it holds that

$$\int \mathrm{d}^3x S_{\mathrm{F}}^{(0)}(x'-x)\gamma^0\psi(x)$$
$$= -\mathrm{i}\Theta(x'^0 - x^0)\int \mathrm{d}^3x \int \mathrm{d}^3p \int \mathrm{d}^3p'$$
$$\times \sum_{r=1}^{2}\sum_{r'=1}^{4} \psi_{\boldsymbol{p}}^{(r)}(x')\psi_{\boldsymbol{p}}^{(r)\dagger}(x)\psi_{\boldsymbol{p}'}^{(r')}(x)a^{(r')}(\boldsymbol{p}')$$
$$+\mathrm{i}\Theta(x^0 - x'^0)\int \mathrm{d}^3x \int \mathrm{d}^3p \int \mathrm{d}^3p'$$
$$\times \sum_{r=3}^{4}\sum_{r'=1}^{4} \psi_{\boldsymbol{p}}^{(r)}(x')\psi_{\boldsymbol{p}}^{(r)\dagger}(x)\psi_{\boldsymbol{p}'}^{(r')}(x)a^{(r')}(\boldsymbol{p}')$$
$$= -\mathrm{i}\Theta(x'^0 - x^0)\int \mathrm{d}^3p \int \mathrm{d}^3p' \sum_{r=1}^{2}\sum_{r'=1}^{4} \delta_{rr'}\delta(\boldsymbol{p}-\boldsymbol{p}')\psi_{\boldsymbol{p}}^{(r)}(x')a^{(r')}(\boldsymbol{p}')$$
$$+\mathrm{i}\Theta(x^0 - x'^0)\int \mathrm{d}^3p \int \mathrm{d}^3p' \sum_{r=3}^{4}\sum_{r'=1}^{4} \delta_{rr'}\delta(\boldsymbol{p}-\boldsymbol{p}')\psi_{\boldsymbol{p}}^{(r)}(x')a^{(r')}(\boldsymbol{p}')$$
$$= -\mathrm{i}\Theta(x'^0 - x^0)\int \mathrm{d}^3p \sum_{r=1}^{2} \psi_{\boldsymbol{p}}^{(r)}(x')a^{(r)}(\boldsymbol{p})$$
$$+\mathrm{i}\Theta(x^0 - x'^0)\int \mathrm{d}^3p \sum_{r=3}^{4} \psi_{\boldsymbol{p}}^{(r)}(x')a^{(r)}(\boldsymbol{p})$$
$$= -\mathrm{i}\Theta(x'^0 - x^0)\psi^{(+)}(x') + \mathrm{i}\Theta(x^0 - x'^0)\psi^{(-)}(x') \; .$$

Similarly, we have

$$
\begin{aligned}
&\int \mathrm{d}^3x \bar{\psi}(x)\gamma^0 S_{\mathrm{F}}^{(0)}(x-x') \\
&= -\mathrm{i}\Theta(x^0 - x'^0)\int \mathrm{d}^3x \int \mathrm{d}^3p' \int \mathrm{d}^3p \\
&\quad \times \sum_{r'=1}^{4}\sum_{r=1}^{2} a^{(r')*}(\boldsymbol{p}')\psi_{\boldsymbol{p}'}^{(r')\dagger}(x)\psi_{\boldsymbol{p}}^{(r)}(x)\bar{\psi}_{\boldsymbol{p}}^{(r)}(x') \\
&\quad +\mathrm{i}\Theta(x'^0 - x^0)\int \mathrm{d}^3x \int \mathrm{d}^3p' \int \mathrm{d}^3p \\
&\quad \times \sum_{r'=1}^{4}\sum_{r=3}^{4} a^{(r')*}(\boldsymbol{p}')\psi_{\boldsymbol{p}'}^{(r')\dagger}(x)\psi_{\boldsymbol{p}}^{(r)}(x)\bar{\psi}_{\boldsymbol{p}}^{(r)}(x') \\
&= -\mathrm{i}\Theta(x^0 - x'^0)\int \mathrm{d}^3p' \int \mathrm{d}^3p \sum_{r'=1}^{4}\sum_{r=1}^{2} a^{(r')*}(\boldsymbol{p}')\bar{\psi}_{\boldsymbol{p}}^{(r)}(x')\delta_{rr'}\delta(\boldsymbol{p}-\boldsymbol{p}') \\
&\quad +\mathrm{i}\Theta(x'^0 - x^0)\int \mathrm{d}^3p' \int \mathrm{d}^3p \sum_{r'=1}^{4}\sum_{r=3}^{4} a^{(r')*}(\boldsymbol{p}')\bar{\psi}_{\boldsymbol{p}}^{(r)}(x')\delta_{rr'}\delta(\boldsymbol{p}-\boldsymbol{p}') \\
&= -\mathrm{i}\Theta(x^0 - x'^0)\int \mathrm{d}^3p \sum_{r=1}^{2} a^{(r)*}(\boldsymbol{p})\bar{\psi}_{\boldsymbol{p}}^{(r)}(x') \\
&\quad +\mathrm{i}\Theta(x'^0 - x^0)\int \mathrm{d}^3p \sum_{r=3}^{4} a^{(r)*}(\boldsymbol{p})\bar{\psi}_{\boldsymbol{p}}^{(r)}(x') \\
&= -\mathrm{i}\Theta(x^0 - x'^0)\bar{\psi}^{(+)}(x') + \mathrm{i}\Theta(x'^0 - x^0)\bar{\psi}^{(-)}(x') \; .
\end{aligned}
$$

3.3 Spin-1/2 Scattering Processes

After the preparatory considerations of the preceding two sections, we now carry out concrete calculations of spin-1/2 scattering processes to the lowest orders of the scattering theory. As the simplest example, we first consider the Coulomb scattering of electrons. Thereafter, we discuss the more realistic case of electron scattering against freely moving protons. Here we encounter a close correspondence between scattering processes, Feynman diagrams, and scattering amplitudes that can be cast into a simple set of rules, the so-called *Feynman rules*. Furthermore, we address the processes of electron-electron and electron-positron scattering as well as Compton scattering against electrons, electron-positron creation by two photons, and electron-positron annihilation into two photons. The first two and the last three processes are interconnected via the principle of *crossing symmetry*. This section ends with a complete compilation of the Feynman rules.

Before we start, we highlight some points that are essential for the correct understanding of the whole third chapter.

- The Dirac equation deals with the movement of (anti)fermions within an external classical background potential. The same is true for the propagator formalism (Theorem 3.5) as it is merely based upon a perturbation theoretical expansion of the Dirac equation.
- In this respect, the scattering of an electron against an external classical Coulomb potential is a concrete and totally legitimate application of this formalism.
- Strictly speaking, the electron-proton scattering lies outside the range of our formalism. However, it can be integrated into it by a plausible extension, namely that the external potential is considered to be created by the proton or electron current. In this way, any two-particle scatterings can generally be described as current-current interactions within this formalism, whereas the interactions themselves, i.e. the A^μ-fields in the scattering series (3.61), can be interpreted as the exchange of n virtual photons between both particles to n-th order.
- The other processes, Compton scattering, electron-positron annihilation, and electron-positron creation, clearly go beyond the scope of our formalism since they involve photonic initial and final states. Fortunately, even here one can find a reasonable way of integration by considering the A^μ-fields in the term of the lowest (here: second) order of (3.61) as the incoming and outgoing photons. However, in this case the question naturally arises how the A^μ-fields of higher-order terms are to be interpreted.
- As we will see, the scattering amplitudes of all of these processes can be graphically depicted and calculated using the Feynman rules. However, the Feynman rules are much more general than what follows from our scattering formalism. Besides *tree diagrams*, they also allow the presence of *loop diagrams* corresponding to the creation and subsequent annihilation of virtual particles. Strictly speaking, those effects (*radiation corrections*) cannot be explained by our formalism and are purely of quantum field theoretical nature.
- In other words, the Feynman rules are the quantum electrodynamical directives for the construction of scattering amplitudes. They can be deduced from the propagator scattering formalism with some additional, not necessarily obvious generalizations. In the narrow sense, the propagator scattering formalism only provides their tree level part.
- Apart from the presentation of the complete Feynman rules, this section deals exclusively with scattering processes on tree level. Quantum electrodynamical corrections (loop level) are the subject of Section 3.4.

Further notes. From now on we use the *natural unit system* throughout in order to keep our calculations as clear as possible. In this system we have

$$\hbar = c = 1 \ .$$

Consequently, there is only one energy or mass unit left which is typically measured in electron volt (eV). Furthermore, length and time are proportional to each other and possess the unit 1/eV. The conversion from the MKS to the natural system is

$$1\mathrm{s} = \frac{1.519 \cdot 10^{15}}{\mathrm{eV}} \ , \ 1\mathrm{m} = \frac{5.068 \cdot 10^{6}}{\mathrm{eV}} \ . \tag{3.69}$$

Finally, note that we will use two different notions of order in the following. Firstly, "order" refers to the number of the term within the series of multiple scatterings from Theorem 3.5 starting from zero. Secondly, it means the order in the coupling constant e.

3.3.1 Coulomb Scattering of Electrons

First, we consider the scattering of electrons against a Coulomb potential of the form

$$eA^0(x) = V(x) = \frac{\alpha}{|\boldsymbol{x}|} \ , \ \boldsymbol{A}(x) = \boldsymbol{0} \ , \ \alpha = -Ze^2 \ ,$$

where we proceed similarly to the nonrelativistic calculation in Subsection 3.1.4. Since incoming and scattered particles are electrons, we choose for $\Psi_{i,f}$ positive plane Dirac waves normalized to the volume V with electron mass m_0, energies $E_{i,f}$, four-momentum indices $p_{i,f}$, and four-polarization indices $s_{i,f}$ [compare to (3.62)]:

$$\begin{aligned}
\Psi_i(x) &= \sqrt{\frac{m_0}{E_i V}} u(p_i, s_i) \mathrm{e}^{-\mathrm{i}E_i t} \mathrm{e}^{\mathrm{i}\boldsymbol{p}_i \boldsymbol{x}} \quad \text{(in the limit } t \to -\infty) \\
\Psi_f(x) &= \sqrt{\frac{m_0}{E_f V}} u(p_f, s_f) \mathrm{e}^{-\mathrm{i}E_f t} \mathrm{e}^{\mathrm{i}\boldsymbol{p}_f \boldsymbol{x}} \ \text{(in the limit } t \to +\infty) \ .
\end{aligned}$$

According to Theorem 3.5, the corresponding scattering amplitude to first order ($\epsilon_f = +1$, $f \neq i$) is

$$\begin{aligned}
S_{fi} &= -\mathrm{i}e \int \mathrm{d}^4 x \bar{\Psi}_f(x) \gamma^\mu A_\mu(x) \Psi_i(x) \\
&= -\frac{\mathrm{i}\alpha}{V} \sqrt{\frac{m_0^2}{E_f E_i}} \bar{u}(p_f, s_f) \gamma^0 u(p_i, s_i) \\
&\quad \times \int\limits_{-T/2}^{T/2} \mathrm{d}t \mathrm{e}^{\mathrm{i}(E_f - E_i)t} \int\limits_V \mathrm{d}^3 x \mathrm{e}^{-\mathrm{i}\boldsymbol{q}\boldsymbol{x}} \frac{1}{|\boldsymbol{x}|} \ , \ \boldsymbol{q} = \boldsymbol{p}_f - \boldsymbol{p}_i \\
&= -\frac{\mathrm{i}[2\pi\delta(E_f - E_i)]}{V} \sqrt{\frac{m_0^2}{E_i E_f}} M_{fi} \ , \ M_{fi} = \frac{4\pi\alpha}{\boldsymbol{q}^2} \bar{u}(p_f, s_f) \gamma^0 u(p_i, s_i) \ ,
\end{aligned}$$

where, in the last step, the relations

$$\int_{-T/2}^{T/2} \mathrm{d}t \mathrm{e}^{\mathrm{i}(E_f - E_i)t} \overset{T\to\infty}{=} 2\pi\delta(E_f - E_i) \; , \; \int_V \mathrm{d}^3x \frac{\mathrm{e}^{-\mathrm{i}\boldsymbol{q}\boldsymbol{x}}}{|\boldsymbol{x}|} = \frac{4\pi}{\boldsymbol{q}^2}$$

[see (3.25) and (3.26)] have been used. Due to Theorem 3.3, the differential cross section follows as

$$\mathrm{d}\sigma = \frac{|S_{fi}|^2}{T|\boldsymbol{j}_i|} \frac{V\mathrm{d}^3p_f}{(2\pi)^3} = \frac{m_0}{E_i} \frac{[2\pi\delta(E_f - E_i)]^2}{TV|\boldsymbol{j}_i|} |M_{fi}|^2 \frac{m_0\mathrm{d}^3p_f}{(2\pi)^3 E_f}$$

or, after the replacement $[2\pi\delta(E_f - E_i)]^2 \to 2\pi T\delta(E_f - E_i)$ with finite T,

$$\mathrm{d}\sigma = \frac{m_0}{E_i} \frac{1}{V|\boldsymbol{j}_i|} |M_{fi}|^2 (2\pi)\delta(E_f - E_i) \frac{m_0\mathrm{d}^3p_f}{(2\pi)^3 E_f} \; . \tag{3.70}$$

To determine the current density $|\boldsymbol{j}_i|$ we assume the velocity of the incoming particle to be oriented toward the z-direction. It then follows that

$$|\boldsymbol{j}_i| = \frac{m_0}{E_i V} \left| u^\dagger(p_i, s_i)\alpha_3 u(p_i, s_i) \right|$$

and, in the Dirac representation,

$$u(p_i, s_i) = \sqrt{\frac{E_i + m_0}{2m_0}} \begin{pmatrix} \chi_{s_i} \\ \dfrac{\sigma_3|\boldsymbol{p}_i|}{E_i + m_0}\chi_{s_i} \end{pmatrix} \; , \; \chi_{s_i}^\dagger \chi_{s_i} = 1 \; ,$$

where the concrete form of the spinors χ_{s_i} depends on the direction of polarization s_i. From this follows the polarization-independent and intuitively expected relation

$$|\boldsymbol{j}_i| = \frac{|\boldsymbol{p}_i|}{E_i V} \; .$$

Inserting this into (3.70), the differential cross section becomes

$$\mathrm{d}\sigma = \frac{m_0}{|\boldsymbol{p}_i|} |M_{fi}|^2 (2\pi)\delta(E_f - E_i) \frac{m_0\mathrm{d}^3p_f}{(2\pi)^3 E_f} \; ,$$

where, as desired, all dependencies of T and V are removed. Taking into account

$$\mathrm{d}^3p_f = \boldsymbol{p}_f^2 \mathrm{d}|\boldsymbol{p}_f|\mathrm{d}\Omega \; , \; E_f^2 = \boldsymbol{p}_f^2 + m_0^2 \Longrightarrow \mathrm{d}|\boldsymbol{p}_f| = \frac{E_f\mathrm{d}E_f}{|\boldsymbol{p}_f|} \; ,$$

we finally get

$$\begin{aligned} \frac{\mathrm{d}\sigma}{\mathrm{d}\Omega} &= \frac{m_0^2}{(2\pi)^2|\boldsymbol{p}_i|} \int \mathrm{d}E_f |\boldsymbol{p}_f| |M_{fi}|^2 \delta(E_f - E_i) \\ &= \frac{m_0^2}{(2\pi)^2} |M_{fi}|^2_{|\boldsymbol{p}_f| = |\boldsymbol{p}_i|} \\ &= \frac{4\alpha^2 m_0^2}{\boldsymbol{q}^4} \left| \bar{u}(p_f, s_f)\gamma^0 u(p_i, s_i) \right|^2_{|\boldsymbol{p}_f| = |\boldsymbol{p}_i|} \; . \end{aligned} \tag{3.71}$$

As expected, in the nonrelativistic limit this expression turns into Rutherford's scattering formula (3.29):

$$\left|\bar{u}(p_f, s_f)\gamma^0 u(p_i, s_i)\right|^2 = \left|(1,0)\begin{pmatrix}1 & 0\\ 0 & 1\end{pmatrix}\begin{pmatrix}1\\ 0\end{pmatrix}\right|^2 = 1 .$$

Unpolarized cross section. To calculate (3.71) further, we initially assume that in the scattering experiment neither the polarization of the incoming particle beam is prepared nor the polarization of the scattered particles is measured – quite a typical practical situation. This implies that in (3.71) the average over all possible initial polarizations s_i and the sum over all possible final polarizations s_f must be taken (every possible s_i occurs with the same probability and every possible s_f is measured):

$$\overline{\frac{\mathrm{d}\sigma}{\mathrm{d}\Omega}} = \frac{4\alpha^2 m_0^2}{\boldsymbol{q}^4}\frac{1}{2}\sum_{s_f,s_i}\left|\bar{u}(p_f, s_f)\gamma^0 u(p_i, s_i)\right|^2_{|\boldsymbol{p}_f|=|\boldsymbol{p}_i|} . \tag{3.72}$$

The appearance of such double spin sums within cross sections allows a comfortable evaluation using Theorem 3.6 where the concrete form of the involved bispinors does not matter. Therefore, here and in all subsequent scattering problems, we will strive to cast the corresponding cross section into a form similar to (3.72). In the case of (3.72), it follows from (3.64) and the second equation of (3.65) that

$$\begin{aligned}
\left|\bar{u}(p_f, s_f)\gamma^0 u(p_i, s_i)\right|^2 &= \left[\bar{u}(p_f, s_f)\gamma^0 u(p_i, s_i)\right]\left[\bar{u}(p_f, s_f)\gamma^0 u(p_i, s_i)\right]^\dagger \\
&= \left[\bar{u}(p_f, s_f)\gamma^0 u(p_i, s_i)\right]\left[\bar{u}(p_i, s_i)\gamma^0 u(p_f, s_f)\right]
\end{aligned}$$

$$\begin{aligned}
\Longrightarrow \sum_{s_f,s_i}\left|\bar{u}(p_f, s_f)\gamma^0 u(p_i, s_i)\right|^2 &= \mathrm{tr}\left[\Lambda_+(p_f)\gamma^0\Lambda_+(p_i)\gamma^0\right] \\
&= \mathrm{tr}\left(\frac{\not{p}_f + m_0}{2m_0}\gamma^0\frac{\not{p}_i + m_0}{2m_0}\gamma^0\right) \\
&= \frac{1}{4m_0^2}\mathrm{tr}(\not{p}_f\gamma^0\not{p}_i\gamma^0) + \frac{1}{4m_0}\mathrm{tr}(\not{p}_f) \\
&\quad + \frac{1}{4m_0}\mathrm{tr}(\not{p}_i) + \frac{1}{4}\mathrm{tr}(1) \\
&= \frac{1}{4m_0^2}\mathrm{tr}(\not{p}_f\gamma^0\not{p}_i\gamma^0) + 1 .
\end{aligned}$$

Introducing the four-vector $(a^\mu) = (1, 0, 0, 0)$ and using the fourth and first equations of (3.65), the remaining trace becomes

$$\begin{aligned}
\mathrm{tr}\left(\not{p}_f\gamma^0\not{p}_i\gamma^0\right) &= \mathrm{tr}\left(\not{p}_f\not{a}\not{p}_i\not{a}\right) \\
&= 2(p_f\cdot a)\mathrm{tr}(\not{p}_i\not{a}) - (p_f\cdot p_i)\mathrm{tr}(\not{a}\not{a}) \\
&= 8(p_f\cdot a)(p_i\cdot a) - 4(p_f\cdot p_i)(a\cdot a) \\
&= 8E_iE_f - 4\left(E_iE_f - \boldsymbol{p}_i\boldsymbol{p}_f\right) \\
&= 4E_iE_f + 4\boldsymbol{p}_i\boldsymbol{p}_f .
\end{aligned} \tag{3.73}$$

All in all, (3.72) turns into *Mott's scattering formula*

$$\begin{aligned}\overline{\frac{\mathrm{d}\sigma}{\mathrm{d}\Omega}} &= \left(\frac{\mathrm{d}\sigma}{\mathrm{d}\Omega}\right)_{\mathrm{Mott}} = \frac{2\alpha^2 m_0^2}{\boldsymbol{q}^4}\left(1+\frac{E_iE_f+\boldsymbol{p}_i\boldsymbol{p}_f}{m_0^2}\right)_{|\boldsymbol{p}_f|=|\boldsymbol{p}_i|} \\ &= \frac{4\alpha^2}{\boldsymbol{q}^4}\left(E_i^2\cos^2\frac{\theta}{2}+m_0^2\sin^2\frac{\theta}{2}\right) = \frac{\alpha^2\left(1-v_i^2\sin^2\frac{\theta}{2}\right)}{4v_i^4E_i^2\sin^4\frac{\theta}{2}}\;,\end{aligned} \tag{3.74}$$

where, in the last row, the identities

$$\boldsymbol{p}_i\boldsymbol{p}_f\big|_{|\boldsymbol{p}_f|=|\boldsymbol{p}_i|} = \boldsymbol{p}_i^2\cos\theta\;,\;\; \cos\theta = \cos^2\frac{\theta}{2}-\sin^2\frac{\theta}{2}$$

$$\boldsymbol{q}^2\big|_{|\boldsymbol{p}_f|=|\boldsymbol{p}_i|} = 4\boldsymbol{p}_i^2\sin^2\frac{\theta}{2}\;,\;\; \boldsymbol{p}_i^2 = v_i^2E_i^2$$

have been used.

Partially polarized cross section. Next we suppose that the polarization s_f of the scattered particles is measured while the incoming particle beam is still unpolarized. Instead of (3.72), we now have to consider

$$\frac{\mathrm{d}\sigma}{\mathrm{d}\Omega}(s_f) = \frac{4\alpha^2m_0^2}{\boldsymbol{q}^4}\frac{1}{2}\sum_{s_i}\left|\bar{u}(p_f,s_f)\gamma^0u(p_i,s_i)\right|^2_{|\boldsymbol{p}_f|=|\boldsymbol{p}_i|} \tag{3.75}$$

resulting from (3.71) by averaging over all possible initial polarizations s_i. Obviously, this expression does not possess the desired form of a double spin sum. However, using the spin projectors $\Sigma(s)$ of Theorem 2.4, we can transform the single spin sum in (3.75) into a double spin sum which, as before, can be easily evaluated further with the help of Theorem 3.6. Taking into account

$$\Sigma(s)u(p,s) = u(p,s)\;,\;\; \Sigma(s)u(p,-s) = 0\;,$$

we have

$$\begin{aligned}&\sum_{s_i}\left|\bar{u}(p_f,s_f)\gamma^0u(p_i,s_i)\right|^2 \\ &\quad= \sum_{s_i}\left[\bar{u}(p_f,s_f)\gamma^0u(p_i,s_i)\right]\left[\bar{u}(p_f,s_f)\gamma^0u(p_i,s_i)\right]^\dagger \\ &\quad= \sum_{s_i}\left[\bar{u}(p_f,s_f)\gamma^0u(p_i,s_i)\right]\left[\bar{u}(p_i,s_i)\gamma^0u(p_f,s_f)\right] \\ &\quad= \sum_{s_f',s_i}\left[\bar{u}(p_f,s_f')\gamma^0u(p_i,s_i)\right]\left[\bar{u}(p_i,s_i)\gamma^0\Sigma(s_f)u(p_f,s_f')\right] \\ &\quad= \mathrm{tr}\left[\Lambda_+(p_f)\gamma^0\Lambda_+(p_i)\gamma^0\Sigma(s_f)\right] \\ &\quad= \frac{1}{8m_0^2}\mathrm{tr}\left[(\not{p}_f+m_0)\gamma^0(\not{p}_i+m_0)\gamma^0(1+\gamma^5\not{s}_f)\right]\end{aligned}$$

$$= \frac{1}{8m_0^2} \mathrm{tr}\left(\not{p}_f \gamma^0 \not{p}_i \gamma^0\right) + \frac{1}{2}$$

$$= \frac{1}{2}\left(1 + \frac{E_i E_f + \boldsymbol{p}_i \boldsymbol{p}_f}{m_0^2}\right) . \tag{3.76}$$

The last-but-one step results from the fact that traces consisting of an odd number of $\not{a}$-multiplications (with or without an additional γ^5) vanish and that terms with an even number of $\not{a}$-multiplications and one additional γ^5 can be related back to terms of the form $\mathrm{tr}(\gamma^5 \not{a} \not{b}) = 0$ by anticommuting with γ^0. The last step follows from (3.73). Inserting the last relation into (3.75) and comparing this with (3.74) we obtain the final result

$$\frac{\mathrm{d}\sigma}{\mathrm{d}\Omega}(s_f) = \frac{\alpha^2 m_0^2}{\boldsymbol{q}^4}\left(1 + \frac{E_i E_f + \boldsymbol{p}_i \boldsymbol{p}_f}{m_0^2}\right)_{|\boldsymbol{p}_f| = |\boldsymbol{p}_i|} = \frac{1}{2}\left(\overline{\frac{\mathrm{d}\sigma}{\mathrm{d}\Omega}}\right)_{\mathrm{Mott}} . \tag{3.77}$$

Accordingly, $\mathrm{d}\sigma(s_f)/\mathrm{d}\Omega$ is independent from the measured spin of the scattered particles – an effect which is only true in the lowest order of the scattering theory.

The cross section of the complementary situation where the incoming particle beam is polarized while the polarization of the scattered particles is not measured can be obtained analogously. To do this, one has to consider the equation [see (3.71)]

$$\frac{\mathrm{d}\sigma}{\mathrm{d}\Omega}(s_i) = \frac{4\alpha^2 m_0^2}{\boldsymbol{q}^4} \sum_{s_f} \left|\bar{u}(p_f, s_f)\gamma^0 u(p_i, s_i)\right|^2_{|\boldsymbol{p}_f| = |\boldsymbol{p}_i|} ,$$

in which all possible final polarizations are summed up. Performing a calculation similar to (3.76) where the spin projector $\Sigma(s_i)$ [instead of $\Sigma(s_f)$] is inserted appropriately, one finds

$$\sum_{s_f} \left|\bar{u}(p_f, s_f)\gamma^0 u(p_i, s_i)\right|^2 = \frac{1}{2}\left(1 + \frac{E_i E_f + \boldsymbol{p}_i \boldsymbol{p}_f}{m_0^2}\right)$$

and hence

$$\frac{\mathrm{d}\sigma}{\mathrm{d}\Omega}(s_i) = \frac{2\alpha^2 m_0^2}{\boldsymbol{q}^4}\left(1 + \frac{E_i E_f + \boldsymbol{p}_i \boldsymbol{p}_f}{m_0^2}\right)_{|\boldsymbol{p}_f| = |\boldsymbol{p}_i|} = \left(\overline{\frac{\mathrm{d}\sigma}{\mathrm{d}\Omega}}\right)_{\mathrm{Mott}} . \tag{3.78}$$

Totally polarized cross section. Let us now consider the remaining case where the incoming particle beam is polarized and the polarization of the scattered particles is measured. Here we have [see (3.71)]

$$\frac{\mathrm{d}\sigma}{\mathrm{d}\Omega}(s_i, s_f) = \frac{4\alpha^2 m_0^2}{\boldsymbol{q}^4} \left|\bar{u}(p_f, s_f)\gamma^0 u(p_i, s_i)\right|^2_{|\boldsymbol{p}_f| = |\boldsymbol{p}_i|} .$$

Similarly to the earlier cases, this expression can also be written as a double spin sum by inserting the spin projectors $\Sigma(s_i)$ and $\Sigma(s_f)$ and subsequently simplifying the result using Theorem 3.6:

$$\begin{aligned}
&\left|\bar{u}(p_f, s_f)\gamma^0 u(p_i, s_i)\right|^2 \\
&\quad = \left[\bar{u}(p_f, s_f)\gamma^0 u(p_i, s_i)\right]\left[\bar{u}(p_i, s_i)\gamma^0 u(p_f, s_f)\right] \\
&\quad = \sum_{s'_f, s'_i}\left[\bar{u}(p_f, s'_f)\gamma^0 \Sigma(s_i) u(p_i, s'_i)\right]\left[\bar{u}(p_i, s'_i)\gamma^0 \Sigma(s_f) u(p_f, s'_f)\right] \\
&\quad = \mathrm{tr}\left[\Lambda_+(p_f)\gamma^0 \Sigma(s_i)\Lambda_+(p_i)\gamma^0 \Sigma(s_f)\right] \\
&\quad = \mathrm{tr}\left(\frac{\not{p}_f + m_0}{2m_0}\gamma^0 \frac{1+\gamma^5 \not{s}_i}{2}\frac{\not{p}_i + m_0}{2m_0}\gamma^0 \frac{1+\gamma^5 \not{s}_f}{2}\right) \\
&\quad = \frac{1}{16m_0^2}\left\{\mathrm{tr}\left[(\not{p}_f + m_0)\gamma^0 (\not{p}_i + m_0)\gamma^0\right]\right. \\
&\qquad + \mathrm{tr}\left[(\not{p}_f + m_0)\gamma^0 \gamma^5 \not{s}_i (\not{p}_i + m_0)\gamma^0\right] \\
&\qquad + \mathrm{tr}\left[(\not{p}_f + m_0)\gamma^0 (\not{p}_i + m_0)\gamma^0 \gamma^5 \not{s}_f\right] \\
&\qquad \left. + \mathrm{tr}\left[(\not{p}_f + m_0)\gamma^0 \gamma^5 \not{s}_i (\not{p}_i + m_0)\gamma^0 \gamma^5 \not{s}_f\right]\right\} .
\end{aligned}$$

Here the second and third traces vanish due to the same line of argument as in (3.76). Therefore, it follows that

$$\begin{aligned}
\frac{\mathrm{d}\sigma}{\mathrm{d}\Omega}(s_i, s_f) = \frac{\alpha^2}{4\boldsymbol{q}^4}&\left\{\mathrm{tr}\left[(\not{p}_f + m_0)\gamma^0 (\not{p}_i + m_0)\gamma^0\right]\right. \\
&\left. + \mathrm{tr}\left[(\not{p}_f + m_0)\gamma^0 \gamma^5 \not{s}_i (\not{p}_i + m_0)\gamma^0 \gamma^5 \not{s}_f\right]\right\}_{|\boldsymbol{p}_f| = |\boldsymbol{p}_i|} .
\end{aligned} \tag{3.79}$$

Contrary to (3.77) and (3.78), the polarization dependencies are still present. Thus, for further evaluation of this expression, the four-polarizations s_i and s_f need to be concretized. For this purpose, consider an electron whose rest spin is directed toward $\boldsymbol{s}^{(0)}$, $|\boldsymbol{s}^{(0)}| = 1$. Then, according to Theorem 2.3, the electron's four-polarization in a system where it is moving with velocity $\boldsymbol{v} = \boldsymbol{p}/E$ is ($c = 1$)

$$(s^\mu) = [\Lambda_{-\boldsymbol{v}}]^\mu{}_\nu \begin{pmatrix} 0 \\ \boldsymbol{s}^{(0)} \end{pmatrix} = \left(\frac{\boldsymbol{p}\boldsymbol{s}^{(0)}}{m_0}, \boldsymbol{s}^{(0)} + \frac{\boldsymbol{p}\boldsymbol{s}^{(0)}}{m_0(E+m_0)}\boldsymbol{p}\right) , \tag{3.80}$$

where $\Lambda_{-\boldsymbol{v}}$ denotes the Lorentz transformation for the transition from the electron's rest system to a system moving with velocity $-\boldsymbol{v}$ relative to it. If we now assume that the electron's rest spin is parallel or antiparallel to its direction of motion, i.e. that it has positive or negative helicity

$$\boldsymbol{s}^{(0)} = \frac{\lambda \boldsymbol{p}}{|\boldsymbol{p}|} , \ \lambda = \pm 1 ,$$

then (3.80) turns into

$$(s^\mu) = \lambda\left(\frac{|\boldsymbol{p}|}{m_0}, \frac{E}{m_0}\frac{\boldsymbol{p}}{|\boldsymbol{p}|}\right) .$$

Hence, for the case of incoming and scattered particles each with positive or negative helicity, (3.79) can be rewritten as

$$\frac{\mathrm{d}\sigma}{\mathrm{d}\Omega}(\lambda_i,\lambda_f) = \frac{\alpha^2}{4\boldsymbol{q}^4}\Big\{\mathrm{tr}\left[(\not{p}_f+m_0)\gamma^0(\not{p}_i+m_0)\gamma^0\right] \\ +\lambda_i\lambda_f\,\mathrm{tr}\left[(\not{p}_f+m_0)\gamma^0\gamma^5\not{s}_i(\not{p}_i+m_0)\gamma^0\gamma^5\not{s}_f\right]\Big\}_{|\boldsymbol{p}_f|=|\boldsymbol{p}_i|} ,$$

with

$$(s_i^\mu) = \lambda_i s_i \ , \ s_i = \left(\frac{|\boldsymbol{p}_i|}{m_0}, \frac{E_i}{m_0}\frac{\boldsymbol{p}_i}{|\boldsymbol{p}_i|}\right)$$

$$(s_f^\mu) = \lambda_f s_f \ , \ s_f = \left(\frac{|\boldsymbol{p}_f|}{m_0}, \frac{E_f}{m_0}\frac{\boldsymbol{p}_f}{|\boldsymbol{p}_f|}\right) .$$

Taking into account $|\boldsymbol{p}_f| = |\boldsymbol{p}_i|$, $\cos\theta = \boldsymbol{p}_i\boldsymbol{p}_f/|\boldsymbol{p}_i|^2$, examination of the remaining traces according to Theorem 3.6 yields

$$\mathrm{tr}\left[(\not{p}_f+m_0)\gamma^0(\not{p}_i+m_0)\gamma^0\right] = 8\left(E_i^2\cos^2\frac{\theta}{2}+m_0^2\sin^2\frac{\theta}{2}\right)$$

$$\mathrm{tr}\left[(\not{p}_f+m_0)\gamma^0\gamma^5\not{s}_i(\not{p}_i+m_0)\gamma^0\gamma^5\not{s}_f\right] = 8\left(E_i^2\cos^2\frac{\theta}{2}-m_0^2\sin^2\frac{\theta}{2}\right) .$$

Therefore, the totally polarized differential cross section is

$$\frac{\mathrm{d}\sigma}{\mathrm{d}\Omega}(\lambda_i,\lambda_f) = \frac{2\alpha^2}{\boldsymbol{q}^4}\left[E_i^2\cos^2\frac{\theta}{2}+m_0^2\sin^2\frac{\theta}{2} \right. \\ \left. +\lambda_i\lambda_f\left(E_i^2\cos^2\frac{\theta}{2}-m_0^2\sin^2\frac{\theta}{2}\right)\right] .$$

As expected, after averaging over λ_i and/or summing over λ_f, this turns into the expressions (3.74), (3.77), and (3.78) for the unpolarized and partially polarized cross sections. Thus, even in first order, the counting rates of the scattered particles with a particular spin orientation depend on the polarization of the incoming particle beam.

Besides the cross section, the *degree of polarization* is also of interest in spin-sensitive scattering experiments. It is defined as the difference between the counting rates for positive and negative helicity divided by the total counting rate:

$$P(\lambda_i) = \frac{\mathrm{d}\sigma(\lambda_f=+1)-\mathrm{d}\sigma(\lambda_f=-1)}{\mathrm{d}\sigma(\lambda_f=+1)+\mathrm{d}\sigma(\lambda_f=-1)} .$$

If the initial state is totally polarized, e.g. $\lambda_i = +1$, the corresponding degree of polarization is

$$P(\lambda_i=+1) = 1-\frac{2m_0^2\sin^2\frac{\theta}{2}}{E_i^2\cos^2\frac{\theta}{2}+m_0^2\sin^2\frac{\theta}{2}} .$$

In the nonrelativistic limit $E\to m_0$, it becomes

$$P(\lambda_i=+1)\approx 1-2\sin^2\frac{\theta}{2} = \cos\theta ,$$

which is simply the geometric overlap of the quantization axes of the initial and final states. This implies that, in this limit and seen from a fixed system, the spin is not influenced by the scattering at all.

Theorem 3.7: Coulomb scattering of electrons to leading order

The scattering amplitude for the scattering of electrons against a fixed Coulomb potential of the form

$$eA^0(x) = \frac{\alpha}{|\boldsymbol{x}|} \ , \ \boldsymbol{A}(x) = \boldsymbol{0}$$

is given in leading order by ($f \neq i$)

$$S_{fi} = -\mathrm{i}\frac{2\pi\delta(E_f - E_i)}{V}\sqrt{\frac{m_0^2}{E_iE_f}}M_{fi} \ ,$$

with

$$M_{fi} = \frac{4\pi\alpha}{\boldsymbol{q}^2}\bar{u}(p_f, s_f)\gamma^0 u(p_i, s_i) \ , \ \boldsymbol{q} = \boldsymbol{p}_f - \boldsymbol{p}_i \ .$$

The differential cross section follows as

$$\begin{aligned}
\mathrm{d}\sigma &= \frac{m_0}{E_i}\frac{1}{V|\boldsymbol{j}_i|}|M_{fi}|^2(2\pi)\delta(E_f - E_i)\frac{m_0\mathrm{d}^3p_f}{(2\pi)^3E_f} \\
&= \frac{m_0}{|\boldsymbol{p}_i|}|M_{fi}|^2(2\pi)\delta(E_f - E_i)\frac{m_0\mathrm{d}^3p_f}{(2\pi)^3E_f} \\
\Longrightarrow \frac{\mathrm{d}\sigma}{\mathrm{d}\Omega} &= \frac{m_0^2}{(2\pi)^2}|M_{fi}|^2_{|\boldsymbol{p}_f|=|\boldsymbol{p}_i|} \ ,
\end{aligned}$$

where, in the last equation, all scattering momenta $\boldsymbol{p}_f$ directed toward $\mathrm{d}\Omega$ have been integrated out. From this one obtains the unpolarized cross section (average over s_i and sum over s_f, $v_i = |\boldsymbol{p}_i|/E_i$)

$$\begin{aligned}
\left(\frac{\mathrm{d}\sigma}{\mathrm{d}\Omega}\right)_{\mathrm{Mott}} &= \frac{2\alpha^2m_0^2}{\boldsymbol{q}^4}\left(1 + \frac{E_iE_f + \boldsymbol{p}_i\boldsymbol{p}_f}{m_0^2}\right)_{|\boldsymbol{p}_f|=|\boldsymbol{p}_i|} \\
&= \frac{4\alpha^2}{\boldsymbol{q}^4}\left(E_i^2\cos^2\frac{\theta}{2} + m_0^2\sin^2\frac{\theta}{2}\right) = \frac{\alpha^2\left(1 - v_i^2\sin^2\frac{\theta}{2}\right)}{4v_i^4E_i^2\sin^4\frac{\theta}{2}} \ ,
\end{aligned}$$

the partially polarized cross sections (average over s_i or sum over s_f)

$$\frac{\mathrm{d}\sigma}{\mathrm{d}\Omega}(s_f) = \frac{1}{2}\left(\frac{\mathrm{d}\sigma}{\mathrm{d}\Omega}\right)_{\mathrm{Mott}} \ , \ \frac{\mathrm{d}\sigma}{\mathrm{d}\Omega}(s_i) = \left(\frac{\mathrm{d}\sigma}{\mathrm{d}\Omega}\right)_{\mathrm{Mott}} \ ,$$

and the totally polarized cross section (with initial and final helicities $\lambda_{i,f}$) ▷

$$\frac{d\sigma}{d\Omega}(\lambda_i, \lambda_f) = \frac{2\alpha^2}{\boldsymbol{q}^4}\left[E_i^2 \cos^2\frac{\theta}{2} + m_0^2 \sin^2\frac{\theta}{2} + \lambda_i\lambda_f\left(E_i^2\cos^2\frac{\theta}{2} - m_0^2\sin^2\frac{\theta}{2}\right)\right] .$$

Let us now turn shortly to the Coulomb scattering of positrons to leading order. In this case we have to describe the outgoing positron (with p_f, s_f) by a temporally backward directed negative plane Dirac wave ψ_i (with $-p_f$, $-s_f$) moving into the scattering area and, accordingly, the incoming positron (with p_i, s_i) by a temporally backward directed negative plane Dirac wave ψ_f (with $-p_i$, $-s_i$) moving out of the scattering area. We therefore have [compare to (3.63)]

$$\psi_i(x) = \sqrt{\frac{m_0}{E_f V}} v(p_f, s_f) \mathrm{e}^{+\mathrm{i}E_f t}\mathrm{e}^{-\mathrm{i}\boldsymbol{p}_f \boldsymbol{x}} \quad \text{(in the limit } t \to +\infty)$$

$$\psi_f(x) = \sqrt{\frac{m_0}{E_i V}} v(p_i, s_i) \mathrm{e}^{+\mathrm{i}E_i t}\mathrm{e}^{-\mathrm{i}\boldsymbol{p}_i \boldsymbol{x}} \quad \text{(in the limit } t \to -\infty) .$$

Analogously to the electron case, Theorem 3.5 ($\epsilon_f = -1$, $f \neq i$) yields the scattering amplitude

$$S_{fi} = +\frac{\mathrm{i}[2\pi\delta(E_f - E_i)]}{V}\sqrt{\frac{m_0^2}{E_i E_f}} M_{fi} \ , \ M_{fi} = \frac{4\pi\alpha}{\boldsymbol{q}^2}\bar{v}(p_i, s_i)\gamma^0 v(p_f, s_f) \ ,$$

which differs from that of the electron case only by the overall sign (due to ϵ_f) and the involved bispinors. Due to the kinematically equal situations, the calculation of the differential cross section leads again to the formula

$$\frac{d\sigma}{d\Omega} = \frac{m_0^2}{(2\pi)^2}|M_{fi}|^2_{|\boldsymbol{p}_f|=|\boldsymbol{p}_i|} .$$

Depending on the considered situation (unpolarized, partially or totally polarized particle beams), we again have to consider different sums of $|M_{fi}|^2$. They all contain traces differing from those of the electron case only by the replacement $\Lambda_+(p) \to \Lambda_-(p)$. Therefore, due to 3.6, it follows that in the positron case, we get the same $|M_{fi}|^2$-sums and hence the same cross sections as in the electron case. However, this coincidence is only true in the lowest order.

3.3.2 Electron-Proton Scattering (I)

The Coulomb scattering of electrons discussed in the preceding subsection is equivalent to the electron scattering against a fixed, infinitely heavy, spinless, and structureless proton. In this and the next subsection we extend this scenario and deal with the more realistic scattering of electrons against freely moving, finitely heavy protons. These protons are now considered to be spin-1/2 particles, whereas their internal structure is still ignored. Thus, particularly due to repulsion effects, we expect some differences compared to the Coulomb scattering.

The starting point of our discussion is the scattering amplitude to first order (see Theorem 3.5, $\epsilon_f = +1, f \neq i$)

$$S_{fi} = -\mathrm{i}e \int \mathrm{d}^4 x \bar{\Psi}_f(x) \not{A} \Psi_i(x) \ . \tag{3.81}$$

As before, Ψ_i and Ψ_f denote positive plane Dirac waves for the initial and final electron states normalized to the volume V. A^μ is the four-potential generated by the proton whose form is a priori not clear and needs to be determined. We assume that the proton's electric current (more precisely: electric current density) $J^{(\mathrm{p})\mu}$ is known. The corresponding electromagnetic *radiation field* A^μ can then be calculated via the Maxwell equation[16]

$$\partial_\mu F^{\mu\nu}(x) = 4\pi J^{(\mathrm{p})\nu}(x) \ , \ F^{\mu\nu} = \partial^\mu A^\nu - \partial^\nu A^\mu$$

or, using the Lorentz gauge $\partial_\mu A^\mu = 0$, via

$$\partial_\mu \partial^\mu A^\nu(x) = 4\pi J^{(\mathrm{p})\nu}(x) \ .$$

For our purposes, it it advantageous to use the Green function calculus and write the solution to the last equation as

$$A^\mu(x) = \int \mathrm{d}^4 y D_{\mathrm{F}}^{(0)}(x-y) J^{(\mathrm{p})\mu}(y) \ . \tag{3.82}$$

$D_{\mathrm{F}}^{(0)}$ is the *free photon propagator* which, in turn, has to fulfill the equation

$$\partial_\mu \partial^\mu D_{\mathrm{F}}^{(0)}(x-y) = 4\pi\delta(x-y) \ . \tag{3.83}$$

For its solution, we proceed similarly to Exercise 31 and Subsection 3.2.2. Using the fourdimensional Fourier representations

$$\begin{aligned} D_{\mathrm{F}}^{(0)}(x-y) &= \int \frac{\mathrm{d}^4 q}{(2\pi)^4} \mathrm{e}^{-\mathrm{i}q\cdot(x-y)} \tilde{D}_{\mathrm{F}}^{(0)}(q) \\ \delta(x-y) &= \int \frac{\mathrm{d}^4 q}{(2\pi)^4} \mathrm{e}^{-\mathrm{i}q\cdot(x-y)} \end{aligned}$$

and inserting them into (3.83), it follows that

$$\begin{aligned} \tilde{D}_{\mathrm{F}}^{(0)}(q) &= -\frac{4\pi}{q^2} \ , \ q^2 = q_\mu q^\mu \neq 0 \\ \Longrightarrow D_{\mathrm{F}}^{(0)}(x-y) &= \int \frac{\mathrm{d}^4 q}{(2\pi)^4} \frac{-4\pi}{q^2 + \mathrm{i}\epsilon} \mathrm{e}^{-\mathrm{i}q\cdot(x-y)} \ . \end{aligned} \tag{3.84}$$

Based on our conclusions in Subsection 3.2.2, we have added a small imaginary part to the denominator from the beginning, which guarantees the desired causal behavior of A^μ, namely that only electromagnetic radiation with

---

[16] The factor $4\pi/c = 4\pi$ results from the use of Gaussian units ($c = 1$).

positive frequency, i.e. positive energy, occurs. Of course, we also have contributions with negative energy moving backward in time. However, since the photon carries no charge and is therefore its own antiparticle, both processes are physically identical.

Combining the two equations (3.81) and (3.82), the scattering amplitude follows as

$$\begin{aligned} S_{fi} &= -\mathrm{i}e \int \mathrm{d}^4x \bar{\Psi}_f(x)\gamma_\mu A^\mu(x)\Psi_i(x) \\ &= -\mathrm{i} \int \mathrm{d}^4x \int \mathrm{d}^4y [e\bar{\Psi}_f(x)\gamma_\mu \Psi_i(x)] A^\mu(x) \\ &= -\mathrm{i} \int \mathrm{d}^4x \int \mathrm{d}^4y \left[e\bar{\Psi}_f(x)\gamma_\mu \Psi_i(x)\right] D_{\mathrm{F}}^{(0)}(x-y) J^{(\mathrm{p})\mu}(y) , \end{aligned} \tag{3.85}$$

where the proton current $J^{(\mathrm{p})\mu}$ is still undetermined. Obviously, the square bracket can be identified (to first order) with the current of the electrons:

$$J_\mu(x) = e\bar{\Psi}_f(x)\gamma_\mu\Psi_i(x) .$$

Since the electronic and protonic currents should be physically on equal footing (electron scattering within the proton field $\Longleftrightarrow$ proton scattering within the electron field), it makes sense to choose the protonic current (to first order) as[17]

$$J^{(\mathrm{p})\mu}(y) = e_{\mathrm{p}}\bar{\Psi}_f^{(\mathrm{p})}(y)\gamma^\mu\Psi_i^{(\mathrm{p})}(y) . \tag{3.86}$$

Here $e_{\mathrm{p}} = -e$ denotes the proton charge and $\Psi_{i,f}^{(\mathrm{p})}$ the proton wave functions in the initial and final states, i.e. positive plane Dirac waves normalized to the volume V. Understandably, both currents are also called *transition currents.* Inserting the electron and proton wave functions

$$\begin{aligned} \Psi_i(x) &= \sqrt{\frac{m_0}{E_i V}} u(p_i, s_i) \mathrm{e}^{-\mathrm{i}p_i \cdot x} \\ \Psi_f(x) &= \sqrt{\frac{m_0}{E_f V}} u(p_f, s_f) \mathrm{e}^{-\mathrm{i}p_f \cdot x} \\ \Psi_i^{(\mathrm{p})}(y) &= \sqrt{\frac{M_0}{E_i^{(\mathrm{p})} V}} u(P_i, S_i) \mathrm{e}^{-\mathrm{i}P_i \cdot y} \\ \Psi_f^{(\mathrm{p})}(y) &= \sqrt{\frac{M_0}{E_f^{(\mathrm{p})} V}} u(P_f, S_f) \mathrm{e}^{-\mathrm{i}P_f \cdot y} \end{aligned}$$

[17] Strictly speaking, this choice of the proton current along with the resulting symmetric current-current interaction is an extension of our scattering formalism, since now the A^μ-field is no longer an external background potential that cannot be influenced by the scattering. A formal justification of this procedure can only be found within the framework of quantum electrodynamics (compare to the introductory notes in this section).

as well as the photon propagator (3.84) and the proton current (3.86), the scattering amplitude (3.85) now becomes

$$\begin{aligned}
S_{fi} &= -\frac{\mathrm{i}}{V^2}\sqrt{\frac{m_0^2}{E_iE_f}}\sqrt{\frac{M_0^2}{E_i^{(\mathrm{p})}E_f^{(\mathrm{p})}}}\int \mathrm{d}^4x\int \mathrm{d}^4y\int\frac{\mathrm{d}^4q}{(2\pi)^4}\\
&\quad\times\left[\bar{u}(p_f,s_f)\gamma_\mu u(p_i,s_i)\frac{-4\pi e e_{\mathrm{p}}}{q^2+\mathrm{i}\epsilon}\bar{u}(P_f,S_f)\gamma^\mu u(P_i,S_i)\right]\\
&\quad\times \mathrm{e}^{\mathrm{i}(p_f-p_i)\cdot x}\mathrm{e}^{-\mathrm{i}q\cdot(x-y)}\mathrm{e}^{\mathrm{i}(P_f-P_i)\cdot y}\ ,
\end{aligned}$$

where M_0, $E_{i,f}^{(\mathrm{p})}$, $P_{i,f}$, $S_{i,f}$ denote the mass, energies, four-momentum, and four-polarization indices of the proton. The x and y integrations can be carried out immediately,

$$\begin{aligned}
\int \mathrm{d}^4x\mathrm{e}^{\mathrm{i}(p_f-p_i-q)\cdot x} &= (2\pi)^4\delta(p_f-p_i-q)\\
\int \mathrm{d}^4y\mathrm{e}^{\mathrm{i}(P_f-P_i+q)\cdot y} &= (2\pi)^4\delta(P_f-P_i+q)\ ,
\end{aligned}$$

and lead to the q integration

$$\begin{aligned}
&(2\pi)^4\int \mathrm{d}^4q\delta(p_f-p_i-q)\delta(P_f-P_i+q)\frac{-4\pi e e_{\mathrm{p}}}{q^2+\mathrm{i}\epsilon}\\
&\qquad = (2\pi)^4\delta(p_f+P_f-p_i-P_i)\frac{-4\pi e e_{\mathrm{p}}}{(p_f-p_i)^2+\mathrm{i}\epsilon}\ .
\end{aligned}$$

All in all, we end up with the expression

$$\left.\begin{aligned}
S_{fi} &= -\frac{\mathrm{i}\left[(2\pi)^4\delta(p_f+P_f-p_i-P_i)\right]}{V^2}\sqrt{\frac{m_0^2}{E_iE_f}}\sqrt{\frac{M_0^2}{E_i^{(\mathrm{p})}E_f^{(\mathrm{p})}}}M_{fi}\\
M_{fi} &= \bar{u}(p_f,s_f)\gamma_\mu u(p_i,s_i)\frac{-4\pi e e_{\mathrm{p}}}{q^2+\mathrm{i}\epsilon}\bar{u}(P_f,S_f)\gamma^\mu u(P_i,S_i)\\
q &= p_f-p_i\ .
\end{aligned}\right\}\quad(3.87)$$

Note that the amplitude M_{fi} is manifestly Lorentz-invariant. Furthermore, it displays a perfect symmetry with respect to the electron and proton variables, which justifies our choice of (3.86) for the protonic transition current. Finally, the fourdimensional δ-function expresses the fact that, contrary to the Coulomb scattering, not only energy but also momentum is conserved (four-momentum conservation).

Cross section. Coming from (3.87), we can now use Theorem 3.3 to calculate the cross section for the electron-proton scattering. Here we have to keep in mind that the integration needs to be carried out over all possible final states of both electrons and protons. This means that, in Theorem 3.3, we have to consider not only the number of final electron states within the momentum interval $[\boldsymbol{p}_f:\boldsymbol{p}_f+\mathrm{d}^3p_f]$ but also the number of final proton states within the momentum interval $[\boldsymbol{P}_f : \boldsymbol{P}_f + \mathrm{d}^3 P_f]$, which, in total, leads to the phase space factor

$$\frac{V\mathrm{d}^3 p_f}{(2\pi)^3}\frac{V\mathrm{d}^3 P_f}{(2\pi)^3} .$$

Therefore, the sixfold differential cross section is

$$\begin{aligned}
\mathrm{d}\sigma &= \frac{|S_{fi}|^2}{T|\boldsymbol{j}_i|}\frac{V\mathrm{d}^3 p_f}{(2\pi)^3}\frac{V\mathrm{d}^3 P_f}{(2\pi)^3} \\
&= \frac{m_0}{E_i}\frac{M_0}{E_i^{(\mathrm{p})}}\frac{\left[(2\pi)^4\delta(p_f+P_f-p_i-P_i)\right]^2}{TV^2|\boldsymbol{j}_i|}|M_{fi}|^2\frac{m_0\mathrm{d}^3 p_f}{(2\pi)^3 E_f}\frac{M_0\mathrm{d}^3 P_f}{(2\pi)^3 E_f^{(\mathrm{p})}} \\
&= \frac{m_0}{E_i}\frac{M_0}{E_i^{(\mathrm{p})}}\frac{1}{|\boldsymbol{j}_i|V}|M_{fi}|^2(2\pi)^4\delta(p_f+P_f-p_i-P_i) \\
&\quad\times\frac{m_0\mathrm{d}^3 p_f}{(2\pi)^3 E_f}\frac{M_0\mathrm{d}^3 P_f}{(2\pi)^3 E_f^{(\mathrm{p})}} . \qquad (3.88)
\end{aligned}$$

In the last step, the mathematically ill-defined δ-square has been replaced by the appropriate generalization of (3.28) for a finite T and V:

$$\left[(2\pi)^4\delta(p_f+P_f-p_i-P_i)\right]^2 \longrightarrow TV(2\pi)^4\delta(p_f+P_f-p_i-P_i) .$$

Next we need the current density $|\boldsymbol{j}_i|$ resulting from the relative motion of the mutually approaching electrons and protons ($\rho, \boldsymbol{v}_i$ = particle density and velocity of the electrons; $\rho^{(\mathrm{p})}, \boldsymbol{V}_i$ = particle density and velocity of the protons):

$$|\boldsymbol{j}_i| = |\rho\boldsymbol{v}_i - \rho^{(\mathrm{p})}\boldsymbol{V}_i| = \frac{|\boldsymbol{v}_i - \boldsymbol{V}_i|}{V} = \frac{\sqrt{\left(E_i^{(\mathrm{p})}\boldsymbol{p}_i - E_i\boldsymbol{P}_i\right)^2}}{VE_iE_i^{(\mathrm{p})}} .$$

Since in the following we focus on collinear currents $[(\boldsymbol{j}_i \| \boldsymbol{j}_i^{(\mathrm{p})} \Longrightarrow (\boldsymbol{p}_i\boldsymbol{P}_i)^2 = \boldsymbol{p}_i^2\boldsymbol{P}_i^2]$, this equation can be rewritten as

$$|\boldsymbol{j}_i| = \frac{\sqrt{(p_i\cdot P_i)^2 - m_0^2M_0^2}}{VE_iE_i^{(\mathrm{p})}} . \qquad (3.89)$$

Inserting this into (3.88), the differential cross section follows as

$$\begin{aligned}
\mathrm{d}\sigma &= \frac{m_0M_0}{\sqrt{(p_i\cdot P_i)^2 - m_0^2M_0^2}}|M_{fi}|^2(2\pi)^4\delta(p_f+P_f-p_i-P_i) \\
&\quad\times\frac{m_0\mathrm{d}^3 p_f}{(2\pi)^3 E_f}\frac{M_0\mathrm{d}^3 P_f}{(2\pi)^3 E_f^{(\mathrm{p})}} . \qquad (3.90)
\end{aligned}$$

Note that the last two factors and hence the whole expression are Lorentz-invariant. This can be seen by considering the identity

$$\delta\left[f(x)\right] = \sum_k \frac{\delta(x-x_k)}{\left|\frac{\mathrm{d}f}{\mathrm{d}x}\right|_{x_k}} \ , \ x_k = \text{zeros of } f \ , \tag{3.91}$$

from which follows that (integration only over p_0):

$$\frac{\mathrm{d}^3p}{2E} = \int\limits_0^\infty \mathrm{d}p_0\delta(p^2-m_0^2)\mathrm{d}^3p = \int\limits_{-\infty}^{+\infty} \mathrm{d}^4p\delta(p^2-m_0^2)\Theta(p_0) \ .$$

Since p^μ is a time-like four-vector in every Lorentz system, it generally holds that $p^2 = m_0^2 \Longrightarrow p_0^2 > \boldsymbol{p}^2 > 0$. The Lorentz invariance of the Θ-function and of d^3p/E follows from this immediately.

Contrary to $\mathrm{d}\sigma$, the quantity $\mathrm{d}\sigma/\mathrm{d}\Omega$ is not Lorentz-invariant so that now we need to specify the reference frame. Since electron-proton scattering experiments are usually performed against a fixed proton target, we choose the laboratory system where the proton is initially at rest. Taking into account

$$p_i = (E_i, \boldsymbol{p}_i) \ , \ p_f = (E_f, \boldsymbol{p}_f) \ , \ P_i = (M_0, \boldsymbol{0}) \ , \ \boldsymbol{p}_i\boldsymbol{p}_f = |\boldsymbol{p}_i||\boldsymbol{p}_f|\cos\theta$$

as well as

$$\begin{aligned}
\frac{m_0M_0}{\sqrt{(p_i\cdot P_i)^2 - m_0^2M_0^2}} &= \frac{m_0M_0}{\sqrt{E_i^2M_0^2 - m_0^2M_0^2}} = \frac{m_0}{\sqrt{E_i^2-m_0^2}} = \frac{m_0}{|\boldsymbol{p}_i|} \\
\mathrm{d}^3p_f &= \boldsymbol{p}_f^2\mathrm{d}|\boldsymbol{p}_f|\mathrm{d}\Omega = |\boldsymbol{p}_f|E_f\mathrm{d}E_f\mathrm{d}\Omega \\
\frac{\mathrm{d}^3P_f}{E_f^{(\mathrm{p})}} &= 2\int \mathrm{d}^4P_f\delta(P_f^2-M_0^2)\Theta(P_f^0) \ ,
\end{aligned}$$

(3.90) leads to

$$\begin{aligned}
\frac{\mathrm{d}\sigma}{\mathrm{d}\Omega} &= \frac{m_0^2M_0}{2\pi^2|\boldsymbol{p}_i|}\int \mathrm{d}E_f \int \mathrm{d}^4P_f|\boldsymbol{p}_f||M_{fi}|^2\delta(p_f+P_f-p_i-P_i) \\
&\quad \times\delta(P_f^2-M_0^2)\Theta(P_f^0) \\
&= \frac{m_0^2M_0}{2\pi^2|\boldsymbol{p}_i|}\int \mathrm{d}E_f|\boldsymbol{p}_f||M_{fi}|^2_{P_f=P_i+p_i-p_f} \\
&\quad \times\delta\left[(P_i+p_i-p_f)^2-M_0^2\right]\Theta(M_0+E_i-E_f) \\
&= \frac{m_0^2M_0}{2\pi^2|\boldsymbol{p}_i|}\int\limits_{m_0}^{M_0+E_i} \mathrm{d}E_f|\boldsymbol{p}_f||M_{fi}|^2_{P_f=P_i+p_i-p_f} \\
&\quad \times\delta\left[(P_i+p_i-p_f)^2-M_0^2\right] \\
&= \frac{m_0^2M_0}{2\pi^2|\boldsymbol{p}_i|}\int\limits_{m_0}^{M_0+E_i} \mathrm{d}E_f|\boldsymbol{p}_f||M_{fi}|^2_{P_f=P_i+p_i-p_f} \\
&\quad \times\delta\left[2m_0^2-2M_0(E_f-E_i)-2E_iE_f+2|\boldsymbol{p}_i||\boldsymbol{p}_f|\cos\theta\right] \ .
\end{aligned}$$

The integration limits result from the fact that, on the one hand, $E_f \geq m_0$ is required (lower limit) and, on the one hand, $\Theta(M_0+E_i-E_f)$ yields a contribution only for $E_f < M_0 + E_i$ (upper limit). The remaining integral can again be evaluated using the identity (3.91) so that, finally, we get

$$\left.\begin{aligned} \frac{\mathrm{d}\sigma}{\mathrm{d}\Omega} &= \frac{m_0^2 M_0 |\boldsymbol{p}_f|}{4\pi^2 |\boldsymbol{p}_i|} \frac{|M_{fi}|^2_{\mathrm{co}}}{M_0 + E_i - \frac{|\boldsymbol{p}_i| E_f}{|\boldsymbol{p}_f|}\cos\theta} \\ |M_{fi}|^2_{\mathrm{co}} &= |M_{fi}|^2_{P_f = P_i + p_i - p_f} , \end{aligned}\right\} \quad (3.92)$$

with the secondary condition

$$2m_0^2 - 2M_0(E_f - E_i) - 2E_iE_f + 2|\boldsymbol{p}_i||\boldsymbol{p}_f|\cos\theta = 0$$

connecting E_f or $|\boldsymbol{p}_f|$ with θ and E_i or $|\boldsymbol{p}_i|$. Note the index co (=conservation) that was introduced to abbreviate four-momentum conservation.

Amplitude square. With the last equations, we already have a compact representation of the differential cross section of electron-proton scattering to first order where, however, the amplitude square $|M_{fi}|^2_{\mathrm{co}}$ needs to be evaluated further. In order to keep the calculations simple, we do not study polarization effects. Therefore, instead of (3.92), we consider the unpolarized cross section

$$\overline{\frac{\mathrm{d}\sigma}{\mathrm{d}\Omega}} = \frac{m_0^2 M_0 |\boldsymbol{p}_f|}{4\pi^2 |\boldsymbol{p}_i|} \frac{\overline{|M_{fi}|^2_{\mathrm{co}}}}{M_0 + E_i - \frac{|\boldsymbol{p}_i| E_f}{|\boldsymbol{p}_f|}\cos\theta} , \quad (3.93)$$

with the amplitude

$$\overline{|M_{fi}|^2} = \frac{1}{4} \sum_{\substack{s_f, s_i \\ S_f, S_i}} \left| \bar{u}(p_f, s_f)\gamma_\mu u(p_i, s_i) \frac{4\pi e e_{\mathrm{p}}}{q^2 + \mathrm{i}\epsilon} \bar{u}(P_f, S_f\gamma^\mu u(P_i, S_i) \right|^2 . \quad (3.94)$$

The latter results from $|M_{fi}|^2$ in (3.87) by averaging over all incoming and summing over all outgoing electronic and protonic spin states. Taking into account that terms of the form $\bar{u}\gamma^\mu u$ are $\mathbb{C}$-numbers, we can rewrite (3.94) as a product of two double spin sums that, in turn, can be calculated further with the help of Theorem 3.6:

$$\begin{aligned} \overline{|M_{fi}|^2} &= \frac{(4\pi)^2 e^2 e_{\mathrm{p}}^2}{4(q^2)^2} \sum_{\substack{s_f, s_i \\ S_f, S_i}} [\bar{u}(p_f, s_f)\gamma_\mu u(p_i, s_i)] [\bar{u}(P_f, S_f)\gamma^\mu u(P_i, S_i)] \\ &\quad \times [\bar{u}(p_f, s_f)\gamma_\nu u(p_i, s_i)]^\dagger [\bar{u}(P_f, S_f)\gamma^\nu u(P_i, S_i)]^\dagger \\ &= \frac{(4\pi)^2 e^2 e_{\mathrm{p}}^2}{4(q^2)^2} \sum_{s_f, s_i} [\bar{u}(p_f, s_f)\gamma_\mu u(p_i, s_i)] [\bar{u}(p_i, s_i)\gamma_\nu u(p_f, s_f)] \\ &\quad \times \sum_{S_f, S_i} [\bar{u}(P_f, S_f)\gamma_\mu u(P_i, S_i)] [\bar{u}(P_i, S_i)\gamma_\nu u(P_f, S_f)] \\ &= \frac{(4\pi)^2 e^2 e_{\mathrm{p}}^2}{4(q^2)^2} \mathrm{tr}\left[\Lambda_+(p_f)\gamma_\mu \Lambda_+(p_i)\gamma_\nu\right] \mathrm{tr}\left[\Lambda_+(P_f)\gamma^\mu \Lambda_+(P_i)\gamma^\nu\right] . \end{aligned} \quad (3.95)$$

Next we introduce the auxiliary quantities

$$(a^\mu) = \begin{pmatrix} 0 \\ 1 \\ \cdot \\ 0 \end{pmatrix} \leftarrow \mu\text{-th position} \quad , \quad (b^\nu) = \begin{pmatrix} 0 \\ \cdot \\ 1 \\ 0 \end{pmatrix} \leftarrow \nu\text{-th position} \; ,$$

make the replacements $\gamma_\mu \to \not{a}$, $\gamma_\nu \to \not{b}$, and calculate as follows:

$$\begin{aligned}
\operatorname{tr}\left[\Lambda_+(p_f)\gamma_\mu \Lambda_+(p_i)\gamma_\nu\right]
&= \frac{1}{4m_0^2}\operatorname{tr}\left[(\not{p}_f + m_0)\not{a}(\not{p}_i + m_0)\not{b}\right] \\
&= \frac{1}{4m_0^2}\left[\operatorname{tr}(\not{p}_f\not{a}\not{p}_i\not{b}) + m_0^2\operatorname{tr}(\not{a}\not{b})\right] \\
&= \frac{1}{4m_0^2}\left[(p_f\cdot a)\operatorname{tr}(\not{p}_i\not{b}) - (p_f\cdot p_i)\operatorname{tr}(\not{a}\not{b}) + (p_f\cdot b)\operatorname{tr}(\not{a}\not{p}_i) + m_0^2\operatorname{tr}(\not{a}\not{b})\right] \\
&= \frac{4}{4m_0^2}\left[(p_f)_\mu(p_i)_\nu - (p_f\cdot p_i)g_{\mu\nu} + (p_f)_\nu(p_i)_\mu + m_0^2 g_{\mu\nu}\right] \\
&= \frac{1}{m_0^2}\left[(p_f)_\mu(p_i)_\nu + (p_i)_\mu(p_f)_\nu - g_{\mu\nu}(p_f\cdot p_i - m_0^2)\right] \; .
\end{aligned}$$

Correspondingly, the second trace yields

$$\operatorname{tr}\left[\Lambda_+(P_f)\gamma^\mu\Lambda_+(P_i)\gamma^\nu\right] = \frac{1}{M_0^2}\left[P_f^\mu P_i^\nu + P_i^\mu P_f^\nu - g^{\mu\nu}(P_f\cdot P_i - M_0^2)\right] \; .$$

Hence, after expanding the two traces, (3.95) turns into

$$\begin{aligned}
\overline{|M_{fi}|^2} = \frac{(4\pi)^2 e^2 e_\mathrm{p}^2}{2m_0^2 M_0^2 (q^2)^2}\big[&(p_i\cdot P_i)(p_f\cdot P_f) + (p_i\cdot P_f)(p_f\cdot P_i) \\
&- (p_i\cdot p_f)M_0^2 - (P_i\cdot P_f)m_0^2 + 2m_0^2 M_0^2\big] \; .
\end{aligned} \tag{3.96}$$

If we now replace the four-momenta by the kinematic quantities $p_i = (E_i, \boldsymbol{p}_i)$, $p_f = (E_f, \boldsymbol{p}_f)$, $P_i = (M_0, \mathbf{0})$ in the laboratory system and take into account the four-momentum conservation $P_f = P_i + p_i - p_f$, we arrive at the final result

$$\begin{aligned}
\overline{|M_{fi}|^2_\mathrm{co}} &= \frac{(4\pi)^2 e^2 e_\mathrm{p}^2}{2m_0^2 M_0^2 (q^2)^2}\Big\{2M_0^2 E_i E_f + 2M_0 m_0^2 (E_f - E_i) \\
&\qquad - (p_i\cdot p_f)\left[M_0^2 + M_0(E_f - E_i)\right] + m_0^2 M_0^2\Big\} \\
&= \frac{(4\pi)^2 e^2 e_\mathrm{p}^2}{2m_0^2 M_0^2 (q^2)^2}\Big\{2M_0^2 E_i E_f + M_0 m_0^2 (E_f - E_i) \\
&\qquad + \frac{q^2}{2}\left[M_0^2 + M_0(E_f - E_i)\right]\Big\} \; ,
\end{aligned}$$

which has to be inserted into (3.93). In the last step, the scalar product $p_i \cdot p_f$ was expressed by the quadratic four-momentum transfer

$$q^2 = (p_f - p_i)^2 = p_f^2 + p_i^2 - 2p_i\cdot p_f = 2m_0^2 - 2p_i\cdot p_f \; .$$

Theorem 3.8: Electron-proton scattering to leading order

The scattering amplitude of the electron-proton scattering to leading order is $(f \neq i)$

$$S_{fi} = -\mathrm{i}\frac{(2\pi)^4\delta(p_f + P_f - p_i - P_i)}{V^2}\sqrt{\frac{m_0^2}{E_i E_f}}\sqrt{\frac{M_0^2}{E_i^{(\mathrm{p})}E_f^{(\mathrm{p})}}}M_{fi} ,$$

with the Lorentz-invariant amplitude $(q = p_f - p_i)$

$$M_{fi} = \bar{u}(p_f, s_f)\gamma_\mu u(p_i, s_i)\frac{-4\pi e e_{\mathrm{p}}}{q^2 + \mathrm{i}\epsilon}\bar{u}(P_f, S_f)\gamma^\mu u(P_i, S_i) .$$

The differential cross section follows as

$$\begin{aligned} \mathrm{d}\sigma &= \frac{m_0}{E_i}\frac{M_0}{E_i^{(\mathrm{p})}}\frac{1}{|\boldsymbol{j}_i|V}|M_{fi}|^2(2\pi)^4\delta(p_f + P_f - p_i - P_i) \\ &\quad\times\frac{m_0\mathrm{d}^3p_f}{(2\pi)^3E_f}\frac{M_0\mathrm{d}^3P_f}{(2\pi)^3E_f^{(\mathrm{p})}} \\ &= \frac{m_0M_0}{\sqrt{(p_i\cdot P_i)^2 - m_0^2M_0^2}}|M_{fi}|^2(2\pi)^4\delta(p_f + P_f - p_i - P_i) \\ &\quad\times\frac{m_0\mathrm{d}^3p_f}{(2\pi)^3E_f}\frac{M_0\mathrm{d}^3P_f}{(2\pi)^3E_f^{(\mathrm{p})}} \quad \text{(collinear currents)} . \end{aligned}$$

In the laboratory system where the proton is initially at rest, this becomes

$$\left.\begin{aligned} &\frac{\mathrm{d}\sigma}{\mathrm{d}\Omega} = \frac{m_0^2M_0|\boldsymbol{p}_f|}{4\pi^2|\boldsymbol{p}_i|}\frac{|M_{fi}|^2_{\mathrm{co}}}{M_0 + E_i - \frac{|\boldsymbol{p}_i|E_f}{|\boldsymbol{p}_f|}\cos\theta} \\ &|M_{fi}|^2_{\mathrm{co}} = |M_{fi}|^2_{P_f = P_i + p_i - p_f} \\ &2m_0^2 - 2M_0(E_f - E_i) - 2E_iE_f + 2|\boldsymbol{p}_i||\boldsymbol{p}_f|\cos\theta = 0 , \end{aligned}\right\} \quad (3.97)$$

where, in $\mathrm{d}\sigma/\mathrm{d}\Omega$, all electronic scattering momenta $\boldsymbol{p}_f$ directed toward $\mathrm{d}\Omega$ and all protonic scattering momenta $\boldsymbol{P}_f$ have been integrated out. Disregarding electronic and protonic polarization effects, the unpolarized amplitude square is obtained as

$$\begin{aligned} \overline{|M_{fi}|^2_{\mathrm{co}}} = \frac{(4\pi)^2e^2e_{\mathrm{p}}^2}{2m_0^2M_0^2(q^2)^2}\Big\{&2M_0^2E_iE_f + M_0m_0^2(E_f - E_i) \\ &+ \frac{q^2}{2}\left[M_0^2 + M_0(E_f - E_i)\right]\Big\} . \end{aligned}$$

Low-energy and ultrarelativistic limits. We can assure ourselves of the correctness of this theorem, for example, by proving it for the low-energy limit, which should lead us back to the laws for the Coulomb scattering of electrons. In this limit we have $E_{i,f}, |\boldsymbol{p}_{i,f}| \ll m_0 < M_0$, and the secondary condition in (3.97) reduces to $E_f \approx E_i \Longleftrightarrow |\boldsymbol{p}_f| \approx |\boldsymbol{p}_i|$ (no repulsion effects, totally elastic electron scattering). Taking into account

$$q^2 \approx -\boldsymbol{q}^2 = -2(E_i^2 - m_0^2) + 2\boldsymbol{p}_i\boldsymbol{p}_f \ ,$$

the mean amplitude square now becomes

$$\begin{aligned}\overline{|M_{fi}|^2_{\rm co}} &\approx \frac{(4\pi)^2 e^2 e_{\rm p}^2}{2m_0^2\boldsymbol{q}^4}\left(2E_i^2 - \frac{\boldsymbol{q}^2}{2}\right)_{|\boldsymbol{p}_f|=|\boldsymbol{p}_i|} \\ &= \frac{(4\pi)^2 e^2 e_{\rm p}^2}{2\boldsymbol{q}^4}\left(1 + \frac{E_i^2 + \boldsymbol{p}_i\boldsymbol{p}_f}{m_0^2}\right)_{|\boldsymbol{p}_f|=|\boldsymbol{p}_i|} ,\end{aligned}$$

from which follows the expected Mott scattering formula:

$$\overline{\frac{\mathrm{d}\sigma}{\mathrm{d}\Omega}} \approx \frac{m_0^2\overline{|M_{fi}|^2_{\rm co}}}{4\pi^2} \approx \frac{2m_0^2 e^2 e_{\rm p}^2}{\boldsymbol{q}^4}\left(1 + \frac{E_i^2 + \boldsymbol{p}_i\boldsymbol{p}_f}{m_0^2}\right)_{|\boldsymbol{p}_f|=|\boldsymbol{p}_i|} = \left(\overline{\frac{\mathrm{d}\sigma}{\mathrm{d}\Omega}}\right)_{\rm Mott} .$$

The other extremum is the ultrarelativistic limit. It is defined by $E_{i,f}/m_0 \gg 1$. Together with $E_{i,f} \approx |\boldsymbol{p}_{i,f}|$ and

$$q^2 \approx -2E_iE_f(1-\cos\theta) = -4E_iE_f\sin^2\frac{\theta}{2}$$

$$\Longrightarrow 1 + \frac{q^2}{4E_iE_f} \approx 1 - \sin^2\frac{\theta}{2} = \cos^2\frac{\theta}{2}$$

as well as the secondary condition

$$M_0(E_f - E_i) \approx m_0^2 - E_iE_f + E_iE_f(1-\cos\theta) = -2E_iE_f\sin^2\frac{\theta}{2} ,$$

the mean amplitude square becomes

$$\begin{aligned}\overline{|M_{fi}|^2_{\rm co}} &= \frac{(4\pi)^2 e^2 e_{\rm p}^2 E_iE_f}{m_0^2(q^2)^2} \\ &\quad\times\left[1 + \frac{q^2}{4E_iE_f}\left(1 + \frac{E_f - E_i}{M_0}\right) + \underbrace{\frac{m_0^2}{2E_iE_f}}_{\approx 0}\frac{E_f - E_i}{M_0}\right] \\ &\approx \frac{\pi^2 e^2 e_{\rm p}^2}{m_0^2 E_iE_f\sin^4\frac{\theta}{2}}\left(1 + \frac{q^2}{4E_iE_f} - \frac{q^2}{4E_iE_f}\frac{2E_iE_f\sin^2\frac{\theta}{2}}{M_0^2}\right) \\ &= \frac{\pi^2 e^2 e_{\rm p}^2}{m_0^2 E_iE_f\sin^4\frac{\theta}{2}}\left(\cos^2\frac{\theta}{2} - \frac{q^2}{2M_0^2}\sin^2\frac{\theta}{2}\right) ,\end{aligned}$$

and we obtain the unpolarized differential cross section as

$$\overline{\frac{\mathrm{d}\sigma}{\mathrm{d}\Omega}} \approx \frac{m_0^2 E_f}{4\pi^2 E_i}\frac{\overline{|M_{fi}|^2_{\rm co}}}{1 + \frac{2E_i}{M_0}\sin^2\frac{\theta}{2}} \approx \frac{e^2 e_{\rm p}^2}{4E_i^2\sin^4\frac{\theta}{2}}\frac{\cos^2\frac{\theta}{2} - \frac{q^2}{2M_0^2}\sin^2\frac{\theta}{2}}{1 + \frac{2E_i}{M_0}\sin^2\frac{\theta}{2}} .$$

Note that, according to our initial presupposition, this equation disregards the internal structure of the proton and its anomalous magnetic moment. In this respect, it does not provide a realistic description of electron-proton scattering under extremely high energies. A more realistic description is given by the *Rosenbluth formula* where the proton's internal structure is taken into account by so-called *electric* and *magnetic form factors.* Nonetheless, the above formula describes the scattering of electrons and myons with high accuracy as both behave like structureless Dirac particles.

Feynman diagrams and characteristic factors. After these many and sometimes lengthy calculations, it is instructive to highlight their results with respect to their systematics as well as their relationship to the Feynman diagrams. Having identified the electronic and protonic transition currents, we initially found the scattering amplitude of the electron-proton scattering to be [see (3.85) and (3.86)]

$$S_{fi} = -\mathrm{i}\int \mathrm{d}^4x \int \mathrm{d}^4y \left[e\bar{\Psi}_f(x)\gamma_\mu\Psi_i(x)\right] D_{\mathrm{F}}^{(0)}(x-y) \times \left[e_{\mathrm{p}}\bar{\Psi}_f^{(\mathrm{p})}(y)\gamma^\mu\Psi_i^{(\mathrm{p})}(y)\right] . \tag{3.98}$$

With reference to our general considerations in Subsection 3.2.3, this expression can be depicted in a Feynman diagram in coordinate space as shown in Figure 3.9**a**. The left-hand thin line with a positive temporal direction represents the propagation of the electron (electronic transition current). Correspondingly, the right-hand thick and likewise temporally forward directed line represents the proton's propagation (protonic transition current). The influence of the electromagnetic interaction (photon propagator) is visualized as a wavy line. It can be viewed as a virtual photon which is exchanged

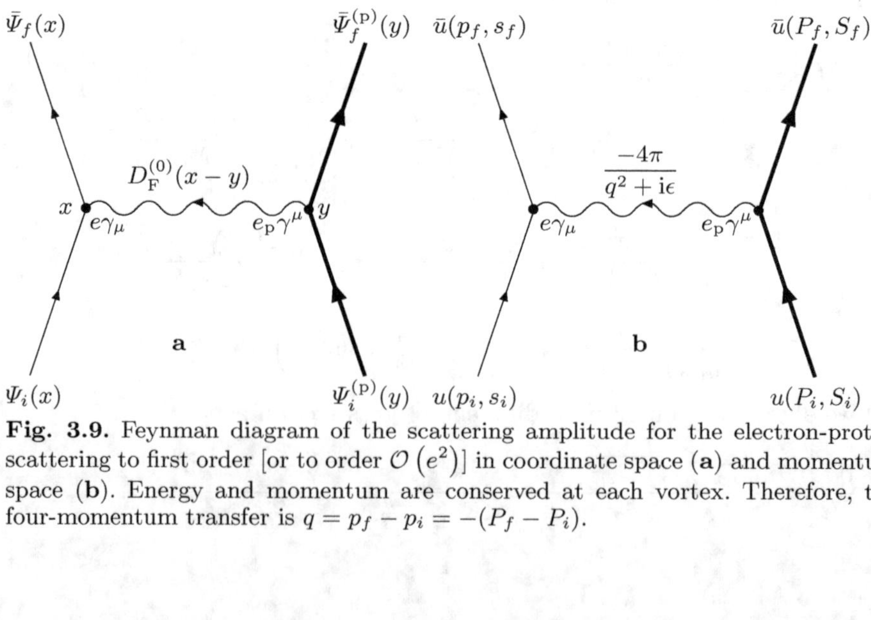

Fig. 3.9. Feynman diagram of the scattering amplitude for the electron-proton scattering to first order [or to order $\mathcal{O}\left(e^2\right)$] in coordinate space (**a**) and momentum space (**b**). Energy and momentum are conserved at each vortex. Therefore, the four-momentum transfer is $q = p_f - p_i = -(P_f - P_i)$.

between the electron and proton, and causes a scattering at both vortices x and y. The correspondence is completed by assigning certain factors to the four external fermion lines (with one open end), the closed photon line (with a start and an end point) as well as to the vortices. Later on we see that these factor assignments are characteristic and remain valid for other scattering processes, too. Thus, after some practice, one should be able to deduce the scattering amplitude directly from the corresponding Feynman diagram.

After inserting the explicit expressions for the wave functions and the photon propagator into (3.98) and then integrating out the position and photon momentum variables, we arrived at

$$\left.\begin{aligned} S_{fi} &= -\mathrm{i}\frac{(2\pi)^4\delta(p_f+P_f-p_i-P_i)}{V^2}\sqrt{\frac{m_0^2}{E_iE_f}}\sqrt{\frac{M_0^2}{E_i^{(\mathrm{p})}E_f^{(\mathrm{p})}}}M_{fi} \\ M_{fi} &= \bar{u}(p_f,s_f)\gamma_\mu u(p_i,s_i)\frac{-4\pi e e_\mathrm{p}}{q^2+\mathrm{i}\epsilon}\bar{u}(P_f,S_f)\gamma^\mu u(P_i,S_i) \\ q &= p_f-p_i\;, \end{aligned}\right\} \quad (3.99)$$

where the fourdimensional δ-function ensures energy and momentum conservation. The Feynman diagram in momentum space belonging to M_{fi} is depicted in Figure 3.9**b**. It is obviously connected to the diagram in position space (Figure 3.9**a**) through the replacements

$$\Psi_{i,f}(x) \longrightarrow u(p_{i,f},s_{i,f})\;,\;\Psi_{i,f}^{(\mathrm{p})}(y) \longrightarrow u(P_{i,f},S_{i,f})$$

$$D_\mathrm{F}^{(0)}(x-y) \longrightarrow \tilde{D}_\mathrm{F}^{(0)}(q) = \frac{-4\pi}{q^2+\mathrm{i}\epsilon}\;,\; q=p_f-p_i\;,$$

where the four-momentum transfer q guarantees energy and momentum conservation at each vortex. All in all, we see a close correspondence between scattering processes, Feynman diagrams, and scattering amplitudes, to which we will often return in the following.

Let us, at the end, consider the formula for the differential cross section

$$\begin{aligned} \mathrm{d}\sigma &= \left(\frac{m_0}{E_i}\frac{M_0}{E_i^{(\mathrm{p})}}\frac{1}{|\boldsymbol{j}_i|V}\right)|M_{fi}|^2(2\pi)^4\delta(p_f+P_f-p_i-P_i) \\ &\quad\times\frac{m_0\mathrm{d}^3p_f}{(2\pi)^3E_f}\frac{M_0\mathrm{d}^3P_f}{(2\pi)^3E_f^{(\mathrm{p})}}\;, \end{aligned}$$

which also exhibits some interesting characteristics: apart from the δ-function and the amplitude square $|M_{fi}|^2$, there occurs a factor of m_0/E for each external fermion line. Furthermore, each outgoing particle yields an additional phase space factor of $\mathrm{d}^3p/(2\pi)^3$. In the case of collinear currents the bracket term can be expressed by the four-momenta of the incoming particles via $m_0M_0/\sqrt{(p_i\cdot P_i)-m_0^2M_0^2}$.

3.3.3 Electron-Proton Scattering (II)

Now, we extend our considerations on the electron-proton scattering even further and discuss the corrections of second order. First we try to develop the corresponding Feynman diagrams on the basis of the above correspondences in order to derive the correct form of the scattering amplitude $S_{fi}^{(2)}$. Then we verify our reasoning by analytical calculations.

Direct scattering amplitude.[18] The scattering amplitude of second order is characterized by the fact that both particles, electron and proton, suffer two scatterings caused by the exchange of two virtual photons. Between the scatterings the fermions and photons move undisturbed with a probability amplitude proportional to the propagator of the respective particle. Therefore, it is sensible to draw the corresponding Feynman diagram in coordinate space as shown in Figure 3.10, where the end point of each photon line is

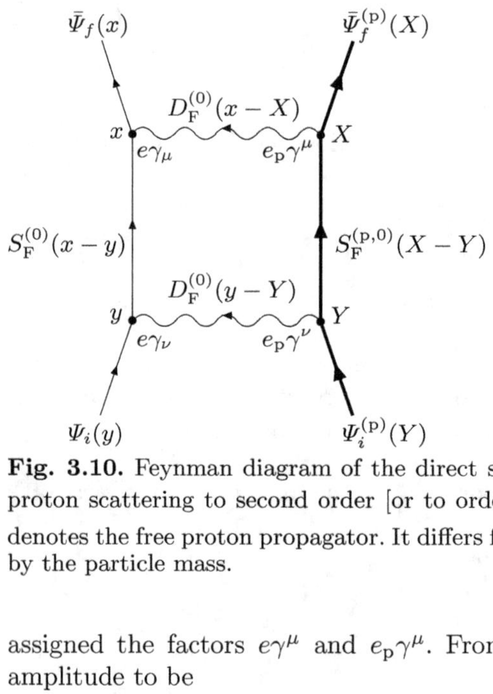

Fig. 3.10. Feynman diagram of the direct scattering amplitude for the electron-proton scattering to second order [or to order $\mathcal{O}\left(e^4\right)$] in coordinate space. $S_{\mathrm{F}}^{(\mathrm{p},0)}$ denotes the free proton propagator. It differs from the electron propagator $S_{\mathrm{F}}^{(0)}$ only by the particle mass.

assigned the factors $e\gamma^\mu$ and $e_{\mathrm{p}}\gamma^\mu$. From this we conclude the scattering amplitude to be

$$\begin{aligned}
S_{fi}^{(2)}(\mathrm{dir}) = -\mathrm{i}\int \mathrm{d}^4x \int \mathrm{d}^4y \int \mathrm{d}^4X \int \mathrm{d}^4Y \\
\times \left[e^2\bar{\Psi}_f(x)\gamma_\mu S_{\mathrm{F}}^{(0)}(x-y)\gamma_\nu\Psi_i(y)\right] \\
\times D_{\mathrm{F}}^{(0)}(x-X)D_{\mathrm{F}}^{(0)}(y-Y)
\end{aligned}$$

[18] In addition to the *direct scattering amplitude*, there exists the so-called *exchange scattering amplitude* that will be discussed later.

$$\times \left[e_{\mathrm{p}}^2 \bar{\Psi}_f^{(\mathrm{p})}(X) \gamma^\mu S_{\mathrm{F}}^{(\mathrm{p},0)}(X-Y) \gamma^\nu \Psi_i^{(\mathrm{p})}(Y) \right] , \tag{3.100}$$

with an inserted factor of $-\mathrm{i}$ following (3.98). Of course, Figure 3.10 shows only one representative of all the 4! possible temporal arrangements of the vortices (see Figure 3.11). They are automatically taken into account in (3.100) by the four time integrations over x^0, y^0, X^0, and Y^0.

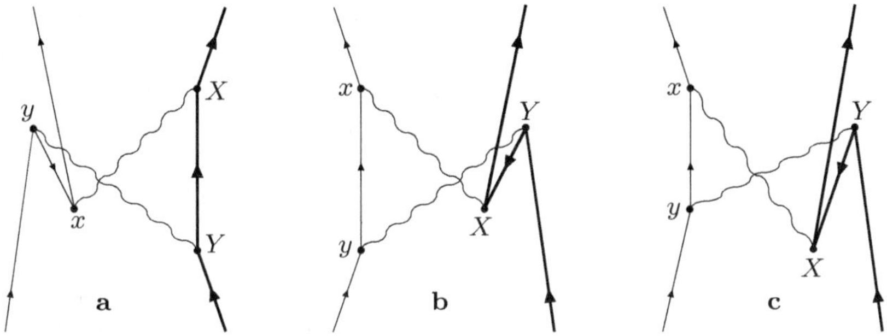

Fig. 3.11. Three of the 4! possible relative temporal arrangements of the vortices in Figure 3.10. In **a** a virtual positron along with the outgoing electron is created at x. In **b** a virtual antiproton along with the outgoing proton is created at X.

If we now perform the corresponding replacements in Figure 3.10 we are led to the Feynman diagram in momentum space shown in Figure 3.12. Here energy and momentum conservation is postulated at each vortex, and the circulating four-momentum q_1 remains as a degree of freedom. Together with

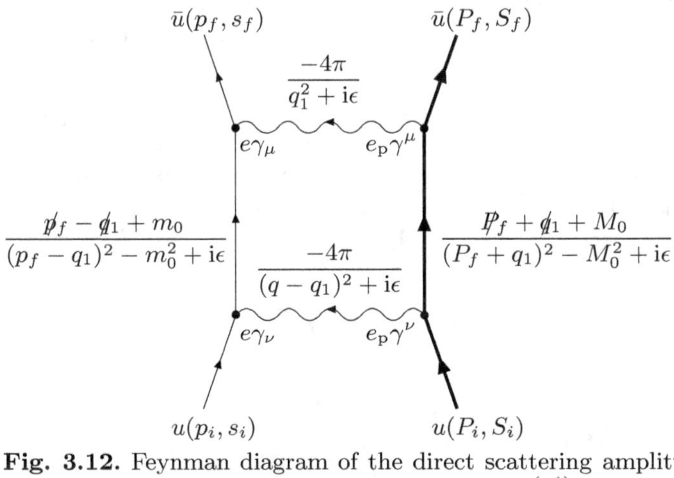

Fig. 3.12. Feynman diagram of the direct scattering amplitude for the electron-proton scattering to second order [or to order $\mathcal{O}\left(e^4\right)$] in momentum space. Energy and momentum are conserved at each vortex. Therefore, the four-momentum transfer is $q = p_f - p_i = -(P_f - P_i)$.

the energy and momentum conserving δ-function as well as the normalization factors $\sqrt{m_0/E_{i,f}V}$ and $\sqrt{M_0/E_{i,f}^{(\mathrm{p})}V}$ for the incoming and outgoing fermions we should finally obtain the integrated scattering amplitude

$$
\begin{aligned}
S_{fi}^{(2)}(\mathrm{dir}) = & -\mathrm{i}\frac{(2\pi)^4\delta(p_f+P_f-p_i-P_i)}{V^2}\sqrt{\frac{m_0^2}{E_iE_f}}\sqrt{\frac{M_0^2}{E_i^{(\mathrm{p})}E_f^{(\mathrm{p})}}} \\
& \times\int\frac{\mathrm{d}^4q_1}{(2\pi)^4}\frac{-4\pi ee_{\mathrm{p}}}{q_1^2+\mathrm{i}\epsilon}\frac{-4\pi ee_{\mathrm{p}}}{(q-q_1)^2+\mathrm{i}\epsilon} \\
& \times\left[\bar{u}(p_f,s_f)\gamma_\mu\frac{\not{p}_f-\not{q}_1+m_0}{(p_f-q_1)^2-m_0^2+\mathrm{i}\epsilon}\gamma_\nu u(p_i,s_i)\right] \\
& \times\left[\bar{u}(P_f,S_f)\gamma^\mu\frac{\not{P}_f+\not{q}_1+M_0}{(P_f+q_1)^2-M_0^2+\mathrm{i}\epsilon}\gamma^\nu u(P_i,S_i)\right] \qquad (3.101)
\end{aligned}
$$

with $q=p_f-p_i=-(P_f-P_i)$ and an inserted factor of $-\mathrm{i}$ as in (3.99).

Let us now compare the heuristically derived expressions (3.100) and (3.101) with the scattering amplitudes calculated on the basis of Theorem 3.5. Our starting point is ($\epsilon_f=+1, f\neq i$)

$$
\begin{aligned}
S_{fi}^{(2)}(\mathrm{dir}) &= -\mathrm{i}e^2\int\mathrm{d}^4x\int\mathrm{d}^4y\bar{\Psi}_f(x)\not{A}(x)S_{\mathrm{F}}^{(0)}(x-y)\not{A}(y)\Psi_i(y) \\
&= -\int\mathrm{d}^4x\int\mathrm{d}^4y \\
&\quad\times\left[\mathrm{i}e^2\bar{\Psi}_f(x)\gamma_\mu S_{\mathrm{F}}^{(0)}(x-y)\gamma_\nu\Psi_i(y)\right]A^\mu(x)A^\nu(y)\,. \qquad (3.102)
\end{aligned}
$$

Similarly to the preceding subsection, we first aim to identify the electronic and protonic transition currents (to second order) in a way that the scattering amplitude is symmetric under both of them. For the electronic current, the square bracket term

$$
J_{\mu\nu}^{(2)}(x)=\mathrm{i}e^2\bar{\Psi}_f(x)\gamma_\mu S_{\mathrm{F}}^{(0)}(x-y)\gamma_\nu\Psi_i(y) \qquad (3.103)
$$

seems to be a good choice. The entrainment of the factor i ensures that $J_{\mu\nu}^{(2)}$ can be written as a product of first-order transition currents, because, using the wave decomposition (3.58) of $S_{\mathrm{F}}^{(0)}$, we have

$$
\begin{aligned}
J_{\mu\nu}^{(2)}(x) &= e^2\Theta(x^0-y^0)\bar{\Psi}_f(x)\gamma_\mu\sum_{\boldsymbol{p},r=1}^{2}\Psi_{\boldsymbol{p}}^{(r)}(x)\bar{\Psi}_{\boldsymbol{p}}^{(r)}(y)\gamma_\nu\Psi_i(y) \\
&\quad -e^2\Theta(y^0-x^0)\bar{\Psi}_f(x)\gamma_\mu\sum_{\boldsymbol{p},r=3}^{4}\Psi_{\boldsymbol{p}}^{(r)}(x)\bar{\Psi}_{\boldsymbol{p}}^{(r)}(y)\gamma_\nu\Psi_i(y) \\
&= e^2\Theta(x^0-y^0)\sum_{\boldsymbol{p},r=1}^{2}\left[\bar{\Psi}_f(x)\gamma_\mu\Psi_{\boldsymbol{p}}^{(r)}(x)\right]\left[\bar{\Psi}_{\boldsymbol{p}}^{(r)}(y)\gamma_\nu\Psi_i(y)\right]
\end{aligned}
$$

$$-e^2\Theta(y^0 - x^0) \sum_{\boldsymbol{p},r=3}^{4} \left[\bar{\Psi}_f(x)\gamma_\mu \Psi_{\boldsymbol{p}}^{(r)}(x)\right] \left[\bar{\Psi}_{\boldsymbol{p}}^{(r)}(y)\gamma_\nu \Psi_i(y)\right]$$

$$= \Theta(x^0 - y^0) \sum_{\boldsymbol{p},r=1}^{2} \left[J_\mu(x)\right]_{f,(\boldsymbol{p},r)} \left[J_\nu(y)\right]_{(\boldsymbol{p},r),i}$$

$$-\Theta(y^0 - x^0) \sum_{\boldsymbol{p},r=3}^{4} \left[J_\mu(x)\right]_{f,(\boldsymbol{p},r)} \left[J_\nu(y)\right]_{(\boldsymbol{p},r),i} \ .$$

Since, according to (3.82), each first-order current causes an electromagnetic field of the form

$$A^\mu(x) = \int \mathrm{d}^4X D_{\mathrm{F}}^{(0)}(x - X) J^{(\mathrm{p})\mu}(X) \ ,$$

it is reasonable to assume that the combination $A^\mu A^\nu$ in (3.102) is connected to the second-order protonic current $J^{(\mathrm{p},2)\mu\nu}$ via

$$A^\mu(x)A^\nu(y) = \int \mathrm{d}^4X \int \mathrm{d}^4Y D_{\mathrm{F}}^{(0)}(x - X) D_{\mathrm{F}}^{(0)}(y - Y) J^{(\mathrm{p},2)\mu\nu}(X,Y) \ ,$$

which, in turn, has to be chosen as [see (3.103)]

$$J^{(\mathrm{p},2)\mu\nu}(X,Y) = \mathrm{i}e_{\mathrm{p}}^2 \bar{\Psi}_f^{(\mathrm{p})}(X)\gamma^\mu S_{\mathrm{F}}^{(\mathrm{p},0)}(X - Y)\gamma^\nu \Psi_i^{(\mathrm{p})}(Y) \ .$$

Consequently, (3.102) turns into

$$\begin{aligned} S_{fi}^{(2)}(\mathrm{dir}) = & \int \mathrm{d}^4x \int \mathrm{d}^4y \int \mathrm{d}^4X \int \mathrm{d}^4Y \\ & \times \left[e^2 \bar{\Psi}_f(x)\gamma_\mu S_{\mathrm{F}}^{(0)}(x - y)\gamma_\nu \Psi_i(y)\right] \\ & \times D_{\mathrm{F}}^{(0)}(x - X) D_{\mathrm{F}}^{(0)}(y - Y) \\ & \times \left[e_{\mathrm{p}}^2 \bar{\Psi}_f^{(\mathrm{p})}(X)\gamma^\mu S_{\mathrm{F}}^{(\mathrm{p},0)}(X - Y)\gamma^\nu \Psi_i^{(\mathrm{p})}(Y)\right] \ . \end{aligned} \tag{3.104}$$

Apart from a factor of $-\mathrm{i}$, this equation does indeed coincide with the heuristically derived expression (3.100). In order to verify (3.101), we now insert the known expressions for the electron and proton wave functions as well as the Fourier representations of the electron, proton, and photon propagators. This leads to

$$\begin{aligned} S_{fi}^{(2)}(\mathrm{dir}) = & \frac{1}{V^2}\sqrt{\frac{m_0^2}{E_i E_f}}\sqrt{\frac{M_0^2}{E_i^{(\mathrm{p})} E_f^{(\mathrm{p})}}} \int \mathrm{d}^4x \int \mathrm{d}^4y \int \mathrm{d}^4X \int \mathrm{d}^4Y \\ & \times \int \frac{\mathrm{d}^4q_1}{(2\pi)^4} \int \frac{\mathrm{d}^4q_2}{(2\pi)^4} \int \frac{\mathrm{d}^4p}{(2\pi)^4} \int \frac{\mathrm{d}^4P}{(2\pi)^4} \\ & \times \frac{-4\pi e e_{\mathrm{p}}}{q_1^2 + \mathrm{i}\epsilon} \frac{-4\pi e e_{\mathrm{p}}}{q_2^2 + \mathrm{i}\epsilon} \left[\bar{u}(p_f, s_f)\gamma_\mu \frac{(\not{p} + m_0)}{p^2 - m_0^2 + \mathrm{i}\epsilon}\gamma_\nu u(p_i, s_i)\right] \end{aligned}$$

$$\times \left[\bar{u}(P_f, S_f)\gamma^\mu \frac{(\not{P} + M_0)}{P^2 - M_0^2 + \mathrm{i}\epsilon} \gamma^\nu u(P_i, S_i) \right]$$
$$\times \mathrm{e}^{-\mathrm{i}q_1\cdot(x-X)} \mathrm{e}^{-\mathrm{i}q_2\cdot(y-Y)} \mathrm{e}^{\mathrm{i}p_f\cdot x} \mathrm{e}^{-\mathrm{i}p\cdot(x-y)}$$
$$\times \mathrm{e}^{-\mathrm{i}p_i\cdot y} \mathrm{e}^{\mathrm{i}P_f\cdot X} \mathrm{e}^{-\mathrm{i}P\cdot(X-Y)} \mathrm{e}^{-\mathrm{i}P_i\cdot Y} .$$

Here we first perform the coordinate integrations,

$$\int \mathrm{d}^4x \int \mathrm{d}^4y \int \mathrm{d}^4X \int \mathrm{d}^4Y \mathrm{e}^{-\mathrm{i}q_1\cdot(x-X)} \mathrm{e}^{-\mathrm{i}q_2\cdot(y-Y)} \mathrm{e}^{\mathrm{i}p_f\cdot x} \mathrm{e}^{-\mathrm{i}p\cdot(x-y)}$$
$$\times \mathrm{e}^{-\mathrm{i}p_i\cdot y} \mathrm{e}^{\mathrm{i}P_f\cdot X} \mathrm{e}^{-\mathrm{i}P\cdot(X-Y)} \mathrm{e}^{-\mathrm{i}P_i\cdot Y}$$
$$= (2\pi)^4\delta(q_1 + p - p_f)(2\pi)^4\delta(q_2 - p + p_i)(2\pi)^4\delta(-q_1 + P - P_f)$$
$$\times (2\pi)^4\delta(-q_2 - P + P_i) ,$$

and then the momentum integrations over p, P, and q_2:

$$\int \frac{\mathrm{d}^4q_1}{(2\pi)^4} \int \frac{\mathrm{d}^4q_2}{(2\pi)^4} \int \frac{\mathrm{d}^4p}{(2\pi)^4} \int \frac{\mathrm{d}^4P}{(2\pi)^4} (2\pi)^4\delta(q_1 + p - p_f)$$
$$\times (2\pi)^4\delta(q_2 - p + p_i)(2\pi)^4\delta(-q_1 + P - P_f)(2\pi)^4\delta(-q_2 - P + P_i)$$
$$\times \frac{-4\pi e e_\mathrm{p}}{q_1^2 + \mathrm{i}\epsilon} \frac{-4\pi e e_\mathrm{p}}{q_2^2 + \mathrm{i}\epsilon} \left[\bar{u}(p_f, s_f)\gamma_\mu \frac{\not{p} + m_0}{p^2 - m_0^2 + \mathrm{i}\epsilon} \gamma_\nu u(p_i, s_i) \right]$$
$$\times \left[\bar{u}(P_f, S_f)\gamma^\mu \frac{\not{P} + M_0}{P^2 - M_0^2 + \mathrm{i}\epsilon} \gamma^\nu u(P_i, S_i) \right]$$
$$= (2\pi)^4\delta(p_f + P_f - p_i - P_i) \int \frac{\mathrm{d}^4q_1}{(2\pi)^4} \frac{-4\pi e e_\mathrm{p}}{q_1^2 + \mathrm{i}\epsilon} \frac{-4\pi e e_\mathrm{p}}{(p_f - p_i - q_1)^2 + \mathrm{i}\epsilon}$$
$$\times \left[\bar{u}(p_f, s_f)\gamma_\mu \frac{\not{p}_f - \not{q}_1 + m_0}{(p_f - q_1)^2 - m_0^2 + \mathrm{i}\epsilon} \gamma_\nu u(p_i, s_i) \right]$$
$$\times \left[\bar{u}(P_f, S_f)\gamma^\mu \frac{\not{P}_f + \not{q}_1 + M_0}{(P_f + q_1)^2 - M_0^2 + \mathrm{i}\epsilon} \gamma^\nu u(P_i, S_i) \right] .$$

Note that the four δ-functions stemming from the coordinate integrations express energy and momentum conservation that was postulated at each vortex in Figure 3.12. For the integrated scattering amplitude we finally obtain

$$S_{fi}^{(2)}(\mathrm{dir}) = \frac{(2\pi)^4\delta(p_f + P_f - p_i - P_i)}{V^2} \sqrt{\frac{m_0^2}{E_i E_f}} \sqrt{\frac{M_0^2}{E_i^{(\mathrm{p})} E_f^{(\mathrm{p})}}}$$
$$\times \int \frac{\mathrm{d}^4q_1}{(2\pi)^4} \frac{-4\pi e e_\mathrm{p}}{q_1^2 + \mathrm{i}\epsilon} \frac{-4\pi e e_\mathrm{p}}{(q - q_1)^2 + \mathrm{i}\epsilon}$$
$$\times \left[\bar{u}(p_f, s_f)\gamma_\mu \frac{\not{p}_f - \not{q}_1 + m_0}{(p_f - q_1)^2 - m_0^2 + \mathrm{i}\epsilon} \gamma_\nu u(p_i, s_i) \right]$$
$$\times \left[\bar{u}(P_f, S_f)\gamma^\mu \frac{\not{P}_f + \not{q}_1 + M_0}{(P_f + q_1)^2 - M_0^2 + \mathrm{i}\epsilon} \gamma^\nu u(P_i, S_i) \right] , \quad (3.105)$$

with $q = p_f - p_i = -(P_f - P_i)$. Happily this equation also coincides with the corresponding heuristic expression (3.101) but, again, except for a factor of $-\mathrm{i}$.

Overall, this example demonstrates very nicely the correspondence between scattering processes, Feynman diagrams, and scattering amplitudes. In the next subsection we form this correspondence into some simple rules that especially remove the observed ambiguities with respect to the i-factors.

Exchange amplitude. Apart from the direct scattering amplitude, there is another amplitude which contributes to the electron-proton scattering to second order since the two photons emitted by the proton current cannot be distinguished.[19] This means, for example, that the electron which is interacting with a photon at the space-time point x cannot know whether this photon comes from the vortex X or Y. The complete scattering amplitude is therefore given by the addition

$$S_{fi}^{(2)} = S_{fi}^{(2)}(\mathrm{dir}) + S_{fi}^{(2)}(\mathrm{ex}) \ .$$

$S_{fi}^{(2)}(\mathrm{ex})$ denotes the *exchange scattering amplitude* and is represented by the Feynman diagram of Figure 3.13. It differs from the direct scattering amplitude $S_{fi}^{(2)}(\mathrm{dir})$ by the fact that the end points of the photon line on one side (here: on the proton's side) together with the associated γ-factors are

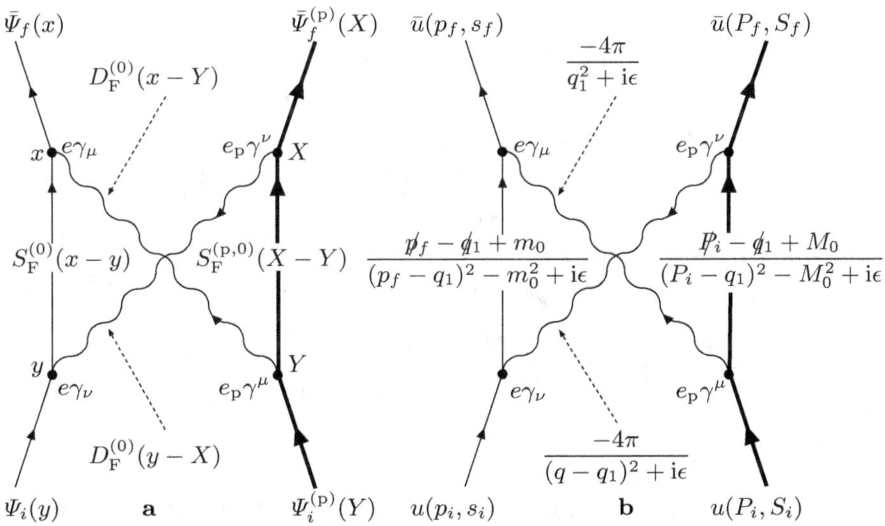

Fig. 3.13. Feynman diagram of the exchange scattering amplitude for the electron-proton scattering to second order [or to order $\mathcal{O}\left(e^4\right)$] in coordinate space (**a**) and momentum space (**b**). Energy and momentum are conserved at each vortex. Therefore, the four-momentum transfer is $q = p_f - p_i = -(P_f - P_i)$.

[19] Additionally, in this order there also exist diagrams that are related to the production and absorption of virtual particles. They are not considered here.

swapped. Therefore, making the corresponding replacements in (3.105), the integrated expression for the exchange scattering amplitude can be determined immediately:

$$\begin{aligned} S_{fi}^{(2)}(\mathrm{ex}) &= \frac{(2\pi)^4\delta(p_f+P_f-p_i-P_i)}{V^2}\sqrt{\frac{m_0^2}{E_iE_f}}\sqrt{\frac{M_0^2}{E_i^{(\mathrm{p})}E_f^{(\mathrm{p})}}} \\ &\quad\times\int\frac{\mathrm{d}^4q_1}{(2\pi)^4}\frac{-4\pi ee_\mathrm{p}}{q_1^2+\mathrm{i}\epsilon}\frac{-4\pi ee_\mathrm{p}}{(q-q_1)^2+\mathrm{i}\epsilon} \\ &\quad\times\left[\bar{u}(p_f,s_f)\gamma_\mu\frac{\not{p}_f-\not{q}_1+m_0}{(p_f-q_1)^2-m_0^2+\mathrm{i}\epsilon}\gamma_\nu u(p_i,s_i)\right] \\ &\quad\times\left[\bar{u}(P_f,S_f)\gamma^\nu\frac{\not{P}_i-\not{q}_1+M_0}{(P_i-q_1)^2-M_0^2+\mathrm{i}\epsilon}\gamma^\mu u(P_i,S_i)\right] . \end{aligned}$$

In total, we have

Theorem 3.9: Electron-proton scattering next to leading order

The scattering amplitude for the electron-proton scattering next to leading order is $(f \neq i)$

$$S_{fi}^{(2)} = \frac{(2\pi)^4\delta(p_f+P_f-p_i-P_i)}{V^2}\sqrt{\frac{m_0^2}{E_iE_f}}\sqrt{\frac{M_0^2}{E_i^{(\mathrm{p})}E_f^{(\mathrm{p})}}}M_{fi}^{(2)} ,$$

with the Lorentz-invariant amplitude $(q = p_f - p_i)$

$$\begin{aligned} M_{fi}^{(2)} &= \int\frac{\mathrm{d}^4q_1}{(2\pi)^4}\frac{-4\pi ee_\mathrm{p}}{q_1^2+\mathrm{i}\epsilon}\frac{-4\pi ee_\mathrm{p}}{(q-q_1)^2+\mathrm{i}\epsilon} \\ &\quad\times\left[\bar{u}(p_f,s_f)\gamma_\mu\frac{\not{p}_f-\not{q}_1+m_0}{(p_f-q_1)^2-m_0^2+\mathrm{i}\epsilon}\gamma_\nu u(p_i,s_i)\right]P^{\mu\nu} \end{aligned}$$

and the proton tensor

$$\begin{aligned} P^{\mu\nu} &= \bar{u}(P_f,S_f)\left[\gamma^\mu\frac{\not{P}_f+\not{q}_1+M_0}{(P_f+q_1)^2-M_0^2+\mathrm{i}\epsilon}\gamma^\nu\right. \\ &\quad\left.+\gamma^\nu\frac{\not{P}_i-\not{q}_1+M_0}{(P_i-q_1)^2-M_0^2+\mathrm{i}\epsilon}\gamma^\mu\right]u(P_i,S_i) . \end{aligned}$$

Static limit. Generally, due to the fourdimensional integral, a further evaluation of this theorem is difficult and nontrivial. However, the calculation can be carried out a little further in the limit of an infinitely heavy point-like proton at rest. In this case, taking into account

$$M_0 \to \infty \Longrightarrow \begin{cases} P_i \approx P_f \approx (M_0, \mathbf{0}) \\ \dfrac{M_0^2}{E_i^{(\mathrm{p})} E_f^{(\mathrm{p})}} \approx 1 \\ \delta(E_f^{(\mathrm{p})} - E_i^{(\mathrm{p})} + E_f - E_i) \approx \delta(E_f - E_i) \\ u(P_{i,f}, S_{i,f}) \approx u(0, S_{i,f}) \end{cases}$$

and

$$\gamma^\mu u(0,S) = g^{\mu 0} u(0,S) \; , \quad \frac{1}{q_1^0 + \mathrm{i}\epsilon} - \frac{1}{q_1^0 - \mathrm{i}\epsilon} = -2\pi \mathrm{i} \delta(q_1^0) \; ,$$

the proton tensor is simplified to

$$\begin{aligned} P^{\mu\nu} &\approx \bar{u}(0,S_f) \left[\gamma^\mu \frac{M_0 \gamma^0 + \not{q}_1 + M_0}{(P_f + q_1)^2 - M_0^2 + \mathrm{i}\epsilon} \gamma^\nu \right. \\ &\quad \left. + \gamma^\nu \frac{M_0 \gamma^0 - \not{q}_1 + M_0}{(P_i - q_1)^2 - M_0^2 + \mathrm{i}\epsilon} \gamma^\mu \right] u(0,S_i) \\ &\approx \bar{u}(0,S_f) \left[\gamma^\mu \frac{M_0(\gamma^0 + 1)}{M_0^2 + 2M_0 q_1^0 - M_0^2 + \mathrm{i}\epsilon} \gamma^\nu \right. \\ &\quad \left. + \gamma^\nu \frac{M_0(\gamma^0 + 1)}{M_0^2 - 2M_0 q_1^0 - M_0^2 + \mathrm{i}\epsilon} \gamma^\mu \right] u(0,S_i) \\ &= \bar{u}(0,S_f) \left[\gamma^\mu \frac{\gamma^0 + 1}{2q_1^0 + \mathrm{i}\epsilon} \gamma^\nu + \gamma^\nu \frac{\gamma^0 + 1}{-2q_1^0 + \mathrm{i}\epsilon} \gamma^\mu \right] u(0,S_i) \\ &= g^{\mu 0} g^{\nu 0} u^\dagger(0,S_f) u(0,S_i) \left[\frac{1}{q_1^0 + \mathrm{i}\epsilon} - \frac{1}{q_1^0 - \mathrm{i}\epsilon} \right] \\ &= -2\pi \mathrm{i} g^{\mu 0} g^{\nu 0} \delta_{S_f S_i} \delta(q_1^0) \; . \end{aligned}$$

The scattering amplitude follows from this as

$$\begin{aligned} S_{fi}^{(2)} &\approx -\mathrm{i} \frac{(2\pi)^4 \delta(E_f - E_i) \delta(\boldsymbol{p}_f + \boldsymbol{P}_f - \boldsymbol{p}_i - \boldsymbol{P}_i) \delta_{S_f,S_i}}{V^2} \sqrt{\frac{m_0^2}{E_i E_f}} \\ &\quad \times 2\pi \int \frac{\mathrm{d}^4 q_1}{(2\pi)^4} \frac{-4\pi e e_\mathrm{p}}{q_1^2 + \mathrm{i}\epsilon} \frac{-4\pi e e_\mathrm{p}}{(q - q_1)^2 + \mathrm{i}\epsilon} \delta(q_1^0) \\ &\quad \times \left[\bar{u}(p_f, s_f) \gamma_0 \frac{\not{p}_f - \not{q}_1 + m_0}{(p_f - q_1)^2 - m_0^2 + \mathrm{i}\epsilon} \gamma_0 u(p_i, s_i) \right] \; . \end{aligned}$$

Now we assume that momentum and polarization effects of the proton are not measured. This implies that in the last equation we can make the replacement

$$(2\pi)^3 \delta(\boldsymbol{p}_f + \boldsymbol{P}_f - \boldsymbol{p}_i - \boldsymbol{P}_i) \delta_{S_f S_i} \longrightarrow V$$

since in the cross section we have (integration over $\boldsymbol{P}_f$, average over S_i and sum over S_f):

$$\frac{1}{2} \sum_{S_f, S_i} \int \frac{V \mathrm{d}^3 P_f}{(2\pi)^3} \left[(2\pi)^3 \delta(\boldsymbol{p}_f + \boldsymbol{P}_f - \boldsymbol{p}_i - \boldsymbol{P}_i) \right]^2 \delta_{S_f S_i}$$

$$= \int \frac{V^2 \mathrm{d}^3 P_f}{(2\pi)^3} (2\pi)^3 \delta(\boldsymbol{p}_f + \boldsymbol{P}_f - \boldsymbol{p}_i - \boldsymbol{P}_i) = V^2 ,$$

where, again, $[(2\pi)^3 \delta(\boldsymbol{p}_f + \ldots)]^2 \to V(2\pi)^3 \delta(\boldsymbol{p}_f + \ldots)$ has been used. Thus, all in all, we obtain

$$\begin{aligned}
S_{fi}^{(2)} \approx & -\mathrm{i} \frac{2\pi\delta(E_f - E_i)}{V} \sqrt{\frac{m_0^2}{E_i E_f}} \\
& \times \int \frac{\mathrm{d}^3 q_1}{(2\pi)^3} \int \mathrm{d}q_1^0 \frac{-4\pi e e_\mathrm{p}}{q_1^2 + \mathrm{i}\epsilon} \frac{-4\pi e e_\mathrm{p}}{(q - q_1)^2 + \mathrm{i}\epsilon} \delta(q_1^0) \\
& \times \left[\bar{u}(p_f, s_f) \gamma_0 \frac{\not{p}_f - \not{q}_1 + m_0}{(p_f - q_1)^2 - m_0^2 + \mathrm{i}\epsilon} \gamma_0 u(p_i, s_i) \right] \\
= & -\mathrm{i} \frac{2\pi\delta(E_f - E_i)}{V} \sqrt{\frac{m_0^2}{E_i E_f}} \int \frac{\mathrm{d}^3 q_1}{(2\pi)^3} \frac{-4\pi e e_\mathrm{p}}{\boldsymbol{q}_1^2} \frac{-4\pi e e_\mathrm{p}}{(\boldsymbol{q} - \boldsymbol{q}_1)^2} \\
& \times \left[\bar{u}(p_f, s_f) \frac{\gamma_0 E_i + \boldsymbol{\gamma}(\boldsymbol{p}_f - \boldsymbol{q}_1) + m_0}{\boldsymbol{p}_f^2 - (\boldsymbol{p}_f - \boldsymbol{q}_1)^2 + \mathrm{i}\epsilon} u(p_i, s_i) \right] .
\end{aligned}$$

As can be shown, this corresponds exactly to the scattering amplitude for the Coulomb scattering of electrons to second order. The remaining threedimensional integral is divergent, which is due to the long reach of the Coulomb potential.

3.3.4 Preliminary Feynman Rules in Momentum Space

Before we discuss more scattering processes in the subsequent subsections, we bring together the correspondences observed so far between scattering processes, Feynman diagrams, and scattering amplitudes, and cast them into a simple set of rules in momentum space that significantly simplifies the calculation of scattering amplitudes and cross sections. However, these *Feynman rules* in their present form are still incomplete and have to be completed in appropriate places in the following. We will provide the complete set of Feynman rules in Subsection 3.3.9 after we have discussed other types of scattering processes, particularly those involving real photons.

1. The scattering amplitude for a scattering process of the kind

   $$I + I' \longrightarrow F + F' \quad (I\text{=incoming, } F\text{=outgoing particle})$$

   is given by

   $$S_{fi} = \frac{(2\pi)^4 \delta(p_f + p_f' - p_i - p_i')}{V^2} \sqrt{\frac{N_i}{E_i}} \sqrt{\frac{N_i'}{E_i'}} \sqrt{\frac{N_f}{E_f}} \sqrt{\frac{N_f'}{E_f'}} M_{fi} ,$$

   where $N_{i,f}^{(\ldots)} = m_{0_{i,f}}$ are the *fermion factors*. Each incoming antifermion (outgoing fermion wave function with negative energy) yields an additional factor of (-1).

2. In the case of collinear currents the corresponding differential cross section is

$$\mathrm{d}\sigma = \frac{N_i N_i'}{\sqrt{(p_i \cdot p_i')^2 - m_{0,i}^2 m_{0,i}'^2}} |M_{fi}|^2 (2\pi)^4 \delta(p_f + p_f' - p_i - p_i') \times \frac{N_f \mathrm{d}^3 p_f}{(2\pi)^3 E_f} \frac{N_f' \mathrm{d}^3 p_f'}{(2\pi)^3 E_f'} \,.$$

3. The Lorentz-invariant amplitude M_{fi} can be expanded in powers of the coupling constant e. The expansion terms of order $\mathcal{O}(e^n)$ are obtained from the Feynman rules in momentum space containing all topological constellations of fermion lines, photon lines, and n vortices that are consistent with the scattering process.

4. In the Feynman diagrams, all vortices, fermion lines, and photon lines are assigned the factors given in Figure 3.14.

5. Four-momentum conservation holds at each vortex. All remaining (undetermined) momenta p are subject to integration with $\int \mathrm{d}^4p/(2\pi)^4$ in the amplitude M_{fi}.

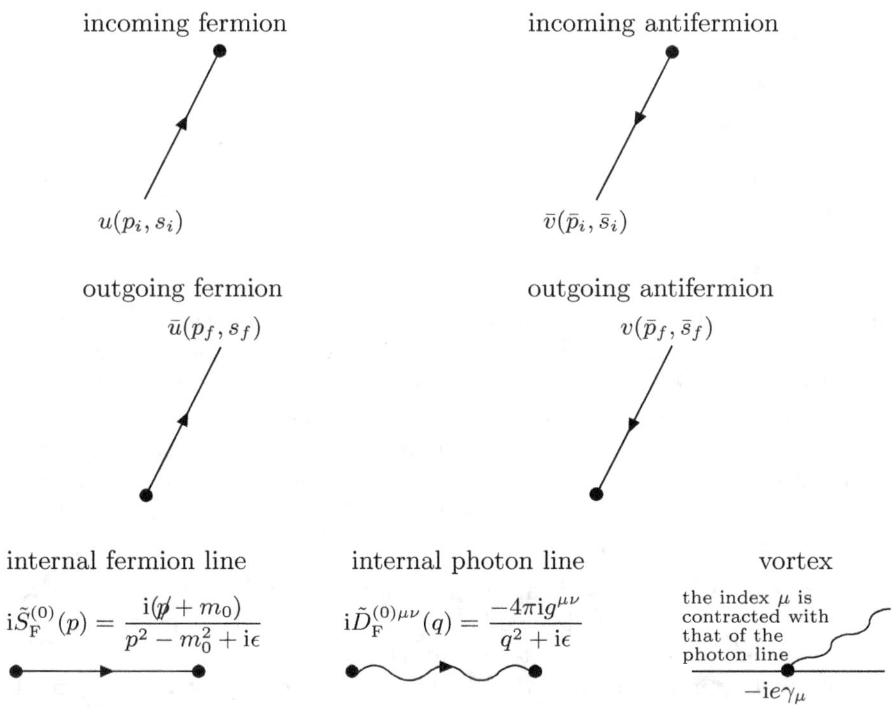

Fig. 3.14. Feynman diagram elements and characteristic factors in momentum space.

To 1. and 2. So far, the validity of these rules has only been shown to first and partially second order (with respect to the number of the scattering series term). However, they turn out to be true to any arbitrary order. The factor (-1) results from $\epsilon_f = -1$ in the case of an incoming antifermion.

To 3. (Tree diagrams and loop diagrams). Up to now, we have exclusively studied scattering processes where the electromagnetic potential is either a classical background field (Coulomb scattering of electrons, Subsection 3.3.1) or created by the transition currents of mutually scattered particles (electron-proton scattering, Subsections 3.3.2 and 3.3.3). In the latter case the electromagnetic interaction can be viewed as an exchange of virtual photons between both particles. In Feynman diagrams this is reflected by one internal photon line to first order and two internal photon lines to second order, each connecting one vortex of the first particle with one vortex of the second one. If, for example, we calculated the electron-proton scattering in the same way to higher orders, it would lead graphically to an increasing number of internal photon lines between the electron and the proton. Those diagrams are called *tree diagrams* and are, of course, covered by the 3. rule.

The decisive point is that, due to the combinatory diversity of vortices and lines, the 3. rule also allows the construction of *loop diagrams* in higher orders, as shown, for example, in Figure 3.15. These kinds of diagrams clearly

Fig. 3.15. Possible loop diagrams resulting from the 3. Feynman rule: vacuum polarization (**a**) and self-energy (**b**).

lie outside of our scattering formalism (with its view of classical background fields or its modification in terms of current-current interactions) and can only be justified within quantum field theory. Physically, loop diagrams correspond to *radiation corrections* caused by *quantum fluctuations of the vacuum.* Those corrections have an impact, for example, on the gyromagnetic ratio of the electron and the bound spectrum of atomic systems. They will be the subject of Section 3.4. In the current section we further concentrate on the tree diagrams of scattering processes in the lowest orders (compare to the introductory notes in this section).

To 4. Principally, one has to bear in mind that the Feynman diagrams are to be constructed on the level of wave functions. Therefore, the direction of the four-momentum arrows of the external antifermion lines follow the Feynman-Stückelberg interpretation, according to which a temporally forward incoming [outgoing] antifermion is described by a temporally backward outgoing [incoming] fermionic wave function with negative energy. On the other hand, the momenta and spins in the bispinors refer to the particle level; in the antifermion case they are reversed to the respective arrow directions.

Obviously, the vortices and internal fermion and photon lines are assigned slightly different factors than the ones used earlier. This removes uncertainties in the scattering amplitude with respect to i-factors that we encountered in the preceding subsection. There we defined the electronic and protonic transition currents for the electron-proton scattering to first order without a factor of i but to second order with i. Putting it differently, to first order we arrived at (3.85) that can be schematically written as

$$S_{fi} \sim -\mathrm{i}J_\mu(x)D_{\mathrm{F}}^{(0)}(x-y)J^{(\mathrm{p})\mu}(y) \ .$$

By contrast, to second order we came to (3.104), i.e.

$$S_{fi}^{(2)} \sim -J_{\mu\nu}^{(2)}(x,y)D_{\mathrm{F}}^{(0)}(x-X)D_{\mathrm{F}}^{(0)}(y-Y)J^{(\mathrm{p},2)\mu\nu}(X,Y) \ ,$$

where an i-factor was included in both transition currents due to reasons of factorizability. As this factorizability is supposed to hold to every order, we obtain an unambiguous treatment of i-factors, if we assign each occurring fermion propagator the factor +i, each occurring field A^μ the factor −i, and, in return, drop the inserted factor −i in $S_{fi}^{(n)}$ (1. rule) since we have

$$S_{fi}^{(n)} \sim -\mathrm{i}\not{A}(+\mathrm{i}S_{\mathrm{F}}^{(0)})\cdots\not{A} = (-\mathrm{i}\not{A})(+\mathrm{i}S_{\mathrm{F}}^{(0)})\cdots(-\mathrm{i}\not{A}) \ .$$

As can easily be seen, this corresponds exactly to the factor assignment of internal fermion and photon lines, and vortices shown in Figure 3.14.

3.3.5 Electron-Electron Scattering

We now turn to the process of electron-electron scattering to leading order, which we will describe using the rules presented immediately above. The kinematic situation where the electrons fly past each other is shown in Figure 3.16**a**. The corresponding Feynman diagram is given in Figure 3.16**b** and leads to the Lorentz-invariant amplitude

$$\begin{aligned} M_{fi}(\mathrm{dir}) &= \bar{u}(p_f,s_f)(-\mathrm{i}e)\gamma_\mu u(p_i,s_i)\frac{-4\pi\mathrm{i}}{q^2+\mathrm{i}\epsilon}\bar{u}(p_f',s_f')(-\mathrm{i}e)\gamma^\mu u(p_i',s_i') \\ q &= p_f - p_i \ . \end{aligned}$$

Obviously, it has the same structure as the $\mathcal{O}\left(e^2\right)$-amplitude of the electron-proton scattering in Theorem 3.8, which, of course, is due to the kinematic similarity of both processes. However, apart from this direct scattering, we have to consider another type of scattering as well since here, contrary to the electron-proton scattering, we are dealing with identical particles. This means that in the scattering experiment we are not able to distinguish the kinematic situation of Figure 3.16**a** from that of Figure 3.17**a** where both particles reflect each other. In addition to the direct amplitude $M_{fi}(\mathrm{dir})$, we

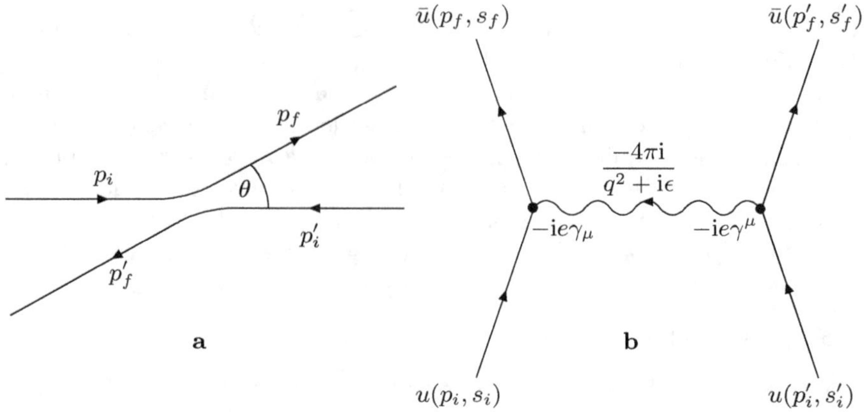

Fig. 3.16. Electron-electron scattering. **a** depicts the kinematic situation of direct scattering in the center of mass system and **b** the Feynman diagram of the direct $\mathcal{O}\left(e^2\right)$-scattering amplitude in momentum space (compare to Figure 3.9**b**). Energy and momentum are conserved at each vortex. Therefore, the four-momentum transfer is $q = p_f - p_i = -(p'_f - p'_i)$.

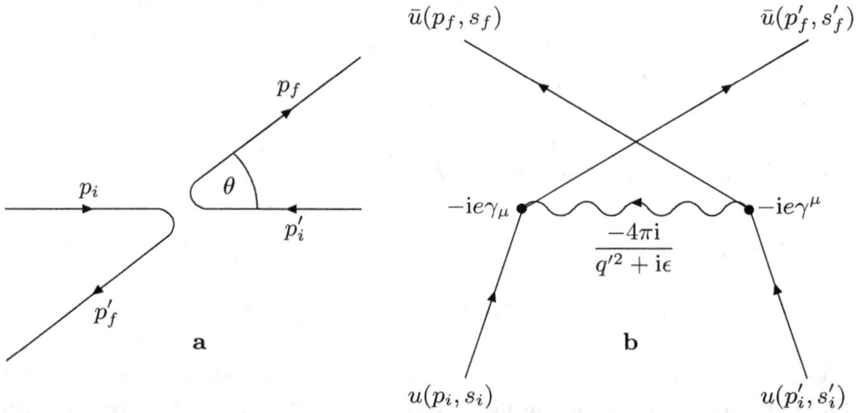

Fig. 3.17. Electron-electron scattering. **a** depicts the kinematic situation of exchange scattering in the center of mass system and **b** the Feynman diagram of the exchange $\mathcal{O}\left(e^2\right)$-scattering amplitude in momentum space. Energy and momentum are conserved at each vortex. Therefore, the four-momentum transfer is $q' = p'_f - p_i = -(p_f - p'_i)$.

therefore have to bring along the exchange amplitude $M_{fi}(\mathrm{ex})$ that results from $M_{fi}(\mathrm{dir})$ after the replacement $p_f \leftrightarrow p'_f$. Thus, overall, we obtain the scattering amplitude $(f \neq i)$

$$\left.\begin{aligned} S_{fi} &= \frac{(2\pi)^4\delta(p_f+p'_f-p_i-p'_i)}{V^2}\sqrt{\frac{m_0^2}{E_iE_f}}\sqrt{\frac{m_0^2}{E'_iE'_f}}M_{fi} \\ M_{fi} &= M_{fi}(\mathrm{dir}) - M_{fi}(\mathrm{ex}) \\ M_{fi}(\mathrm{dir}) &= \bar{u}(p_f,s_f)\gamma_\mu u(p_i,s_i)\frac{4\pi\mathrm{i}e^2}{q^2+\mathrm{i}\epsilon}\bar{u}(p'_f,s'_f)\gamma^\mu u(p'_i,s'_i) \\ M_{fi}(\mathrm{ex}) &= \bar{u}(p'_f,s'_f)\gamma_\mu u(p_i,s_i)\frac{4\pi\mathrm{i}e^2}{q'^2+\mathrm{i}\epsilon}\bar{u}(p_f,s_f)\gamma^\mu u(p'_i,s'_i) \\ q &= p_f - p_i \ , \ q' = p'_f - p_i \ . \end{aligned}\right\} \quad (3.106)$$

The relative sign between $M_{fi}(\mathrm{dir})$ and $M_{fi}(\mathrm{ex})$ accounts for the Fermi-Dirac statistics according to which, in the case of identical fermions, the whole scattering amplitude must be symmetric under the exchange of both fermions in the initial state ($p_i \leftrightarrow p'_i$) or final state ($p_f \leftrightarrow p'_f$).

Cross section. For the calculation of the differential cross section

$$\begin{aligned} \mathrm{d}\sigma &= \frac{m_0^2}{\sqrt{(p_i\cdot p'_i)^2-m_0^4}}|M_{fi}|^2(2\pi)^4\delta(p_f+p'_f-p_i-p'_i) \\ &\quad\times\frac{m_0\mathrm{d}^3p_f}{(2\pi)^3E_f}\frac{m_0\mathrm{d}^3p'_f}{(2\pi)^3E'_f} \ , \end{aligned} \quad (3.107)$$

we can proceed similarly to the calculations of the electron-proton case that lead to (3.92). However, in this case it is more realistic to work in the center of mass system instead of the laboratory system. In the former we have due to momentum conservation

$$\boldsymbol{p}_i + \boldsymbol{p}'_i = \boldsymbol{0} = \boldsymbol{p}_f + \boldsymbol{p}'_f \Longrightarrow \begin{cases} \boldsymbol{p}_i = -\boldsymbol{p}'_i \ , \ \boldsymbol{p}_f = -\boldsymbol{p}'_f \\ E_i = E'_i \ , \ E_f = E'_f \end{cases}$$

and due to energy conservation

$$E_i + E'_i = E_f + E'_f \Longrightarrow \begin{cases} E_i = E'_i = E_f = E'_f \\ |\boldsymbol{p}_i| = |\boldsymbol{p}'_i| = |\boldsymbol{p}_f| = |\boldsymbol{p}'_f| \ . \end{cases}$$

Using these and the identities

$$\frac{m_0^2}{\sqrt{(p_i\cdot p'_i)^2-m_0^4}} = \frac{m_0^2}{\sqrt{(E_i^2+\boldsymbol{p}_i^2)^2-m_0^4}} = \frac{m_0^2}{2E_i|\boldsymbol{p}_i|}$$

$$\mathrm{d}^3p_f = |\boldsymbol{p}_f|E_f\mathrm{d}E_f\mathrm{d}\Omega \ , \ \frac{\mathrm{d}^3p'_f}{E'_f} = 2\int \mathrm{d}^4p'_f\delta(p_f'^2-m_0^2)\Theta(p_f'^0) \ ,$$

(3.107) can be rewritten as (cm=center of mass system)

$$
\begin{aligned}
\left(\frac{\mathrm{d}\sigma}{\mathrm{d}\Omega}\right)_{\mathrm{cm}} &= \frac{m_0^4}{(2\pi)^2 E_i|\boldsymbol{p}_i|}\int \mathrm{d}E_f|\boldsymbol{p}_f|\int \mathrm{d}^4p_f'|M_{fi}|^2\delta(p_f+p_f'-p_i-p_i') \\
&\quad\times\delta(p_f'^2-m_0^2)\Theta(p_f'^0) \\
&= \frac{m_0^4}{(2\pi)^2 E_i|\boldsymbol{p}_i|}\int \mathrm{d}E_f|\boldsymbol{p}_f||M_{fi}|^2_{p_f'=p_i'+p_i-p_f} \\
&\quad\times\delta\left[(p_i'+p_i-p_f)^2-m_0^2\right]\Theta(p_i'^0+p_i^0-p_f^0) \\
&= \frac{m_0^4}{(2\pi)^2 E_i|\boldsymbol{p}_i|}\int \mathrm{d}E_f|\boldsymbol{p}_f||M_{fi}|^2_{p_f'=p_i'+p_i-p_f}\delta([4E_i(E_i-E_f)] \\
&\quad\times\Theta(2E_i-E_f) \\
&= \frac{m_0^4}{(2\pi)^2 E_i|\boldsymbol{p}_i|}\int\limits_{m_0}^{2E_i} \mathrm{d}E_f|\boldsymbol{p}_f||M_{fi}|^2_{p_f'=p_i'+p_i-p_f}\delta[4E_i(E_i-E_f)] \\
&= \frac{m_0^4}{(2\pi)^2 E_i|\boldsymbol{p}_i|}\int\limits_{m_0}^{2E_i} \mathrm{d}E_f|\boldsymbol{p}_f||M_{fi}|^2_{p_f'=p_i'+p_i-p_f}\frac{\delta(E_f-E_i)}{4E_i} \\
&= \frac{m_0^4}{4(2\pi)^2E_i^2}|M_{fi}|^2_{\mathrm{cm}}\ . \qquad (3.108)
\end{aligned}
$$

Amplitude square. To determine $|M_{fi}|^2_{\mathrm{cm}}$ we assume, as in the electron-proton case, that polarization effects do not play any role and consider the amplitude square

$$
\overline{|M_{fi}|^2} = \overline{|M_{fi}(\mathrm{dir})|^2} + \overline{|M_{fi}(\mathrm{ex})|^2} - 2\mathrm{Re}\overline{\left[M_{fi}(\mathrm{dir})M_{fi}^\dagger(\mathrm{ex})\right]}\ ,
$$

where the average over all incoming spins s_i, s_i' (a factor of 1/4) and the sum over all outgoing spins s_f, s_f' are taken. Comparing (3.106) with (3.94), we can immediately read off the square of the direct and exchange amplitudes from (3.95) and (3.96) with the corresponding replacements. This yields

$$
\begin{aligned}
\overline{|M_{fi}(\mathrm{dir})|^2} &= \frac{(4\pi)^2e^4}{4(q^2)^2}\mathrm{tr}\left[\Lambda_+(p_f)\gamma_\mu\Lambda_+(p_i)\gamma_\nu\right]\mathrm{tr}\left[\Lambda_+(p_f')\gamma^\mu\Lambda_+(p_i')\gamma^\nu\right] \\
&= \frac{(4\pi)^2e^4}{2m_0^4(q^2)^2}\left[(p_i\cdot p_i')(p_f\cdot p_f')+(p_i\cdot p_f')(p_f\cdot p_i')\right. \\
&\quad\left.-m_0^2(p_i\cdot p_f)-m_0^2(p_i'\cdot p_f')+2m_0^4\right] \qquad (3.109)
\end{aligned}
$$

$$\begin{aligned}
\overline{|M_{fi}(\text{ex})|^2} &= \frac{(4\pi)^2 e^4}{4(q'^2)^2} \text{tr}\left[\Lambda_+(p'_f)\gamma_\mu \Lambda_+(p_i)\gamma_\nu\right] \text{tr}\left[\Lambda_+(p_f)\gamma^\mu \Lambda_+(p'_i)\gamma^\nu\right] \\
&= \frac{(4\pi)^2 e^4}{2m_0^4(q'^2)^2}\left[(p_i \cdot p'_i)(p_f \cdot p'_f) + (p_i \cdot p_f)(p'_f \cdot p'_i)\right. \\
&\quad \left. -m_0^2(p_i \cdot p'_f) - m_0^2(p'_i \cdot p_f) + 2m_0^4\right] . \qquad (3.110)
\end{aligned}$$

With the help of $\sum\limits_s u(p,s)\bar{u}(p,s) = \Lambda_+(p)$, the interference term can be related back to a double spin sum and subsequently simplified further using Theorem 3.6:

$$\begin{aligned}
2\text{Re}\overline{\left[M_{fi}(\text{dir})M^\dagger_{fi}(\text{ex})\right]} &= 2\overline{\left[M_{fi}(\text{dir})M^\dagger_{fi}(\text{ex})\right]} \\
&= \frac{1}{2}\sum_{\substack{s_f, s_i \\ s'_f, s'_i}} \left[\bar{u}(p_f,s_f)\gamma_\mu u(p_i,s_i)\frac{4\pi \mathrm{i} e^2}{q^2}\bar{u}(p'_f,s'_f)\gamma^\mu u(p'_i,s'_i)\right] \\
&\quad \times \left[\bar{u}(p'_f,s'_f)\gamma_\nu u(p_i,s_i)\frac{4\pi \mathrm{i} e^2}{q'^2}\bar{u}(p_f,s_f)\gamma^\nu u(p'_i,s'_i)\right]^\dagger \\
&= \frac{(4\pi)^2 e^4}{2q^2 q'^2}\sum_{\substack{s_f, s_i \\ s'_f, s'_i}} \left[\bar{u}(p_f,s_f)\gamma_\mu u(p_i,s_i)\bar{u}(p'_f,s'_f)\gamma^\mu u(p'_i,s'_i)\right] \\
&\quad \times \left[\bar{u}(p'_i,s'_i)\gamma^\nu u(p_f,s_f)\bar{u}(p_i,s_i)\gamma_\nu u(p'_f,s'_f)\right] \\
&= \frac{(4\pi)^2 e^4}{2q^2 q'^2}\sum_{\substack{s_f, s_i \\ s'_f, s'_i}} \left[\bar{u}(p_f,s_f)\gamma_\mu u(p_i,s_i)\right]\left[\bar{u}(p_i,s_i)\gamma_\nu u(p'_f,s'_f)\right] \\
&\quad \times \left[\bar{u}(p'_f,s'_f)\gamma^\mu u(p'_i,s'_i)\right]\left[\bar{u}(p'_i,s'_i)\gamma^\nu u(p_f,s_f)\right] \\
&= \frac{(4\pi)^2 e^4}{2q^2 q'^2}\sum_{s_f, s'_f}\left[\bar{u}(p_f,s_f)\gamma_\mu \Lambda_+(p_i)\gamma_\nu u(p'_f,s'_f)\right] \\
&\quad \times \left[\bar{u}(p'_f,s'_f)\gamma^\mu \Lambda_+(p'_i)\gamma^\nu u(p_f,s_f)\right] \\
&= \frac{(4\pi)^2 e^4}{2q^2 q'^2}\text{tr}\left[\Lambda_+(p_f)\gamma_\mu \Lambda_+(p_i)\gamma_\nu \Lambda_+(p'_f)\gamma^\mu \Lambda_+(p'_i)\gamma^\nu\right] \\
&= \frac{(4\pi)^2 e^4}{2m_0^4 q^2 q'^2}\left[-2(p_i \cdot p'_i)(p_f \cdot p'_f) + m_0^2(p_i \cdot p'_i + p_i \cdot p_f\right. \\
&\quad \left. + p_i \cdot p'_f + p_f \cdot p'_i + p_f \cdot p'_f + p'_i \cdot p'_f) - 2m_0^4\right] . \qquad (3.111)
\end{aligned}$$

Putting all three amplitude square contributions together and replacing the scalar products with the corresponding relations in the center of mass system,

$$\left.\begin{aligned} p_i \cdot p_i &= p_i' \cdot p_i' = p_f \cdot p_f = p_f' \cdot p_f' = m_0^2 \\ p_i \cdot p_i' &= p_f \cdot p_f' = 2E_i^2 - m_0^2 \\ p_i \cdot p_f &= p_i' \cdot p_f' = 2E_i^2 \sin^2\frac{\theta}{2} + m_0^2 \cos\theta \\ p_i \cdot p_f' &= p_i' \cdot p_f = 2E_i^2 \cos^2\frac{\theta}{2} - m_0^2 \cos\theta \\ q^2 &= (p_f - p_i)^2 = -4(E_i^2 - m_0^2)\sin^2\frac{\theta}{2} \\ q'^2 &= (p_f' - p_i)^2 = -4(E_i^2 - m_0^2)\cos^2\frac{\theta}{2} \ , \end{aligned}\right\} \tag{3.112}$$

we arrive, after some algebraic manipulations, at

$$\overline{|M_{fi}|^2_{\mathrm{cm}}} = \frac{(2\pi)^2 e^4}{m_0^4}\left[\frac{4(2E_i^2 - m_0^2)^2}{\boldsymbol{p}_i^4 \sin^4\theta} - \frac{4E_i^2(E_i^2 + \boldsymbol{p}_i^2) - m_0^4}{\boldsymbol{p}_i^4 \sin^2\theta} + 1\right] .$$

The fact that this expression contains only trigonometric powers of $\sin^2\theta$ is plausible since, due to the identity of the particles, the differential cross section has to be symmetric under $\theta \to \pi - \theta$.

Theorem 3.10: Electron-electron scattering to leading order

The scattering amplitude for the electron-electron scattering (*Møller scattering*) to leading order is ($f \neq i$)

$$S_{fi} = \frac{(2\pi)^4 \delta(p_f + p_f' - p_i - p_i')}{V^2}\sqrt{\frac{m_0^2}{E_i E_f}}\sqrt{\frac{m_0^2}{E_i' E_f'}} M_{fi} \ ,$$

with the Lorentz-invariant amplitude ($q = p_f - p_i$, $q' = p_f' - p_i$)

$$\begin{aligned} M_{fi} &= M_{fi}(\mathrm{dir}) - M_{fi}(\mathrm{ex}) \\ M_{fi}(\mathrm{dir}) &= \bar{u}(p_f, s_f)\gamma_\mu u(p_i, s_i)\frac{4\pi \mathrm{i} e^2}{q^2 + \mathrm{i}\epsilon}\bar{u}(p_f', s_f')\gamma^\mu u(p_i', s_i') \\ M_{fi}(\mathrm{ex}) &= \bar{u}(p_f', s_f')\gamma_\mu u(p_i, s_i)\frac{4\pi \mathrm{i} e^2}{q'^2 + \mathrm{i}\epsilon}\bar{u}(p_f, s_f)\gamma^\mu u(p_i', s_i') \ . \end{aligned}$$

The differential cross section follows as

$$\begin{aligned} \mathrm{d}\sigma &= \frac{m_0^2}{\sqrt{(p_i \cdot p_i')^2 - m_0^4}} |M_{fi}|^2 (2\pi)^4 \delta(p_f + p_f' - p_i - p_i') \\ &\quad \times \frac{m_0 \mathrm{d}^3 p_f}{(2\pi)^3 E_f}\frac{m_0 \mathrm{d}^3 p_f'}{(2\pi)^3 E_f'} \end{aligned}$$

and, particularly in the center of mass system,

$$\left(\frac{\mathrm{d}\sigma}{\mathrm{d}\Omega}\right)_{\mathrm{cm}} = \frac{m_0^4}{4(2\pi)^2 E_i^2}|M_{fi}|^2_{\mathrm{cm}} \ .$$

▷

Here all electronic scattering momenta $\boldsymbol{p}_f$ directed toward $\mathrm{d}\Omega$ as well as all electronic scattering momenta $\boldsymbol{p}_f'$ have been integrated out. The unpolarized amplitude square is evaluated as

$$\overline{|M_{fi}|^2_{\mathrm{cm}}} = \frac{(2\pi)^2 e^4}{m_0^4}\left[\frac{4(2E_i^2 - m_0^2)^2}{\boldsymbol{p}_i^4 \sin^4\theta} - \frac{4E_i^2(E_i^2 + \boldsymbol{p}_i^2) - m_0^4}{\boldsymbol{p}_i^4 \sin^2\theta} + 1\right] .$$

This example demonstrates nicely how effective and time-saving the consistent use of the Feynman rules is. Note that there is no additional factor 1/2 or $\sqrt{1/2}$ in the scattering amplitude as may be expected due to the particles' identity. And also the rules for calculating the differential cross section are not altered at the presence of identical particles. However, we have to attach a factor of 1/2 to the total cross section in order to avoid double counting the identical particles in the final state.

3.3.6 Electron-Positron Scattering

Next we discuss the electron-positron scattering to leading order following the rules of Subsection 3.3.4. In doing so, we will make an interesting discovery, namely that the corresponding scattering amplitude is directly connected to that of the electron-electron case – a phenomenon which is generally true when comparing particle-particle and particle-antiparticle scattering processes.

Analogously to the electron-electron case, the most obvious kinematic constellation of the electron-positron scattering is the direct scattering, i.e. the fly-by of both particles as shown in Figure 3.18**a**. We have to take into account at the construction of the corresponding Feynman diagram that the temporally forward incoming [outgoing] positron is related to a temporally backward outgoing [incoming] electronic Dirac wave with negative energy. Therefore, on the right hand (positronic) side of Figure 3.18**b** the time arrows are directed backward and the vortex is assigned a factor of $-\mathrm{i}e\gamma^\mu$ (and not $+\mathrm{i}e\gamma^\mu$ as might be expected due to the positron's charge sign). By contrast, the momenta and spins in the v-spinors refer to the particle level, i.e. to the temporally forward moving positron. Overall, we obtain the amplitude

$$\begin{aligned} M_{fi}(\mathrm{dir}) &= \bar{u}(p_f, s_f)(-\mathrm{i}e)\gamma_\mu u(p_i, s_i)\frac{-4\pi\mathrm{i}}{q^2 + \mathrm{i}\epsilon}\bar{v}(\bar{p}_i, \bar{s}_i)(-\mathrm{i}e)\gamma^\mu v(\bar{p}_f, \bar{s}_f) \\ q &= p_f - p_i \ . \end{aligned}$$

Another scattering constellation is that the incoming electron and positron are annihilated during their "collision" and a new outgoing electron-positron pair is created (see Figure 3.19**a**). The corresponding annihilation diagram is shown in Figure 3.19**b** and leads to the amplitude

$$\begin{aligned} M_{fi}(\mathrm{ex}) &= \bar{v}(\bar{p}_i, \bar{s}_i)(-\mathrm{i}e)\gamma_\mu u(p_i, s_i)\frac{-4\pi\mathrm{i}}{q'^2 + \mathrm{i}\epsilon}\bar{u}(p_f, s_f)(-\mathrm{i}e)\gamma^\mu v(\bar{p}_f, \bar{s}_f) \\ q' &= p_i + \bar{p}_i \ . \end{aligned}$$

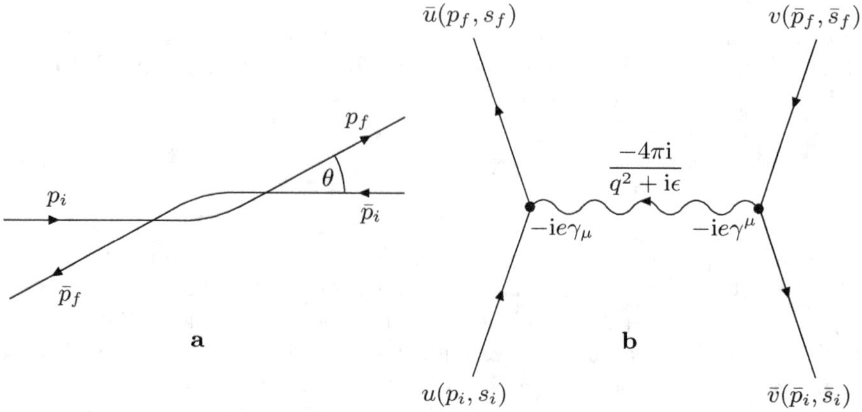

Fig. 3.18. Electron-positron scattering. **a** depicts the kinematic situation of direct scattering in the center of mass system and **b** the Feynman diagram of the direct $\mathcal{O}\left(e^2\right)$-scattering amplitude (compare to Figure 3.16). The unbarred quantities refer to the electron, the over-barred quantities to the positron. Energy and momentum are conserved at each vortex. Therefore the four-momentum transfer is $q = p_f - p_i = -(\bar{p}_f - \bar{p}_i)$.

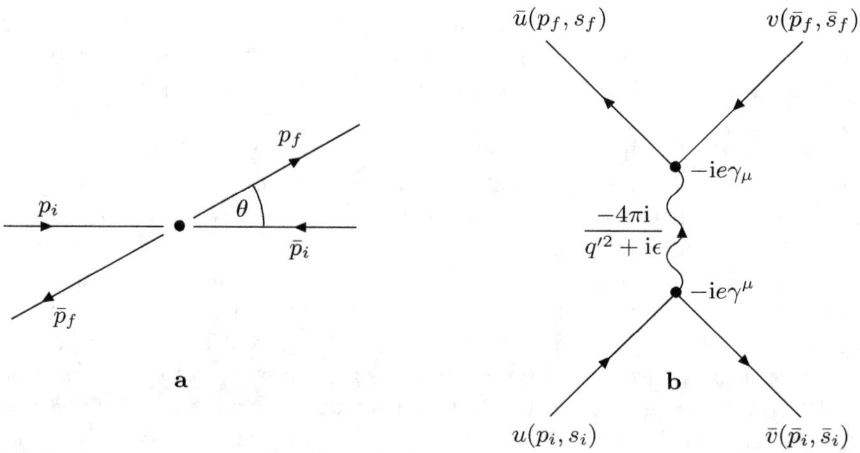

Fig. 3.19. Electron-positron scattering. **a** depicts the kinematic situation of annihilation scattering in the center of mass system and **b** the Feynman diagram of the annihilation $\mathcal{O}\left(e^2\right)$-scattering amplitude in momentum space (compare to Figure 3.17) for which we keep the symbol "ex" due to convenience. Energy and momentum are conserved at each vortex. Therefore, the four-momentum transfer is $q' = p_i + \bar{p}_i = p_f + \bar{p}_f$.

Here the peculiarity is that, contrary to all processes considered so far, the four-momentum transfer q' is time-like, for which reason the photon line is drawn vertically in Figure 3.19**b**. This can best be seen in the center of mass system, where $p_i = (E_i, \boldsymbol{p})$, $\bar{p}_i = (E_i, -\boldsymbol{p})$, and hence $q' = (2E_i, \boldsymbol{0})$, $q'^2 > 0$.

Combining the two amplitudes, we finally obtain $(f \neq i)$

$$\left.\begin{aligned}
S_{fi} &= -\frac{(2\pi)^4\delta(p_f+\bar{p}_f-p_i-\bar{p}_i)}{V^2}\sqrt{\frac{m_0^2}{E_iE_f}}\sqrt{\frac{m_0^2}{\bar{E}_i\bar{E}_f}}M_{fi}\\
M_{fi} &= M_{fi}(\mathrm{dir})-M_{fi}(\mathrm{ex})\\
M_{fi}(\mathrm{dir}) &= \bar{u}(p_f,s_f)\gamma_\mu u(p_i,s_i)\frac{4\pi\mathrm{i}e^2}{q^2+\mathrm{i}\epsilon}\bar{v}(\bar{p}_i,\bar{s}_i)\gamma^\mu v(\bar{p}_f,\bar{s}_f)\\
M_{fi}(\mathrm{ex}) &= \bar{v}(\bar{p}_i,\bar{s}_i)\gamma_\mu u(p_i,s_i)\frac{4\pi\mathrm{i}e^2}{q'^2+\mathrm{i}\epsilon}\bar{u}(p_f,s_f)\gamma^\mu v(\bar{p}_f,\bar{s}_f)\\
q &= p_f-p_i\ ,\ q'=p_i+\bar{p}_i\ .
\end{aligned}\right\}\quad(3.113)$$

The relative sign between $M_{fi}(\mathrm{dir})$ and $M_{fi}(\mathrm{ex})$ is, again, a consequence of the Fermi-Dirac statistics on the level of wave functions. It expresses the necessary antisymmetry between the incoming electron of positive energy (p_i) and the temporally backward incoming electron with negative energy ($-\bar{p}_f$) or, equally, between the outgoing electron with positive energy (p_f) and the temporally backward outgoing electron with negative energy ($-\bar{p}_i$).

Cross section. Since, on the particle level, the electron-positron and electron-electron scatterings are equal with respect to their energy and momentum aspects, we can entirely carry over the calculation of the differential cross section from the preceding subsection with the replacement $p'_{i,f}\to\bar{p}_{i,f}$. Thus, in the center of mass, we again obtain

$$\left(\frac{\mathrm{d}\sigma}{\mathrm{d}\Omega}\right)_{\mathrm{cm}}=\frac{m_0^4}{4(2\pi)^2E_i}|M_{fi}|^2_{\mathrm{cm}}\ .$$

Amplitude square. By contrast, the further calculation of $|M_{fi}|^2$ proceeds differently. Disregarding any polarization effects, the amplitude square is

$$\overline{|M_{fi}|^2}=\overline{|M_{fi}(\mathrm{dir})|^2}+\overline{|M_{fi}(\mathrm{ex})|^2}-2\mathrm{Re}\overline{\left[M_{fi}(\mathrm{dir})M^\dagger_{fi}(\mathrm{ex})\right]}\ ,$$

where the average over all incoming spins s_i, $\bar{s}_i$ (a factor of 1/4) and the sum over all outgoing spins s_f, $\bar{s}_f$ have been taken. Using Theorem 3.6, the individual terms are evaluated as

$$\begin{aligned}
\overline{|M_{fi}(\mathrm{dir})|^2} &= \frac{(4\pi)^2e^4}{4(q^2)^2}\sum_{\substack{s_f,s_i\\ \bar{s}_f,\bar{s}_i}}[\bar{u}(p_f,s_f)\gamma_\mu u(p_i,s_i)][\bar{v}(\bar{p}_i,\bar{s}_i)\gamma^\mu v(\bar{p}_f,\bar{s}_f)]\\
&\quad\times[\bar{u}(p_f,s_f)\gamma_\nu u(p_i,s_i)]^\dagger[\bar{v}(\bar{p}_i,\bar{s}_i)\gamma^\nu v(\bar{p}_f,\bar{s}_f)]^\dagger\\
&= \frac{(4\pi)^2e^4}{4(q^2)^2}\sum_{\substack{s_f,s_i\\ \bar{s}_f,\bar{s}_i}}[\bar{u}(p_f,s_f)\gamma_\mu u(p_i,s_i)][\bar{u}(p_i,s_i)\gamma_\nu u(p_f,s_f)]\\
&\quad\times[\bar{v}(\bar{p}_i,\bar{s}_i)\gamma^\mu v(\bar{p}_f,\bar{s}_f)][\bar{v}(\bar{p}_f,\bar{s}_f)\gamma^\nu v(\bar{p}_i,\bar{s}_i)]\\
&= \frac{(4\pi)^2e^4}{4(q^2)^2}\mathrm{tr}\left[\Lambda_+(p_f)\gamma_\mu\Lambda_+(p_i)\gamma_\nu\right]\mathrm{tr}\left[\Lambda_-(\bar{p}_i)\gamma^\mu\Lambda_-(\bar{p}_f)\gamma^\nu\right]
\end{aligned}$$

$$
\begin{aligned}
\overline{|M_{fi}(\mathrm{ex})|^2} &= \frac{(4\pi)^2 e^4}{4(q'^2)^2} \sum_{\substack{s_f,\, s_i \\ \bar{s}_f,\, \bar{s}_i}} [\bar{v}(\bar{p}_i, \bar{s}_i)\gamma_\mu u(p_i, s_i)][\bar{u}(p_f, s_f)\gamma^\mu v(\bar{p}_f, \bar{s}_f)] \\
&\quad \times [\bar{v}(\bar{p}_i, \bar{s}_i)\gamma_\nu u(p_i, s_i)]^\dagger [\bar{u}(p_f, s_f)\gamma^\nu v(\bar{p}_f, \bar{s}_f)]^\dagger \\
&= \frac{(4\pi)^2 e^4}{4(q'^2)^2} \sum_{\substack{s_f,\, s_i \\ \bar{s}_f,\, \bar{s}_i}} [\bar{v}(\bar{p}_i, \bar{s}_i)\gamma_\mu u(p_i, s_i)][\bar{u}(p_i, s_i)\gamma_\nu v(\bar{p}_i, \bar{s}_i)] \\
&\quad \times [\bar{u}(p_f, s_f)\gamma^\mu v(\bar{p}_f, \bar{s}_f)][\bar{v}(\bar{p}_f, \bar{s}_f)\gamma^\nu u(p_f, s_f)] \\
&= \frac{(4\pi)^2 e^4}{4(q'^2)^2} \operatorname{tr}\left[\Lambda_-(\bar{p}_i)\gamma_\mu \Lambda_+(p_i)\gamma_\nu\right] \operatorname{tr}\left[\Lambda_+(p_f)\gamma^\mu \Lambda_-(\bar{p}_f)\gamma^\nu\right]
\end{aligned}
$$

$$
\begin{aligned}
2\mathrm{Re}\overline{\left[M_{fi}(\mathrm{dir})M^\dagger_{fi}(\mathrm{ex})\right]} &= 2\overline{\left[M_{fi}(\mathrm{dir})M^\dagger_{fi}(\mathrm{ex})\right]} \\
&= \frac{(4\pi)^2 e^4}{2q^2 q'^2} \sum_{\substack{s_f,\, s_i \\ \bar{s}_f,\, \bar{s}_i}} [\bar{u}(p_f, s_f)\gamma_\mu u(p_i, s_i)][\bar{v}(\bar{p}_i, \bar{s}_i)\gamma^\mu v(\bar{p}_f, \bar{s}_f)] \\
&\quad \times [\bar{v}(\bar{p}_i, \bar{s}_i)\gamma_\nu u(p_i, s_i)]^\dagger [\bar{u}(p_f, s_f)\gamma^\nu v(\bar{p}_f, \bar{s}_f)]^\dagger \\
&= \frac{(4\pi)^2 e^4}{2q^2 q'^2} \sum_{\substack{s_f,\, s_i \\ \bar{s}_f,\, \bar{s}_i}} [\bar{u}(p_f, s_f)\gamma_\mu u(p_i, s_i)][\bar{u}(p_i, s_i)\gamma_\nu v(\bar{p}_i, \bar{s}_i)] \\
&\quad \times [\bar{v}(\bar{p}_i, \bar{s}_i)\gamma^\mu v(\bar{p}_f, \bar{s}_f)][\bar{v}(\bar{p}_f, \bar{s}_f)\gamma^\nu u(p_f, s_f)] \\
&= -\frac{(4\pi)^2 e^4}{2q^2 q'^2} \sum_{s_f, \bar{s}_i} [\bar{u}(p_f, s_f)\gamma_\mu \Lambda_+(p_i)\gamma_\nu v(\bar{p}_i, \bar{s}_i)] \\
&\quad \times [\bar{v}(\bar{p}_i, \bar{s}_i)\gamma^\mu \Lambda_-(\bar{p}_f)\gamma^\nu u(p_f, s_f)] \\
&= \frac{(4\pi)^2 e^4}{2q^2 q'^2} \operatorname{tr}\left[\Lambda_+(p_f)\gamma_\mu \Lambda_+(p_i)\gamma_\nu \Lambda_-(\bar{p}_i)\gamma^\mu \Lambda_-(\bar{p}_f)\gamma^\nu\right] \ .
\end{aligned}
$$

Comparing these expressions with the corresponding relations (3.109), (3.110), and (3.111) for the electron-electron scattering, we find that the amplitude squares of both processes emerge from each other if the four-momenta are replaced as shown in Figure 3.20. Obviously, this is due to the fact that the scattering amplitude of the electron-positron scattering in (3.113) results from that of the electron-electron scattering (Theorem 3.10) via exactly these replacements. This *crossing symmetry* turns out to be generally true – exactly and to every order of the scattering theory – for S-matrix elements of processes where incoming particles are replaced by the antiversions of the respective outgoing particles and vice versa. This means, for example, that the scattering amplitude of the particle-particle reaction $A + B \to C + D$ follows from that of the particle-antiparticle reaction $A + \bar{D} \to C + \bar{B}$ by simply replacing the momentum variables $\bar{p}_B \to -p_D$ and $\bar{p}_D \to -p_B$. And even processes with a different grouping of incoming and outgoing particles,

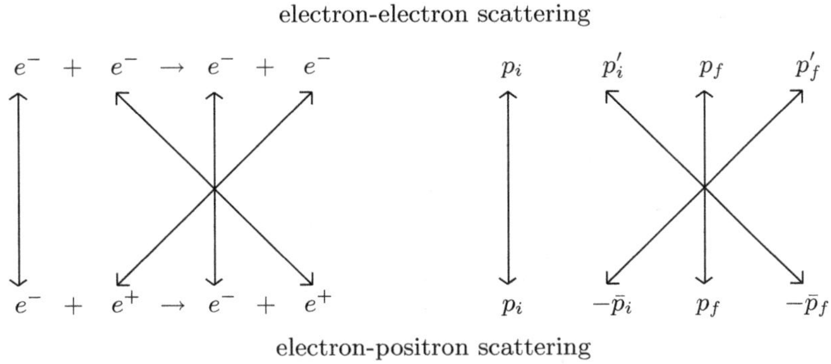

Fig. 3.20. Crossing symmetry between the electron-electron and electron-positron scatterings.

for example $A \rightarrow \bar{B} + C + D$ and $A + B \rightarrow C + D$, are interrelated through the crossing symmetry.

If we now make the replacements $p_i' \rightarrow -\bar{p}_f$, $p_f' \rightarrow -\bar{p}_i$ in (3.109), (3.110), and (3.111) and then evaluate the scalar products in the center of mass system, we finally obtain the result

$$\overline{|M_{fi}|^2_{\text{cm}}} = \frac{(2\pi)^2 e^4}{4m_0^4} \left[\frac{m_0^4 + 4\boldsymbol{p}_i^2 m_0^2 \cos^2\frac{\theta}{2} + 2\boldsymbol{p}_i^4 \left(1 + \cos^4\frac{\theta}{2}\right)}{\boldsymbol{p}_i^4 \sin^4\frac{\theta}{2}} \right.$$
$$+ \frac{3m_0^4 + 4\boldsymbol{p}_i^2 m_0^2 + \boldsymbol{p}_i^4(1+\cos^2\theta)}{E_i^4}$$
$$\left. - \frac{3m_0^4 + 8\boldsymbol{p}_i^2 m_0^2 \cos^2\frac{\theta}{2} + 4\boldsymbol{p}_i^4 \cos^4\frac{\theta}{2}}{E_i^2 \boldsymbol{p}_i^2 \sin^2\frac{\theta}{2}} \right] .$$

Contrary to the electron-electron case, this formula cannot be expressed by powers of $1/\sin^2\theta$ since now it is possible to distinguish between forward scattering ($\theta < \pi/2$) and backward scattering ($\theta > \pi/2$).

Theorem 3.11: Electron-positron scattering to leading order

The scattering amplitude for the electron-positron scattering (*Bhabba scattering*) to leading order is ($f \neq i$)

$$S_{fi} = -\frac{(2\pi)^4 \delta(p_f + \bar{p}_f - p_i - \bar{p}_i)}{V^2} \sqrt{\frac{m_0^2}{E_i E_f}} \sqrt{\frac{m_0^2}{\bar{E}_i \bar{E}_f}} M_{fi} ,$$

with the Lorentz-invariant amplitude ($q = p_f - p_i$, $q' = p_i + \bar{p}_i$) ▷

$$M_{fi} = M_{fi}(\mathrm{dir}) - M_{fi}(\mathrm{ex})$$

$$M_{fi}(\mathrm{dir}) = \bar{u}(p_f, s_f)\gamma_\mu u(p_i, s_i)\frac{4\pi \mathrm{i}e^2}{q^2 + \mathrm{i}\epsilon}\bar{v}(\bar{p}_i, \bar{s}_i)\gamma^\mu v(\bar{p}_f, \bar{s}_f)$$

$$M_{fi}(\mathrm{ex}) = \bar{v}(\bar{p}_i, \bar{s}_i)\gamma_\mu u(p_i, s_i)\frac{4\pi \mathrm{i}e^2}{q'^2 + \mathrm{i}\epsilon}\bar{u}(p_f, s_f)\gamma^\mu v(\bar{p}_f, \bar{s}_f)\ .$$

The differential cross section follows as

$$\mathrm{d}\sigma = \frac{m_0^2}{\sqrt{(p_i \cdot \bar{p}_i)^2 - m_0^4}}|M_{fi}|^2(2\pi)^4\delta(p_f + \bar{p}_f - p_i - \bar{p}_i) \times \frac{m_0 \mathrm{d}^3 p_f}{(2\pi)^3 E_f}\frac{m_0 \mathrm{d}^3 \bar{p}_f}{(2\pi)^3 \bar{E}_f}$$

and, particularly in the center of mass system,

$$\left(\frac{\mathrm{d}\sigma}{\mathrm{d}\Omega}\right)_{\mathrm{cm}} = \frac{m_0^4}{4(2\pi)^2 E_i^2}|M_{fi}|^2_{\mathrm{cm}}\ .$$

Here all electronic scattering momenta $\boldsymbol{p}_f$ directed toward $\mathrm{d}\Omega$ as well as all positronic scattering momenta $\bar{\boldsymbol{p}}_f$ have been integrated out. The unpolarized amplitude square is evaluated as

$$\overline{|M_{fi}|^2_{\mathrm{cm}}} = \frac{(2\pi)^2 e^4}{4m_0^4}\left[\frac{m_0^4 + 4\boldsymbol{p}_i^2 m_0^2 \cos^2\frac{\theta}{2} + 2\boldsymbol{p}_i^4\left(1 + \cos^4\frac{\theta}{2}\right)}{\boldsymbol{p}_i^4 \sin^4\frac{\theta}{2}} + \frac{3m_0^4 + 4\boldsymbol{p}_i^2 m_0^2 + \boldsymbol{p}_i^4(1 + \cos^2\theta)}{E_i^4} - \frac{3m_0^4 + 8\boldsymbol{p}_i^2 m_0^2 \cos^2\frac{\theta}{2} + 4\boldsymbol{p}_i^4 \cos^4\frac{\theta}{2}}{E_i^2 \boldsymbol{p}_i^2 \sin^2\frac{\theta}{2}}\right]\ .$$

The electron-positron and electron-electron scatterings are connected via the crossing symmetry.

3.3.7 Compton Scattering against Electrons

Up to now, we have discussed exclusively scattering processes where virtual photons have mediated the electromagnetic interaction between real fermions. Accordingly, the photonic four-momenta were space- or time-like and have been represented in the Feynman diagrams by closed lines with a start and an end point. However, there also exist processes with real photons whose four-momenta satisfy the Einstein condition $k_\mu k^\mu = 0$. Three such processes are the Compton scattering, the electron-positron annihilation and the electron-positron creation, all of which we discuss in this and the next subsection as well as in Exercises 38 and 39.[20]

Description of real photons. For the description of real photons, we start from the four-potential A^μ that, in the Lorentz gauge $\partial_\mu A^\mu = 0$, fulfills the Maxwell equation

$$\partial_\mu \partial^\mu A^\nu = 0 \ .$$

Similarly to fermions, we assume for A^μ a plane wave ($\hbar = c = 1$),

$$A_k^\mu(x) = \epsilon^\mu N_k \left(\mathrm{e}^{-\mathrm{i}k\cdot x} + \mathrm{e}^{\mathrm{i}k\cdot x}\right) \ , \ k^\mu = \begin{pmatrix} \omega \\ \boldsymbol{k} \end{pmatrix} \ , \ k \cdot k = 0 \ ,$$

with the normalization constant N_k, the polarization vector ϵ^μ, and the conditions

$$k \cdot \epsilon = 0 \ , \ \epsilon \cdot \epsilon = -1 \ . \tag{3.114}$$

The first condition follows from the Lorentz gauge and reflects the transversal nature of A^μ. Since further gauge transformations of the kind

$$A^\mu(x) \longrightarrow A^\mu(x) - \chi(x) \ , \ \partial_\mu \partial^\mu \chi = 0$$

do not affect the Lorentz gauge, we can turn to the *radiation gauge* by choosing $\chi(x) = A^0(x)$ so that

$$A^0(x) = 0 \Longrightarrow \boldsymbol{\nabla} \boldsymbol{A}(x) = 0 \ .$$

In this particular Lorentz system the polarization vectors are space-like and there remain two transversal, linearly independent three-polarization vectors:

$$(\epsilon^\mu) = \begin{pmatrix} 0 \\ \boldsymbol{\epsilon}(\boldsymbol{k}, \lambda) \end{pmatrix} \ , \ \boldsymbol{k}\boldsymbol{\epsilon}(\boldsymbol{k}, \lambda) = 0 \ , \ \boldsymbol{\epsilon}(\boldsymbol{k}, \lambda)\boldsymbol{\epsilon}(\boldsymbol{k}, \lambda) = 1 \ , \ \lambda = 1, 2 \ .$$

The normalization constant N_k can be determined by the constraint that the mean energy of the wave A_k^μ,

$$E_k = \frac{1}{8\pi} \int\limits_V \mathrm{d}^3x \left\langle \boldsymbol{E}_k^2 + \boldsymbol{B}_k^2 \right\rangle \ , \ \left\langle \boldsymbol{E}_k^2 \right\rangle = \left\langle \boldsymbol{B}_k^2 \right\rangle = \frac{1}{T} \int\limits_0^T \mathrm{d}t \boldsymbol{B}_k^2 \ , \ T = \frac{2\pi}{\omega} \ ,$$

is just the energy ω of a single photon. Taking into account

$$\boldsymbol{B}_k = \boldsymbol{\nabla} \times \boldsymbol{A}_k = \mathrm{i} N_k \boldsymbol{k} \times \boldsymbol{\epsilon} \left(\mathrm{e}^{-\mathrm{i}k\cdot x} - \mathrm{e}^{\mathrm{i}k\cdot x}\right) = 2N_k \boldsymbol{k} \times \boldsymbol{\epsilon} \sin k \cdot x$$

[20] This clearly implies that we have to modify our scattering formalism in Theorem 3.5 a second time (the first modification was the current-current interaction, see footnote 17 on page 234), since now we are dealing with emitted or absorbed photon quanta that are incompatible with a classical background field. However, at least to leading order $\mathcal{O}\left(e^2\right)$, it seems to be plausible to interpret the two A^μ-fields of the second scattering series term as incoming and outgoing photons. By contrast, in higher orders one could not get around a quantum field theoretical treatment (compare to the introductory notes in this section).

and

$$(\boldsymbol{k}\times\boldsymbol{\epsilon})^2 = \boldsymbol{k}^2\boldsymbol{\epsilon}^2 - (\boldsymbol{k}\boldsymbol{\epsilon})^2 = \boldsymbol{k}^2 = \omega^2 ,$$

it follows that

$$E_k = \frac{\omega^2 N_k^2}{\pi}\int\limits_V \mathrm{d}^3x \left\langle \sin^2(\omega t - \boldsymbol{k}\boldsymbol{x})\right\rangle = \frac{\omega^2 N_k^2 V}{2\pi} \Longrightarrow N_k = \sqrt{\frac{2\pi}{\omega V}} .$$

After these preliminary considerations, we now turn to the Compton scattering, where a photon is scattered against a free electron as shown in Figure 3.21, and start directly from Theorem 3.5.

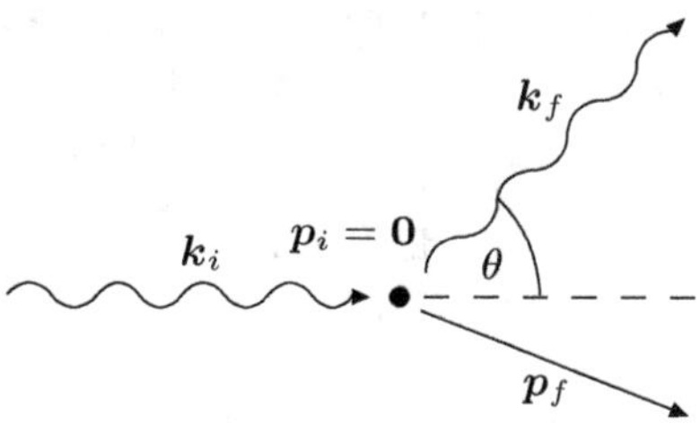

Fig. 3.21. Kinematic situation of the Compton scattering in the laboratory system where the electron is initially at rest.

Direct scattering amplitude. The leading term of the scattering series for the Compton scattering is number two. Thus, we start with the equation (the index 2 is suppressed)

$$\begin{aligned} S_{fi}(\mathrm{dir}) &= -\mathrm{i}e^2\int \mathrm{d}^4x\int \mathrm{d}^4y \bar{\Psi}_f(x)\not{A}_f(x)S_\mathrm{F}^{(0)}(x-y)\not{A}_i(y)\Psi_i(y) \\ &= \int \mathrm{d}^4x\int \mathrm{d}^4y\bar{\Psi}_f(x)(-\mathrm{i}e)\not{A}_f(x) \\ &\quad \times(+\mathrm{i})S_\mathrm{F}^{(0)}(x-y)(-\mathrm{i}e)\not{A}_i(y)\Psi_i(y) , \end{aligned} \quad (3.115)$$

where it is obvious to identify the four-potentials with the incoming (i) and outgoing (f) photons. Within our nomenclature, this corresponds to a direct scattering amplitude. Inserting the known expressions for the electron and photon wave functions as well as the electron propagator, we obtain

$$\begin{aligned} S_{fi}(\mathrm{dir}) &= -\frac{\mathrm{i}e^2}{V^2}\sqrt{\frac{m_0}{E_iE_f}}\sqrt{\frac{(2\pi)^2}{\omega_i\omega_f}}\int \mathrm{d}^4x\int \mathrm{d}^4y\int\frac{\mathrm{d}^4p}{(2\pi)^4} \\ &\quad\times\left[\bar{u}(p_f,s_f)\not{\epsilon}(k_f,\lambda_f)\frac{(\not{p}+m_0)}{p^2-m_0^2+\mathrm{i}\epsilon}\not{\epsilon}(k_i,\lambda_i)u(p_i,s_i)\right] \\ &\quad\times\mathrm{e}^{\mathrm{i}p_f\cdot x}\left(\mathrm{e}^{-\mathrm{i}k_f\cdot x}+\mathrm{e}^{\mathrm{i}k_f\cdot x}\right)\mathrm{e}^{-\mathrm{i}p\cdot(x-y)}\left(\mathrm{e}^{-\mathrm{i}k_i\cdot y}+\mathrm{e}^{\mathrm{i}k_i\cdot y}\right)\mathrm{e}^{-\mathrm{i}p_i\cdot y} . \end{aligned}$$

First we carry out the coordinate integration,

$$\int \mathrm{d}^4x \int \mathrm{d}^4y \mathrm{e}^{\mathrm{i}p_f\cdot x}\left(\mathrm{e}^{-\mathrm{i}k_f\cdot x}+\mathrm{e}^{\mathrm{i}k_f\cdot x}\right)\mathrm{e}^{-\mathrm{i}p\cdot(x-y)}\left(\mathrm{e}^{-\mathrm{i}k_i\cdot y}+\mathrm{e}^{\mathrm{i}k_i\cdot y}\right)\mathrm{e}^{-\mathrm{i}p_i\cdot y}$$

$$\begin{aligned}
&= \int \mathrm{d}^4x \left[\mathrm{e}^{\mathrm{i}(p_f-p-k_f)\cdot x}+\mathrm{e}^{\mathrm{i}(p_f-p+k_f)\cdot x}\right] \\
&\quad \times \int \mathrm{d}^4y \left[\mathrm{e}^{-\mathrm{i}(p_i-p+k_i)\cdot y}+\mathrm{e}^{-\mathrm{i}(p_i-p-k_i)\cdot y}\right] \\
&= (2\pi)^8\left[\delta(p_f-p-k_f)+\delta(p_f-p+k_f)\right] \\
&\quad \times \left[\delta(p_i-p+k_i)+\delta(p_i-p-k_i)\right] \\
&= (2\pi)^8\delta(p_f-p+k_f)\delta(p_i-p+k_i)\ ,
\end{aligned} \tag{3.116}$$

where, in the last step, it was taken into account that three of the four $\delta()\delta()$-combinations correspond to other or not realizable kinematic situations (see Exercise 37). The momentum integration now yields

$$\begin{aligned}
&\int \frac{\mathrm{d}^4p}{(2\pi)^4}(2\pi)^8\delta(p_f-p+k_f)\delta(p_i-p+k_i)\frac{\not{p}+m_0}{p^2-m_0^2+\mathrm{i}\epsilon} \\
&\quad = (2\pi)^4\delta(p_f+k_f-p_i-k_i)\frac{\not{p}_i+\not{k}_i+m_0}{(p_i+k_i)^2-m_0^2+\mathrm{i}\epsilon}\ ,
\end{aligned}$$

so that we end up with the integrated scattering amplitude

$$\begin{aligned}
S_{fi}(\mathrm{dir}) &= \frac{(2\pi)^4\delta(p_f+k_f-p_i-k_i)}{V^2}\sqrt{\frac{m_0^2}{E_iE_f}}\sqrt{\frac{(2\pi)^2}{\omega_i\omega_f}}M_{fi}(\mathrm{dir}) \\
M_{fi}(\mathrm{dir}) &= u(p_f,s_f)(-\mathrm{i}e)\not{\epsilon}(k_f,\lambda_f)\frac{(+\mathrm{i})(\not{p}_i+\not{k}_i+m_0)}{(p_i+k_i)^2-m_0^2+\mathrm{i}\epsilon} \\
&\quad \times(-\mathrm{i}e)\not{\epsilon}(k_i,\lambda_i)u(p_i,s_i)\ .
\end{aligned}$$

Comparing these expressions as well as the Feynman diagram in Figure 3.22a belonging to $M_{fi}(\mathrm{dir})$ with our Feynman rules in Subsection 3.3.4, we see that these rules can easily be extended to include processes involving real photons:

1', 2'. Each incoming or outgoing photon yields a *photon factor* (Gaussian unit system) $N_{i,f}^{(\dots)}=2\pi$.

4'. Within Feynman diagrams in momentum space, incoming and outgoing photons are represented by lines and factors as given in Figure 3.23.

3', 5'. Unchanged compared to 3. and 5.

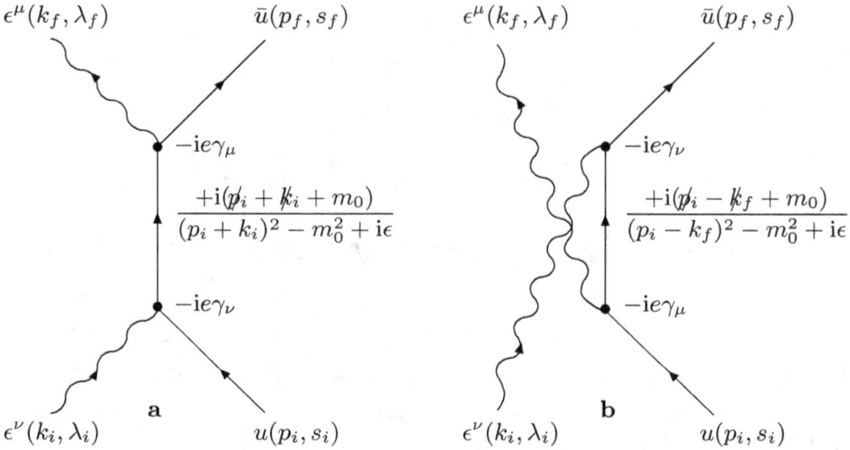

Fig. 3.22. Feynman diagrams of the direct $\mathcal{O}\left(e^2\right)$-scattering amplitude (**a**) and the exchange $\mathcal{O}\left(e^2\right)$-scattering amplitude (**b**) for the Compton scattering against electrons in momentum space. Energy and momentum are conserved at each vortex.

Fig. 3.23. Supplementary Feynman diagram elements and characteristic factors in momentum space (see Figure 3.14).

Exchange scattering amplitude. Apart from the direct scattering, we also have to take into account the constellation where the assignment of incoming and outgoing or, likewise, absorbed and emitted photons in (3.115) is reversed. After some calculations, similar to the above ones, this leads to the integrated exchange scattering amplitude

$$
\begin{aligned}
S_{fi}(\mathrm{ex}) &= \frac{(2\pi)^4\delta(p_f+k_f-p_i-k_i)}{V^2}\sqrt{\frac{m_0^2}{E_iE_f}}\sqrt{\frac{(2\pi)^2}{\omega_i\omega_f}}M_{fi}(\mathrm{ex})\\
M_{fi}(\mathrm{ex}) &= u(p_f,s_f)(-\mathrm{i}e)\not{\epsilon}(k_i,\lambda_i)\frac{(+\mathrm{i})(\not{p}_i-\not{k}_f+m_0)}{(p_i-k_f)^2-m_0^2+\mathrm{i}\epsilon}\\
&\quad\times(-\mathrm{i}e)\not{\epsilon}(k_f,\lambda_f)u(p_i,s_i)
\end{aligned}
$$

together with its graphical representation in Figure 3.22**b**, both being consistent with our modified rules 1' to 5'.

All in all, we obtain the scattering amplitude

$$\left.\begin{aligned} S_{fi} &= \frac{(2\pi)^4\delta(p_f+k_f-p_i-k_i)}{V^2}\sqrt{\frac{m_0^2}{E_iE_f}}\sqrt{\frac{(2\pi)^2}{\omega_i\omega_f}}M_{fi} \\ M_{fi} &= M_{fi}(\mathrm{dir})+M_{fi}(\mathrm{ex}) \\ &= -\mathrm{i}e^2\bar{u}(p_f,s_f)\left[\frac{\not{\epsilon}(k_f,\lambda_f)(\not{p}_i+\not{k}_i+m_0)\not{\epsilon}(k_i,\lambda_i)}{(p_i+k_i)^2-m_0^2+\mathrm{i}\epsilon}\right. \\ &\quad+\left.\frac{\not{\epsilon}(k_i,\lambda_i)(\not{p}_i-\not{k}_f+m_0)\not{\epsilon}(k_f,\lambda_f)}{(p_i-k_f)^2-m_0^2+\mathrm{i}\epsilon}\right]u(p_i,s_i)\ . \end{aligned}\right\} \tag{3.117}$$

Its invariance under the exchange $k_i \leftrightarrow -k_f$ is another example of the crossing symmetry that we encountered in the preceding subsection. In this case it means that the scattering amplitude for the absorption and emission of photons with momenta k_i and k_f is equal to the scattering amplitude for the absorption and emission of antiphotons with momenta k_f and k_i. However, the differentiation between photon and antiphoton is irrelevant as the photon is its own antiparticle.

Cross section. The differential cross section is best evaluated in the laboratory system where the electron is initially at rest, $p_i = (m_0, \mathbf{0})$, so that

$$\mathrm{d}\sigma = \frac{2\pi m_0}{\sqrt{\omega_i^2 m_0^2}}|M_{fi}|^2(2\pi)^4\delta(p_f+k_f-p_i-k_i)\frac{m_0\mathrm{d}^3p_f}{(2\pi)^3E_f}\frac{2\pi\mathrm{d}^3k_f}{(2\pi)^3\omega_f}\ .$$

Using

$$\frac{\mathrm{d}^3p_f}{E_f} = 2\int \mathrm{d}^4p_f\delta(p_f^2-m_0^2)\Theta(p_f^0)\ ,\quad \mathrm{d}^3k_f = \omega_f^2\mathrm{d}\omega_f\mathrm{d}\Omega\ ,$$

we now calculate as follows:

$$\begin{aligned} \frac{\mathrm{d}\sigma}{\mathrm{d}\Omega} &= \frac{2m_0}{\omega_i}\int\mathrm{d}\omega_f\omega_f\int\mathrm{d}^4p_f|M_{fi}|^2\delta(p_f+k_f-p_i-k_i) \\ &\quad\times\delta(p_f^2-m_0^2)\Theta(p_f^0) \\ &= \frac{2m_0}{\omega_i}\int\mathrm{d}\omega_f\omega_f|M_{fi}|^2_{p_f=p_i+k_i-k_f} \\ &\quad\times\delta[(p_i+k_i-k_f)^2-m_0^2]\Theta(m_0+\omega_i-\omega_f) \\ &= \frac{2m_0}{\omega_i}\int\limits_0^{m_0+\omega_i}\mathrm{d}\omega_f\omega_f|M_{fi}|^2_{p_f=p_i+k_i-k_f} \\ &\quad\times\delta[2m_0(\omega_i-\omega_f)-2\omega_i\omega_f(1-\cos\theta)] \\ &= \frac{2m_0}{\omega_i}\int\limits_0^{m_0+\omega_i}\mathrm{d}\omega_f\omega_f|M_{fi}|^2_{p_f=p_i+k_i-k_f} \\ &\quad\times\frac{\delta\left[\omega_f-\dfrac{\omega_i}{1+\frac{\omega_i}{m_0}(1-\cos\theta)}\right]}{2m_0+2\omega_i(1-\cos\theta)} \end{aligned}$$

$$= \frac{\omega_f^2}{\omega_i^2}|M_{fi}|^2_{\rm co}\ ,\ |M_{fi}|^2_{\rm co} = |M_{fi}|^2_{p_f=p_i+k_i-k_f}\ , \quad (3.118)$$

with the secondary condition

$$\omega_f = \frac{\omega_i}{1+\frac{\omega_i}{m_0}(1-\cos\theta)}\ .$$

As a consequence of energy and momentum conservation, the last equation interrelates the energies of incoming and outgoing photons. With $\lambda = 2\pi/\omega$, it can be cast into the *Compton formula*

$$\lambda_f = \lambda_i + 2\pi\frac{1}{m_0}(1-\cos\theta)\ .$$

Accordingly, the wave length of the scattered photon is raised by an amount comparable to the electron's Compton wave length $\lambda_{\rm c} = \hbar/m_0c$.

Amplitude square. If we are interested in the case of unpolarized electrons but keep the photon polarizations $\lambda_{i,f}$, we are led from (3.117) to (average over the initial and summation over the final polarizations of the electron)

$$\begin{aligned}\overline{|M_{fi}|^2}(\lambda_i,\lambda_f) &= \frac{e^4}{2}\sum_{s_f,s_i}\left[\bar{u}(p_f,s_f)\Gamma_1 u(p_i,s_i)\right]\left[\bar{u}(p_f,s_f)\Gamma_1 u(p_i,s_i)\right]^\dagger \\ &= \frac{e^4}{2}\sum_{s_f,s_i}\left[\bar{u}(p_f,s_f)\Gamma_1 u(p_i,s_i)\right]\left[\bar{u}(p_i,s_i)\Gamma_2 u(p_f,s_f)\right] \\ &= \frac{e^4}{2}\mathrm{tr}\left[\Lambda_+(p_f)\Gamma_1\Lambda_+(p_i)\Gamma_2\right]\ , \end{aligned} \quad (3.119)$$

with the operators

$$\left.\begin{aligned}\Gamma_1 &= \frac{\not\epsilon_f(\not p_i+\not k_i+m_0)\not\epsilon_i}{2p_i\cdot k_i} - \frac{\not\epsilon_i(\not p_i-\not k_f+m_0)\not\epsilon_f}{2p_i\cdot k_f} \\ \Gamma_2 &= \gamma^0\Gamma_1^\dagger\gamma^0 = \frac{\not\epsilon_i(\not p_i+\not k_i+m_0)\not\epsilon_f}{2p_i\cdot k_i} - \frac{\not\epsilon_f(\not p_i-\not k_f+m_0)\not\epsilon_i}{2p_i\cdot k_f} \\ &= \Gamma_1(\epsilon_i\leftrightarrow\epsilon_f)\end{aligned}\right\} \quad (3.120)$$

and the abbreviations $\epsilon_{i,f} = \epsilon(k_{i,f},\lambda_{i,f})$. Due to the many γ-matrix combinations, the evaluation of the trace according to Theorem 3.6 is much more complicated than in all the preceding examples. However, we can initially simplify the operators $\Gamma_{1,2}$ through the following reasoning: commuting $\not p_i$ in Γ_1 to the right and in Γ_2 to the left, we obtain

$$\begin{aligned}\Gamma_1 &= \frac{2p_i\cdot\epsilon_i\not\epsilon_f + \not\epsilon_f\not k_i\not\epsilon_i - \not\epsilon_f\not\epsilon_i(\not p_i - m_0)}{2p_i\cdot k_i} \\ &\quad - \frac{2p_i\cdot\epsilon_f\not\epsilon_i - \not\epsilon_i\not k_f\not\epsilon_f - \not\epsilon_i\not\epsilon_f(\not p_i - m_0)}{2p_i\cdot k_f} \\ \Gamma_2 &= \frac{2p_i\cdot\epsilon_i\not\epsilon_f + \not\epsilon_i\not k_i\not\epsilon_f - (\not p_i - m_0)\not\epsilon_i\not\epsilon_f}{2p_i\cdot k_i}\end{aligned}$$

$$-\frac{2p_i\cdot\epsilon_f\not\epsilon_i-\not\epsilon_f\not k_f\not\epsilon_i-(\not p_i-m_0)\not\epsilon_f\not\epsilon_i}{2p_i\cdot k_f}\,.$$

Here we can ignore the $(\not p_i - m_0)$-terms because they are orthogonal to the energy projection operators $\Lambda_+(p_i)$. Furthermore, we can always find a gauge where the photon polarizations $\epsilon_{i,f}$ are perpendicular to p_i. In the laboratory system this is the radiation gauge where $\epsilon^0 = 0$. Thus, overall, we can replace $\Gamma_{1,2}$ with

$$\Gamma_1 \longrightarrow \frac{\not\epsilon_f\not k_i\not\epsilon_i}{2p_i\cdot k_i}+\frac{\not\epsilon_i\not k_f\not\epsilon_f}{2p_i\cdot k_f}\,,\quad \Gamma_2 \longrightarrow \frac{\not\epsilon_i\not k_i\not\epsilon_f}{2p_i\cdot k_i}+\frac{\not\epsilon_f\not k_f\not\epsilon_i}{2p_i\cdot k_f}\,.$$

Despite this simplification, the determination of the trace remains quite complicated due to the presence of products with up to eight γ-matrices. Having struggled through these calculations and taking four-momentum conservation into account, one arrives at

$$\begin{aligned}\overline{|M_{fi}|^2_{\mathrm{co}}}(\lambda_i,\lambda_f) &= \frac{e^4}{4m_0^2}\left\{\frac{p_i\cdot k_f}{p_i\cdot k_i}+\frac{p_i\cdot k_i}{p_i\cdot k_f}\right.\\ &\quad\left.+\,4[\epsilon(k_i,\lambda_i)\cdot\epsilon(k_f,\lambda_f)]^2-2\right\}\,. \end{aligned}\tag{3.121}$$

Incorporating the laboratory conditions $k_i\cdot p_i=\omega_i m_0$, $k_f\cdot p_i=\omega_f m_0$ leads to the *Klein-Nishina formula*

$$\overline{|M_{fi}|^2_{\mathrm{co}}}(\lambda_i,\lambda_f)=\frac{e^4}{4m_0^2}\left\{\frac{\omega_f}{\omega_i}+\frac{\omega_i}{\omega_f}+4[\epsilon(k_i,\lambda_i)\cdot\epsilon(k_f,\lambda_f)]^2-2\right\}\,.$$

To determine the totally unpolarized cross section, we now have to average over the photon's initial polarizations and sum over its final polarizations:

$$\overline{|M_{fi}|^2_{\mathrm{co}}}=\frac{e^4}{2m_0^2}\left\{\frac{\omega_f}{\omega_i}+\frac{\omega_i}{\omega_f}+\sum_{\lambda_i,\lambda_f}[\epsilon(k_i,\lambda_i)\cdot\epsilon(k_f,\lambda_f)]^2-2\right\}\,. \tag{3.122}$$

Here the use of the radiation gauge presents itself where, without restricting generality, we can choose the three-vectors $\boldsymbol{\epsilon}(\boldsymbol{k}_i,1)$ and $\boldsymbol{\epsilon}(\boldsymbol{k}_f,1)$ in such a way that they lie in the plane spanned by $\boldsymbol{k}_i$ and $\boldsymbol{k}_f$. Consequently, the angle between $\boldsymbol{\epsilon}(\boldsymbol{k}_i,1)$ and $\boldsymbol{\epsilon}(\boldsymbol{k}_f,1)$ is equal to the scattering angle θ. Furthermore, $\boldsymbol{\epsilon}(\boldsymbol{k}_i,2)$ and $\boldsymbol{\epsilon}(\boldsymbol{k}_f,2)$ are perpendicular to the plane and therefore identical:

$$\left.\begin{aligned}\boldsymbol{\epsilon}(\boldsymbol{k}_i,1)\boldsymbol{\epsilon}(\boldsymbol{k}_f,1) &= \cos\theta\,,\ \boldsymbol{\epsilon}(\boldsymbol{k}_i,2)\boldsymbol{\epsilon}(\boldsymbol{k}_f,2)=1\\ \boldsymbol{\epsilon}(\boldsymbol{k}_i,1)\boldsymbol{\epsilon}(\boldsymbol{k}_f,2) &= \boldsymbol{\epsilon}(\boldsymbol{k}_i,2)\boldsymbol{\epsilon}(\boldsymbol{k}_f,1)=0\\ \sum_{\lambda_i,\lambda_f}[\epsilon(k_i,\lambda_i)\cdot\epsilon(k_f,\lambda_f)]^2 &= \sum_{\lambda_i,\lambda_f}[\boldsymbol{\epsilon}(\boldsymbol{k}_i,\lambda_i)\boldsymbol{\epsilon}(\boldsymbol{k}_f,\lambda_f)]^2\\ &= 1+\cos^2\theta\,.\end{aligned}\right\} \tag{3.123}$$

With this (3.122) finally turns into

$$\overline{|M_{fi}|^2_{\mathrm{co}}}=\frac{e^4}{2m_0^2}\left(\frac{\omega_f}{\omega_i}+\frac{\omega_i}{\omega_f}-\sin^2\theta\right)\,.$$

Theorem 3.12: Compton scattering against electrons to leading order

The scattering amplitude for the Compton scattering against electrons to leading order is $(f \neq i)$

$$S_{fi} = \frac{(2\pi)^4\delta(p_f + k_f - p_i - k_i)}{V^2}\sqrt{\frac{m_0^2}{E_i E_f}}\sqrt{\frac{(2\pi)^2}{\omega_i\omega_f}}M_{fi} ,$$

with the Lorentz-invariant amplitude

$$\begin{aligned}
M_{fi} &= M_{fi}(\mathrm{dir}) + M_{fi}(\mathrm{ex}) \\
M_{fi}(\mathrm{dir}) &= -\mathrm{i}e^2\bar{u}(p_f, s_f)\frac{\not\epsilon(k_f,\lambda_f)(\not p_i + \not k_i + m_0)\not\epsilon(k_i,\lambda_i)}{(p_i + k_i)^2 - m_0^2 + \mathrm{i}\epsilon}u(p_i, s_i) \\
M_{fi}(\mathrm{ex}) &= -\mathrm{i}e^2\bar{u}(p_f, s_f)\frac{\not\epsilon(k_i,\lambda_i)(\not p_i - \not k_f + m_0)\not\epsilon(k_f,\lambda_f)}{(p_i - k_f)^2 - m_0^2 + \mathrm{i}\epsilon}u(p_i, s_i) .
\end{aligned}$$

The differential cross section follows as

$$\mathrm{d}\sigma = \frac{2\pi m_0}{\sqrt{(p_i \cdot k_i)^2}}|M_{fi}|^2(2\pi)^4\delta(p_f + k_f - p_i - k_i)\frac{m_0\mathrm{d}^3p_f}{(2\pi)^3E_f}\frac{2\pi\mathrm{d}^3k_f}{(2\pi)^3\omega_f}$$

and, particularly in the laboratory system where the electron is initially at rest,

$$\begin{aligned}
\frac{\mathrm{d}\sigma}{\mathrm{d}\Omega} &= \frac{\omega_f^2}{\omega_i^2}|M_{fi}|^2_{\mathrm{co}} ,\quad |M_{fi}|^2_{\mathrm{co}} = |M_{fi}|^2_{p_f = p_i + k_i - k_f} \\
\omega_f &= \frac{\omega_i}{1 + \frac{\omega_i}{m_0}(1 - \cos\theta)} .
\end{aligned}$$

Here all photonic scattering momenta $\boldsymbol{k}_f$ directed toward $\mathrm{d}\Omega$ as well as all electronic scattering momenta $\boldsymbol{p}_f$ have been integrated out. Disregarding electronic polarization effects while keeping the photon polarizations leads to the Klein-Nishina formula

$$\overline{|M_{fi}|^2_{\mathrm{co}}}(\lambda_i, \lambda_f) = \frac{e^4}{4m_0^2}\left\{\frac{\omega_f}{\omega_i} + \frac{\omega_i}{\omega_f} + 4[\epsilon(k_i,\lambda_i)\cdot\epsilon(k_f,\lambda_f)]^2 - 2\right\} .$$

The totally unpolarized amplitude square is

$$\overline{|M_{fi}|^2_{\mathrm{co}}} = \frac{e^4}{2m_0^2}\left(\frac{\omega_f}{\omega_i} + \frac{\omega_i}{\omega_f} - \sin^2\theta\right) .$$

3.3.8 Electron-Positron Annihilation

Now we apply the extended set of rules 1' to 5' from the preceding subsection to the process of electron-positron annihilation where an electron and a positron collide and radiate into two photons (see Figure 3.24). The cor-

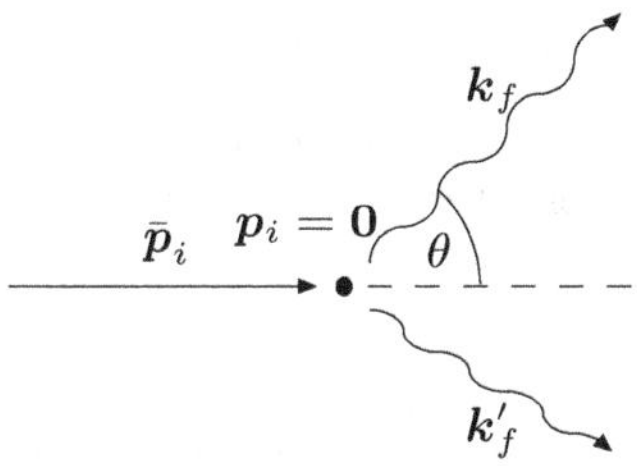

Fig. 3.24. Kinematic situation of the electron-positron annihilation in the laboratory system where the electron is initially at rest.

responding Feynman diagrams for the direct and exchange scatterings to leading order in momentum space are depicted in Figure 3.25. They lead to the scattering amplitude

$$\left.\begin{aligned}
S_{fi} &= \frac{(2\pi)^4\delta(k_f + k'_f - p_i - \bar{p}_i)}{V^2}\sqrt{\frac{m_0^2}{E_i\bar{E}_i}}\sqrt{\frac{(2\pi)^2}{\omega_f\omega'_f}}M_{fi} \\
M_{fi} &= M_{fi}(\mathrm{dir}) + M_{fi}(\mathrm{ex}) \\
&= -\mathrm{i}e^2\bar{v}(\bar{p}_i, \bar{s}_i)\left[\frac{\not\epsilon(k'_f, \lambda'_f)(\not p_i - \not k_f + m_0)\not\epsilon(k_f, \lambda_f)}{(p_i - k_f)^2 - m_0^2 + \mathrm{i}\epsilon}\right. \\
&\quad \left. + \frac{\not\epsilon(k_f, \lambda_f)(\not p_i - \not k'_f + m_0)\not\epsilon(k'_f, \lambda'_f)}{(p_i - k'_f)^2 - m_0^2 + \mathrm{i}\epsilon}\right] u(p_i, s_i)\ .
\end{aligned}\right\} \tag{3.124}$$

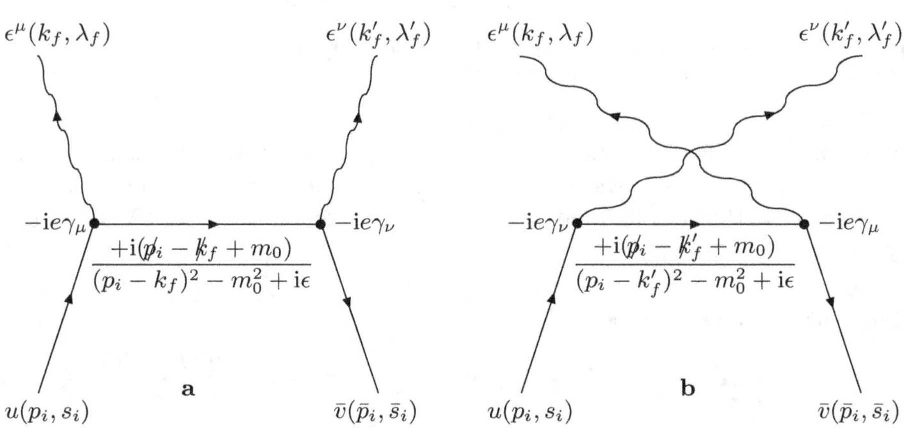

Fig. 3.25. Feynman diagrams of the direct $\mathcal{O}\left(e^2\right)$-scattering amplitude (**a**) and the exchange or annihilation $\mathcal{O}\left(e^2\right)$-scattering amplitude (**b**) for the electron-positron annihilation in momentum space. Energy and momentum are conserved at each vortex.

The following points have to be kept in mind:

- The leading order of this process is $\mathcal{O}\left(e^2\right)$ since the radiation of a (free!) electron-positron pair into a single photon is kinematically not possible. This means that the condition
$$k_f^2 = (p_i + \bar{p}_i)^2 = (p_i^0 + \bar{p}_i^0)^2 - (\boldsymbol{p}_i + \bar{\boldsymbol{p}}_i)^2 = 0$$
cannot be satisfied, as can easily be seen in the center of mass system where $\boldsymbol{p}_i = -\bar{\boldsymbol{p}}_i$.
- In accordance with the Bose-Einstein statistics, the scattering amplitude is symmetric under the exchange of the photons in the final state ($k_f \leftrightarrow k_f'$).
- Obviously, the Feynman diagrams of electron-positron annihilation and Compton scattering emerge from each other by 90°-rotations. Furthermore, comparing the scattering amplitude (3.124) with that in Theorem 3.12, we encounter, once again, an example of the crossing symmetry according to which both processes are connected via the substitution rule of Figure 3.26.

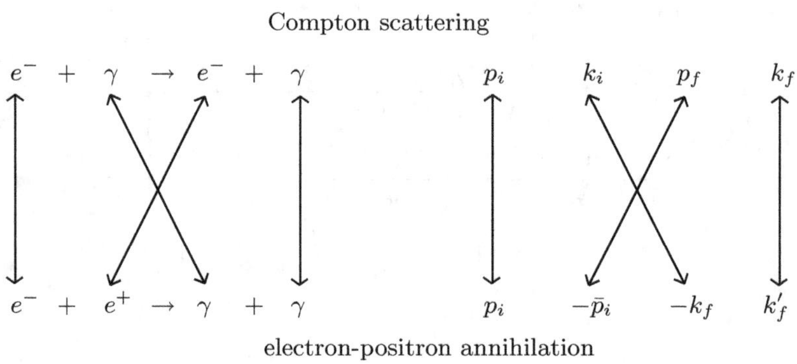

Fig. 3.26. Crossing symmetry between the Compton scattering against two electrons and the electron-positron annihilation.

A similar relationship exists between the Compton scattering and the process $\gamma + \gamma \to e^- + e^+$, i.e. the electron-positron creation by two photons (see Exercise 39). Hence, all three processes are interconnected via the crossing symmetry.

Cross section. In the laboratory system we have $p_i = (m_0, \mathbf{0})$ and the differential cross section is

$$\mathrm{d}\sigma = \frac{m_0^2}{\sqrt{m_0^2 \bar{E}_i^2 - m_0^4}} |M_{fi}|^2 (2\pi)^4 \delta(k_f + k_f' - p_i - \bar{p}_i) \frac{2\pi \mathrm{d}^3 k_f}{(2\pi)^3 \omega_f} \frac{2\pi \mathrm{d}^3 k_f'}{(2\pi)^3 \omega_f'} \ .$$

With the help of

$$\mathrm{d}^3 k_f = \omega_f^2 \mathrm{d}\omega_f \mathrm{d}\Omega \ , \ \frac{\mathrm{d}^3 k_f'}{\omega_f'} = 2 \int \mathrm{d}^4 k_f' \delta(k_f'^2) \Theta(k_f'^0) \ ,$$

this can be rewritten in a familiar way as

$$\begin{aligned}
\frac{\mathrm{d}\sigma}{\mathrm{d}\Omega} &= \frac{2m_0}{|\bar{\boldsymbol{p}}_i|}\int \mathrm{d}\omega_f\omega_f \int \mathrm{d}^4k'_f|M_{fi}|^2\delta(k_f+k'_f-p_i-\bar{p}_i)\delta(k'^2_f)\Theta(k'^0_f)\\
&= \frac{2m_0}{|\bar{\boldsymbol{p}}_i|}\int \mathrm{d}\omega_f\omega_f|M_{fi}|^2_{k'_f=\bar{p}_i+p_i-k_f}\\
&\quad\times\delta[(\bar{p}_i+p_i-k_f)^2]\Theta(\bar{E}_i+m_0-\omega_f)\\
&= \frac{2m_0}{|\bar{\boldsymbol{p}}_i|}\int\limits_0^{m_0+\bar{E}_i} \mathrm{d}\omega_f\omega_f|M_{fi}|^2_{k'_f=\bar{p}_i+p_i-k_f}\\
&\quad\times\delta[2m_0^2+2m_0\bar{E}_i-2\omega_f(m_0+\bar{E}_i-|\bar{\boldsymbol{p}}_i|\cos\theta)]\\
&= \frac{2m_0}{|\bar{\boldsymbol{p}}_i|}\int\limits_0^{m_0+\bar{E}_i} \mathrm{d}\omega_f\omega_f|M_{fi}|^2_{k'_f=\bar{p}_i+p_i-k_f}\\
&\quad\times\frac{\delta\left[\omega_f-\frac{m_0(m_0+\bar{E}_i)}{m_0+\bar{E}_i-|\bar{\boldsymbol{p}}_i|\cos\theta}\right]}{2(m_0+\bar{E}_i-|\bar{\boldsymbol{p}}_i|\cos\theta)}\\
&= \frac{\omega_f^2}{|\bar{\boldsymbol{p}}_i|(m_0+\bar{E}_i)}|M_{fi}|^2_{\mathrm{co}}\ ,\ |M_{fi}|^2_{\mathrm{co}}=|M_{fi}|^2_{k'_f=\bar{p}_i+p_i-k_f}\ ,
\end{aligned}$$

where the photon energy ω_f is related to the energy of the outgoing positron via

$$\omega_f=\frac{m_0(m_0+\bar{E}_i)}{m_0+\bar{E}_i-|\bar{\boldsymbol{p}}_i|\cos\theta}\ .$$

Amplitude square. Similarly to the Compton scattering, we assume to have unpolarized fermions while keeping the photon polarizations λ_f and λ'_f. Thus, starting from (3.124), we consider the equation (average over all initial polarizations of the electrons and positrons)

$$\begin{aligned}
\overline{|M_{fi}|^2}(\lambda_f,\lambda'_f) &= \frac{e^4}{4}\sum_{s_i,\bar{s}_i}[\bar{v}(\bar{p}_i,\bar{s}_i)\tilde{\Gamma}_1u(p_i,s_i)][\bar{v}(\bar{p}_i,\bar{s}_i)\tilde{\Gamma}_1u(p_i,s_i)]^\dagger\\
&= \frac{e^4}{4}\sum_{s_i,\bar{s}_i}[\bar{v}(\bar{p}_i,\bar{s}_i)\tilde{\Gamma}_1\dot{u}(p_i,s_i)][\bar{u}(p_i,s_i)\tilde{\Gamma}_2v(\bar{p}_i,\bar{s}_i)]\\
&= -\frac{e^4}{4}\mathrm{tr}[\Lambda_-(\bar{p}_i)\tilde{\Gamma}_1\Lambda_+(p_i)\tilde{\Gamma}_2]\ ,
\end{aligned}\tag{3.125}$$

with the operators

$$\begin{aligned}
\tilde{\Gamma}_1 &= -\frac{\not{\epsilon}(k'_f,\lambda'_f)(\not{p}_i-\not{k}_f+m_0)\not{\epsilon}(k_f,\lambda_f)}{2p_i\cdot k_f}\\
&\quad-\frac{\not{\epsilon}(k_f,\lambda_f)(\not{p}_i-\not{k}'_f+m_0)\not{\epsilon}(k'_f,\lambda'_f)}{2p_i\cdot k'_f}
\end{aligned}$$

$$\tilde{\Gamma}_2 = \gamma^0 \tilde{\Gamma}_1^\dagger \gamma^0 = -\frac{\not\epsilon(k_f, \lambda_f)(\not p_i - \not k_f + m_0)\not\epsilon(k'_f, \lambda'_f)}{2p_i \cdot k_f} - \frac{\not\epsilon(k'_f, \lambda'_f)(\not p_i - \not k'_f + m_0)\not\epsilon(k_f, \lambda_f)}{2p_i \cdot k'_f} .$$

A further evaluation of these expressions is easy if we bear in mind that they are connected with the corresponding equations (3.119) and (3.120) of the Compton scattering via the substitution rule of the crossing symmetry in Figure 3.26. The additional factor $-1/2$ in (3.125) results from the facts that there the average is taken over both fermion polarizations and that a v-bispinor is involved in the trace (compare to Theorem 3.6). Overall, we can therefore carry over (3.121) with the corresponding replacements to obtain

$$\overline{|M_{fi}|^2_{\mathrm{co}}}(\lambda_f, \lambda'_f) = \frac{e^4}{8m_0^2} \left\{ \frac{p_i \cdot k'_f}{p_i \cdot k_f} + \frac{p_i \cdot k_f}{p_i \cdot k'_f} + 2 - 4[\epsilon(k_f \lambda_f) \cdot \epsilon(k'_f, \lambda'_f)]^2 \right\} \quad (3.126)$$

and, after inserting the laboratory conditions $p_i \cdot k_f = \omega_f m_0$, $p_i \cdot k'_f = \omega'_f m_0$,

$$\overline{|M_{fi}|^2_{\mathrm{co}}}(\lambda_f, \lambda'_f) = \frac{e^4}{8m_0^2} \left\{ \frac{\omega'_f}{\omega_f} + \frac{\omega_f}{\omega'_f} + 2 - 4[\epsilon(k_f \lambda_f) \cdot \epsilon(k'_f, \lambda'_f)]^2 \right\} .$$

Energy conservation determines the value of ω'_f to be

$$\omega'_f = m_0 + \bar{E}_i - \omega_f = (m_0 + \bar{E}_i) \left(1 - \frac{m_0}{m_0 + \bar{E}_i - |\bar{\boldsymbol{p}}_i| \cos\theta} \right) .$$

Summing over the photon polarizations finally yields the totally unpolarized amplitude square

$$\overline{|M_{fi}|^2_{\mathrm{co}}} = \frac{e^4}{2m_0^2} \left(\frac{\omega'_f}{\omega_f} + \frac{\omega_f}{\omega'_f} + \sin^2\tilde{\theta} \right) ,$$

where $\tilde{\theta}$ denotes the angle between the photon momenta $\boldsymbol{k}_f$ and $\boldsymbol{k}'_f$ (in the case of the Compton scattering $\tilde{\theta}$ was identical to the scattering angle θ).

Theorem 3.13: Electron-positron annihilation to leading order

The scattering amplitude for the electron-positron annihilation to leading order is $(f \neq i)$

$$S_{fi} = \frac{(2\pi)^4 \delta(k_f + k'_f - p_i - \bar{p}_i)}{V^2} \sqrt{\frac{m_0^2}{E_i \bar{E}_i}} \sqrt{\frac{(2\pi)^2}{\omega_f \omega'_f}} M_{fi} ,$$

▷

with the Lorentz-invariant amplitude

$$\begin{aligned}
M_{fi} &= M_{fi}(\mathrm{dir}) + M_{fi}(\mathrm{ex}) \\
M_{fi}(\mathrm{dir}) &= -\mathrm{i}e^2 \bar{v}(\bar{p}_i, \bar{s}_i) \frac{\not{\epsilon}(k'_f, \lambda'_f)(\not{p}_i - \not{k}_f + m_0)\not{\epsilon}(k_f, \lambda_f)}{(p_i - k_f)^2 - m_0^2 + \mathrm{i}\epsilon} u(p_i, s_i) \\
M_{fi}(\mathrm{ex}) &= -\mathrm{i}e^2 \bar{v}(\bar{p}_i, \bar{s}_i) \frac{\not{\epsilon}(k_f, \lambda_f)(\not{p}_i - \not{k}'_f + m_0)\not{\epsilon}(k'_f, \lambda'_f)}{(p_i - k'_f)^2 - m_0^2 + \mathrm{i}\epsilon} u(p_i, s_i) .
\end{aligned}$$

The differential cross section follows as

$$\begin{aligned}
\mathrm{d}\sigma &= \frac{m_0^2}{\sqrt{(p_i \cdot \bar{p}_i)^2 - m_0^4}} |M_{fi}|^2 (2\pi)^4 \delta(k_f + k'_f - p_i - \bar{p}_i) \\
&\quad \times \frac{2\pi \mathrm{d}^3 k_f}{(2\pi)^3 \omega_f} \frac{2\pi \mathrm{d}^3 k'_f}{(2\pi)^3 \omega'_f}
\end{aligned}$$

and, particularly in the laboratory system where the electron is initially at rest,

$$\begin{aligned}
\frac{\mathrm{d}\sigma}{\mathrm{d}\Omega} &= \frac{\omega_f^2}{|\bar{\boldsymbol{p}}_i|(m_0 + \bar{E}_i)} |M_{fi}|^2_{\mathrm{co}} , \quad |M_{fi}|^2_{\mathrm{co}} = |M_{fi}|^2_{k'_f = \bar{p}_i + p_i - k_f} \\
\omega_f &= \frac{m_0(m_0 + \bar{E}_i)}{m_0 + \bar{E}_i - |\bar{\boldsymbol{p}}_i| \cos\theta} .
\end{aligned}$$

Here all photonic scattering momenta $\boldsymbol{k}_f$ directed toward $\mathrm{d}\Omega$ as well as all photonic scattering momenta $\boldsymbol{k}'_f$ have been integrated out. Disregarding fermionic polarization effects while keeping the photon polarizations yields

$$\begin{aligned}
\overline{|M_{fi}|^2_{\mathrm{co}}}(\lambda_f, \lambda'_f) &= \frac{e^4}{8m_0^2} \left\{ \frac{\omega'_f}{\omega_f} + \frac{\omega_f}{\omega'_f} + 2 - 4[\epsilon(k_f \lambda_f) \cdot \epsilon(k'_f, \lambda'_f)]^2 \right\} \\
\omega'_f &= (m_0 + \bar{E}_i) \left(1 - \frac{m_0}{m_0 + \bar{E}_i - |\bar{\boldsymbol{p}}_i| \cos\theta} \right) .
\end{aligned}$$

The totally unpolarized amplitude square is

$$\overline{|M_{fi}|^2_{\mathrm{co}}} = \frac{e^4}{2m_0^2} \left(\frac{\omega'_f}{\omega_f} + \frac{\omega_f}{\omega'_f} + \sin^2 \tilde{\theta} \right) , \quad \tilde{\theta} = \sphericalangle(\boldsymbol{k}_f, \boldsymbol{k}'_f) .$$

The processes of Compton scattering, electron-positron annihilation and creation are connected via the crossing symmetry.

3.3.9 Conclusion: Feynman Diagrams in Momentum Space

In the preceding subsections we have studied several relativistic spin-1/2 scattering processes. First, we considered pure fermion processes such as Coulomb, electron-proton, electron-electron, and electron-positron scatterings, where only virtual photons are involved as mediators of the electromagnetic interaction. There it turned out that the corresponding scattering amplitudes follow particular characteristics that can be formalized in some simple rules. Subsequently, we discussed the Compton scattering as well as the electron-positron annihilation and found that also those processes encompassing real photons can be described by the discovered rules with a few extensions. As was pointed out several times in this section, this modus operandi comprises two important aspects:

- Enhancements to the original scattering formalism in Theorem 3.5. They allow the description of two-particle scatterings via current-current interactions as well as of processes with real photons and lie beyond the treatment of A^μ as a classical background field.
- Derivation of the Feynman rules that can only be fully explained within quantum field theory. In this section we have only considered their tree level part to the lowest orders.

In the following section we see that it is the purely quantum field theoretically motivated loop diagrams in higher orders that raise difficulties for the construction of scattering amplitudes.

At the end of this section we present, once again, the complete set of Feynman rules in momentum space. Here the extensions, compared to the rules in Subsections 3.3.4 and 3.3.7, refer mainly to the generalization of scattering processes with more than two scattering products (1. rule), the incorporation of the quantum mechanical indistinguishability of identical particles (2. rule) as well as the Fermi statistics (4. rule).

1. The scattering amplitude of an elastic or inelastic scattering process of the kind

$$I + I' \longrightarrow F + F' + F'' + \ldots + F^{(m)} \quad \begin{pmatrix} I\text{=incoming particle} \\ F\text{=outgoing particle} \end{pmatrix}$$

is given by

$$S_{fi} = \frac{(2\pi)^4 \delta\left(p_i + p_i' - \sum\limits_{k=1}^{m} p_f^{(k)}\right)}{V^2} \prod_{j=1}^{2} \sqrt{\frac{N_i^{(j)}}{E_i^{(j)}}} \prod_{k=1}^{m} \sqrt{\frac{N_f^{(k)}}{E_f^{(k)}}} M_{fi} ,$$

with a fermion factor of $N_{i,f}^{(\ldots)} = m_0$ for each (anti)fermion and a photon factor of $N_{i,f}^{(\ldots)} = 2\pi$ for each photon. S_{fi} receives an additional sign for each incoming antifermion (outgoing fermion wave function with negative energy).

2. The corresponding differential cross section for collinear currents is

$$\mathrm{d}\sigma = \frac{N_i N_i'}{\sqrt{(p_i \cdot p_i')^2 - m_{0,i}^2 m_{0,i}'^2}} |M_{fi}|^2 (2\pi)^4 \delta\left(p_i + p_i' - \sum_{k=1}^{m} p_f^{(k)}\right)$$

$$\times \prod_{k=1}^{m} \frac{N_f^{(k)} \mathrm{d}^3 p_f^{(k)}}{(2\pi)^3 E_f^{(k)}} \ .$$

When calculating the total cross section, the additional *degeneracy factor*

$$\prod_{k=1}^{m} \frac{1}{g^{(k)}!}$$

has to be taken into account for $g^{(k)}$ identical particles of the kind $F^{(k)}$ in the final state.

3. The Lorentz-invariant amplitude M_{fi} can be expanded in powers of the coupling constant e. The expansion terms of order $\mathcal{O}(e^n)$ are obtained from the Feynman diagrams in momentum space containing all topological constellations of fermion lines, photon lines, and n vortices that are consistent with the scattering process.

4. All vortices, fermion lines, and photon lines in the Feynman diagrams are assigned the factors given in Figure 3.27. Furthermore, the following factors have to be taken into account:

   i) a relative sign when two Feynman diagrams differ only by the exchange of two fermion lines (of the same fermion type).

   ii) a factor of (-1) for each closed fermion loop.

5. Four-momentum conservation holds at each vortex. All remaining (undetermined) momenta p are subject to integration with $\int \mathrm{d}^4p/(2\pi)^4$ in the amplitude M_{fi}.

To 2. The degeneracy factor takes into account the quantum mechanical indistinguishability of the trajectories of two identical outgoing particles, for example, as in the electron-electron scattering (see Subsection 3.3.5).

To 4. The relative sign in i) is a consequence of the necessary antisymmetrization of the whole scattering amplitude on the level of wave functions due to the Fermi statistics. Therefore, it takes effect also at the exchange of one incoming [outgoing] particle line with one outgoing [incoming] antiparticle line (see electron-electron scattering, Subsection 3.3.5 and electron-positron scattering, Subsection 3.3.6). The rule ii) is nothing more than a special case of rule i) as can be seen immediately by means of Figure 3.28. If we exchange the two marked fermion lines in **a**, we obtain diagram **b**, which can equally be drawn as in **c**. Thus, compared to **c**, diagram **a** contains a phase factor of (-1).[21]

[21] As regards the equivalence of the diagrams **b** and **c**, recall that, due to the integration over all space-time points, a Feynman diagram can be arbitrarily deformed as long as the order of vortices and lines remains unchanged.

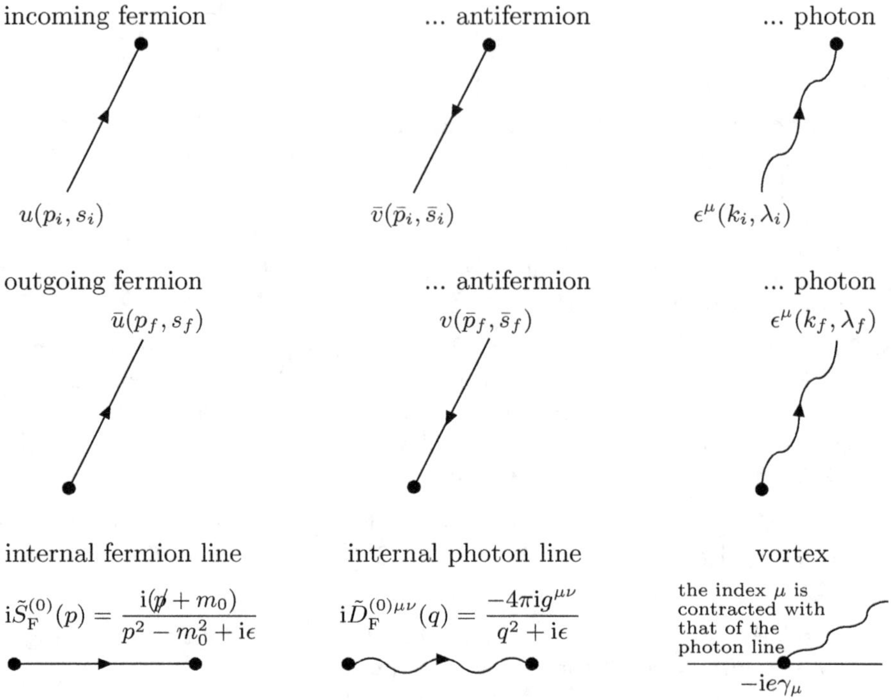

Fig. 3.27. Complete set of Feynman diagram elements and characteristic factors in momentum space (see Figures 3.14 and 3.23).

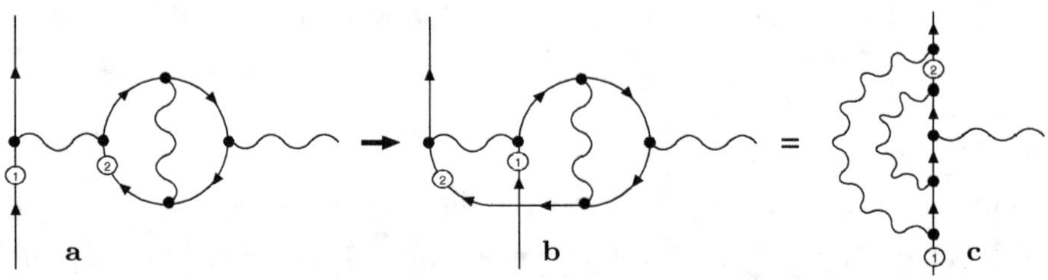

Fig. 3.28. Fermion loop (**a**), exchange of two fermion lines (**b**), and topologically equivalent deformation (**c**).

Summary

- The treatment of relativistic spin-1/2 scattering processes essentially contains the following steps:
  - Constructing the scattering amplitude S_{fi} or M_{fi} up to the desired order of the coupling constant e.

▷

- Taking the absolute square $|S_{fi}|^2$ or $|M_{fi}|^2$ and, if required, averaging over the initial polarizations and/or summing over the final polarizations.
- Inserting $|S_{fi}|^2$ or $|M_{fi}|^2$ into the formula for the differential cross section while taking into account the phase factors of all scattering products involved.

- The first and third steps can be formalized and simplified in a generally valid way with the help of the **Feynman rules**. These rules contain a **tree level** and a **loop level**. The latter is of purely quantum field theoretical nature and goes beyond relativistic quantum mechanics "in the narrow sense".
- The second step is best carried out by relating $|M_{fi}|^2$ back to a double spin sum (if required by inserting appropriate projection operators) and subsequently evaluating this expression using the trace rules from Theorem 3.6.
- Scattering processes with initial and final fermion products are the **Coulomb**, **electron-proton**, **electron-electron**, and **electron-positron scattering**. Exclusively virtual photons occur here as mediators of the electromagnetic interaction.
- Furthermore, there also exist scattering processes with initial and final photonic states such as the **Compton scattering**, **electron-positron annihilation**, and **electron-positron creation**.
- Electron-electron and electron-positron scatterings on the one side and Compton scattering, electron-positron annihilation, and electron-positron creation on the other side are interconnected via the principle of **crossing symmetry**.

Exercises

37. Kinematic constellations at the Compton scattering. Show that only one of the four $\delta()\delta()$-combinations in (3.116) yields a contribution to the considered scattering process.

Solution. Resolving (3.116) yields the following combinations and momentum balances:

$$\begin{aligned}
\mathrm{A} &: \delta(p_f - p + k_f)\delta(p_i - p + k_i) \Longrightarrow p_i + k_i = p_f + k_f \\
\mathrm{B} &: \delta(p_f - p - k_f)\delta(p_i - p - k_i) \Longrightarrow p_i - k_i = p_f - k_f \\
\mathrm{C} &: \delta(p_f - p - k_f)\delta(p_i - p + k_i) \Longrightarrow p_i + k_i = p_f - k_f \\
\mathrm{D} &: \delta(p_f - p + k_f)\delta(p_i - p - k_i) \Longrightarrow p_i - k_i = p_f + k_f \ .
\end{aligned}$$

To A. This balance reflects the correct constellation of the Compton scattering as presupposed: an incoming photon with momentum $+k_i$ and an outgoing photon with momentum $+k_f$.

To B. This case describes the Compton scattering, too, but with reversed photon momenta: an outgoing photon with momentum $-k_i$ and an incoming photon with momentum $-k_f$.

To C and D. Physically these two balances correspond respectively to the absorption and emission of two photons via a free electron, which is kinematically forbidden. In case C this can be seen by considering the equation

$$p_f - p_i = k_i + k_f \ . \tag{3.127}$$

Taking into account the mass shell conditions $p_{i,f}^2 = m_0^2$, $k_{i,f}^2 = 0$, its square is

$$\omega_i\omega_f + E_iE_f - m_0^2 = \boldsymbol{p}_i\boldsymbol{p}_f + \boldsymbol{k}_i\boldsymbol{k}_f \ .$$

With the inequality

$$\begin{aligned}\boldsymbol{p}_i\boldsymbol{p}_f + \boldsymbol{k}_i\boldsymbol{k}_f \leq |\boldsymbol{p}_i\boldsymbol{p}_f + \boldsymbol{k}_i\boldsymbol{k}_f| &\leq |\boldsymbol{p}_i||\boldsymbol{p}_f| + |\boldsymbol{k}_i||\boldsymbol{k}_f| \\ &= \sqrt{E_i^2 - m_0^2}\sqrt{E_f^2 - m_0^2} + \omega_i\omega_f \ ,\end{aligned}$$

it follows that ($E_{i,f} \geq m_0$)

$$E_iE_f - m_0^2 \leq \sqrt{E_i^2 - m_0^2}\sqrt{E_f^2 - m_0^2} \Longrightarrow (E_i - E_f)^2 \leq 0 \Longrightarrow E_i = E_f$$

and, due to (3.127),

$$\omega_i = -\omega_f \Longleftrightarrow |\boldsymbol{k}_i| = -|\boldsymbol{k}_f| \Longrightarrow \omega_i = \omega_f = 0 \ .$$

Thus, balance C only possesses the trivial solution of a noninteracting electron. The impossibility of balance D is shown similarly.

38. Electron-positron annihilation in the center of mass system. Use the results in Subsection 3.3.8 to calculate the totally unpolarized differential and total cross section of the electron-positron annihilation to leading order in the center of mass system.
Tip: due to the gauge invariance of the electromagnetic field, the completeness relations

$$\sum_{\lambda=1}^{2} \epsilon_\mu(k,\lambda)\epsilon_\nu(k,\lambda) = -g_{\mu\nu} \tag{3.128}$$

for the polarization vectors ϵ (perpendicular to k) can be used.

Solution. Our starting point is the differential cross section

$$\mathrm{d}\sigma = \frac{m_0^2}{\sqrt{(p_i\cdot\bar{p}_i)^2 - m_0^4}}|M_{fi}|^2(2\pi)^4\delta(k_f + k'_f - p_i - \bar{p}_i)\frac{2\pi\mathrm{d}^3k_f}{(2\pi)^3\omega_f}\frac{2\pi\mathrm{d}^3k'_f}{(2\pi)^3\omega'_f},$$

with M_{fi} from (3.124), which has to be evaluated in the center of mass system (see Figure 3.29). Taking into account

$$\frac{m_0^2}{\sqrt{(p_i\cdot\bar{p}_i)^2 - m_0^4}} = \frac{m_0^2}{\sqrt{(E_i^2 + \boldsymbol{p}_i^2)^2 - m_0^4}} = \frac{m_0^2}{2E_i|\boldsymbol{p}_i|}$$

$$\mathrm{d}^3k_f = \omega_f^2\mathrm{d}\omega_f\mathrm{d}\Omega\ ,\quad \frac{\mathrm{d}^3k'_f}{\omega'_f} = 2\int \mathrm{d}^4k'_f\delta(k'^2_f)\Theta(k'^0_f)\ ,$$

it follows that

$$\begin{aligned}\left(\overline{\frac{\mathrm{d}\sigma}{\mathrm{d}\Omega}}\right)_{\mathrm{cm}} &= \frac{m_0^2}{E_i|\boldsymbol{p}_i|}\int \mathrm{d}\omega_f\omega_f\int \mathrm{d}^4k'_f\overline{|M_{fi}|^2}\delta(k_f + k'_f - p_i - \bar{p}_i)\\ &\quad\times\delta(k'^2_f)\theta(k'^0_f)\\ &= \frac{m_0^2}{E_i|\boldsymbol{p}_i|}\int \mathrm{d}\omega_f\omega_f\overline{|M_{fi}|^2_{k'_f=\bar{p}_i+p_i-k_f}}\\ &\quad\times\delta[(\bar{p}_i + p_i - k_f)^2]\Theta(\bar{p}_i^0 + p_i^0 - k_f^0)\\ &= \frac{m_0^2}{E_i|\boldsymbol{p}_i|}\int \mathrm{d}\omega_f\omega_f\overline{|M_{fi}|^2_{k'_f=\bar{p}_i+p_i-k_f}}\\ &\quad\times\delta[4E_i(E_i - \omega_f)]\Theta(2E_i - \omega_f)\\ &= \frac{m_0^2}{E_i|\boldsymbol{p}_i|}\int_0^{2E_i} \mathrm{d}\omega_f\omega_f\overline{|M_{fi}|^2_{k'_f=\bar{p}_i+p_i-k_f}}\frac{\delta(\omega_f - E_i)}{4E_i}\\ &= \frac{m_0^2}{4E_i|\boldsymbol{p}_i|}\overline{|M_{fi}|^2_{\mathrm{cm}}}\ . \end{aligned}\qquad(3.129)$$

To calculate the amplitude square, we can resort to (3.126),

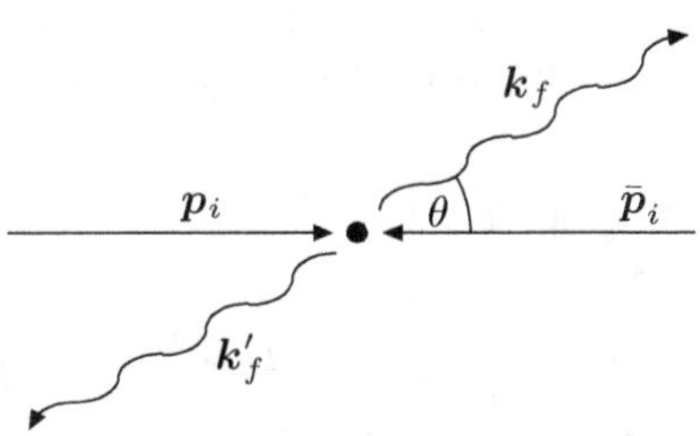

Fig. 3.29. Kinematic situation of the electron-positron annihilation in the center of mass system. Because of energy and momentum conservation, $E_i = \bar{E}_i = \omega_f = \omega'_f$ and $\boldsymbol{p}_i = -\bar{\boldsymbol{p}}_i\ ,\ \boldsymbol{k}_f = -\boldsymbol{k}'_f$.

$$\overline{|M_{fi}|^2_{\rm co}}(\lambda_f,\lambda'_f) = \frac{e^4}{8m_0^2}\left\{\frac{p_i\cdot k'_f}{p_i\cdot k_f}+\frac{p_i\cdot k_f}{p_i\cdot k'_f}\right.$$
$$\left. +2-4[\epsilon(k_f\lambda_f)\cdot\epsilon(k'_f,\lambda'_f)]^2\right\} , \qquad (3.130)$$

where four-momentum conservation as well as the average over the fermions' initial polarizations are already taken into account. From that we obtain the totally unpolarized amplitude square (sum over the photons' final polarizations)

$$\overline{|M_{fi}|^2_{\rm co}} = \frac{e^4}{2m_0^2}\left\{\frac{p_i\cdot k'_f}{p_i\cdot k_f}+\frac{p_i\cdot k_f}{p_i\cdot k'_f}+2-\sum_{\lambda_f,\lambda'_f}[\epsilon(k_f\lambda_f)\cdot\epsilon(k'_f,\lambda'_f)]^2\right\} .$$

Note that (3.130) was derived under the assumption that the polarization vectors ϵ_f, ϵ'_f are perpendicular to p_i (see Compton scattering, Subsection 3.3.7). In the laboratory system ($\boldsymbol{p}_i = \mathbf{0}$) this was easily achieved by choosing the purely space-like photon polarizations $\epsilon = (0, \boldsymbol{\epsilon})$. By contrast, in the more general case in hand it is advantageous to incorporate the conditions $\epsilon_f\cdot p_i = \epsilon'_f\cdot p_i = 0$ by using the approach or gauge transformation

$$\epsilon_f \to \tilde{\epsilon}_f = \epsilon_f - \frac{\epsilon_f\cdot p_i}{k_f\cdot p_i}k_f \ , \ \epsilon'_f \to \tilde{\epsilon}'_f = \epsilon'_f - \frac{\epsilon'_f\cdot p_i}{k'_f\cdot p_i}k'_f \qquad (3.131)$$

that does not affect the other orthogonality and transversality conditions (3.114):

$$\epsilon\cdot k = 0 \ , \ \epsilon\cdot\epsilon = -1 \Longrightarrow \tilde{\epsilon}\cdot k = 0 \ , \ \tilde{\epsilon}\cdot\tilde{\epsilon} = -1 \ .$$

Thus, the polarization sum in $\overline{|M_{fi}|^2_{\rm co}}$ can be rewritten as

$$\sum_{\lambda_f,\lambda'_f}[\tilde{\epsilon}_f\cdot\tilde{\epsilon}'_f]^2 = \sum_{\lambda_f,\lambda'_f}\tilde{\epsilon}^\mu_f\tilde{\epsilon}'_{f,\mu}\tilde{\epsilon}^\nu_f\tilde{\epsilon}'_{f,\nu} = A^{\mu\nu}B_{\mu\nu} \ ,$$

with

$$A^{\mu\nu} = \sum_{\lambda_f}\tilde{\epsilon}^\mu_f\tilde{\epsilon}^\nu_f \ , \ B_{\mu\nu} = \sum_{\lambda'_f}\tilde{\epsilon}_{f,\mu}\tilde{\epsilon}_{f,\nu} \ .$$

If we now use the completeness relations (3.128) both tensors can be simplified to

$$\begin{aligned} A^{\mu\nu} &= \sum_{\lambda_f}\left(\epsilon_f - \frac{\epsilon_f\cdot p_i}{k_f\cdot p_i}k_f\right)^\mu\left(\epsilon_f - \frac{\epsilon_f\cdot p_i}{k_f\cdot p_i}k_f\right)^\nu \\ &= \sum_{\lambda_f}\left(\epsilon^\mu_f\epsilon^\nu_f - \epsilon^\nu_f\epsilon^\alpha_f\frac{p_{i,\alpha}k^\mu_f}{k_f\cdot p_i} - \epsilon^\mu_f\epsilon^\alpha_f\frac{p_{i,\alpha}k^\nu_f}{k_f\cdot p_i} + \epsilon^\alpha_f\epsilon^\beta_f\frac{p_{i,\alpha}p_{i,\beta}k^\mu_f k^\nu_f}{(k_f\cdot p_i)^2}\right) \\ &= -g^{\mu\nu} + \frac{p^\nu_i k^\mu_f + p^\mu_i k^\nu_f}{k_f\cdot p_i} - \frac{m_0^2 k^\mu_f k^\nu_f}{(k_f\cdot p_i)^2} \end{aligned}$$

$$B_{\mu\nu} = \sum_{\lambda_f} \epsilon_{f,\mu}\epsilon_{f,\nu} = \sum_{\lambda'_f} \left(\epsilon'_f - \frac{\epsilon'_f \cdot p_i}{k'_f \cdot p_i} k'_f\right)_\mu \left(\epsilon'_f - \frac{\epsilon'_f \cdot p_i}{k'_f \cdot p_i} k'_f\right)_\nu$$
$$= -g_{\mu\nu} + \frac{p_{i,\nu}k'_{f,\mu} + p_{i,\mu}k'_{f,\nu}}{k'_f \cdot p_i} - \frac{m_0^2 k'_{f,\mu}k'_{f,\nu}}{(k_f \cdot p_i)^2} \ .$$

Contracting $A^{\mu\nu}$ and $B_{\mu\nu}$ and taking into account the mass shell conditions $k_f^2 = k_f'^2 = 0$ finally yields

$$\sum_{\lambda_f,\lambda'_f} [\tilde{\epsilon}_f \cdot \tilde{\epsilon}'_f]^2 = 2 - \frac{2m_0^2 k_f \cdot k'_f}{(k_f \cdot p_i)(k'_f \cdot p_i)} + \frac{m_0^4 (k_f \cdot k'_f)^2}{(k_f \cdot p_i)^2 (k'_f \cdot p_i)^2} \ , \tag{3.132}$$

and the totally unpolarized amplitude square follows as

$$\overline{|M_{fi}|^2_{\mathrm{co}}} = \frac{e^4}{2m_0^2}\left[\frac{p_i \cdot k'_f}{p_i \cdot k_f} + \frac{p_i \cdot k_f}{p_i \cdot k'_f} + \frac{2m_0^2 k_f \cdot k'_f}{(k_f \cdot p_i)(k'_f \cdot p_i)} - \frac{m_0^4 (k_f \cdot k'_f)^2}{(k_f \cdot p_i)^2 (k'_f \cdot p_i)^2}\right] \ . \tag{3.133}$$

In order to get $\overline{|M_{fi}|^2_{\mathrm{cm}}}$, which has to be inserted into (3.129), we still have to incorporate the center of mass conditions. Taking into account $|\boldsymbol{p}_i| = v_i E_i$, we have

$$\begin{aligned} p_i \cdot k_f &= E_i^2 - |\boldsymbol{p}_i||\boldsymbol{k}_f| \cos\theta = E_i^2 (1 - v_i \cos\theta) \\ p_i \cdot k'_f &= E_i^2 + |\boldsymbol{p}_i||\boldsymbol{k}_f| \cos\theta = E_i^2 (1 + v_i \cos\theta) \\ k_f \cdot k'_f &= E_i^2 + |\boldsymbol{k}_f||\boldsymbol{k}'_f| = 2E_i^2 \ . \end{aligned}$$

Thus, the final result for the differential cross section in the center of mass system is

$$\begin{aligned} \left(\overline{\frac{\mathrm{d}\sigma}{\mathrm{d}\Omega}}\right)_{\mathrm{cm}} &= \frac{e^4}{8E_i^2 v_i}\left[\frac{1 + v_i\cos\theta}{1 - v_i\cos\theta} + \frac{1 - v_i\cos\theta}{1 + v_i\cos\theta} + \frac{4m_0^2}{E_i^2(1 - v_i^2\cos^2\theta)} - \frac{4m_0^4}{E_i^4(1 - v_i^2\cos^2\theta)^2}\right] \\ &= \frac{e^4}{4E_i^2 v_i}\frac{1 + 2v_i^2(1 - v_i^2) - 2v_i^2(1 - v_i^2)\cos^2\theta - v_i^4\cos^4\theta}{(1 - v_i^2\cos^2\theta)^2} \ , \end{aligned}$$

where $m_0^2/E_i^2 = 1 - v_i^2$ has been used in the last step. The calculation of the total cross section, particularly the integration over $\mathrm{d}\cos\theta$, can be carried out easily:

$$\bar{\sigma} = \frac{1}{2}\int \left(\overline{\frac{\mathrm{d}\sigma}{\mathrm{d}\Omega}}\right)_{\mathrm{cm}} \mathrm{d}\Omega = \frac{\pi e^4 (1 - v_i^2)}{4m_0^2 v_i^2}\left[(3 - v_i^4)\ln\frac{1 + v_i}{1 - v_i} - 2v_i(2 - v_i^2)\right] \ .$$

The factor 1/2 accounts for the indistinguishability of both photons in the final state and prevents their double counting.

39. Electron-positron creation in the center of mass system. Use the crossing symmetry between the electron-positron creation and annihilation to calculate the totally unpolarized differential and total cross section of the electron-positron creation to leading order in the center of mass system.

Solution. Figure 3.30 depicts the kinematic situation of the electron-positron creation in the center of mass system. Analogously to the preceding exercise,

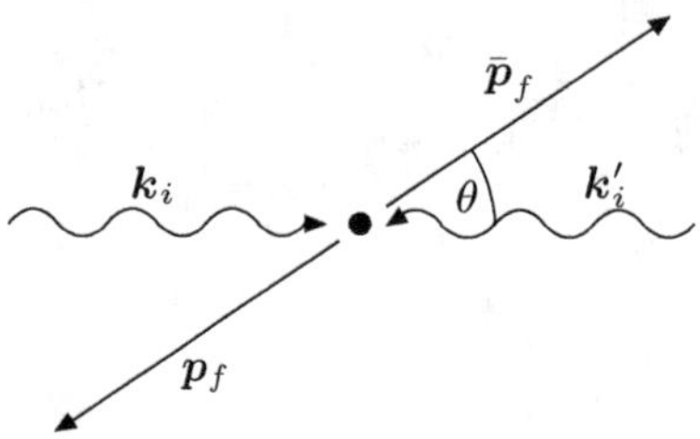

Fig. 3.30. Kinematic situation of the electron-positron creation in the center of mass system. Because of energy and momentum conservation, $\omega_i = \omega_i' = E_f = \bar{E}_f$ and $\boldsymbol{k}_i = -\boldsymbol{k}_i'$, $\boldsymbol{p}_f = -\bar{\boldsymbol{p}}_f$.

we start from the differential cross section

$$\mathrm{d}\sigma = \frac{(2\pi)^2}{\sqrt{(k_i \cdot k_i')^2}} |M_{fi}|^2 (2\pi)^4 \delta(p_f + \bar{p}_f - k_i - k_i') \frac{m_0 \mathrm{d}^3 p_f}{(2\pi)^3 E_f} \frac{m_0 \mathrm{d}^3 \bar{p}_f}{(2\pi)^3 \bar{E}_f}$$

and calculate, with the help of

$$\frac{(2\pi)^2}{\sqrt{(k_i \cdot k_i')^2}} = \frac{(2\pi)^2}{2\omega_i \omega_i'} = \frac{(2\pi)^2}{2\omega_i^2}$$

$$\frac{\mathrm{d}^3 p_f}{E_f} = 2\int \mathrm{d}^4 p_f \delta(p_f^2 - m_0^2)\Theta(p_f^0) \ , \ \mathrm{d}^3\bar{p}_f = |\bar{\boldsymbol{p}}_f| \bar{E}_f \mathrm{d}\bar{E}_f \mathrm{d}\Omega \ ,$$

as follows:

$$\begin{aligned}
\left(\overline{\frac{\mathrm{d}\sigma}{\mathrm{d}\Omega}}\right)_{\mathrm{cm}} &= \frac{m_0^2}{\omega_i^2} \int \mathrm{d}\bar{E}_f |\bar{\boldsymbol{p}}_f| \int \mathrm{d}^4 p_f \overline{|M_{fi}|^2} \delta(p_f + \bar{p}_f - k_i - k_i') \\
&\quad \times \delta(p_f^2 - m_0^2)\Theta(p_f^0) \\
&= \frac{m_0^2}{\omega_i^2} \int \mathrm{d}\bar{E}_f |\bar{\boldsymbol{p}}_f| \overline{|M_{fi}|^2}_{p_f = k_i + k_i' - \bar{p}_f} \\
&\quad \times \delta[(k_i + k_i' - \bar{p}_f)^2 - m_0^2] \Theta(k_i^0 + k_i'^0 - \bar{p}_f^0) \\
&= \frac{m_0^2}{\omega_i^2} \int \mathrm{d}\bar{E}_f |\bar{\boldsymbol{p}}_f| \overline{|M_{fi}|^2}_{p_f = k_i + k_i' - \bar{p}_f} \\
&\quad \times \delta[4\omega_i(\omega_i - \bar{E}_f)] \Theta(2\omega_i - \bar{E}_f)
\end{aligned}$$

$$= \frac{m_0^2}{\omega_i^2} \int_{m_0}^{2\omega_i} \mathrm{d}\bar{E}_f |\bar{\boldsymbol{p}}_f| \overline{|M_{fi}|^2}_{p_f = k_i + k_i' - \bar{p}_f} \frac{\delta(\bar{E}_f - \omega_i)}{4\omega_i}$$

$$= \frac{m_0^2 \sqrt{\omega_i^2 - m_0^2}}{4\omega_i^3} \overline{|M_{fi}|^2_{\mathrm{cm}}} \, .$$

As expected, this formula makes sense only if the energy of each photon is at least equal to the rest energy of the created electron or positron. Without constructing the amplitude M_{fi} explicitly we can derive its square directly from the corresponding expression (3.133) of the electron-positron annihilation by considering the relationship of the crossing symmetry in Figure 3.31. Making the appropriate replacements in (3.133) leads to

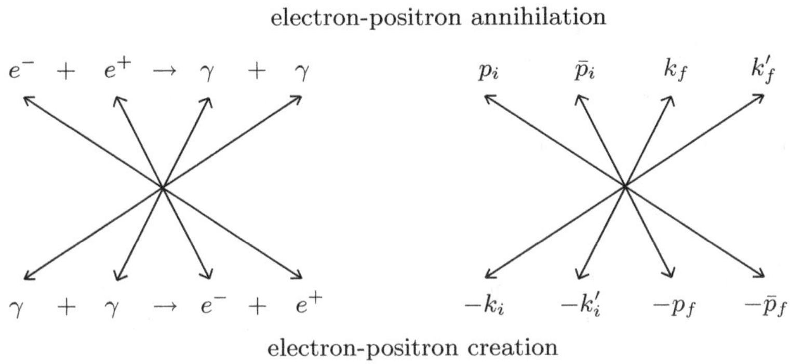

Fig. 3.31. Crossing symmetry between the electron-positron annihilation and creation.

$$\overline{|M_{fi}|^2_{\mathrm{co}}} = \frac{e^4}{2m_0^2} \left[\frac{\bar{p}_f \cdot k_i}{\bar{p}_f \cdot k_i'} + \frac{\bar{p}_f \cdot k_i'}{\bar{p}_f \cdot k_i} + \frac{2m_0^2 k_i \cdot k_i'}{(k_i \cdot \bar{p}_f)(k_i' \cdot \bar{p}_f)} - \frac{m_0^4 (k_i \cdot k_i')^2}{(k_i \cdot \bar{p}_f)^2 (k_i' \cdot \bar{p}_f)^2} \right] .$$

$\overline{|M_{fi}|^2_{\mathrm{cm}}}$ follows from this by evaluating the scalar products in the center of mass system:

$$\begin{aligned} \bar{p}_f \cdot k_i &= \omega_i^2 - |\boldsymbol{p}_f||\boldsymbol{k}_i| \cos\theta = \omega_i^2 (1 - v_f \cos\theta) \, , \; |\boldsymbol{p}_f| = v_f E_f = v_f \omega_i \\ \bar{p}_f \cdot k_i' &= \omega_i^2 + |\boldsymbol{p}_f||\boldsymbol{k}_i| \cos\theta = \omega_i^2 (1 + v_f \cos\theta) \\ \bar{k}_f \cdot k_f' &= \omega_i^2 - |\boldsymbol{k}_f||\boldsymbol{k}_f| = 2\omega_i^2 \, . \end{aligned}$$

In total, we obtain for the differential cross section of the electron-positron creation in the center of mass system:

$$\left(\overline{\frac{\mathrm{d}\sigma}{\mathrm{d}\Omega}} \right)_{\mathrm{cm}} = \frac{e^4 v_f}{8\omega_i^2} \left[\frac{1 + v_f \cos\theta}{1 - v_f \cos\theta} + \frac{1 - v_f \cos\theta}{1 + v_f \cos\theta} \right.$$

$$+\frac{4m_0^2}{\omega_i^2(1-v_f^2\cos^2\theta)}-\frac{4m_0^4}{\omega_i^4(1-v_f^2\cos^2\theta)^2}\Bigg]$$

$$=\frac{e^4 v_f}{4\omega_i}\frac{1+2v_f^2(1-v_f^2)-2v_f^2(1-v_f^2)\cos^2\theta-v_f^4\cos^4\theta}{(1-v_f^2\cos^2\theta)^2}\;.$$

For the calculation of the total cross section we can resort to the corresponding integration of the electron-positron annihilation to find

$$\bar{\sigma}=\int\left(\frac{\mathrm{d}\sigma}{\mathrm{d}\Omega}\right)_{\mathrm{cm}}\mathrm{d}\Omega=\frac{\pi e^4(1-v_f^2)}{2m_0^2}\left[(3-v_f^4)\ln\frac{1+v_f}{1-v_f}-2v_f(2-v_f^2)\right]\;.$$

Compared to the annihilation case, a factor of 1/2 is missing here, since now the scattering products can be distinguished.

40. Furry theorem. Prove the Furry theorem which states the following: two identical Feynman diagrams, each with one n-vortex fermion loop differing from the other just by the direction of the circulating fermion,

- yield the same contribution if n is even,
- cancel each other if n is odd.

What about the cases where $n=1$ and $n=2$?

Solution. Figure 3.32 shows both n-vortex fermion loops with opposite directions of circulation as parts of otherwise identical Feynman diagrams. The

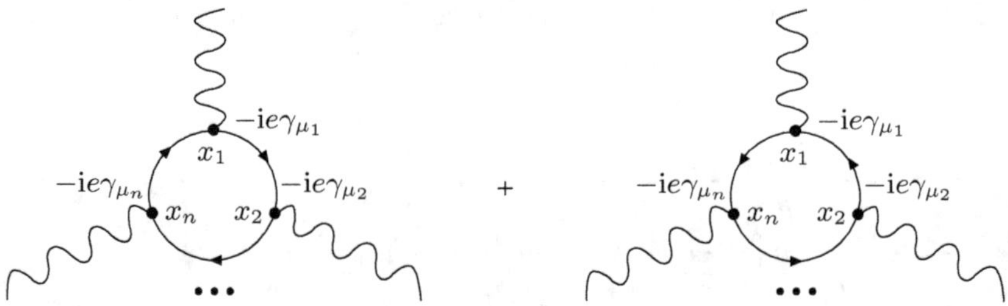

Fig. 3.32. n-vortex fermion loops with opposite directions of the virtual fermion.

corresponding amplitude can be written in the form

$$M_{fi}=[\ldots]\left[M^{(a)}+M^{(b)}\right][\ldots]\;,$$

where $M^{(a)}$ and $M^{(b)}$ denote the contributions of both parts and are given by

$$M^{(a)}=\mathrm{tr}\Big[(-\mathrm{i}e)\gamma_{\mu_n}\mathrm{i}S_{\mathrm{F}}^{(0)}(x_n-x_{n-1})(-\mathrm{i}e)\gamma_{\mu_{n-1}}\mathrm{i}S_{\mathrm{F}}^{(0)}(x_{n-1}-x_{n-2})\cdots$$
$$\times(-\mathrm{i}e)\gamma_{\mu_1}\mathrm{i}S_{\mathrm{F}}^{(0)}(x_1-x_n)\Big]$$

$$
\begin{aligned}
&= e^n \mathrm{tr}\left[\gamma_{\mu_n} S_{\mathrm{F}}^{(0)}(x_n - x_{n-1})\gamma_{\mu_{n-1}} S_{\mathrm{F}}^{(0)}(x_{n-1} - x_{n-2}) \cdots \right.\\
&\quad \left. \times\, \gamma_{\mu_1} S_{\mathrm{F}}^{(0)}(x_1 - x_n)\right]\\
M^{(b)} &= \mathrm{tr}\left[(-\mathrm{i}e)\gamma_{\mu_1} \mathrm{i}S_{\mathrm{F}}^{(0)}(x_1 - x_2)(-\mathrm{i}e)\gamma_{\mu_2}\mathrm{i}S_{\mathrm{F}}^{(0)}(x_2 - x_3) \cdots \right.\\
&\quad \left. \times\, (-\mathrm{i}e)\gamma_{\mu_n}\mathrm{i}S_{\mathrm{F}}^{(0)}(x_n - x_1)\right]\\
&= e^n \mathrm{tr}\left[\gamma_{\mu_1} S_{\mathrm{F}}^{(0)}(x_1 - x_2)\gamma_{\mu_2} S_{\mathrm{F}}^{(0)}(x_2 - x_3) \cdots \gamma_{\mu_n} S_{\mathrm{F}}^{(0)}(x_n - x_1)\right] .
\end{aligned}
$$

The trace results from the cyclic multiplication of the vortex factors and fermion propagators along the loop. Taking into account the charge conjugation transformation C from Subsection 2.1.6 and using the relations [see (2.40)]

$$
\begin{aligned}
C^{-1}\gamma_\mu C &= -\gamma_\mu^* = -\gamma^0\gamma_\mu^T\gamma^0\\
C^{-1}S_{\mathrm{F}}^{(0)}(x)C &= \int \frac{\mathrm{d}^4p}{(2\pi)^4}\mathrm{e}^{-\mathrm{i}p\cdot x}\frac{p^\mu C^{-1}\gamma_\mu C + m_0}{p^2 - m_0^2 + \mathrm{i}\epsilon}\\
&= \int \frac{\mathrm{d}^4p}{(2\pi)^4}\mathrm{e}^{-\mathrm{i}p\cdot x}\frac{-p^\mu \gamma^0\gamma_\mu^T\gamma^0 + m_0}{p^2 - m_0^2 + \mathrm{i}\epsilon}\\
&= \gamma^0 S_{\mathrm{F}}^{(0)T}(-x)\gamma^0 ,
\end{aligned}
$$

$M^{(b)}$ can be transformed into

$$
\begin{aligned}
M^{(b)} &= e^n \mathrm{tr}\left[CC^{-1}\gamma_{\mu_1} CC^{-1}S_{\mathrm{F}}^{(0)}(x_1 - x_2)\right.\\
&\quad \times CC^{-1}\gamma_{\mu_2}CC^{-1}S_{\mathrm{F}}^{(0)}(x_2 - x_3)CC^{-1}\cdots\\
&\quad \left.\times\, CC^{-1}\gamma_{\mu_n}CC^{-1}S_{\mathrm{F}}^{(0)}(x_n - x_1)\right]\\
&= e^n \mathrm{tr}\left[C^{-1}\gamma_{\mu_1} CC^{-1}S_{\mathrm{F}}^{(0)}(x_1 - x_2)\right.\\
&\quad \times CC^{-1}\gamma_{\mu_2}CC^{-1}S_{\mathrm{F}}^{(0)}(x_2 - x_3)CC^{-1}\cdots\\
&\quad \left.\times\, CC^{-1}\gamma_{\mu_n}CC^{-1}S_{\mathrm{F}}^{(0)}(x_n - x_1)C\right]\\
&= e^n(-1)^n \mathrm{tr}\left[\gamma_{\mu_1}^T S_{\mathrm{F}}^{(0)T}(x_2 - x_1)\gamma_{\mu_2}^T S_{\mathrm{F}}^{(0)T}(x_3 - x_2)\cdots\right.\\
&\quad \left.\times\, \gamma_{\mu_n}^T S_{\mathrm{F}}^{(0)T}(x_1 - x_n)\right]\\
&= e^n(-1)^n \mathrm{tr}\left[S_{\mathrm{F}}^{(0)}(x_1 - x_n)\gamma_{\mu_n} S_{\mathrm{F}}^{(0)}(x_n - x_{n-1})\gamma_{\mu_{n-1}}\cdots\right.\\
&\quad \left.\times\, S_{\mathrm{F}}^{(0)}(x_3 - x_2)\gamma_{\mu_2} S_{\mathrm{F}}^{(0)}(x_2 - x_1)\gamma_{\mu_1}\right]\\
&= (-1)^n M^{(a)} ,
\end{aligned}
$$

from which the statement follows. The cyclic permutability of the traces is used in the second and the last steps.

$n = 1$ and $n = 2$ are the only cases where the Furry theorem cannot be applied. Here the two corresponding subdiagrams with opposite directions of circulation are topologically equivalent so that, in fact, only one subdiagram remains. Because of its specific shape, the $n = 1$-subdiagram is also called *tadpole diagram.* Due to four-momentum conservation, it can only be connected to the rest by a virtual photon (with $k = 0$), thus contributing to the self-energy of the electron (or positron). However, contrary to the self-energy contribution discussed in the following section, this contribution is not observable because it can be totally absorbed into a (divergent) *renormalization constant.* Therefore, it is justified to omit Feynman diagrams with 1-vortex fermion loops right from the beginning.

3.4 Higher Order Corrections

In the preceding section we dealt with concrete examples of relativistic spin-1/2 scattering processes to the lowest orders of the scattering theory and derived the complete set of Feynman rules for the construction of scattering amplitudes and cross sections. Principally, this system of rules is valid to all orders. Apart from the tree diagrams, in higher orders it also contains the quantum field theoretically motivated loop diagrams. Now we turn to the Feynman rules in higher orders and discuss the new problems arising connected with them. As we will see, these problems result from the fact that some particular corrections of higher orders, namely those containing loop diagrams, lead to infinities that put the reasonableness of the whole formalism into question. Fortunately, this divergence problem can be removed with the program of *renormalization.* The decisive argument in this context is that parameters like the electric charge e or mass m_0 occurring in the Dirac equation are to be regarded rather as bookkeeping quantities that do not possess any physical significance since they do not account for certain experimental effects. Therefore, those quantities themselves contain divergences that have to be compensated for by the above mentioned infinities. In other words, if, instead of the bare quantities e and m_0, the physical ones are used any divergences are removed from the scattering formalism.

As a prominent example, we consider the $\mathcal{O}\left(e^4\right)$-corrections to the electron-positron scattering (compare to Subsection 3.3.6) whose Feynman diagrams – 18 in total – contain all the topological and kinematically possible constellations with four external fermion lines and four vortices, as shown in Figure 3.33.

- Diagrams **a** and **b** represent the two-photon exchange of the direct and exchange scatterings similar to the $\mathcal{O}\left(e^4\right)$-corrections of the electron-proton scattering in Figures 3.12 and 3.13. Diagrams **c** and **d** are annihilation diagrams. Writing down the amplitudes belonging to **a** to **d** and counting the powers of the integration momentum, one realizes that they can be

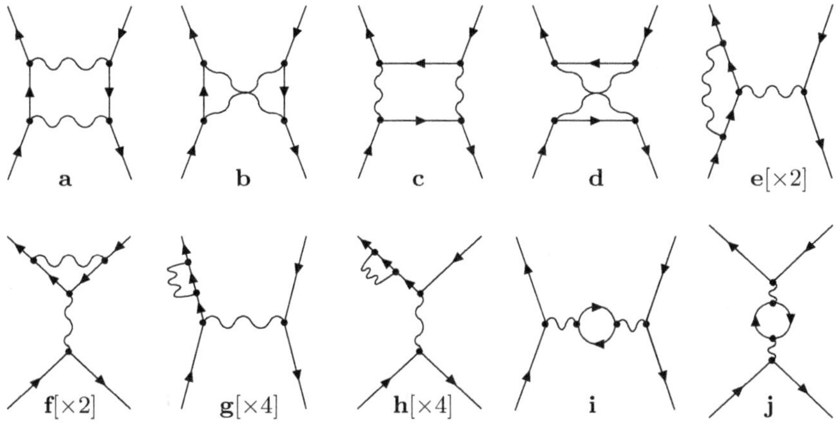

Fig. 3.33. All Feynman diagrams of the electron-positron scattering amplitude to order $\mathcal{O}\left(e^4\right)$. **a** to **d** are tree diagrams and **e** to **j** are loop diagrams.

effortlessly calculated with our existing methods. Therefore, they are not considered any more.

- In diagram **e** the electron (positron) emits a photon before the exchange scattering and absorbs it afterward. Diagram **f** corresponds to a process where the pair created by the annihilation photon scatters once again before reaching its final state. Both diagrams belong to the class of *vortex corrections* and yield divergent contributions.
- Contrary to **e** and **f**, in diagrams **g** and **h** the electron (positron) emits a photon and absorbs it immediately afterward without interacting otherwise. These divergent diagrams are called *self-energy corrections* as they describe the interaction of the electron (positron) with its own radiation field.
- In diagrams **i** and **j** a virtual electron-positron pair is created by the annihilation photon and destroyed immediately afterward. In view of the fluctuating dipole moment of the virtual pair that may by polarized by the electric field, those diagrams are termed *vacuum polarization*. They also lead to divergent contributions.

Obviously, in Feynman diagrams only loops cause divergences in the corresponding scattering amplitudes. They are called *radiation corrections*. In Figure 3.35 these components are depicted, once again, together with another two that occur, for example, in the $\mathcal{O}\left(e^4\right)$-electron-positron annihilation diagrams (see Figure 3.34, compare to Subsection 3.3.8). All of them imply four-dimensional momentum integrals diverging differently at $k \to \infty$ (*ultraviolet divergence*). While the volume elements behave as k^4, the integrands of the three processes from Figure 3.35 are proportional to k^{-2}, k^{-3}, and k^{-4} respectively. We therefore expect the strongest divergence for the vacuum

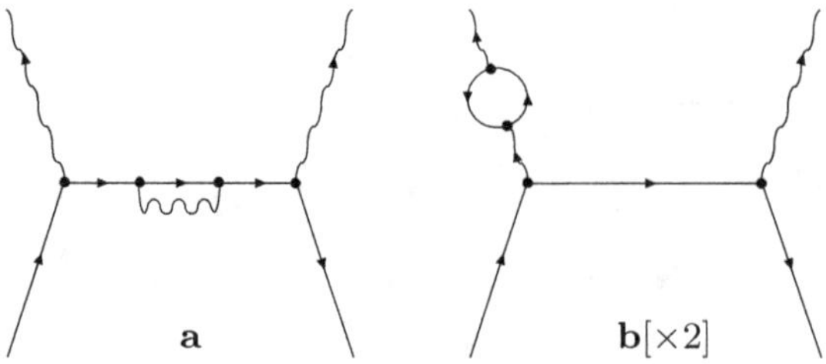

Fig. 3.34. Two Feynman diagrams of the scattering amplitude for electron-positron annihilation to order $\mathcal{O}\left(e^4\right)$: internal self-energy (**a**) and external vacuum polarization (**b**).

polarization, namely a quadratic one, a linear one for the self-energy, and a logarithmic one for the vortex correction.

In the following three subsections we discuss the radiation corrections in more detail and show how they can be absorbed into the bare parameters e and m_0 in a physically sensible way by means of the renormalization program. The fourth subsection deals with some physical consequences connected with the radiation corrections.

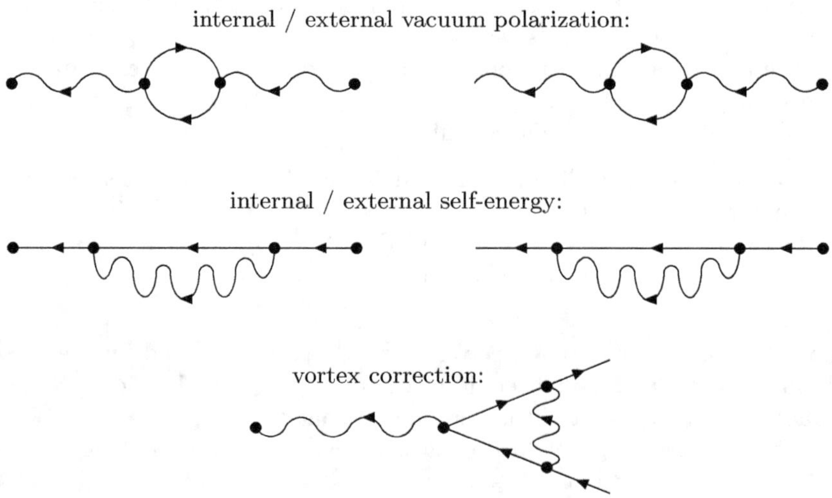

Fig. 3.35. Vacuum polarization, self-energy, and vortex correction as elements causing divergences in $\mathcal{O}\left(e^4\right)$-Feynman diagrams.

Note. In Figure 3.33, we have defalcated some $\mathcal{O}\left(e^4\right)$-Feynman diagrams, namely those consisting of unconnected parts. One example of this kind is shown in Figure 3.36. It represents the electron-positron scattering to lowest order whereas, independently from it, a virtual electron-positron pair is created out of the vacuum and immediately destroyed thereafter. In the

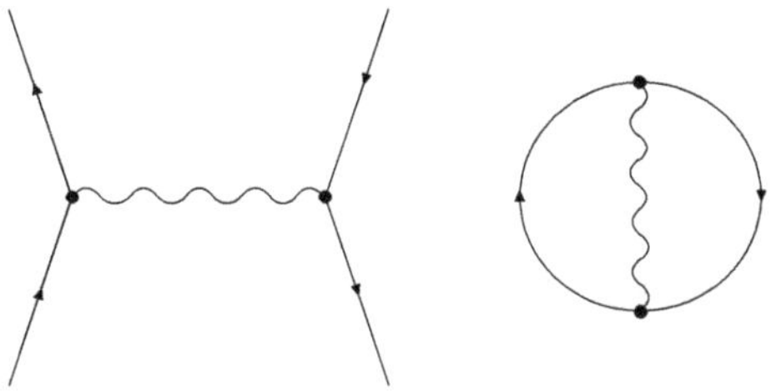

Fig. 3.36. Example of an unconnected $\mathcal{O}\left(e^4\right)$-Feynman diagram of electron-positron scattering.

scattering amplitudes the impact of those *vacuum fluctuations* appears as a multiplicative factor in the part of the connected diagrams with external lines. However, since we are interested in scattering amplitudes relative to the omnipresent vacuum fluctuations, we can simply divide this factor out or disregard all unconnected diagrams right from the beginning.

3.4.1 Vacuum Polarization

We start with the internal vacuum polarization and consider the modification of the free photon propagator[22]

$$D_{\mathrm{F}}^{(0)\mu\nu}(q) = \frac{-4\pi g^{\mu\nu}}{q^2 + \mathrm{i}\epsilon}$$

by a $\mathcal{O}\left(e^2\right)$-fermion loop. According to Figure 3.37, this leads to the replacement

$$D_{\mathrm{F}}^{(0)\mu\nu}(q) \longrightarrow D_{\mathrm{F}}^{\mu\nu}(q) = D_{\mathrm{F}}^{(0)\mu\nu}(q) + D_{\mathrm{F}}^{(0)\mu\alpha}(q) P_{\alpha\beta}(q) D_{\mathrm{F}}^{(0)\beta\nu}(q) \;, \quad (3.134)$$

with the *polarization tensor*

$$P_{\mu\nu}(q) = -\mathrm{i}e^2 \int \frac{\mathrm{d}^4k}{(2\pi)^4} \mathrm{tr}\left[\gamma_\mu \frac{\not{k} + m_0}{k^2 - m_0^2 + \mathrm{i}\epsilon} \gamma_\nu \frac{\not{k} - \not{q} + m_0}{(k-q)^2 - m_0^2 + \mathrm{i}\epsilon}\right] . \quad (3.135)$$

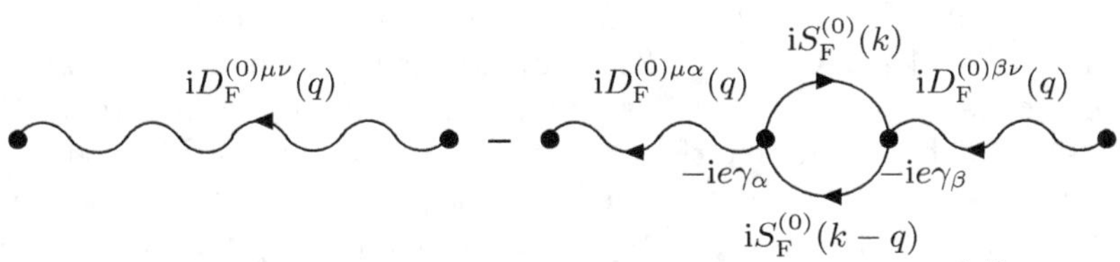

Fig. 3.37. Modification of the free photon propagator by the $\mathcal{O}\left(e^2\right)$-vacuum polarization. The relative sign results from the fermion loop (fourth rule).

---

[22] The $\sim$-symbol above the propagators in momentum space is suppressed here and in the following.

The trace results from the cyclic multiplication of the γ-matrices by the propagators along the loop. Before we calculate this quadratically ultraviolet-divergent expression, it is instructive to take a closer look at its general properties. First, $P_{\mu\nu}$ is a Lorentz tensor and can therefore be written as

$$P_{\mu\nu}(q) = Dg_{\mu\nu} + g_{\mu\nu}q^2 \Pi^{(1)}(q^2) + q_\mu q_\nu \Pi^{(2)}(q^2) \; , \tag{3.136}$$

with the constant D and the scalar functions $\Pi^{(1,2)}(q^2)$. Furthermore, gauge invariance considerations yield the constraints $P_{\mu\nu}q^\nu = q^\mu P_{\mu\nu} = 0$ so that (3.136) is reduced to

$$P_{\mu\nu}(q) = \left(q^2 g_{\mu\nu} - q_\mu q_\nu\right) \Pi(q^2) \; , \; D = 0 \; , \tag{3.137}$$

where $\Pi(q^2)$ denotes the *polarization function.* That the constant D must vanish can also be shown through another reasoning: going beyond the $\mathcal{O}\left(e^2\right)$-fermion loop and taking higher corrections into account, we obtain the expansion

$$\begin{aligned}
D_{\mathrm{F}} &= D_{\mathrm{F}}^{(0)} + D_{\mathrm{F}}^{(0)} P D_{\mathrm{F}}^{(0)} + D_{\mathrm{F}}^{(0)} P D_{\mathrm{F}}^{(0)} P D_{\mathrm{F}}^{(0)} + \ldots \\
&= D_{\mathrm{F}}^{(0)} + D_{\mathrm{F}}^{(0)} P \left(D_{\mathrm{F}}^{(0)} + D_{\mathrm{F}}^{(0)} P D_{\mathrm{F}}^{(0)} + \ldots\right) \\
&= D_{\mathrm{F}}^{(0)} + D_{\mathrm{F}}^{(0)} P D_{\mathrm{F}} \\
&= \frac{1}{\left[D_{\mathrm{F}}^{(0)}\right]^{-1} - P} \quad (\textit{Dyson equation}) \; .
\end{aligned} \tag{3.138}$$

This corresponds graphically to a series of more and more consecutively linked fermion loops within the photon line (however, more complicated constellations such as, for example, interleaved fermion loops remain ignored). Thus, using our general ansatz (3.136), the modified photon propagator in the limit $q^2 \to 0$ and up to order $\mathcal{O}\left(e^2\right)$ can be written as

$$D_{\mathrm{F}}^{\mu\nu}(q) \approx \frac{-4\pi g^{\mu\nu}}{q^2 - D + \mathrm{i}\epsilon} \; .$$

This is exactly the free propagator for a boson with mass $\sqrt{D}$, so that, again, $D = 0$ follows.

However, a direct calculation seems to contradict this result since, as already mentioned, the integral (3.135) is divergent in k and yields an infinite value particularly for $q^2 \to 0$. There exist different approaches to resolve this problem. They all enforce the convergence of the integral through *regularization*, for example by cutting off the k-integration at a particular large momentum or by introducing a damping factor approaching continuously zero for large k. In the following we use the *Pauli-Villars procedure*, where a function with the same asymptotic behavior is subtracted from the integrand in (3.135) so that the resulting integral becomes convergent. One advantage of this method is that the gauge invariance condition (3.137) can be upheld. In practice, this means that, instead of (3.135), we have to consider the *regularized polarization tensor*

$$\bar{P}_{\mu\nu}(p) = -\mathrm{i}e^2 \int \frac{\mathrm{d}^4k}{(2\pi)^4} \sum_{i=0}^{N} C_i \times \mathrm{tr}\left[\gamma_\mu \frac{\not{k} + M_i}{k^2 - M_i^2 + \mathrm{i}\epsilon} \gamma_\nu \frac{\not{k} - \not{q} + M_i}{(k-q)^2 - M_i^2 + \mathrm{i}\epsilon}\right] , \quad (3.139)$$

with $C_0 = 1$, $M_0 = m_0$ (original integrand). $C_{i>0}$, $M_{i>0}$ denote appropriately chosen *cut-off parameters* that ensure the convergence of the integral. Of course, at the end of the calculations, we have to take the limit $M_{i>0} \to \infty$. Since this cutting off procedure (as any other) is arbitrary, the physical observables must not depend on the parameters $C_{i>0}$ and $M_{i>0}$. As we see shortly, this can indeed be achieved.

The concrete calculation of (3.139) requires some considerable effort that we do not present here. It shows that the not gauge invariant terms can be removed using a particular choice of the cut-off parameters. At the end the gauge invariant expression [compare to (3.137)]

$$\bar{P}_{\mu\nu}(q) = \left(g_{\mu\nu}q^2 - q_\mu q_\nu\right) \bar{\Pi}(q^2)$$

remains with the now only logarithmically divergent *regularized polarization function*

$$\bar{\Pi}(q^2) = \frac{e^2}{2\pi^2} \int_0^1 \mathrm{d}\beta\beta(1-\beta) \underbrace{\int_0^\infty \frac{\mathrm{d}\rho}{\rho} \sum_{i=0}^{N} C_i \exp\left\{\mathrm{i}\rho\left[-M_i^2 + \beta(1-\beta)q^2\right]\right\}}_{I} .$$

If we now assume[23] $q^2 < 4m_0^2 \ll M_{i>0}^2$, the I-integral can be rewritten as

$$I = -\lim_{\eta\to 0} \sum_{i=0}^{N} C_i \ln\eta - \sum_{i=0}^{N} C_i \ln\left[M_i^2 - \beta(1-\beta)q^2\right] + \sum_{i=0}^{N} C_i \int_0^\infty \mathrm{d}t \ln t e^{-t}$$

by deforming the integration contour onto the negative imaginary axis. Obviously, the first and the third infinite terms can be eliminated with the additional parameter condition $\sum\limits_{i=0}^{N} C_i = 0$ so that

$$\begin{aligned} I &= -\left\{\ln\left[m_0^2 - \beta(1-\beta)q^2\right] + \sum_{i=1}^{N} C_i \ln\left[M_i^2 - \beta(1-\beta)q^2\right]\right\} \\ &\approx -\left\{\ln\left[m_0^2 - \beta(1-\beta)q^2\right] + \sum_{i=1}^{N} C_i \ln M_i^2\right\} \end{aligned}$$

[23] This ensures a negative imaginary exponent within the β-integration interval $[0:1]$. Beyond this, i.e. above the *threshold energy* $q^2 = (2m_0)^2$, the production of real electron-positron pairs becomes possible.

$$= -\left\{ \ln\left[1-\beta(1-\beta)\frac{q^2}{m_0^2}\right] + \sum_{i=1}^{N} C_i \ln\frac{M_i^2}{m_0^2} + \underbrace{\sum_{i=0}^{N} C_i m_0^2}_{0} \right\}$$

$$= -\left\{ \ln\left[1-\beta(1-\beta)\frac{q^2}{m_0^2}\right] - \ln\frac{\Lambda^2}{m_0^2} \right\} ,$$

where the abbreviation

$$\sum_{i=0}^{N} C_i \ln\frac{M_i^2}{m_0^2} = -\ln\frac{\Lambda^2}{m_0^2} \qquad (\Lambda = \textit{cut-off momentum})$$

has been introduced in the last step. Thus, the regularized polarization tensor finally becomes

$$\left.\begin{aligned}
\bar{P}_{\mu\nu}(q) &= \left(g_{\mu\nu}q^2 - q_\mu q_\nu\right)\bar{\Pi}(q^2) \\
\bar{\Pi}(q^2) &= \frac{1}{4\pi}\left[\frac{e^2}{3\pi}\ln\frac{\Lambda^2}{m_0^2} + \Pi^{(\mathrm{R})}(q^2)\right] \\
\Pi^{(\mathrm{R})}(q^2) &= -\frac{2e^2}{\pi}\int_0^1 \mathrm{d}\beta\,\beta(1-\beta)\ln\left[1-\beta(1-\beta)\frac{q^2}{m_0^2}\right] \\
&\overset{q^2/m_0^2\ll 1}{=} \frac{e^2}{\pi}\frac{q^2}{m_0^2}\left(\frac{1}{15}+\frac{1}{140}\frac{q^2}{m_0^2}+\ldots\right) .
\end{aligned}\right\} \qquad (3.140)$$

As one can see, the effect of the regularization is that now the polarization tensor

- fulfills the gauge invariance condition (3.137),
- shows the correct asymptotic behavior for $q^2 \to 0$,
- possesses no quadratic divergence any more but only a portion that is independent from q and only logarithmically divergent with the cut-off momentum Λ,
- and possesses a well-defined, q-dependent, and finite term $\Pi^{(\mathrm{R})}(q^2)$.

However, even this result seems unsatisfactory because of the presence of a divergent Λ-dependence. This can be understood by turning back to the modified photon propagator and considering its influence on the electron-positron scattering. Taking into account (3.134) and (3.140) as well as Figure 3.38, the part of the direct scattering amplitude up to order $\mathcal{O}\left(e^4\right)$ originating from the vacuum polarization is

$$\begin{aligned}
M_{fi}(\mathrm{dir}) &= \bar{u}(p_f)(-\mathrm{i}e)\gamma_\mu u(p_i)\left[\mathrm{i}D_{\mathrm{F}}^{\mu\nu}(q)\right]\bar{v}(\bar{p}_i)(-\mathrm{i}e)\gamma_\nu v(\bar{p}_f) \\
&= -\mathrm{i}e^2\bar{u}(p_f)\gamma_\mu u(p_i)D_{\mathrm{F}}^{(0)}(q)\left[g^{\mu\nu} + \left(g^{\mu\nu}q^2 - q^\mu q^\nu\right)\bar{\Pi}(q^2)\frac{-4\pi}{q^2}\right]
\end{aligned}$$

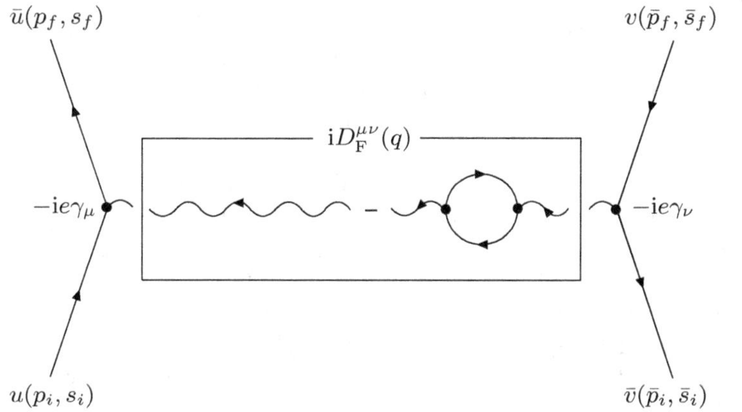

Fig. 3.38. Feynman diagram of the direct electron-positron scattering up to order $\mathcal{O}\left(e^4\right)$ in momentum space resulting from the vacuum polarization. The momentum transfer is $q = p_f - p_i = -(\bar{p}_f - \bar{p}_i)$.

$$
\begin{aligned}
&\times \bar{v}(\bar{p}_i)\gamma_\nu v(\bar{p}_f) \\
&= -\mathrm{i}e^2\bar{u}(p_f)\gamma_\mu u(p_i) D_{\mathrm{F}}^{(0)}(q)\left[1 - 4\pi\bar{\Pi}(q^2)\right]\bar{v}(\bar{p}_i)\gamma^\mu v(\bar{p}_f) \\
&= -\mathrm{i}e^2\bar{u}(p_f)\gamma_\mu u(p_i) D_{\mathrm{F}}^{(0)}(q)\left[1 - \frac{e^2}{3\pi}\ln\frac{\Lambda^2}{m_0^2} - \Pi^{(\mathrm{R})}(q^2)\right] \\
&\times \bar{v}(\bar{p}_i)\gamma^\mu v(\bar{p}_f) ,
\end{aligned} \tag{3.141}
$$

with $D_{\mathrm{F}}^{(0)}(q) = -4\pi/q^2$. In the third step, the fact was used that for free(!) positrons, we have

$$
\left(\bar{\not{p}}_f + m_0\right) v(\bar{p}_f) = 0 \; , \; \bar{v}(\bar{p}_i)\left(\bar{\not{p}}_i + m_0\right) = 0
$$

so that the term proportional to $q^\mu q^\nu$ vanishes:

$$
\bar{v}(\bar{p}_i)\left(\bar{\not{p}}_i - \bar{\not{p}}_f\right) v(\bar{p}_f) = \bar{v}(\bar{p}_i)\not{q} v(\bar{p}_f) = q^\nu \bar{v}(\bar{p}_i)\gamma_\nu v(\bar{p}_f) = 0 \; .
$$

Up to an error of order $\mathcal{O}\left(e^6\right)$, we can now prefix multiplicatively the logarithmically divergent part in (3.141) and finally obtain [compare to the $\mathcal{O}\left(e^2\right)$-scattering amplitude $M_{fi}(\mathrm{dir})$ in Theorem 3.11]

$$
\begin{aligned}
M_{fi}(\mathrm{dir}) &= -\mathrm{i}e^2\bar{u}(p_f)\gamma_\mu u(p_i) D_{\mathrm{F}}^{(0)}(q) Z_3\left[1 - \Pi^{(\mathrm{R})}(q^2) + \mathcal{O}\left(e^4\right)\right] \\
&\times \bar{v}(\bar{p}_i)\gamma^\mu v(\bar{p}_f) \\
&= -\mathrm{i}e_{\mathrm{R}}^2\bar{u}(p_f)\gamma_\mu u(p_i) D_{\mathrm{F}}^{(0)}(q)\left[1 - \Pi^{(\mathrm{R})}(q^2) + \mathcal{O}\left(e^4\right)\right] \\
&\times \bar{v}(\bar{p}_i)\gamma^\mu v(\bar{p}_f) ,
\end{aligned}
$$

with the *renormalized charge*

$$
e_{\mathrm{R}} = \sqrt{Z_3}e \; , \; Z_3 = 1 - \frac{e^2}{3\pi}\ln\frac{\Lambda^2}{m_0^2} \; . \tag{3.142}
$$

At this stage the renormalization argument comes into play. Experimentally, the charge of a particle is determined by its interaction with another charged particle. An inseparable part of this interaction is the vacuum polarization which is not taken into account by using the bare charge e and therefore causes the divergent factor Z_3. Physically, only the renormalized charge e_{R} is relevant that encompasses the fermion-photon interaction. Its value $e_{\mathrm{R}}^2 \approx 1/137$ is determined by scattering experiments with small q^2, i.e. by slight scatterings of two widely separated charges. With this *charge renormalization* one finally ends up with a well-defined scattering amplitude correct up to order $\mathcal{O}\left(e^4\right)$ where the dependence on the cut-off momentum Λ is completely absorbed in the electron charge.[24] Here the detailed relationship between bare and renormalized charge and, hence, the regularization procedure are irrelevant. In order to calculate the influence of the vacuum polarization one simply uses the renormalized charge e_{R} as well as the *renormalized photon propagator*

$$D_{\mathrm{F}}^{(\mathrm{R})\mu\nu}(q) = D_{\mathrm{F}}^{(0)\mu\nu}(q)\left[1 - \Pi^{(\mathrm{R})}(q^2)\right]$$

and is only faced with finite quantities. Compared to the free photon propagator $D_{\mathrm{F}}^{(0)\mu\nu}(q)$, the actual, physically observable correction is given by the momentum-dependent term $\Pi^{(\mathrm{R})}(q^2)$. Its contribution is finite, independent of the cut-off momentum Λ, and vanishes for $q^2 \to 0$.

External vacuum polarization. Having discussed the correction of internal photon lines by the vacuum polarization, it remains to be clarified how external photon lines are to be modified in the presence of a fermion loop. As can easily be seen, the direct calculation leads to an undefined expression. However, in this case it helps to consider that incoming and outgoing fermions are not actually free but have been emitted sometime from a source and will be absorbed sometime by an observer so that an external photon line with a fermion loop can be viewed as an internal photon line with a fermion loop whose one end is the distant vortex (reduction of the external to the internal vacuum polarization). This corresponds to the modified photon propagator

$$q^2 \approx 0 \Longrightarrow D_{\mathrm{F}}^{\mu\nu}(q) = Z_3 D_{\mathrm{F}}^{(0)\mu\nu}(q)$$

or the known renormalization $e_{\mathrm{R}} = \sqrt{Z_3}e$ at both end vortices. Thus, we obtain the simple rule that in an external photon line the contribution of a

---

[24] Note that Z_3 only deviates substantially from one if $\Lambda \gtrsim m_0 \mathrm{e}^{3\pi/2e^2} \approx 10^{280} m_0$, which corresponds to a length of $\Delta x \lesssim \hbar/\Lambda \approx 10^{-293}$cm. However, in practice, this is completely irrelevant since the presence of other quantum fields delimits the range of validity of quantum electrodynamics to much smaller momenta and much larger distances. $Z_3 \lesssim 1$ means that for a remote observer the physical (renormalized) charge of a fermion seems to be weakened compared to its bare charge due to the cloud of virtual particle-antiparticle pairs surrounding it (*screening*).

fermion loop can be omitted if the renormalized charge $e_{\rm R}$ instead of e is used at the photon line's vortex.

Theorem 3.14: Vacuum polarization

The modification of an internal photon line caused by the vacuum polarization adds a quadratically divergent part to the free photon propagator $D_{\rm F}^{(0)\mu\nu}(q)$. Using the Pauli-Villars procedure, this part can be regularized in such a way that, up to order $\mathcal{O}\left(e^2\right)$, the modified photon propagator can be written as

$$D_{\rm F}^{\mu\nu}(q) = Z_3 D_{\rm F}^{(0)\mu\nu}(q)\left[1 - \Pi^{(\rm R)}(q^2)\right] ,$$

with the q-dependent, finite function

$$\begin{aligned}\Pi^{(\rm R)}(q^2) &= -\frac{2e^2}{\pi}\int\limits_0^1 \mathrm{d}\beta\beta(1-\beta)\ln\left[1-\beta(1-\beta)\frac{q^2}{m_0^2}\right]\\ &\overset{q^2/m_0^2\ll 1}{\approx} \frac{e^2}{\pi}\frac{q^2}{m_0^2}\left(\frac{1}{15}+\frac{1}{140}\frac{q^2}{m_0^2}+\ldots\right)\end{aligned}$$

and the *renormalization constant*

$$Z_3 = 1 - \frac{e^2}{3\pi}\ln\frac{\Lambda^2}{m_0^2} .$$

As a result, the former quadratic divergence is weakened to a logarithmic divergence in the cut-off momentum Λ which is solely contained in Z_3. Due to the renormalization idea, $D_{\rm F}^{\mu\nu}(q)$ can be replaced with the renormalized photon propagator

$$D_{\rm F}^{(\rm R)\mu\nu}(q) = D_{\rm F}^{(0)\mu\nu}(q)\left[1 - \Pi^{(\rm R)}(q^2)\right]$$

if, at the same time, the renormalized charge

$$e_{\rm R} = \sqrt{Z_3}e$$

instead of the bare charge e is used at its end vortices. The contribution of the vacuum polarization to an external photon line can be omitted if the renormalized charge $e_{\rm R}$ is used at its vortex, too.

Note that in $\Pi^{(\rm R)}(q^2)$ up to order $\mathcal{O}\left(e^2\right)$ the bare charge can also be replaced with the renormalized charge. Thus, the bare charge falls completely out of the calculation of scattering amplitudes.

3.4.2 Self-Energy

Next we investigate the impact of the internal $\mathcal{O}\left(e^2\right)$-self-energy on the free fermion propagator

$$S_{\mathrm{F}}^{(0)}(p) = \frac{\not{p} + m_0}{p^2 - m_0^2 + \mathrm{i}\epsilon} \,.$$

According to Figure 3.39, the transition to the modified fermion propagator proceeds along

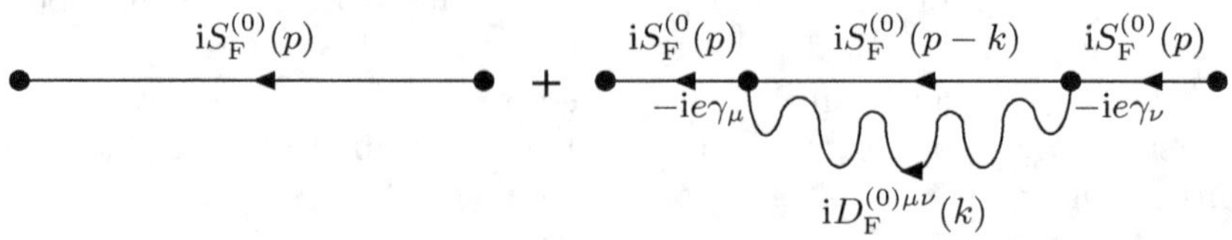

Fig. 3.39. Modification of the free fermion propagator by the $\mathcal{O}\left(e^2\right)$-self-energy.

$$S_{\mathrm{F}}^{(0)}(p) \longrightarrow S_{\mathrm{F}}(p) = S_{\mathrm{F}}^{(0)}(p) + S_{\mathrm{F}}^{(0)}(p)\Sigma(p)S_{\mathrm{F}}^{(0)}(p) \,, \tag{3.143}$$

with the linearly ultraviolet-divergent *self-energy function*

$$\Sigma(p) = -4\pi \mathrm{i}e^2 \int \frac{\mathrm{d}^4k}{(2\pi)^4} \frac{1}{k^2 + \mathrm{i}\epsilon} \gamma_\mu \frac{\not{p} - \not{k} + m_0}{(p-k)^2 - m_0^2 + \mathrm{i}\epsilon} \gamma^\mu \,, \tag{3.144}$$

which, contrary to the polarization tensor $P_{\mu\nu}(p)$, is a 4×4-matrix in spinor space. Using the Dyson equation (3.138), we can formally rewrite $S_{\mathrm{F}}(p)$ as

$$S_{\mathrm{F}}(p) = \frac{1}{\left[S_{\mathrm{F}}^{(0)}(p)\right]^{-1} - \Sigma(p)} = \frac{1}{\not{p} - m_0 - \Sigma(p) + \mathrm{i}\epsilon} \,. \tag{3.145}$$

This expression is correct up to order $\mathcal{O}\left(e^2\right)$ (and contains additional terms in higher orders corresponding to more and more consecutively linked self-energy insertions).

In the following we assume the restriction that $S_{\mathrm{F}}(p)$ is placed between two free electron (and not positron) states, i.e. $\bar{u}(p)S_{\mathrm{F}}(p)u(p)$, that reside "near the mass shell": $(\not{p} - m_0)u(p) \approx 0$, $\bar{u}(p)(\not{p} - m_0) \approx 0$. In this case a reasonable ansatz for the energy function is

$$\Sigma(p) = \delta m + (Z_2 - 1)(\not{p} - m_0) + \Sigma^{(\mathrm{R})}(p)(\not{p} - m_0)^2 \,. \tag{3.146}$$

It corresponds to a Taylor-like expansion around the "point" $\not{p} = m_0$. The quantities δm and $Z_2 - 1$ are to be regarded as small constant $\mathbb{C}$-numbers of order $\mathcal{O}\left(e^2\right)$ that, as we show later on, contain formally the divergence of the integral (3.144), whereas the scalar $\mathcal{O}\left(e^2\right)$-"residue function" $\Sigma^{(\mathrm{R})}(p)$ is finite. Thus, disregarding terms of higher orders, (3.145) turns into

$$\begin{aligned}
S_{\mathrm{F}}(p) &= \frac{1}{\not{p} - m_0 - \delta m - (Z_2 - 1)(\not{p} - m_0) - (\not{p} - m_0)^2\Sigma^{(\mathrm{R})}(p) + \mathrm{i}\epsilon} \\
&\approx \frac{1}{(\not{p} - m_0 - \delta m)[1 - (Z_2 - 1)][1 - (\not{p} - m_0)\Sigma^{(\mathrm{R})}(p)] + \mathrm{i}\epsilon} \\
&\approx \frac{Z_2}{(\not{p} - m_0 - \delta m)[1 - (\not{p} - m_0)\Sigma^{(\mathrm{R})}(p)] + \mathrm{i}\epsilon} \,,
\end{aligned} \tag{3.147}$$

where $[1-(Z_2-1)]^{-1} \approx Z_2$ has been used in the last step. If we now make use of the above restriction concerning quasi-free electron bispinors on the mass shell, the momentum-dependent correction $(\not{p}-m_0)\Sigma^{(\mathrm{R})}(p)$ in (3.147) can be ignored, and we obtain

$$S_\mathrm{F}(p) \approx \frac{Z_2}{\not{p}-m_0-\delta m+\mathrm{i}\epsilon} = \frac{Z_2(\not{p}+m_0+\delta m)}{p^2-(m_0+\delta m)^2+\mathrm{i}\epsilon} \ .$$

Apart from the multiplicative constant Z_2, this corresponds, as expected, to the free electron propagator, where, however, m_0 is replaced by the mass $m_0+\delta m$. Analogously to the charge renormalization in the vacuum polarization, the divergent *self-energy* δm can be removed by the argument that the *renormalized mass*

$$m_\mathrm{R} = m_0 + \delta m$$

is the actual, physically measured electron mass, whereas the bare mass m_0 has no physical meaning since it does not include the interaction of the electron with its own radiation field. A formal procedure for the *mass renormalization* is expressing the Dirac equation through the renormalized mass:

$$(\not{p} - e\not{A} - m_\mathrm{R})\,\psi = -\delta m\psi \ .$$

This implies within our scattering formalism that, on the one hand, m_0 is to be replaced by m_R everywhere. On the other hand, the additional term or *counter term*

$$-S_\mathrm{F}^{(0)}(p)\delta m S_\mathrm{F}^{(0)}(p)$$

arises in (3.143) that eliminates δm in $\Sigma(p)$ and hence also in (3.147). Therefore, the modified (mass-renormalized) electron propagator takes the form

$$S_\mathrm{F}(p) = \frac{Z_2}{(\not{p}-m_\mathrm{R})[1-(\not{p}-m_\mathrm{R})\Sigma^{(\mathrm{R})}(p)]+\mathrm{i}\epsilon} \ .$$

The remaining divergent factor Z_2 can, once again, be absorbed in the bare charge e by performing the charge renormalization

$$e \longrightarrow e_\mathrm{R}' = Z_2 e \ . \tag{3.148}$$

Contrary to (3.142), here is no root present since each vortex is shared by two electron lines. All in all, the impact of the electronic self-energy can be determined in a well-defined way, too, by using the renormalized charge e_R' as well as the *charge- and mass-renormalized electron propagator*

$$S_\mathrm{F}^{(\mathrm{R})}(p) = \frac{1}{(\not{p}-m_\mathrm{R})[1-(\not{p}-m_\mathrm{R})\Sigma^{(\mathrm{R})}(p)]+\mathrm{i}\epsilon} \ ,$$

where the actual, physically relevant correction to the free electron propagator resides solely in the momentum-dependent and finite function $\Sigma^{(\mathrm{R})}(p)$.

External self-energy. Analogously to the vacuum polarization, the effect of a self-energy insertion in an external electron line can be easily understood considering that even a free electron was emitted sometime and will be absorbed sometime. Therefore, it can be regarded as an inner electron on the mass shell within a larger process sequence (reduction of the external to the internal self-energy). Consequently, the combination of an external electron line with or without a self-energy insertion leads to the modified electron propagator

$$\not{p} \approx m_{\rm R} \Longrightarrow S_{\rm F}(p) = Z_2 S_{\rm F}^{(0)}(m_{\rm R}, p) ,$$

which simply corresponds to a factor of $\sqrt{Z_2}$ at both end vortices. Similarly to the external vacuum polarization, we finally obtain the rule that the self-energy contribution to an external electron line can be omitted if the renormalized charge $e'_{\rm R}$ instead of e is used at its vortex.

Determination of δm and Z_2. We now show how the divergences of the self-energy function $\Sigma(p)$ can be shifted to the renormalization constants δm and Z_2. As already mentioned, one ultraviolet divergence of the integral (3.144) lies at $k \to \infty$ so that it needs to be regularized. However, due to the first term in the integrand, the integral also contains an *infrared divergence* at $k \to 0$. This can be circumvented by introducing a photon mass μ, whereas, at the end of our calculations, we have to consider the limit $\mu \to 0$. Let us therefore start with the *regularized self-energy function*

$$\bar{\Sigma}(p,\mu) = -4\pi \mathrm{i} e^2 \int \frac{\mathrm{d}^4 k}{(2\pi)^4} \left[\frac{1}{k^2 - \mu^2 + \mathrm{i}\epsilon} \gamma_\mu \frac{\not{p} - \not{k} + m_0}{(p-k)^2 - m_0^2 + \mathrm{i}\epsilon} \gamma^\mu + \mathrm{Reg} \right] ,$$

where the regularizing Pauli-Villars terms are subsumed within "Reg". After some intermediate calculations, this becomes

$$\begin{aligned} \bar{\Sigma}(p,\mu) &= \frac{e^2}{2\pi} \int_0^1 \mathrm{d}\beta (2m_0 - \beta \not{p}) \\ &\quad \times \left[\int_0^\infty \frac{\mathrm{d}\rho}{\rho} \exp\left\{ \mathrm{i}\rho \left[\beta(1-\beta)p^2 - \beta\mu^2 - (1-\beta)m_0^2 \right] \right\} + \mathrm{Reg} \right] . \end{aligned}$$

As one can see, the ρ-integral is logarithmically divergent at its lower limit. For its regularization it is sufficient to subtract from the integrand one single Pauli-Villars term of the form $\exp(-\mathrm{i}\rho\beta\Lambda^2)/\rho$ with the cut-off momentum Λ. With the assumption[25] $p^2 < (m_0 + \mu)^2$, the ρ-integration can, as in the case of vacuum polarization, be carried out by deforming the integration contour onto the negative imaginary axis. Finally, one obtains

---

[25] For four-momenta above the *threshold energy* $p^2 = (m_0 + \mu)^2$ the virtual photon can decay into a real fermion and a real photon (compare to footnote 23 on page 297).

$$\bar{\Sigma}(p, \mu, \Lambda) = \frac{e^2}{2\pi} \int_0^1 \mathrm{d}\beta (2m_0 - \beta \not{p}) \ln \frac{\beta \Lambda^2}{(1-\beta)m_0^2 + \beta\mu^2 - \beta(1-\beta)p^2} .$$

According to our ansatz (3.146), the self-energy contribution δm follows from calculating $\bar{\Sigma}(p, \mu, \Lambda)$ on the mass shell. The final result is independent of μ and reads

$$\delta m = \bar{\Sigma}(p, \mu, \Lambda)\Big|_{\not{p}=m_0, p^2=m_0^2} = \frac{3e^2 m_0}{4\pi} \ln\left(\frac{\Lambda^2}{m_0^2} + \frac{1}{2}\right) .$$

However, for Z_2 we find the μ-dependent expression

$$Z_2 = 1 + \left.\frac{\partial \bar{\Sigma}(p, \mu, \Lambda)}{\partial \not{p}}\right|_{\not{p}=m_0, p^2=m_0^2} = 1 - \frac{e^2}{2\pi}\left(\frac{1}{2}\ln\frac{\Lambda^2}{m_0^2} + \ln\frac{\mu^2}{m_0^2} + \frac{9}{4}\right) .$$

The determination of the residue function $\Sigma^{(\mathrm{R})}(p)$ is much more complicated and is not pursued here.

It is noticeable that δm as well as Z_2 are logarithmically divergent in Λ whereas an inspection of (3.144) originally led to the more pessimistic prediction of a linear divergence. Furthermore, the quantity Z_2 possesses two awkward properties: firstly, it is infrared-divergent in the photon mass and secondly, it is not gauge invariant. However, as we see shortly, this is irrelevant since Z_2 is exactly canceled out by the renormalization constant stemming from the vortex correction.

Theorem 3.15: Self-energy

The modification of an internal fermion line caused by the self-energy adds a logarithmically ultraviolet- and infrared-divergent part to the free fermion propagator $S_{\mathrm{F}}^{(0)}(p)$. Using the Pauli-Villars procedure, this can be regularized in such a way that, up to order $\mathcal{O}\left(e^2\right)$, the modified fermion propagator can be written as

$$S_{\mathrm{F}}(p) = \frac{Z_2}{(\not{p} - m_{\mathrm{R}})[1 - (\not{p} - m_{\mathrm{R}})\Sigma^{(\mathrm{R})}(p)] + \mathrm{i}\epsilon} ,$$

with the p-dependent, finite function $\Sigma^{(\mathrm{R})}(p)$ and the renormalization constant

$$Z_2 = 1 - \frac{e^2}{2\pi}\left(\frac{1}{2}\ln\frac{\Lambda^2}{m_0^2} + \ln\frac{\mu^2}{m_0^2} + \frac{9}{4}\right) .$$

Here it is assumed that

- $S_{\mathrm{F}}(p)$ is placed between quasi-free electron bispinors and
- the mass renormalization has already been carried out everywhere, i.e. the bare mass m_0 has been replaced by the renormalized (physical) mass m_{R}.

▷

After this Z_2 arises as the only constant from the regularization containing the above mentioned divergences. Due to the renormalization idea, $S_{\rm F}(p)$ can be replaced with the renormalized fermion propagator

$$S_{\rm F}^{(\rm R)}(p) = \frac{1}{(\not p - m_{\rm R})[1-(\not p - m_{\rm R})\Sigma^{(\rm R)}(p)] + {\rm i}\epsilon},$$

if, at the same time, the renormalized charge

$$e'_{\rm R} = Z_2 e$$

instead of the bare charge e is used at its end vortices. The contribution of the self-energy to an external fermion line can be omitted if the renormalized charge $e'_{\rm R}$ is used at its vortex, too.

3.4.3 Vortex Correction

The last radiation correction is the $\mathcal{O}(e^2)$-vortex correction. Here a vortex is modified by an internal photon line according to

$$\gamma_\mu \longrightarrow \Gamma_\mu(p',p) = \gamma_\mu + \Lambda_\mu(p',p) \tag{3.149}$$

(see Figure 3.40) with the logarithmically ultraviolet-divergent *vortex function*[26]

$$\begin{aligned}\Lambda_\mu(p',p) &= -4\pi{\rm i}e^2 \int \frac{{\rm d}^4k}{(2\pi)^4}\frac{1}{k^2-\mu^2+{\rm i}\epsilon} \\ &\quad\times\left[\gamma_\nu \frac{\not p' - \not k + m_{\rm R}}{(p'-k)^2 - m_{\rm R}^2 + {\rm i}\epsilon}\gamma_\mu \frac{\not p - \not k + m_{\rm R}}{(p-k)^2 - m_{\rm R}^2 + {\rm i}\epsilon}\gamma^\nu\right]. \end{aligned} \tag{3.150}$$

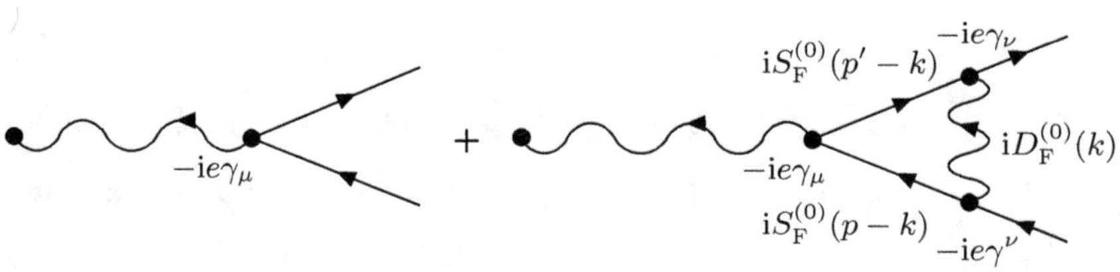

Fig. 3.40. Modification of a vortex by the $\mathcal{O}(e^2)$-vortex correction.

Similarly to the self-energy function, it also possesses an infrared divergence due to the photon propagator term, for which reason the photon mass μ was inserted right from the beginning. As before, our goal is to shift the present divergences into a multiplicative constant that, in turn, leads to a

[26] Remember: $m_{\rm R}$ results from the mass renormalization that has to be performed everywhere.

renormalization of the electric charge. For this purpose we split $\Lambda_\mu(p',p)$ into a part $\Lambda_\mu(p,p)$ representing the "forward scattering" (vanishing momentum transfer: $q = p' - p = 0$) and into a "residue" $\Lambda_\mu^{(\mathrm{R})}(p',p)$:

$$\Lambda_\mu(p',p) = \Lambda_\mu(p,p) + \Lambda_\mu^{(\mathrm{R})}(p',p) \ .$$

Note that the ultraviolet divergence of the integral (3.150) is solely contained by $\Lambda_\mu(p,p)$, whereas $\Lambda_\mu^{(\mathrm{R})}(p',p)$ is well-defined and finite in this respect. This can be seen by expanding the second propagator in (3.150) using the Dyson equation (3.138):

$$\begin{aligned}\frac{\not p' - \not k + m_{\mathrm{R}}}{(p'-k)^2 - m_{\mathrm{R}}^2 + \mathrm{i}\epsilon} &= \frac{1}{\not p' - \not k - m_{\mathrm{R}} + \mathrm{i}\epsilon} \\ &= \frac{1}{\not p - \not k - m_{\mathrm{R}} + \mathrm{i}\epsilon - (\not p - \not p')} \\ &= \frac{1}{\not p - \not k - m_{\mathrm{R}} + \mathrm{i}\epsilon} \\ &\quad + \frac{1}{\not p - \not k - m_{\mathrm{R}} + \mathrm{i}\epsilon}(\not p - \not p')\frac{1}{\not p - \not k - m_{\mathrm{R}} + \mathrm{i}\epsilon} + \dots \ .\end{aligned}$$

For large k the p'-independent term behaves like $1/|k|$ which causes the logarithmic divergence, whereas the p'-dependent terms possess higher k-powers in the denominator thus leaving the integral convergent.

Surprisingly, we can leave out an explicit regularization when calculating the divergent part $\Lambda_\mu(p,p)$ by using the *Ward identity*

$$\Lambda_\mu(p,p) = -\frac{\partial \Sigma(p)}{\partial p^\mu} \ , \tag{3.151}$$

where $\Sigma(p)$ is the known self-energy function from (3.144). This identity follows from the relation[27]

$$\frac{\partial}{\partial p^\mu} S_{\mathrm{F}}^{(0)}(p-k) = -S_{\mathrm{F}}^{(0)}(p-k)\gamma_\mu S_{\mathrm{F}}^{(0)}(p-k)$$

that, in turn, results from the differentiation of $S_{\mathrm{F}}^{(0)}(p-k) S_{\mathrm{F}}^{(0)^{-1}}(p-k) = 1$ using the product rule:

$$\left[\frac{\partial}{\partial p^\mu} S_{\mathrm{F}}^{(0)}(p-k)\right] S_{\mathrm{F}}^{(0)^{-1}}(p-k) + S_{\mathrm{F}}^{(0)}(p-k)\frac{\partial}{\partial p^\mu}(\not p - \not k - m_{\mathrm{R}}) = 0 \ .$$

Due to our ansatz (3.146), the Ward identity yields the simple relationship

$$\Lambda_\mu(p,p) = -(Z_2 - 1)\gamma_\mu + \mathcal{O}(\not p - m_{\mathrm{R}}) \ .$$

If we restrict ourselves once again to the case where the modified vortex $\Gamma_\mu(p',p)$ is placed between two quasi-free electron bispinors near the mass

[27] It means that differentiating the fermion propagator with respect to the momentum corresponds graphically to an insertion of a photon with momentum transfer zero into the fermion line.

shell, i.e. $\bar{u}(p')\Gamma_\mu(p',p)u(p)$, $(\not{p}-m_{\rm R})u(p)\approx 0$, $\bar{u}(p')(\not{p}'-m_0)\approx 0$, then, due to the last relation, we can rewrite (3.149) in the form

$$\begin{aligned}\Gamma_\mu(p',p) &= [1-(Z_2-1)]\,\gamma_\mu + \Lambda_\mu^{(\rm R)}(p',p)\\ &\approx [1-(Z_2-1)]\left[\gamma_\mu + \Lambda_\mu^{(\rm R)}(p',p)\right]\\ &\approx Z_2^{-1}\left[\gamma_\mu + \Lambda_\mu^{(\rm R)}(p',p)\right]\ ,\end{aligned}$$

which is correct up to an error of order $\mathcal{O}\left(e^4\right)$. At this point, our goal is reached: the ultraviolet divergence of the vortex correction, i.e. its dependence on the cut-off momentum, is contained solely in the multiplicative factor Z_2^{-1} and can, as for the vacuum polarization and the self-energy, be removed from the scattering formalism by performing the charge renormalization

$$e \longrightarrow e''_{\rm R} = Z_2^{-1}e \tag{3.152}$$

if, at the same time, the *renormalized vortex*

$$\Gamma_\mu^{(\rm R)}(p',p) = \gamma_\mu + \Lambda_\mu^{(\rm R)}(p',p)$$

instead of γ_μ is used. As before, the actual, physically observable correction is given exclusively by the cut-off momentum-independent residue function $\Lambda_\mu^{(\rm R)}(p',p)$. In the limit $q^2/m_{\rm R}^2 \to 0$ it becomes, after some lengthy calculations (for free electron bispinors!),[28]

$$\Lambda_\mu^{(\rm R)}(p',p) \approx \gamma_\mu \frac{e^2}{3\pi}\frac{q^2}{m_{\rm R}^2}\left(\ln\frac{m_{\rm R}}{\mu} - \frac{3}{8}\right) + \frac{e^2}{2\pi}\frac{\rm i}{2m_{\rm R}}\sigma_{\mu\nu}q^\nu\ ,$$

with $q = p' - p$ and $\sigma_{\mu\nu} = {\rm i}[\gamma_\mu,\gamma_\nu]/2$.

Let us now consider the three fundamental radiation corrections together: vacuum polarization, self-energy, and vortex correction. In calculations of scattering amplitudes this leads, in total, to the charge renormalization

$$e \longrightarrow \sqrt{Z_3}Z_2Z_2^{-1}e = \sqrt{Z_3}e$$

at each vortex correct up to order $\mathcal{O}\left(e^2\right)$ where the renormalizations stemming from the self-energy and the vacuum polarization cancel out each other. This result is satisfactory in many respects. Firstly, contrary to (3.148) and (3.152), the resulting physical charge depends neither on the artificially introduced photon mass nor on the arbitrarily chosen gauge. Secondly, the renormalization of the electric charge is exclusively a consequence of the modified photon propagator caused by the creation of virtual pairs so that two fermions

---

[28] Note that $\Gamma_\mu^{(\rm R)}(p',p)$ still possesses an infrared divergence at vanishing photon mass. However, this so-called *infrared catastrophe* turns out to be not existent if one takes into account that in each calculation of elastic scattering processes one always has to incorporate a "background noise" of inelastic scatterings (e.g. braking radiation, German: Bremsstrahlung) that cannot be switched off in real experiments. Those external radiation corrections exactly compensate for the photon mass dependence of the internal vortex correction (see Exercise 41).

with the same bare charge (e^-, μ^-, τ^-, ...) also possess the same physical charge. Without the above mentioned cancellation or, likewise, without the Ward identity (3.151) this would not be the case due to the fermion mass dependence of each renormalization constant.

Theorem 3.16: Vortex correction

The vortex correction adds a logarithmically ultraviolet- and infrared-divergent part to each vortex factor γ_μ. Using the Ward Identity, the modified vortex can be written up to order $\mathcal{O}\left(e^2\right)$ as

$$\Gamma_\mu(p',p) = Z_2^{-1}\left[\gamma_\mu + \Lambda_\mu^{(\mathrm{R})}(p',p)\right] ,$$

with the momentum-dependent, finite function ($q = p' - p$)

$$\Lambda_\mu^{(\mathrm{R})}(p',p) \overset{q^2/m_\mathrm{R}^2\to 0}{\approx} \gamma_\mu \frac{e^2}{3\pi}\frac{q^2}{m_\mathrm{R}^2}\left(\ln\frac{m_\mathrm{R}}{\mu} - \frac{3}{8}\right) + \frac{e^2}{2\pi}\frac{\mathrm{i}}{2m_\mathrm{R}}\sigma_{\mu\nu}q^\nu$$

and the renormalization constant Z_2 (see Theorem 3.15). Here it is assumed that $\Gamma_\mu(p',p)$ is placed between two quasi-free electron bispinors. Due to the renormalization idea, $\Gamma_\mu(p',p)$ can be replaced with the renormalized vortex

$$\Gamma_\mu^{(\mathrm{R})}(p',p) = \gamma_\mu + \Lambda_\mu^{(\mathrm{R})}(p',p)$$

if, at the same time, the renormalized charge

$$e''_\mathrm{R} = Z_2^{-1}e$$

instead of the bare charge e is used. In total, vacuum polarization, self-energy, and vortex correction lead to the charge renormalization

$$e \to \sqrt{Z_3}e ,$$

which is caused solely by the vacuum polarization. This result is a consequence of the Ward identity which is valid in all orders of the scattering theory.

It is pointed out here, too, that in $\Lambda_\mu^{(\mathrm{R})}$ the bare charge can also be replaced by the renormalized charge up to order $\mathcal{O}\left(e^2\right)$.

After all three $\mathcal{O}\left(e^2\right)$-radiation corrections have been discussed, one may ask what new problems will arise if one goes beyond them. With the renormalization methods presented so far, we already have all the instruments that are necessary to obtain unique, finite, and cut-off parameter-independent physical results in higher orders of the scattering theory. There, however, we should prepare ourselves for much more calculational effort due simply to the much larger diversity of combinations in the construction of Feynman diagrams.

3.4.4 Physical Consequences

At the end of this section we discuss two physical effects caused by the $\mathcal{O}(e^2)$-radiation corrections that are among the most important touchstones of quantum electrodynamics. They are the *electron anomaly*, i.e. the deviation of Landé's g-factor from 2, and the *Lamb shift* in the energy spectra of atomic systems.

Gyromagnetic ratio of the electron. In Subsection 2.4.1 we saw that, in the presence of an external electromagnetic field, the pure Dirac theory leads to an interaction term $-\boldsymbol{MB}$ in the Hamilton operator, where

$$\boldsymbol{M} = g\mu_{\mathrm{B}}\boldsymbol{S} \tag{3.153}$$

denotes the magnetic moment of the electron with

$$g = 2 \quad \text{(Landé factor)} \ , \ \mu_{\mathrm{B}} = \frac{e\hbar}{2m_0c} \quad \text{(Bohr magneton)} \ . \tag{3.154}$$

Now we investigate the impact of radiation corrections on the magnetic moment of the electron. We consider the scattering of an electron against an external electromagnetic potential A^μ_{ext} to lowest order $\mathcal{O}(e)$, include the $\mathcal{O}(e^2)$-radiation corrections and ask for the corresponding interaction energy

$$W = \int \mathrm{d}^3x j_\mu A^\mu_{\mathrm{ext}} \ . \tag{3.155}$$

The Feynman diagrams contributing to this process are depicted in Figure 3.41. The self-energy correction is omitted there since, for free particles, it merely leads to the mass renormalization $m_0 \to m_{\mathrm{R}}$ and, together with the other corrections, to the charge renormalization $e \to e_{\mathrm{R}}$. Diagrams **a** and **b** yield the amplitude (see Theorem 3.14)

$$\begin{aligned} M_{fi}^{(a,b)} &= e_{\mathrm{R}}\bar{u}(p_f, s_f)\gamma_\mu u(p_i, s_i) D_{\mathrm{F}}^{(\mathrm{R})\mu\nu}(q) j_{\nu,\mathrm{ext}} \ , \ q = p_f - p_i \\ &= e_{\mathrm{R}}\bar{u}(p_f, s_f)\gamma_\mu u(p_i, s_i) D_{\mathrm{F}}^{(0)\mu\nu}(q) \left[1 - \Pi^{(\mathrm{R})}(q^2)\right] j_{\nu,\mathrm{ext}} \ , \end{aligned}$$

and diagram **c** leads to (see Theorem 3.16)

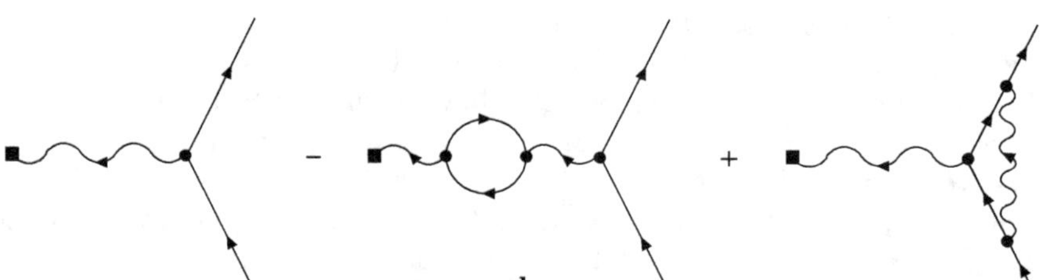

Fig. 3.41. Electron scattering against an external potential to lowest order $\mathcal{O}(e)$ (**a**), including the $\mathcal{O}(e^2)$-radiation corrections of vacuum polarization (**b**), and vortex correction (**c**).

$$M_{fi}^{(c)} = e_{\mathrm{R}}\bar{u}(p_f,s_f)\Lambda_\mu^{(\mathrm{R})}(p_f,p_i)u(p_i,s_i)D_{\mathrm{F}}^{(0)\mu\nu}(q)j_{\nu,\mathrm{ext}} \ .$$

Thus, in total, we obtain the amplitude

$$M_{fi}^{(a,b,c)} = j_\mu D_{\mathrm{F}}^{(0)\mu\nu}(q)j_{\nu,\mathrm{ext}} \ ,$$

where

$$j_\mu = e_{\mathrm{R}}\bar{u}(p_f,s_f)\left\{\gamma_\mu\left[1-\Pi^{(\mathrm{R})}(q^2)\right]+\Lambda_\mu^{(\mathrm{R})}(p_f,p_i)\right\}u(p_i,s_i)$$

is the $\mathcal{O}\left(e^2\right)$-radiation corrected electronic transition current in momentum space, which has to be inserted into (3.155). After this the interaction energy of the electron with an external electromagnetic field becomes

$$W = e_{\mathrm{R}}\int \mathrm{d}^3x\bar{\Psi}_f\left\{\gamma_\mu\left[1-\Pi^{(\mathrm{R})}(q^2)\right]+\Lambda_\mu^{(\mathrm{R})}(p_f,p_i)\right\}\Psi_i A_{\mathrm{ext}}^\mu \ .$$

Using the explicit expressions of $\Pi^{(\mathrm{R})}$ and $\Lambda_\mu^{(\mathrm{R})}$ in the limit $q^2 \to 0$ (see Theorems 3.14 and 3.16), this turns into

$$\begin{aligned}
W &\approx e_{\mathrm{R}}\int \mathrm{d}^3x\bar{\Psi}_f\left\{\gamma_\mu\left[1+\frac{e_{\mathrm{R}}^2q^2}{3\pi m_{\mathrm{R}}^2}\left(\ln\frac{m_{\mathrm{R}}}{\mu}-\frac{3}{8}-\frac{1}{5}\right)\right]\right.\\
&\quad\left.+\frac{\mathrm{i}e_{\mathrm{R}}^2}{4\pi m_{\mathrm{R}}}\sigma_{\mu\nu}q^\nu\right\}\Psi_i A_{\mathrm{ext}}^\mu \qquad (3.156)\\
&\approx e_{\mathrm{R}}\int \mathrm{d}^3x\frac{\mathrm{i}}{2m_{\mathrm{R}}}\left[\bar{\Psi}_f\partial_\mu\Psi_i-(\partial_\mu\bar{\Psi}_f)\Psi_i\right]\\
&\quad\times\left[1+\frac{e_{\mathrm{R}}^2q^2}{3\pi m_{\mathrm{R}}^2}\left(\ln\frac{m_{\mathrm{R}}}{\mu}-\frac{3}{8}-\frac{1}{5}\right)\right]A_{\mathrm{ext}}^\mu\\
&\quad+e_{\mathrm{R}}\int \mathrm{d}^3x\frac{\mathrm{i}}{2m_{\mathrm{R}}}\left(1+\frac{e_{\mathrm{R}}^2}{2\pi}\right)\bar{\Psi}_f\sigma_{\mu\nu}\Psi_i q^\nu A_{\mathrm{ext}}^\mu \ ,
\end{aligned}$$

where in the last step the Gordon decomposition [see (2.86)]

$$\bar{\Psi}_f\gamma_\mu\Psi_i = \frac{\mathrm{i}}{2m_{\mathrm{R}}}\left[\bar{\Psi}_f\partial_\mu\Psi_i-(\partial_\mu\bar{\Psi}_f)\Psi_i\right]+\frac{\mathrm{i}}{2m_{\mathrm{R}}}\bar{\Psi}_f\sigma_{\mu\nu}\Psi_i q^\nu$$

has been used and the $\mathcal{O}\left(q^3\right)$-terms have been omitted. Transforming the q-factors into derivatives in coordinate space, we finally obtain

$$\begin{aligned}
W &\approx e_{\mathrm{R}}\int \mathrm{d}^3x\left\{\frac{\mathrm{i}}{2m_{\mathrm{R}}}\left[\bar{\Psi}_f\partial_\mu\Psi_i-(\partial_\mu\bar{\Psi}_f)\Psi_i\right]\right.\\
&\quad\times\left[1-\frac{e_{\mathrm{R}}^2}{3\pi m_{\mathrm{R}}^2}\left(\ln\frac{m_{\mathrm{R}}}{\mu}-\frac{3}{8}-\frac{1}{5}\right)\partial_\alpha\partial^\alpha\right]A_{\mathrm{ext}}^\mu\\
&\quad\left.-\left(1+\frac{e_{\mathrm{R}}^2}{2\pi}\right)\frac{1}{2m_{\mathrm{R}}}\bar{\Psi}_f\sigma_{\mu\nu}\Psi_i\partial^\nu A_{\mathrm{ext}}^\mu\right\} \ .
\end{aligned}$$

The first term contains a purely convective current that we do not consider any further. Using [see (2.54) in Exercise 15]

$$\begin{aligned}\sigma_{\mu\nu}\partial^\nu A^\mu &= \frac{1}{2}(\sigma_{\mu\nu} - \sigma_{\nu\mu})\partial^\nu A^\mu = \frac{1}{2}(\sigma_{\mu\nu}(\partial^\nu A^\mu - \partial^\mu A^\nu)\\ &= -\frac{1}{2}\sigma_{\mu\nu}F^{\mu\nu} = -(\mathrm{i}\boldsymbol{\alpha}\boldsymbol{E} - \hat{\boldsymbol{\sigma}}\boldsymbol{B}) ,\end{aligned}$$

the second term can be rewritten as

$$\delta W \approx \left(1 + \frac{e_{\mathrm{R}}^2}{2\pi}\right)\frac{1}{2m_{\mathrm{R}}}\int \mathrm{d}^3x\bar{\Psi}_f\left(\mathrm{i}\boldsymbol{\alpha}\Psi_i\boldsymbol{E} - \hat{\boldsymbol{\sigma}}\Psi_i\boldsymbol{B}\right) .$$

Obviously, in the case of a purely magnetic field and of slowly moving electrons where the two upper components of Ψ dominate the lower ones, this expression can be interpreted as magnetic dipole energy since we have

$$\begin{aligned}\delta W &\approx -2\left(1 + \frac{e_{\mathrm{R}}^2}{2\pi}\right)\frac{1}{2m_{\mathrm{R}}}\int \mathrm{d}^3x\bar{\Psi}\boldsymbol{S}\Psi\boldsymbol{B}\\ &\approx -2\left(1 + \frac{e_{\mathrm{R}}^2}{2\pi}\right)\frac{1}{2m_{\mathrm{R}}}\int \mathrm{d}^3x\Psi^\dagger\boldsymbol{S}\Psi\boldsymbol{B}\\ &= -\langle\boldsymbol{M}\rangle\boldsymbol{B} ,\end{aligned}$$

with the magnetic moment

$$\langle\boldsymbol{M}\rangle = 2\left(1 + \frac{e_{\mathrm{R}}^2}{2\pi\hbar c}\right)\frac{e_{\mathrm{R}}\hbar}{2m_{\mathrm{R}}c}\langle\boldsymbol{S}\rangle = g\mu_{\mathrm{B}}\langle\boldsymbol{S}\rangle , \; g = 2\left(1 + \frac{e_{\mathrm{R}}^2}{2\pi\hbar c}\right) .$$

Here the physical constants $\hbar$ and c have exceptionally been written down. Thus, compared to (3.153) and (3.154), the electron's magnetic moment is slightly larger when taking into account the $\mathcal{O}\left(e^2\right)$-radiation corrections. With $\alpha_e = e_{\mathrm{R}}^2/\hbar c = 1/137.03602$, it contains the corrected Landé factor

$$g = 2\left(1 + \frac{\alpha_e}{2\pi}\right) = 2(1 + 0.00116141) .$$

The deviation of this factor from the original value 2 is called *electron anomaly*. It was derived by Julian Schwinger in 1948 for the first time and subsequently verified experimentally by others. A modern experimental value is

$$g_{\mathrm{exp}} = 2[1 + 0.00115965219(\pm 1)] .$$

The remaining difference between g and g_{exp} is explained by the omission of higher order corrections. Taking into account the corrections up to order $\mathcal{O}\left(e^8\right)$ yields a theoretical value coinciding with g_{exp} up to a relative deviation of about 10^{-11}.

Qualitatively, the increase in the magnetic moment can be understood as follows: the electron continually emits and absorbs virtual photons that carry away parts of the electron's energy or mass. Therefore, the ratio of the electron's charge and effective mass is raised, and this has an impact on measurements of the magnetic moment in the presence of a magnetic field.

Lamb shift. Comparing the experimentally measured, highly resolved binding spectrum of hydrogen atoms with theoretical predictions of the pure Dirac theory in Subsection 2.5.4, one encounters differences that, once again, result mainly from radiation corrections. In the presentation of this relationship, we will leave out some concrete calculations and, instead, concentrate on the essentials.

As before, we restrict ourselves to the $\mathcal{O}\left(e^2\right)$-radiation corrections and study their impact on the binding energies of the orbital electron in hydrogen-like atoms. The relevant contributions can again be depicted with the Feynman diagrams of Figure 3.41. However, now the external and internal fermion lines represent respectively bound electron solutions and the full electron propagator of the Dirac equation in the Coulomb field, where the latter encompasses all interactions between electron and nucleus to all orders of e. Unfortunately, the accompanying calculations are very complicated and lengthy. However, taking into account that atomic binding energies typically have a magnitude of $Z^2 e_{\mathrm{R}}^4 m_{\mathrm{R}}$ [see (2.115)] and are nonrelativistic in the case of light atoms, the problem can be solved approximately by splitting it up into two partial problems for high- and low-frequency radiation fields.

High-frequency radiation. For radiation fields with frequencies

$$\omega \geq \omega_{\min} \gg Z^2 e_{\mathrm{R}}^4 m_{\mathrm{R}} \ ,$$

the effect of the Coulomb field on the radiation corrections can be ignored. Therefore, it is justified to consider the term

$$\delta H = e_{\mathrm{R}}\gamma^0 \left[\gamma_\mu \frac{e_{\mathrm{R}}^2 q^2}{3\pi m_{\mathrm{R}}^2}\left(\ln\frac{m_{\mathrm{R}}}{\mu} - \frac{3}{8} - \frac{1}{5}\right) + \frac{\mathrm{i}e_{\mathrm{R}}^2}{4\pi m_{\mathrm{R}}}\sigma_{\mu\nu}q^\nu\right] A_{\mathrm{ext}}^\mu$$

in (3.156) to be a perturbation of the Dirac-Hamilton operator with the Coulomb potential

$$[A_{\mathrm{ext}}^\mu(x)] = \begin{pmatrix} A_{\mathrm{ext}}^0(\boldsymbol{x}) \\ \mathbf{0} \end{pmatrix} \ , \ A_{\mathrm{ext}}^0(\boldsymbol{x}) = -\frac{Ze}{|\boldsymbol{x}|}$$

and to determine the resulting energy shift of the unperturbed Coulomb solutions ψ_ν to first order perturbation theory:

$$\delta E_\nu^> = \int \mathrm{d}^3x \psi_\nu^\dagger \delta H \psi_\nu \ .$$

Here one has to bear in mind that the photon mass μ was originally introduced in δH to remove the infrared divergence within the momentum integration (3.150) of the vortex function. However, due to $\omega \geq \omega_{\min}$, this integration is now cut off at the bottom so that μ is not needed any more. As corresponding calculations show, this results in the replacement

$$\ln\frac{m_{\mathrm{R}}}{\mu} \longrightarrow \ln\frac{m_{\mathrm{R}}}{2\omega_{\min}} + \frac{5}{6}$$

in δH. Converting the momentum factors into derivatives in coordinate space,

$$\gamma_\mu q^2 A^\mu_{\text{ext}} = \gamma_0 \left(-\frac{\partial^2}{\partial t^2} + \boldsymbol{\nabla}^2\right) A^0_{\text{ext}} = \gamma_0 \boldsymbol{\nabla}^2 A^0_{\text{ext}}$$

$$\begin{aligned}\gamma^0\sigma_{\mu\nu}q^\nu A^\mu_{\text{ext}} &= \gamma^0\sigma_{0\nu}q^\nu A^0_{\text{ext}} = \mathrm{i}\gamma^0\sigma_{0\nu}\partial^\nu A^0_{\text{ext}} = \mathrm{i}\gamma^0\sigma_{0k}\partial^k A^0_{\text{ext}} \\ &= -\mathrm{i}\gamma_k\partial^k A^0_{\text{ext}} = -\boldsymbol{\gamma}\boldsymbol{\nabla} A^0_{\text{ext}} \ ,\end{aligned}$$

we finally obtain

$$\begin{aligned}\delta E^>_\nu &= e_{\mathrm{R}} \int \psi^\dagger_\nu \left[\frac{e^2_{\mathrm{R}}}{3\pi m^2_{\mathrm{R}}}\left(\ln\frac{m_{\mathrm{R}}}{2\omega_{\min}} + \frac{5}{6} - \frac{3}{8} - \frac{1}{5}\right)(\boldsymbol{\nabla}^2 A^0_{\text{ext}}) \right. \\ &\quad \left. - \frac{\mathrm{i}e^2_{\mathrm{R}}}{4\pi m_{\mathrm{R}}}\boldsymbol{\gamma}(\boldsymbol{\nabla} A^0_{\text{ext}})\right]\psi_\nu \\ &= \frac{e^3_{\mathrm{R}}}{3\pi m^2_{\mathrm{R}}}\left(\ln\frac{m_{\mathrm{R}}}{2\omega_{\min}} + \frac{5}{6} - \frac{3}{8} - \frac{1}{5}\right)\langle\nu|\,(\boldsymbol{\nabla}^2 A^0_{\text{ext}})\,|\nu\rangle \\ &\quad - \frac{\mathrm{i}e^3_{\mathrm{R}}}{4\pi m_{\mathrm{R}}}\langle\nu|\,\boldsymbol{\gamma}(\boldsymbol{\nabla} A^0_{\text{ext}})\,|\nu\rangle \ .\end{aligned}$$

The expectation values can be calculated using the nonrelativistic approximation. This yields

$$\begin{aligned}\langle\nu|\,(\boldsymbol{\nabla}^2 A^0_{\text{ext}})\,|\nu\rangle &= \frac{4Z^4 e^7_{\mathrm{R}} m^3_{\mathrm{R}}}{n^3}\delta_{l0} \\ \langle\nu|\,\boldsymbol{\gamma}(\boldsymbol{\nabla} A^0_{\text{ext}})\,|\nu\rangle &= \frac{2\mathrm{i}Z^4 e^7_{\mathrm{R}} m^2_{\mathrm{R}}}{n^3}\left[\delta_{l0} \pm \frac{2(1-\delta_{l0})}{(2J+1)(2l+1)}\right] \ ,\end{aligned}$$

with the principal quantum number n, the orbital quantum number l, and the total orbital quantum number $J = l \pm 1/2$ of the Schrödinger-Coulomb solutions.

Low-frequency radiation. If, on the other hand, we assume that the radiation frequency is

$$\omega \le \omega_{\max} \ll m_{\mathrm{R}} \ ,$$

the problem is completely nonrelativistic. In this case we can simply regard the radiation corrections as emission and reabsorption of a virtual photon and treat them perturbation theoretically within the framework of Schrödinger's theory. The corresponding perturbation operator is now

$$\delta H = \frac{\mathrm{i}e_{\mathrm{R}}}{m_{\mathrm{R}}}\boldsymbol{A}\boldsymbol{\nabla} \ ,$$

where $\boldsymbol{A}$ denotes the photon potential in the radiation gauge for which we set, as in Subsection 3.3.7,

$$\boldsymbol{A}_k(x) = \sqrt{\frac{2\pi}{\omega V}}\boldsymbol{\epsilon}(\boldsymbol{k},\lambda)\left(\mathrm{e}^{-\mathrm{i}k\cdot x} + \mathrm{e}^{\mathrm{i}k\cdot x}\right) \ , \quad \begin{cases}\boldsymbol{k}\boldsymbol{\epsilon}(\boldsymbol{k},\lambda) = 0 \\ \boldsymbol{\epsilon}(\boldsymbol{k},\lambda)\boldsymbol{\epsilon}(\boldsymbol{k},\lambda) = 1 \ .\end{cases}$$

With this the energy shift of the electron caused by the emission and reabsorption of a photon to second order perturbation theory becomes

$$\delta E_\nu^< = \sum_{\nu',\lambda} \int_{\omega<\omega_{\max}} \frac{V\mathrm{d}^3k}{(2\pi)^3}\frac{2\pi}{\omega V}\frac{e_\mathrm{R}^2}{m_\mathrm{R}^2}\frac{\left|\langle \nu'|\,\mathrm{e}^{\mathrm{i}\boldsymbol{kx}}\boldsymbol{\epsilon}(\boldsymbol{k},\lambda)\mathrm{i}\boldsymbol{\nabla}\,|\nu\rangle\right|^2}{E_\nu - E_{\nu'} - \omega} .$$

The sum is to be carried out over all electron states and transversal directions of the photon's polarization. Using the dipole approximation $\mathrm{e}^{\mathrm{i}\boldsymbol{kx}} \approx 1$ and taking into account $E_{\nu'} - E_\nu \ll \omega_{\max}$, this becomes

$$\begin{aligned}
\delta E_\nu^< &\approx \frac{2e_\mathrm{R}^2}{3\pi}\int_0^{\omega_{\max}} \mathrm{d}\omega\omega \sum_{\nu'} \frac{|\langle \nu'|\,\boldsymbol{v}\,|\nu\rangle|^2}{E_\nu - E_{\nu'} - \omega} \;,\; \boldsymbol{v} = -\frac{\mathrm{i}\boldsymbol{\nabla}}{m_\mathrm{R}} \\
&\approx \frac{2e_\mathrm{R}^2}{3\pi}\left[-\omega_{\max}\langle \nu|\,\boldsymbol{v}^2\,|\nu\rangle \right. \\
&\quad \left. + \sum_{\nu'} \left|\langle \nu'|\,\boldsymbol{v}^2\,|\nu\rangle\right|^2 (E_{\nu'} - E_\nu)\ln\frac{\omega_{\max}}{|E_{\nu'} - E_\nu|}\right] .
\end{aligned}$$

The first term describes the contribution of the low-frequent photons to the mass renormalization and must be subtracted since m_R is already the physical electron mass. After a series of further manipulations, there finally follows

$$\begin{aligned}
\delta E_\nu^< &\approx \frac{2e_\mathrm{R}^2}{3\pi}\sum_{\nu'} \left|\langle \nu'|\,\boldsymbol{v}^2\,|\nu\rangle\right|^2 (E_{\nu'} - E_\nu)\ln\frac{\omega_{\max}}{|E_{\nu'} - E_\nu|} \\
&= \frac{e_\mathrm{R}^3}{3\pi m_\mathrm{R}^2}\langle \nu|\,(\boldsymbol{\nabla}^2 A_\mathrm{ext}^0)\,|\nu\rangle \left[\ln\frac{2\omega_{\max}}{m_\mathrm{R}} - 2\ln(Ze_\mathrm{R}^2)\right] \\
&\quad + \frac{2e_\mathrm{R}^2}{3\pi}\sum_{\nu'} |\langle \nu'|\,\boldsymbol{v}\,|\nu\rangle|^2 (E_{\nu'} - E_\nu)\ln\frac{Z^2e_\mathrm{R}^4 m_\mathrm{R}/2}{|E_{\nu'} - E_\nu|} .
\end{aligned}$$

Before combining the energy shifts of the high- and low-frequency domains, we note that we cover the whole frequency range with the equalization $\omega_{\min} = \omega_{\max}$ as long as $\omega_{\min}$ can be chosen in such a way that $Z^2e_\mathrm{R}^4 m_\mathrm{R} \ll \omega_{\min} \ll m_\mathrm{R}$. However, for light atoms, this is always possible. Following this assumption, the total energy shift is

$$\begin{aligned}
\delta E_\nu &= \delta E_\nu^> + \delta E_\nu^< \\
&= \frac{4Z^4 e_\mathrm{R}^{10} m_\mathrm{R}}{3\pi n^3}\left\{ L_{nl} + \left[\frac{19}{30} - 2\ln(Ze_\mathrm{R}^2)\right]\delta_{l0} \pm \frac{3(1-\delta_{l0})}{4(2J+1)(2l+1)}\right\} ,
\end{aligned}$$

with

$$L_{nl} = \frac{n^3}{2Z^4 e_\mathrm{R}^8 m_\mathrm{R}}\sum_{\nu'} |\langle \nu'|\,\boldsymbol{v}\,|\nu\rangle|^2 (E_{\nu'} - E_\nu)\ln\frac{Z^2e_\mathrm{R}^4 m_\mathrm{R}/2}{|E_{\nu'} - E_\nu|} .$$

The quantity L_{nl} cannot be calculated further analytically and must be determined numerically.

As one can see, compared to the unperturbed binding energies

$$E_{nl} \approx -\frac{Z^2 e_\mathrm{R}^4 m_\mathrm{R}}{2n^3} ,$$

the Lamb shift is very small and suppressed by the factor $Z^2 e_{\mathrm{R}}^6$. Furthermore, it removes the degeneracy of states with equal principal quantum number n and total orbital quantum number J following from the Dirac theory (with a Coulomb potential). The historically most important example of the Lamb shift is the increase of the difference between the $2s_{1/2}$- and $2p_{1/2}$-levels measured for the first time by Lamb and Rutherford in 1947 (see Figure 2.2 in Subsection 2.5.4). With

$$m_{\mathrm{R}} = 5.11004 \cdot 10^5 \mathrm{eV} \ , \ L_{20} = -2.81177 \ , \ L_{21} = 0.03002 \ ,$$

the theoretical value of this increase is

$$\begin{aligned} \delta E &= \delta E_{2s_{1/2}} - \delta E_{2p_{1/2}} = 4.298 \cdot 10^{-6} \ \mathrm{eV} + 5.328 \cdot 10^{-8} \ \mathrm{eV} \\ &= 1039.3 \ \mathrm{MHz} + 12.9 \ \mathrm{MHz} = 1052.2 \ \mathrm{MHz} \ . \end{aligned}$$

A modern experimental value is

$$\delta E_{\mathrm{exp}} = 1057.845(9) \ \mathrm{MHz} \ .$$

The resulting relative deviation between theory and experiment with respect to the binding energies is 10^{-8}. Taking into account further corrections, repulsion effects, as well as the finite radius of the nucleus, the deviation can be reduced further to $< 10^{-11}$.

Summary

- Going beyond the lowest orders of scattering theory, interactions of particles with themselves become possible. They are called **radiation corrections.**
- Three fundamental $\mathcal{O}\left(e^2\right)$-radiation corrections are **vacuum polarization**, **self-energy**, and **vortex correction**. Within the vacuum polarization a (virtual/real) photon creates a virtual fermion-antifermion pair which is destroyed thereafter. In the self-energy and the vortex correction a (virtual/real) fermion emits a virtual photon and absorbs it afterward.
- All radiation corrections are accompanied by particular **ultraviolet divergences** in the corresponding momentum integrals that can be removed using the program of **renormalization**. Here a divergence is isolated into a multiplicative constant through **regularization** of the respective integral and subsequently absorbed in the **bare quantities** e and m_0. The resulting **renormalized quantities** e_{R} and m_{R} represent the actual, experimentally relevant and finite parameters.
- Considering the three radiation corrections together, the **charge renormalization** correct up to order $\mathcal{O}\left(e^2\right)$ results solely from the vacuum polarization.

- Physically, the radiation corrections appear, for example, in the deviation of the electron's gyromagnetic ratio from the value 2, as well as in the **Lamb shift** of atomic binding energies.
- Besides the ultraviolet divergence, the vortex correction also contains an **infrared divergence**. However, this turns out to be fictitious when all the scattering processes are taken into account that contribute due to the particular experimental setup which is used for counting the scattering products.

Exercises

41. Removal of the infrared catastrophe. The process where an electron is scattered against another charged particle while emitting real photons is called *braking radiation* (German: *Bremsstrahlung*). For the simpler case of electron scattering against a Coulomb potential the corresponding Feynman diagrams to leading order are shown in Figure 3.42. In the limit of soft photons with $0 < \omega \leq \omega_{\max}$ and small electron velocities $|\boldsymbol{v}_i| = v_i \ll 1$ (nonrelativistic limit), the corresponding totally polarized cross section is

$$\left(\frac{\mathrm{d}\sigma}{\mathrm{d}\Omega}\right)_{\mathrm{Brems}} = \left(\frac{\mathrm{d}\sigma}{\mathrm{d}\Omega}\right)_{\mathrm{Mott}} \frac{8e_{\mathrm{R}}^2 v_i^2 \sin^2\frac{\theta}{2}}{3\pi} \ln\frac{\omega_{\max}}{\mu} \tag{3.157}$$

containing an infrared divergence at the *cut-off frequency* $\mu \to 0$.
What is the meaning of this result for the $\mathcal{O}\left(e^2\right)$-radiation corrected cross section of the pure Coulomb scattering (without real photon emissions) to leading order?

Solution. If, for the pure Coulomb case, we consider only the infrared divergence causing vortex correction to leading order we have the amplitude square (see Theorem 3.7)

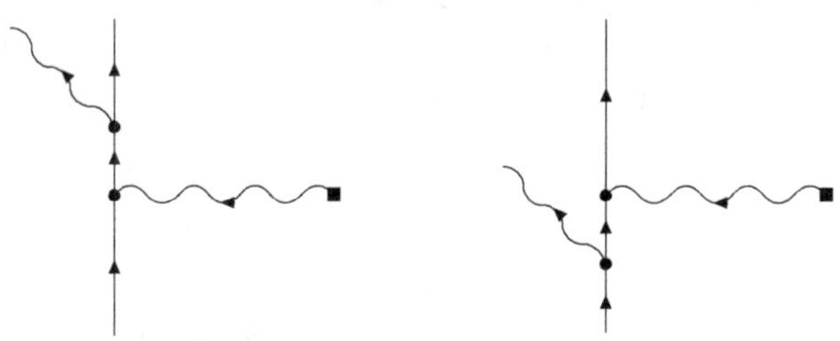

Fig. 3.42. Lowest-order Feynman diagrams of the braking radiation within a Coulomb field.

$$
\begin{aligned}
|M_{fi}|^2 &= \left| \frac{4\pi Z e_{\mathrm{R}}^2}{\boldsymbol{q}^2} \bar{u}(p_f, s_f) \left[\gamma_0 + \Lambda_0^{(\mathrm{R})}(p_f, p_i) \right] u(p_i, s_i) \right|^2 , \quad \boldsymbol{q} = \boldsymbol{p}_f - \boldsymbol{p}_i \\
&= \left| \frac{4\pi Z e_{\mathrm{R}}^2}{\boldsymbol{q}^2} \bar{u}(p_f, s_f) \gamma_0 u(p_i, s_i) \right|^2 \\
&\quad \times \left| \left[1 - \frac{e_{\mathrm{R}}^2 \boldsymbol{q}^2}{3\pi m_{\mathrm{R}}^2} \left(\ln \frac{m_{\mathrm{R}}}{\mu} - \frac{3}{8} \right) + \frac{\mathrm{i} e_{\mathrm{R}}^2}{4\pi m_{\mathrm{R}}} \gamma^0 \sigma_{0k} q^k \right] \right|^2
\end{aligned}
$$

as well as the totally polarized differential cross section

$$
\overline{\frac{\mathrm{d}\sigma}{\mathrm{d}\Omega}} = \left(\overline{\frac{\mathrm{d}\sigma}{\mathrm{d}\Omega}} \right)_{\mathrm{Mott}} \left| \left[1 - \frac{e_{\mathrm{R}}^2 \boldsymbol{q}^2}{3\pi m_{\mathrm{R}}^2} \left(\ln \frac{m_{\mathrm{R}}}{\mu} - \frac{3}{8} \right) + \frac{\mathrm{i} e_{\mathrm{R}}^2}{4\pi m_{\mathrm{R}}} \gamma^0 \sigma_{0k} q^k \right] \right|^2_{|\boldsymbol{p}_f| = |\boldsymbol{p}_i|} .
$$

Here the dominant infrared divergence at $\mu \to 0$ is caused by

$$
A = - \left(\overline{\frac{\mathrm{d}\sigma}{\mathrm{d}\Omega}} \right)_{\mathrm{Mott}} \frac{2 e_{\mathrm{R}}^2 \boldsymbol{q}^2}{3\pi m_{\mathrm{R}}^2} \ln \frac{m_{\mathrm{R}}}{\mu} \bigg|_{|\boldsymbol{p}_f| = |\boldsymbol{p}_i|} .
$$

Taking into account

$$
\boldsymbol{q}^2 \Big|_{|\boldsymbol{p}_f| = |\boldsymbol{p}_i|} = 4 |\boldsymbol{p}_i|^2 \sin^2 \frac{\theta}{2} = \frac{4 v_i^2 m_{\mathrm{R}}^2 \sin^2 \frac{\theta}{2}}{1 - v_i^2}
$$

and considering the nonrelativistic limit $|\boldsymbol{v}_i| = v_i \to 0$, the term A can be rewritten as

$$
A = - \left(\overline{\frac{\mathrm{d}\sigma}{\mathrm{d}\Omega}} \right)_{\mathrm{Mott}} \frac{8 e_{\mathrm{R}}^2 v_i^2 \sin^2 \frac{\theta}{2}}{3\pi} \ln \frac{m_{\mathrm{R}}}{\mu} .
$$

For the physical interpretation of this fact we have to bear in mind that the energy resolution ΔE of every measuring device used for counting the scattered particles is limited. This means that one cannot distinguish between elastically (pure Coulomb scattering) and inelastically (braking radiation) scattered particles in experiments as long as the frequency of the emitted photons fulfills $\omega \leq \Delta E$. Therefore, we have to add to the cross section of the pure Coulomb scattering that of the soft braking radiation (3.157) with $\omega_{\mathrm{max}} = \Delta E$. After this the critical Coulomb term A becomes

$$
\begin{aligned}
A &\longrightarrow \left(\overline{\frac{\mathrm{d}\sigma}{\mathrm{d}\Omega}} \right)_{\mathrm{Mott}} \frac{8 e_{\mathrm{R}}^2 v_i^2 \sin^2 \frac{\theta}{2}}{3\pi} \left(\ln \frac{\Delta E}{\mu} - \ln \frac{m_{\mathrm{R}}}{\mu} \right) \\
&= \left(\overline{\frac{\mathrm{d}\sigma}{\mathrm{d}\Omega}} \right)_{\mathrm{Mott}} \frac{8 e_{\mathrm{R}}^2 v_i^2 \sin^2 \frac{\theta}{2}}{3\pi} \ln \frac{\Delta E}{m_{\mathrm{R}}} .
\end{aligned}
$$

As one can see, the infrared divergence of the braking radiation neutralizes exactly the dominant infrared divergence of the Coulomb scattering – a very satisfying result which also remains valid in higher orders when all relevant internal and external radiation corrections are taken into account properly.

3.5 Scattering of Spin-0 Particles

In this last section we deal with the description of relativistic scattering processes of spin-0 particles on the basis of an appropriate propagator-scattering formalism similar to the one for spin-1/2 particles in the preceding sections. It is not surprising that many of the former concepts can be adopted more or less unchanged. However, at this point it must be stressed that, compared to the spin-1/2 case, the application range of the spin-0 scattering formalism will be severely restricted since in nature there do not exist elementary (point-like) spin-0 particles. Rather, they consist of two spin-1/2 quarks that are subject to the strong interaction. This, in turn, implies that, due to strong (quantum chromodynamical) vacuum polarization effects, each spin-0 particle is inevitably surrounded by a complex cloud of virtual particles that is totally disregarded due to our restriction to the electromagnetic interaction. Furthermore, we have to bear in mind that spin-0 particles are not really stable but decay via the weak interaction. However, with respect to purely electromagnetic scattering processes, some of them, e.g. pions, can be regarded as quasi-stable since their (weak) decay time of 10^{-8} s lies far above the characteristic time unit $\hbar/(m_0c^2) < 10^{-23}$ s.

Thus, due to these restrictions, our discussion of relativistic spin-0 scattering processes is rather of academic and didactic nature and mainly meant to highlight the formal correspondence to the spin-1/2 case. This correspondence is also reflected in the Feynman rules that we derive for the boson case (exclusively from tree level processes) and present in coherence (including the quantum electrodynamical loop level) at the end.

Note. Analogously to the spin-1/2 case, we will often equate spin-0 particles with "pions" and spin-0 antiparticles with "antipions". Furthermore, we will use the natural unit system ($\hbar = c = 1$) throughout the whole section.

3.5.1 Solution of the General Klein-Gordon Equation

As in Subsection 3.2.1, we are initially interested in developing a propagator formalism to solve the general Klein-Gordon equation. It turns out to be useful for this purpose to assign the charge current density and the G-scalar product of the Klein-Gordon theory other prefactors than in Chapter 1 (compare to Theorem 1.2):

$$j^\mu = \mathrm{i}\left[\phi^*\partial^\mu\phi - (\partial^\mu\phi^*)\phi\right] - 2eA^\mu\phi^*\phi = \phi^*\mathrm{i}\overleftrightarrow{\partial^\mu}\phi - 2eA^\mu\phi^*\phi \tag{3.158}$$

$$\langle\phi_1|\,\phi_2\rangle_{\mathrm{G}} = \int \mathrm{d}^3x\left(\phi_1^*\mathrm{i}\overleftrightarrow{\partial_0}\phi_2 - 2eA^0\phi_1^*\phi_2\right) . \tag{3.159}$$

In return, we normalize the free Klein-Gordon solutions as

$$\phi_{\boldsymbol{p}}^{(r)}(x) = \sqrt{\frac{1}{2E(2\pi)^3}}\mathrm{e}^{-\mathrm{i}\epsilon_r p\cdot x}$$

so that they still fulfill the G-continuum normalization condition

$$\left\langle \phi_{\boldsymbol{p}}^{(r)} \middle| \phi_{\boldsymbol{p}'}^{(r')} \right\rangle_{\mathrm{G}} = \epsilon_r \delta_{rr'} \delta(\boldsymbol{p} - \boldsymbol{p}') \ .$$

Consequently, in the case of the G-normalization to a box volume V, which will be important later on, we have (from now on a capital symbol Φ stands for free plane Klein-Gordon waves)

$$\Phi_i^{(r_i)}(x) = \sqrt{\frac{1}{2E_i V}} \mathrm{e}^{-\mathrm{i}\epsilon_i p_i \cdot x} \ , \ \left\langle \Phi_f^{(r_f)} \middle| \Phi_i^{(r_i)} \right\rangle_{\mathrm{G}} = \epsilon_f \delta_{r_f r_i} \delta_{fi} \ . \quad (3.160)$$

All further steps can be carried out quite analogously to the Dirac case in Subsection 3.2.1:

- casting the solutions to the Klein-Gordon equation into an integral form using the Green function calculus,
- calling for an appropriate causality principle for the Green function in accordance with the Feynman-Stückelberg interpretation,
- deriving a differential equation for the *Feynman boson propagator*,
- and finally rewriting it as an iteratively solvable integral equation for the propagator and for the Klein-Gordon wave function itself.

Since these steps are not associated with any new insights, we leave out their concrete execution and, instead, summarize the relevant results in the following theorem:

Theorem 3.17: Solution of the general Klein-Gordon equation in the propagator formalism in consideration of the Feynman-Stückelberg interpretation

The Klein-Gordon equation

$$(p'_\mu p'^\mu - m_0^2)\phi(x') = V(x')\phi(x')$$

with the *modified potential*

$$V(x') = e[p'_\mu A^\mu(x') + A_\mu(x') p'^\mu] - e^2 A_\mu(x') A^\mu(x')$$

is equivalent to the integral equation

$$\phi(x') = \phi_{\text{free}}(x') + \int \mathrm{d}^4 x \Delta_{\mathrm{F}}^{(0)}(x', x) V(x) \phi(x)$$

as long as $\Delta_{\mathrm{F}}^{(0)}$ fulfills the equation

$$(p'_\mu p'^\mu - m_0^2)\Delta_{\mathrm{F}}^{(0)}(x', x) = \delta(x' - x) \ .$$

The quantity $\Delta_{\mathrm{F}}^{(0)}$ is called *free Feynman boson propagator* if it also obeys the causality principles

▷

$$\left\{\begin{array}{l}\Theta(x'^0 - x^0) \\ \Theta(x^0 - x'^0)\end{array}\right\} \phi_{\rm free}^{(\pm)}(x') = \pm \mathrm{i} \int \mathrm{d}^3x \Delta_{\rm F}^{(0)}(x', x) \mathrm{i}\overleftrightarrow{\partial_0} \phi_{\rm free}^{(\pm)}(x)$$

and

$$\left\{\begin{array}{l}\Theta(x^0 - x'^0) \\ \Theta(x'^0 - x^0)\end{array}\right\} \phi_{\rm free}^{(\pm)*}(x') = \pm \mathrm{i} \int \mathrm{d}^3x \phi_{\rm free}^{(\pm)*}(x) \mathrm{i}\overleftrightarrow{\partial_0} \Delta_{\rm F}^{(0)}(x, x') .$$

These principles ensure the temporally forward [backward] propagation of positive [negative] free Klein-Gordon solutions as well as the respective reverse propagation of the complex conjugated solutions. The Fourier decomposition of the free boson propagator is

$$\left.\begin{array}{rl}\Delta_{\rm F}^{(0)}(x', x) = \Delta_{\rm F}^{(0)}(x' - x) = & \displaystyle\int \frac{\mathrm{d}^4p}{(2\pi)^4} \mathrm{e}^{-\mathrm{i}p_\mu(x'^\mu - x^\mu)} \tilde{\Delta}_{\rm F}^{(0)}(p) \\ \tilde{\Delta}_{\rm F}^{(0)}(p) = & \displaystyle\frac{1}{p^2 - m_0^2 + \mathrm{i}\epsilon} .\end{array}\right\} \quad (3.161)$$

As can easily be seen by a calculation similar to that in Subsection 3.2.2, the imaginary part in the denominator of $\tilde{\Delta}_{\rm F}^{(0)}$ is again necessary to ensure the above causality principles. Furthermore, from this calculation follows the decomposition

$$\begin{aligned}\Delta_{\rm F}^{(0)}(x' - x) = & -\mathrm{i}\Theta(x'^0 - x^0) \int \mathrm{d}^3p \phi_{\boldsymbol{p}}^{(1)}(x') \phi_{\boldsymbol{p}}^{(1)*}(x) \\ & -\mathrm{i}\Theta(x^0 - x'^0) \int \mathrm{d}^3p \phi_{\boldsymbol{p}}^{(2)}(x') \phi_{\boldsymbol{p}}^{(2)*}(x)\end{aligned} \quad (3.162)$$

into free Klein-Gordon solutions, which is used in Exercise 42 to show the validity of the causality principles explicitly.

3.5.2 Scattering Formalism

Also with respect to the description of bosonic scattering processes, the same prerequisites (temporal confinement of the scattering interaction, adiabatic approximation etc.) and considerations are valid as presented in Subsection 3.2.3 for the fermion case. Here we are again interested in the projection of the scattered wave ϕ_i onto the free plane Klein-Gordon wave Φ_f long after the scattering, where, long before the scattering, ϕ_i is given by the likewise free plane wave Φ_i. Accordingly, our ansatz for the scattering amplitude S_{fi} in consideration of the G-scalar product (3.159) is

$$S_{fi} = \lim_{t' \to \pm\infty} \epsilon_f \int \mathrm{d}^3x' \Phi_f^*(x') \mathrm{i}\overleftrightarrow{\partial_0'} \phi_i(x') . \quad (3.163)$$

As in the fermion case and according to the Feynman-Stückelberg interpretation, one of two limits has to be considered depending on the particle type in the final state. If we have bosons (pions) Φ_f is a boson wave function with positive energy propagating forward in time so that the upper limit must be taken. In the case of antibosonic (antipionic) scattering states Φ_f represents a temporally backward propagating boson wave function with negative energy and the lower limit becomes relevant. For the scattered wave ϕ_i, originating from Φ_i, we have correspondingly

$$\Phi_i(x) = \lim_{t\to\mp\infty} \phi_i(x) ,$$

with the upper limit for incoming bosons and the lower limit for incoming antibosons. ϵ_f in (3.163) is a pure convention. Now, using the relations

$$\phi_i(x') = \Phi_i(x') + \int \mathrm{d}^4x_1 \Delta_{\mathrm{F}}^{(0)}(x'-x_1)V(x_1)\phi_i(x_1)$$

$$\Phi_f^*(x_1) = \lim_{t'\to\pm\infty} \mathrm{i}\epsilon_f \int \mathrm{d}^3x' \Phi_f^*(x')\mathrm{i}\overleftrightarrow{\partial_0'} \Delta_{\mathrm{F}}^{(0)}(x'-x_1)$$

$$\int \mathrm{d}^3x \Phi_f^{(r_f)*}(x)\mathrm{i}\overleftrightarrow{\partial_0} \Psi_i^{(r_i)}(x) = \epsilon_f \delta_{r_f r_i} \delta(\boldsymbol{p}_f - \boldsymbol{p}_i)$$

[with the upper limit for bosons ($r_f = 1$; $\epsilon_f = +1$) and the lower limit for antibosons ($r_f = 2$; $\epsilon_f = -1$) in the final state], we obtain from (3.163) the expression

$$\begin{aligned} S_{fi} &= \lim_{t'\to\pm\infty} \epsilon_f \left[\int \mathrm{d}^3x' \Phi_f^*(x')\mathrm{i}\overleftrightarrow{\partial_0'} \Phi_i(x') \right. \\ &\quad \left. + \int \mathrm{d}^3x' \int \mathrm{d}^4x_1 \Phi_f^*(x')\mathrm{i}\overleftrightarrow{\partial_0'} \Delta_{\mathrm{F}}^{(0)}(x'-x_1)V(x_1)\phi_i(x) \right] \\ &= \delta(\boldsymbol{p}-\boldsymbol{p}')\delta_{r_f r_i} - \mathrm{i} \int \mathrm{d}^4x_1 \Phi_f^*(x_1)V(x_1)\phi_i(x_1) , \end{aligned}$$

which is equally valid for all four possible scattering constellations

boson or antiboson $\longrightarrow$ boson or antiboson.

Iterating ϕ_i in the way

$$\begin{aligned} \phi_i(x_1) &= \Phi_i(x_1) \\ &\quad + \int \mathrm{d}^4x_2 \Delta_{\mathrm{F}}^{(0)}(x_1-x_2)V(x_2)\Phi_i(x_2) \\ &\quad + \int \mathrm{d}^4x_2 \int \mathrm{d}^4x_3 \Delta_{\mathrm{F}}^{(0)}(x_1-x_2)V(x_2)\Delta_{\mathrm{F}}^{(0)}(x_2-x_3)\Phi_i(x_3) \\ &\quad + \ldots , \end{aligned}$$

we arrive at (compare to Theorem 3.5)

Theorem 3.18: Scattering matrix in the Klein-Gordon theory

The scattering amplitude S_{fi} is defined by the projection of ϕ_i onto Φ_f long after the scattering against a target where ϕ_i is the scattered wave evolving from the free plane wave Φ_i: ▷

$$S_{fi} = \lim_{t'\to\pm\infty} \epsilon_f \int \mathrm{d}^3x' \Phi_f^* \mathrm{i}\overleftrightarrow{\partial_0'} \phi_i(x') \;, \quad \lim_{t\to\mp\infty} \phi_i(x) = \Phi_i(x) \;.$$

In the case of pionic [antipionic] scattering states Φ_f is a plane pion wave with positive [negative] energy propagating forward [backward] in time so that, on the left hand side of this equation, the limit $t' \to +\infty$ [$t' \to -\infty$] has to be considered. If the incident particles are pions [antipions] Φ_i is a plane pion wave with positive [negative] energy and, on the right hand side, the limit $t \to -\infty$ [$t \to +\infty$] is relevant.
With the help of the Feynman propagator formalism, the scattering amplitude S_{fi} can be expanded in a series of multiple scatterings:

$$\begin{aligned} S_{fi} &= \delta(\boldsymbol{p}_f - \boldsymbol{p}_i)\delta_{r_f r_i} - \mathrm{i} \int \mathrm{d}^4x_1 \Phi_f^*(x_1) V(x_1) \phi_i(x_1) \\ &= \delta(\boldsymbol{p}_f - \boldsymbol{p}_i)\delta_{r_f r_i} \\ &\quad - \mathrm{i} \int \mathrm{d}^4x_1 \Phi_f^*(x_1) V(x_1) \Phi_i(x_1) \\ &\quad - \mathrm{i} \int \mathrm{d}^4x_1 \int \mathrm{d}^4x_2 \Phi_f^*(x_2) V(x_2) \Delta_\mathrm{F}^{(0)}(x_2 - x_1) V(x_1) \Phi_i(x_1) \\ &\quad - \mathrm{i} \int \mathrm{d}^4x_1 \int \mathrm{d}^4x_2 \int \mathrm{d}^4x_3 \Phi_f^*(x_3) V(x_3) \Delta_\mathrm{F}^{(0)}(x_3 - x_2) \\ &\quad \times V(x_2) \Delta_\mathrm{F}^{(0)}(x_2 - x_1) V(x_1) \Phi_i(x_1) \\ &\quad - \ldots \;. \end{aligned}$$

V denotes the modified potential of the target (see Theorem 3.17) and $\Delta_\mathrm{F}^{(0)}$ the free boson propagator.
This theorem is based on the adiabatic approximation and the Feynman-Stückelberg interpretation.

All in all, our reasoning leads to a formalism for the relativistic description of bosonic scattering processes which is very similar to that of the fermion case with respect to structure and interpretation. This means that, with an appropriate relabeling of particles and wave functions, we can adopt all considerations employed after Theorem 3.5 about electron and positron scatterings as well as pair creation and annihilation processes, except for the hole theory. However, the boson case possesses one peculiarity: the terms in the scattering series in Theorem 3.18 are no longer identical to the expansion terms of the coupling constant e since the modified potential V contains one linear and one quadratic part in e, which both have to be taken into account adequately.

Analogously to the fermion case, we now apply Theorems 3.17 and 3.18 in combination with Theorem 3.3 to some concrete spin-0 scattering processes.

In doing so, we find that, due to the absence of the spin degree of freedom, some calculations are much simpler than in the fermion case.

3.5.3 Coulomb Scattering of Pions

As the first concrete example we consider the scattering of pions against a Coulomb potential to lowest order of e. If we initially start with an arbitrary background potential A^μ, the corresponding scattering amplitude follows from Theorem 3.18 ($f \neq i$) as

$$\begin{aligned} S_{fi} &= -\mathrm{i}e \int \mathrm{d}^4x \Phi_f^*(x) \left[\mathrm{i}\partial_\mu A^\mu(x) + A^\mu(x)\mathrm{i}\partial_\mu\right] \Phi_i(x) \\ &= -\mathrm{i}e \int \mathrm{d}^4x \left[\Phi_f^*(\mathrm{i}\partial_\mu \Phi_i) A^\mu - (\mathrm{i}\partial_\mu \Phi_f^*)\Phi_i A^\mu\right] \\ &= -\mathrm{i}e \int \mathrm{d}^4x \left(\Phi_f^* \mathrm{i}\overleftrightarrow{\partial}_\mu \Phi_i\right) A^\mu \ , \end{aligned} \tag{3.164}$$

where partial integration has been used in the second step. In the case of pion scattering we have [see (3.160)]

$$\Phi_i(x) = \frac{1}{\sqrt{2E_iV}} \mathrm{e}^{-\mathrm{i}p_i\cdot x} \ , \ \Phi_f(x) = \frac{1}{\sqrt{2E_fV}} \mathrm{e}^{-\mathrm{i}p_f\cdot x}$$

so that (3.164) turns into

$$\begin{aligned} S_{fi} &= \frac{-\mathrm{i}e}{V} \frac{1}{\sqrt{4E_iE_f}} \int \mathrm{d}^4x (p_i + p_f)_\mu A^\mu(x) \mathrm{e}^{i(p_f - p_i)\cdot x} \\ &= \frac{1}{V} \frac{1}{\sqrt{4E_iE_f}} (-\mathrm{i}e)(p_i + p_f)_\mu \tilde{A}^\mu(q) \ , \ q = p_f - p_i \ , \end{aligned} \tag{3.165}$$

with the background potential

$$\tilde{A}^\mu(q) = \int \mathrm{d}^4x \mathrm{e}^{\mathrm{i}q\cdot x} A^\mu(x)$$

in momentum space. If, on the other hand, we are interested in the scattering of antipions we need to choose

$$\Phi_i(x) = \frac{1}{\sqrt{2E_fV}} \mathrm{e}^{+\mathrm{i}p_f\cdot x} \ , \ \Phi_f(x) = \frac{1}{\sqrt{2E_iV}} \mathrm{e}^{+\mathrm{i}p_i\cdot x}$$

so that

$$S_{fi} = \frac{1}{V} \frac{1}{\sqrt{4E_iE_f}} (-\mathrm{i}e)(-p_i - p_f)_\mu \tilde{A}^\mu(q) \ , \ q = p_f - p_i \ . \tag{3.166}$$

Obviously, both expressions (3.165) and (3.166) suggest the *one-photon vortex*[29] in momentum space shown in Figure 3.43, where the vortex factor con-

[29] In the Klein-Gordon case there also exists the *two-photon vortex* to which we return in Subsection 3.5.6.

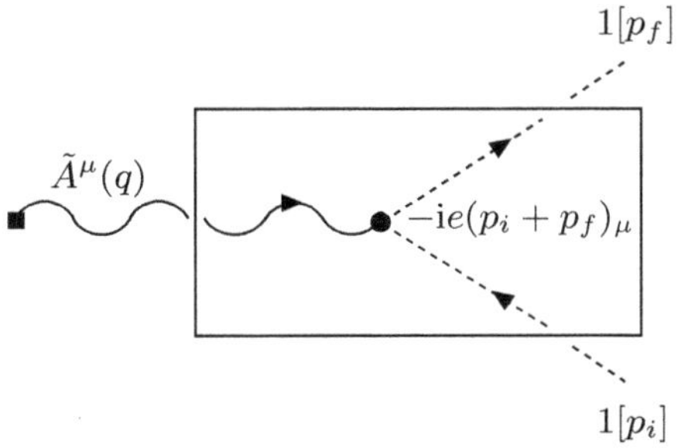

Fig. 3.43. $\mathcal{O}(e)$-Feynman diagram of pion scattering against a background potential in momentum space [see (3.165) and (3.166)]. Inside the box we have the one-photon vortex of the bosonic scattering theory. The vortex factor momenta refer to those of the dashed boson lines on the level of wave functions, whereas the momenta behind the one-factors of the external boson lines represent the particle momenta. The latter were introduced in analogy to the fermion case. They do not occur in matrix elements. The index μ of the vortex factor is contracted with that of the photon line.

tains the momenta of the adjacent boson lines on the level of wave functions, contrary to the provisionally introduced momenta in the square brackets behind the one-factors of the external boson lines referring to the particle level.

Note that in the construction of scattering amplitudes via Feynman diagrams we adopt the i-factor rule of the spin-1/2 scattering theory right from the beginning: a factor of $-\mathrm{i}$ at each one-photon vortex and a factor of $+\mathrm{i}$ at each internal boson line.

Returning to the concrete case of pion scattering against the Coulomb potential

$$eA^\mu(x)=\frac{\alpha g^{\mu 0}}{|\boldsymbol{x}|}\;,\;\alpha=-Ze^2\Longrightarrow e\tilde{A}^\mu(q)=2\pi\delta(q^0)\frac{4\pi\alpha g^{\mu 0}}{\boldsymbol{q}^2}\;,$$

we obtain from (3.165)

$$S_{fi}=\frac{-\mathrm{i}[2\pi\delta(E_f-E_i)]}{V}\frac{1}{\sqrt{4E_iE_f}}M_{fi}\;,\;M_{fi}=\frac{4\pi\alpha}{\boldsymbol{q}^2}(E_i+E_f)\;.$$

Cross section. To determine the differential cross section

$$\mathrm{d}\sigma=\frac{|S_{fi}|^2}{T|\boldsymbol{j}_i|}\frac{V\mathrm{d}^3p_f}{(2\pi)^3}=\frac{1}{2E_i}\frac{[2\pi\delta(E_f-E_i)]^2}{TV|\boldsymbol{j}_i|}|M_{fi}|^2\frac{\mathrm{d}^3p_f}{2E_f(2\pi)^3}$$

we need the current density $|\boldsymbol{j}_i|$ of the incoming pion. Following (3.158), it is

$$|\boldsymbol{j}_i|=\left|-\Phi_i^*\mathrm{i}\overleftrightarrow{\boldsymbol{\nabla}}\Phi_i\right|=\frac{\boldsymbol{p}_i}{2E_iV}\;.$$

Together with the known replacements

$$[2\pi\delta(E_f-E_i)]^2=2\pi T\delta(E_f-E_i)\;,\;\mathrm{d}^3p_f=|\boldsymbol{p}_f|E_f\mathrm{d}E_f\mathrm{d}\Omega\;,$$

follows that

$$\begin{aligned}\frac{\mathrm{d}\sigma}{\mathrm{d}\Omega} &= \frac{1}{4(2\pi)^2|\boldsymbol{p}_i|}\int \mathrm{d}E_f|\boldsymbol{p}_f||M_{fi}|^2\delta(E_f-E_i) = \frac{1}{4(2\pi)^2}|M_{fi}|^2_{|\boldsymbol{p}_f|=|\boldsymbol{p}_i|} \\ &= \left(\frac{4\alpha^2E_i^2}{\boldsymbol{q}^4}\right)_{|\boldsymbol{p}_f|=|\boldsymbol{p}_i|} = \frac{\alpha^2}{4v_i^4E_i^2\sin^4\frac{\theta}{2}} ,\end{aligned}$$

with

$$\boldsymbol{q}^2\Big|_{|\boldsymbol{p}_f|=|\boldsymbol{p}_i|} = 4|\boldsymbol{p}_i|^2\sin^2\frac{\theta}{2} = 4v_i^2E_i^2\sin^2\frac{\theta}{2} .$$

Comparing this with the Mott cross section in Theorem 3.7, one notices an additional factor of $(1-v_i^2\sin^2\theta/2)$ in the Dirac case which is ascribed to the magnetic moment of the electrons. In the limit of small velocities the corresponding magnetic interaction vanishes and both results coincide.

Theorem 3.19: Coulomb scattering of pions to leading order

To leading order, the scattering amplitude for the scattering of pions against a Coulomb potential of the form

$$eA^0(x) = \frac{\alpha}{|\boldsymbol{x}|} , \quad \boldsymbol{A}(x) = \boldsymbol{0}$$

is $(f \neq i)$

$$S_{fi} = -\mathrm{i}\frac{2\pi\delta(E_f-E_i)}{V}\frac{1}{\sqrt{4E_iE_f}}M_{fi} ,$$

with

$$M_{fi} = \frac{4\pi\alpha}{\boldsymbol{q}^2}(E_i+E_f) , \quad \boldsymbol{q} = \boldsymbol{p}_f - \boldsymbol{p}_i .$$

The differential cross section follows as

$$\begin{aligned}\mathrm{d}\sigma &= \frac{1}{2E_i}\frac{1}{V|\boldsymbol{j}_i|}|M_{fi}|^2(2\pi)\delta(E_f-E_i)\frac{\mathrm{d}^3p_f}{2E_f(2\pi)^3} \\ &= \frac{1}{2|\boldsymbol{p}_i|}|M_{fi}|^2(2\pi)\delta(E_f-E_i)\frac{\mathrm{d}^3p_f}{2E_f(2\pi)^3} \\ \Longrightarrow \frac{\mathrm{d}\sigma}{\mathrm{d}\Omega} &= \frac{1}{4(2\pi)^2}|M_{fi}|^2_{|\boldsymbol{p}_f|=|\boldsymbol{p}_i|} = \frac{\alpha^2}{4v_i^4E_i^2\sin^4\frac{\theta}{2}} ,\end{aligned}$$

where all scattering momenta $\boldsymbol{p}_f$ toward $\mathrm{d}\Omega$ have been integrated out in the last relation.

Clearly, the scattering of antipions leads to the same cross section since both amplitudes (3.165) and (3.166) differ only by a sign (compare to the notes after Theorem 3.7).

3.5.4 Pion-Pion Scattering

Similarly to the Dirac case, we are free to consider the background potential A^μ to be created by the current of another particle. In this way, we arrive again at a current-current interaction or at a two-particle scattering. In the following we discuss the mutual scattering of pions following closely our line of argument in Subsections 3.3.2 and 3.3.5. If J'^μ denotes the current of the second pion, then, taking into account (3.82), the scattering amplitude to lowest order in e follows as $(f \neq i)$

$$S_{fi} = -\mathrm{i} \int \mathrm{d}^4x \int \mathrm{d}^4y \left[e\Phi_f^*(x)\mathrm{i}\frac{\overleftrightarrow{\partial}}{\partial x^\mu}\Phi_i(x) \right] D_\mathrm{F}^{(0)}(x-y) J'^\mu(y) ,$$

with the free boson propagator [see (3.84)]

$$D_\mathrm{F}^{(0)}(x-y) = \int \frac{\mathrm{d}^4q}{(2\pi)^4} \frac{-4\pi}{q^2 + \mathrm{i}\epsilon} \mathrm{e}^{-\mathrm{i}q\cdot(x-y)} .$$

As before, it seems plausible to interpret

$$J_\mu(x) = e\Phi_f^*(x)\mathrm{i}\frac{\overleftrightarrow{\partial}}{\partial x^\mu}\Phi_i(x)$$

as the transition current of the first pion (to first order) and, correspondingly,[30]

$$J'^\mu(y) = e\Phi_f'^*(y)\mathrm{i}\frac{\overleftrightarrow{\partial}}{\partial y_\mu}\Phi_i'(y)$$

as the transition current of the second pion (to first order), where $\Phi'_{i,f}$ are the initial and final wave functions of the second pion. Thus, in total, there follows the scattering amplitude

$$\begin{aligned} S_{fi} &= -\mathrm{i} \int \mathrm{d}^4x \int \mathrm{d}^4y \left[e\Phi_f^*(x)\mathrm{i}\frac{\overleftrightarrow{\partial}}{\partial x^\mu}\Phi_i(x) \right] D_\mathrm{F}^{(0)}(x-y) \\ &\quad \times \left[e\Phi_f'^*(y)\mathrm{i}\frac{\overleftrightarrow{\partial}}{\partial y_\mu}\Phi_i'(y) \right] , \end{aligned}$$

which is symmetric in both particles. Since both particles are identical, we have to consider, as for the electron-electron scattering in Subsection 3.3.5, two contributions corresponding to the experimentally indistinguishable situations of the mutual fly-by (direct scattering) and the mutual reflection (exchange scattering). In the case of direct scattering, we have

$$\Phi_i(x) = \frac{1}{\sqrt{2E_iV}}\mathrm{e}^{-\mathrm{i}p_i\cdot x} , \quad \Phi_f(x) = \frac{1}{\sqrt{2E_fV}}\mathrm{e}^{-\mathrm{i}p_f\cdot x}$$

---

[30] See footnote 17 on page 234.

$$\Phi'_i(y) = \frac{1}{\sqrt{2E'_i V}} \mathrm{e}^{-\mathrm{i}p'_i \cdot y} \ , \ \Phi'_f(y) = \frac{1}{\sqrt{2E'_f V}} \mathrm{e}^{-\mathrm{i}p'_f \cdot y} \ ,$$

so that

$$\begin{aligned} S_{fi}(\mathrm{dir}) &= -\frac{\mathrm{i}}{V^2} \frac{1}{\sqrt{4E_i E_f}} \frac{1}{\sqrt{4E'_i E'_f}} \int \mathrm{d}^4x \int \mathrm{d}^4y \int \frac{\mathrm{d}^4q}{(2\pi)^4} \\ &\quad \times (p_i + p_f)_\mu \frac{-4\pi e^2}{q^2 + \mathrm{i}\epsilon} (p'_i + p'_f)^\mu \\ &\quad \times \mathrm{e}^{\mathrm{i}(p_f - p_i)\cdot x} \mathrm{e}^{-\mathrm{i}q\cdot(x-y)} \mathrm{e}^{\mathrm{i}(p'_f - p'_i)\cdot y} \ . \end{aligned}$$

The coordinate and momentum integrations are the same as before (3.87) and lead finally to

$$\begin{aligned} S_{fi}(\mathrm{dir}) &= \frac{(2\pi)^4 \delta(p_f + p'_f - p_i - p'_i)}{V^2} \frac{1}{\sqrt{4E_i E_f}} \frac{1}{\sqrt{4E'_i E'_f}} M_{fi}(\mathrm{dir}) \\ M_{fi}(\mathrm{dir}) &= (-\mathrm{i}e)(p_i + p_f)_\mu \frac{-4\pi\mathrm{i}}{q^2 + \mathrm{i}\epsilon} (-\mathrm{i}e)(p'_i + p'_f)^\mu \ , \ q = p_f - p_i \ . \end{aligned}$$

The exchange scattering amplitude is simply calculated through the replacement $p_f \leftrightarrow p'_f$. All in all, we obtain the scattering amplitude for the pion-pion scattering to lowest order as

$$S_{fi} = \frac{(2\pi)^4 \delta(p_f + p'_f - p_i - p'_i)}{V^2} \frac{1}{\sqrt{4E_i E_f}} \frac{1}{\sqrt{4E'_i E'_f}} M_{fi} \ ,$$

with

$$\left. \begin{aligned} M_{fi} &= M_{fi}(\mathrm{dir}) + M_{fi}(\mathrm{ex}) \\ M_{fi}(\mathrm{dir}) &= (-\mathrm{i}e)(p_i + p_f)_\mu \frac{-4\pi\mathrm{i}}{q^2 + \mathrm{i}\epsilon} (-\mathrm{i}e)(p'_i + p'_f)^\mu \\ &= (p_i + p_f)_\mu \frac{4\pi\mathrm{i}e^2}{q^2 + \mathrm{i}\epsilon} (p'_i + p'_f)^\mu \ , \ q = p_f - p_i \\ M_{fi}(\mathrm{ex}) &= (-\mathrm{i}e)(p_i + p'_f)_\mu \frac{-4\pi\mathrm{i}}{q'^2 + \mathrm{i}\epsilon} (-\mathrm{i}e)(p'_i + p_f)^\mu \\ &= (p_i + p'_f)_\mu \frac{4\pi\mathrm{i}e^2}{q'^2 + \mathrm{i}\epsilon} (p'_i + p_f)^\mu \ , \ q' = p'_f - p_i \ , \end{aligned} \right\} \quad (3.167)$$

where the amplitudes $M_{fi}(\mathrm{dir})$ and $M_{fi}(\mathrm{ex})$ can be represented by the Feynman diagrams of Figure 3.44. Contrary to the electron-electron scattering, there is no relative sign between $M_{fi}(\mathrm{dir})$ and $M_{fi}(\mathrm{ex})$ since, according to the Bose-Einstein statistics, the whole scattering amplitude must be symmetric under the exchange of both bosons in the initial state ($p_i \leftrightarrow p'_i$) or the final state ($p_f \leftrightarrow p'_f$).

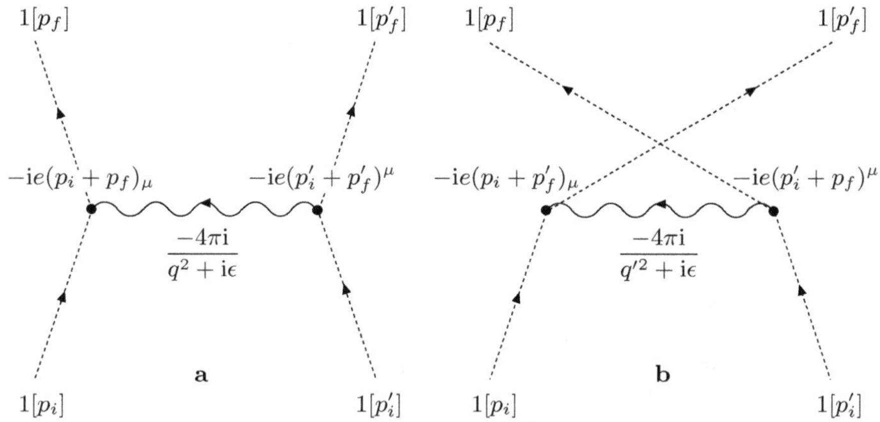

Fig. 3.44. Feynman diagrams of the direct $\mathcal{O}\left(e^2\right)$-scattering amplitude (**a**) and the exchange $\mathcal{O}\left(e^2\right)$-scattering amplitude (**b**) for pion-pion scattering in momentum space (compare to Figures 3.16 and 3.17). As in the Spin-1/2 scattering theory, internal photon lines are represented as wavy lines and assigned the i-fold of the free photon propagator. Energy and momentum are conserved at each vortex. Therefore, the four-momentum transfers are $q = p_f - p_i = -(p'_f - p'_i)$ and $q' = p_f - p'_i = -(p_f - p'_i)$.

Cross section. Next we need to evaluate the sixfold differential cross section

$$\begin{aligned} \mathrm{d}\sigma &= \frac{|S_{fi}|^2}{T|\boldsymbol{j}_i|} \frac{V\mathrm{d}^3 p_f}{(2\pi)^3} \frac{V\mathrm{d}^3 p'_f}{(2\pi)^3} \\ &= \frac{1}{2E_i} \frac{1}{2E'_i} \frac{1}{|\boldsymbol{j}_i|V} |M_{fi}|^2 (2\pi)^4 \delta(p_f + p'_f - p_i - p'_i) \frac{\mathrm{d}^3 p_f}{2E_f(2\pi)^3} \frac{\mathrm{d}^3 p'_f}{2E'_f(2\pi)^3} . \end{aligned}$$

This is done in the center of mass system (as regards the kinematic situation, see Figures 3.16**a** and 3.17**a**). Restricting ourselves again to collinear particle currents, we can adopt formula (3.89) from Subsection 3.3.2 for $|\boldsymbol{j}_i|$,

$$|\boldsymbol{j}_i| = \frac{\sqrt{(p_i \cdot p'_i)^2 - m_0^2 m_0'^2}}{V E_i E'_i} ,$$

in order to obtain the Lorentz-invariant expression

$$\begin{aligned} \mathrm{d}\sigma &= \frac{1}{4\sqrt{(p_i \cdot p'_i)^2 - m_0^2 m_0'^2}} |M_{fi}|^2 (2\pi)^4 \delta(p_f + p'_f - p_i - p'_i) \\ &\quad \times \frac{\mathrm{d}^3 p_f}{2E_f(2\pi)^3} \frac{\mathrm{d}^3 p'_f}{2E'_f(2\pi)^3} . \end{aligned} \tag{3.168}$$

Inserting the center of mass relation $[m_0 = m'_0,\ p_i = (E_i, \boldsymbol{p}_i),\ p'_i = (E_i, -\boldsymbol{p}_i)]$

$$\frac{1}{4\sqrt{(p_i \cdot p'_i)^2 - m_0^4}} = \frac{1}{8E_i|\boldsymbol{p}_i|}$$

as well as the known identities

$$\mathrm{d}^3 p_f = |\boldsymbol{p}_f| E_f \mathrm{d}E_f \mathrm{d}\Omega \ , \quad \frac{\mathrm{d}^3 p'_f}{2E'_f} = \int \mathrm{d}^4 p'_f \delta(p'^2_f - m_0^2)\Theta(p'^0_f) \ ,$$

(3.168) turns into

$$\left(\frac{\mathrm{d}\sigma}{\mathrm{d}\Omega}\right)_{\mathrm{cm}} = \frac{1}{16(2\pi)^2 E_i |\boldsymbol{p}_i|} \int \mathrm{d}E_f |\boldsymbol{p}_f| \int \mathrm{d}^4 p'_f |M_{fi}|^2 \delta(p_f + p'_f - p_i - p'_i) \times \delta(p'^2_f - m_0^2)\Theta(p'^0_f) \ .$$

Obviously, this equation is formally identical to the first row of (3.108) multiplied by $1/(16m_0^4)$. Due to the kinematics being identical to that of the electron-electron scattering, we can read off the final result directly from the last row of (3.108) multiplied by this factor:

$$\left(\frac{\mathrm{d}\sigma}{\mathrm{d}\Omega}\right)_{\mathrm{cm}} = \frac{1}{64(2\pi)^2 E_i^2} |M_{fi}|^2_{\mathrm{cm}} \ .$$

Amplitude square. What remains is the explicit determination of $|M_{fi}|^2_{\mathrm{cm}}$. Starting with (3.167), we have

$$|M_{fi}|^2 = |M_{fi}(\mathrm{dir})|^2 + |M_{fi}(\mathrm{ex})|^2 + 2\mathrm{Re}\left[M_{fi}(\mathrm{dir})M^*_{fi}(\mathrm{ex})\right] \ ,$$

where

$$\left.\begin{aligned} |M_{fi}(\mathrm{dir})|^2 &= \frac{(4\pi)^2 e^4}{(q^2)^2} \left[p_i \cdot p'_i + p_i \cdot p'_f + p_f \cdot p'_i + p_f \cdot p'_f\right]^2 \\ |M_{fi}(\mathrm{ex})|^2 &= \frac{(4\pi)^2 e^4}{(q'^2)^2} \left[p_i \cdot p'_i + p_i \cdot p_f + p'_f \cdot p'_i + p'_f \cdot p_f\right]^2 \\ 2\mathrm{Re}\left[M_{fi}(\mathrm{dir})M^*_{fi}(\mathrm{ex})\right] &= 2M_{fi}(\mathrm{dir})M^*_{fi}(\mathrm{ex}) \\ &= \frac{2(4\pi)^2 e^4}{q^2 q'^2}(p_i \cdot p'_i + p_i \cdot p'_f + p_f \cdot p'_i + p_f \cdot p'_f) \\ &\quad \times (p_i \cdot p'_i + p_i \cdot p_f + p'_f \cdot p'_i + p'_f \cdot p_f) \ . \end{aligned}\right\} \quad (3.169)$$

Using the kinematic relations (3.112) of electron-electron scattering in the center of mass system, this expression can be simplified to

$$|M_{fi}|^2_{\mathrm{cm}} = 4(4\pi)^2 e^4 \left[\frac{4(2E_i^2 - m_0^2)^2}{\boldsymbol{p}_i^4 \sin^4\theta} - \frac{4(2E_i^2 - m_0^2)}{\boldsymbol{p}_i^2 \sin^2\theta} + 1\right] \ .$$

As expected, this result is again symmetric under $\theta \to \pi - \theta$ reflecting the bosons' indistinguishability.

Theorem 3.20: Pion-pion scattering to leading order

The scattering amplitude for pion-pion scattering to leading order is $(f \neq i)$

$$S_{fi} = \frac{(2\pi)^4 \delta(p_f + p'_f - p_i - p'_i)}{V^2} \frac{1}{\sqrt{4E_i E_f}} \frac{1}{\sqrt{4E'_i E'_f}} M_{fi} \ ,$$

with the Lorentz-invariant amplitude $(q = p_f - p_i,\ q' = p'_f - p_i)$ ▷

$$
\begin{aligned}
M_{fi} &= M_{fi}(\mathrm{dir}) + M_{fi}(\mathrm{ex}) \\
M_{fi}(\mathrm{dir}) &= (p_i + p_f)_\mu \frac{4\pi \mathrm{i} e^2}{q^2 + \mathrm{i}\epsilon}(p_i' + p_f')^\mu \\
M_{fi}(\mathrm{ex}) &= (p_i + p_f')_\mu \frac{4\pi \mathrm{i} e^2}{q'^2 + \mathrm{i}\epsilon}(p_i' + p_f)^\mu .
\end{aligned}
$$

The differential cross section follows as

$$
\begin{aligned}
\mathrm{d}\sigma &= \frac{1}{4\sqrt{(p_i \cdot p_i')^2 - m_0^4}} |M_{fi}|^2 (2\pi)^4 \delta(p_f + p_f' - p_i - p_i') \\
&\quad \times \frac{\mathrm{d}^3 p_f}{2E_f(2\pi)^3} \frac{\mathrm{d}^3 p_f'}{2E_f'(2\pi)^3}
\end{aligned}
$$

and, particularly in the center of mass system,

$$
\left(\frac{\mathrm{d}\sigma}{\mathrm{d}\Omega}\right)_{\mathrm{cm}} = \frac{1}{64(2\pi)^2 E_i^2} |M_{fi}|^2_{\mathrm{cm}} ,
$$

where all pionic scattering momenta $\boldsymbol{p}_f$ toward $\mathrm{d}\Omega$ and all pionic scattering momenta $\boldsymbol{p}_f'$ have been integrated out in the last relation. The amplitude square is

$$
|M_{fi}|^2_{\mathrm{cm}} = 4(4\pi)^2 e^4 \left[\frac{4(2E_i^2 - m_0^2)^2}{\boldsymbol{p}_i^4 \sin^4\theta} - \frac{4(2E_i^2 - m_0^2)}{\boldsymbol{p}_i^2 \sin^2\theta} + 1\right] .
$$

Considering the results of this theorem along with Figure 3.44 in view of the spinor Feynman rules in Subsection 3.3.9, we see that, with some small extensions, they also cover the case of bosonic two-particle scatterings. These extensions (presented in totality at the end of this section) are essentially

- a *boson factor* of $N_{i,f}^{(\dots)} = 1/2$ for each (anti)boson (1. and 2. rule),
- a one-factor at each external boson line, the one-photon vortex from Figure 3.43 as well as the *two-photon vortex* (4. rule) that we discuss later on.

Furthermore, due to the Bose-Einstein statistics, the signs in 4.i) and 4.ii) do not apply here.

Using these extensions, we now calculate the bosonic-fermionic mixed process of pion-antipion production by electron-positron annihilation. Subsequently, we draw on the Compton scattering to discuss the treatment of real photons as well as the two-photon vortex originating from the term $-e^2 A_\mu A^\mu$ of the modified potential V in Theorem 3.17.

3.5.5 Pion Production via Electrons

According to our extended Feynman rules, the Feynman diagram for pion-antipion production via electron-positron annihilation to leading order in momentum space is given in Figure 3.45. The scattering amplitude follows from this as

$$\left.\begin{aligned} S_{fi} &= \frac{(2\pi)^4\delta(p_f+\bar{p}_f-p_i-\bar{p}_i)}{V^2}\sqrt{\frac{m_0^2}{E_i\bar{E}_i}}\sqrt{\frac{1}{4E_f\bar{E}_f}}M_{fi} \\ M_{fi} &= -\mathrm{i}e(p_f-\bar{p}_f)_\mu\frac{-4\pi\mathrm{i}}{q^2+\mathrm{i}\epsilon}\bar{v}(\bar{p}_i,\bar{s}_i)(-\mathrm{i}e)\gamma^\mu u(p_i,s_i) \\ &= \frac{4\pi\mathrm{i}e^2}{q^2+\mathrm{i}\epsilon}\bar{v}(\bar{p}_i,\bar{s}_i)(\not{p}_f-\bar{\not{p}}_f)u(p_i,s_i) \\ q &= p_i+\bar{p}_i=p_f+\bar{p}_f\ , \end{aligned}\right\} \quad (3.170)$$

where m_0 denotes the electron mass and M_0 the pion mass.

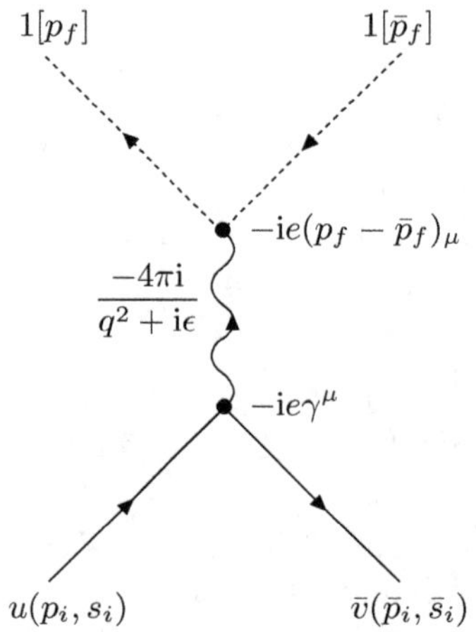

Fig. 3.45. $\mathcal{O}(e^2)$-Feynman diagram for the pion-antipion production in momentum space. Energy and momentum are conserved at each vortex. Therefore, the momentum transfer is given by $q=p_i+\bar{p}_i=p_f+\bar{p}_f$.

Cross section. The starting point for the calculation of the differential cross section is the formula

$$\mathrm{d}\sigma = \frac{m_0^2}{\sqrt{(p_i\cdot\bar{p}_i)^2-m_0^4}}|M_{fi}|^2(2\pi)^4\delta(p_f+\bar{p}_f-p_i-\bar{p}_i) \times\frac{\mathrm{d}^3p_f}{2E_f(2\pi)^3}\frac{\mathrm{d}^3\bar{p}_f}{2\bar{E}_f(2\pi)^3}\ ,$$

which is evaluated in the center of mass system (see Figure 3.46). There we have

$$p_i=(E_i,\boldsymbol{p}_i)\ ,\ \bar{p}_i=(E_i,-\boldsymbol{p}_i)\Longrightarrow\frac{m_0^2}{\sqrt{(p_i\cdot\bar{p}_i)^2-m_0^4}}=\frac{m_0^2}{2E_i|\boldsymbol{p}_i|}\ .$$

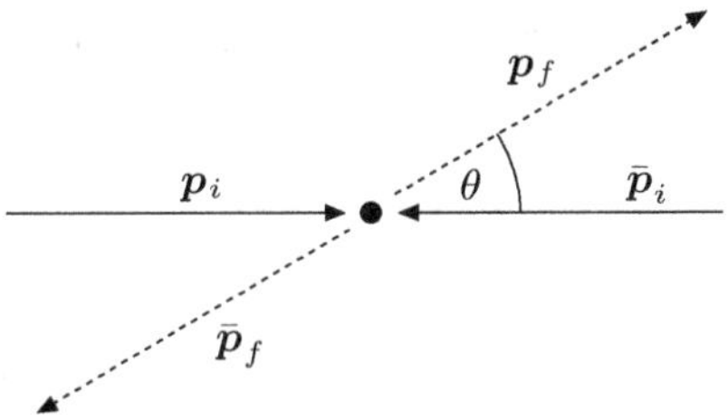

Fig. 3.46. Kinematic situation of pion-antipion production via electron-positron annihilation in the center of mass system. Due to energy and momentum conservation, the relations $E_i = \bar{E}_i = E_f = \bar{E}_f$ and $\boldsymbol{p}_i = -\bar{\boldsymbol{p}}_i$, $\boldsymbol{p}_f = -\bar{\boldsymbol{p}}_f$ hold.

Thus, along with

$$\mathrm{d}^3 p_f = |\boldsymbol{p}_f| E_f \mathrm{d}E_f \mathrm{d}\Omega \ , \quad \frac{\mathrm{d}^3 \bar{p}_f}{2\bar{E}_f} = \int \mathrm{d}^4 \bar{p}_f \delta(\bar{p}_f^2 - M_0^2) \Theta(\bar{p}_f^0) \ ,$$

it follows that

$$\begin{aligned}
\left(\frac{\mathrm{d}\sigma}{\mathrm{d}\Omega}\right)_{\mathrm{cm}} &= \frac{m_0^2}{4(2\pi)^2 E_i |\boldsymbol{p}_i|} \int \mathrm{d}E_f |\boldsymbol{p}_f| \int \mathrm{d}^4 \bar{p}_f |M_{fi}|^2 \delta(p_f + \bar{p}_f - p_i - \bar{p}_i) \\
&\quad \times \delta(\bar{p}_f^2 - M_0^2) \Theta(\bar{p}_f^0) \\
&= \frac{m_0^2}{4(2\pi)^2 E_i |\boldsymbol{p}_i|} \int \mathrm{d}E_f |\boldsymbol{p}_f| |M_{fi}|^2_{\bar{p}_f = p_i + \bar{p}_i - p_f} \\
&\quad \times \delta[(p_i + \bar{p}_i - p_f)^2 - M_0^2] \Theta(p_i^0 + \bar{p}_i^0 - p_f^0) \\
&= \frac{m_0^2}{4(2\pi)^2 E_i |\boldsymbol{p}_i|} \int \mathrm{d}E_f |\boldsymbol{p}_f| |M_{fi}|^2_{\bar{p}_f = p_i + \bar{p}_i - p_f} \\
&\quad \times \delta[4E_i(E_i - E_f)] \Theta(2E_i - E_f) \\
&= \frac{m_0^2}{4(2\pi)^2 E_i |\boldsymbol{p}_i|} \int\limits_{M_0}^{2E_i} \mathrm{d}E_f |\boldsymbol{p}_f| |M_{fi}|_{\bar{p}_f = p_i + \bar{p}_i - p_f} \delta[4E_i(E_i - E_f)] \\
&= \frac{m_0^2}{4(2\pi)^2 E_i |\boldsymbol{p}_i|} \int\limits_{M_0}^{2E_i} \mathrm{d}E_f |\boldsymbol{p}_f| |M_{fi}|^2_{\bar{p}_f = p_i + \bar{p}_i - p_f} \frac{\delta(E_f - E_i)}{4E_i} \\
&= \frac{m_0^2 |\boldsymbol{p}_f|}{16(2\pi)^2 E_i^2 |\boldsymbol{p}_i|} |M_{fi}|^2_{\mathrm{cm}} \ .
\end{aligned}$$

Amplitude square. The calculation of $|M_{fi}|^2$ can be carried out quite easily if we disregard polarization effects and, instead of (3.170), consider the amplitude square

$$\overline{|M_{fi}|^2} = \frac{(4\pi)^2 e^4}{4q^4} \sum_{s_i, \bar{s}_i} \left| \bar{v}(\bar{p}_i, \bar{s}_i)(\not{p}_f - \bar{\not{p}}_f) u(p_i, s_i) \right|^2$$

averaged over the incoming electron spins s_i, $\bar{s}_i$. In this case and taking into account

$$\bar{v}(\bar{p}_i, \bar{s}_i)(\not{p}_f + \bar{\not{p}}_f)u(p_i, s_i) = \bar{v}(\bar{p}_i, \bar{s}_i)(\not{p}_i + \bar{\not{p}}_i)u(p_i, s_i) = 0$$

as well as Theorem 3.6, we obtain

$$\begin{aligned}
\overline{|M_{fi}|^2} &= \frac{(4\pi)^2 e^4}{4q^4} \sum_{s_i, \bar{s}_i} \left|\bar{v}(\bar{p}_i, \bar{s}_i)(\not{p}_f + \bar{\not{p}}_f - 2\bar{\not{p}}_f)u(p_i, s_i)\right|^2 \\
&= \frac{(4\pi)^2 e^4}{q^4} \sum_{s_i, \bar{s}_i} \left|\bar{v}(\bar{p}_i, \bar{s}_i)\bar{\not{p}}_f u(p_i, s_i)\right|^2 \\
&= \frac{(4\pi)^2 e^4}{q^4} \sum_{s_i, \bar{s}_i} [\bar{v}(\bar{p}_i, \bar{s}_i)\bar{\not{p}}_f u(p_i, s_i)][\bar{u}(p_i, s_i)\bar{\not{p}}_f v(\bar{p}_i, \bar{s}_i)] \\
&= -\frac{(4\pi)^2 e^4}{q^4} \mathrm{tr}\left[\Lambda_-(\bar{p}_i)\bar{\not{p}}_f \Lambda_+(p_i)\bar{\not{p}}_f\right] \\
&= -\frac{(4\pi)^2 e^4}{4m_0^2 q^4} \mathrm{tr}\left[(-\bar{\not{p}}_i + m_0)\bar{\not{p}}_f(\not{p}_i + m_0)\bar{\not{p}}_f\right] \\
&= \frac{(4\pi)^2 e^4}{4m_0^2 q^4} \left[\mathrm{tr}(\bar{\not{p}}_i\bar{\not{p}}_f\not{p}_i\bar{\not{p}}_f) - m_0^2 \mathrm{tr}(\bar{\not{p}}_f\bar{\not{p}}_f)\right] \\
&= \frac{(4\pi)^2 e^4}{m_0^2 q^4} \left[2(\bar{p}_i \cdot \bar{p}_f)(p_i \cdot \bar{p}_f) - (p_i \cdot \bar{p}_i)(\bar{p}_f \cdot \bar{p}_f) - m_0^2(\bar{p}_f \cdot \bar{p}_f)\right] .
\end{aligned}$$

Evaluation of the scalar products in the center of mass system,

$$\begin{aligned}
&p_i \cdot \bar{p}_i = E_i^2 + \boldsymbol{p}_i^2 \ , \ p_i \cdot \bar{p}_f = E_i^2 + \boldsymbol{p}_i \boldsymbol{p}_f \ , \ \bar{p}_i \cdot \bar{p}_f = E_i^2 - \boldsymbol{p}_i \boldsymbol{p}_f \\
&\bar{p}_f \cdot \bar{p}_f = E_i^2 - \boldsymbol{p}_f^2 = M_0^2 \ , \ q^2 = 4E_i^2 \ ,
\end{aligned}$$

finally leads to

$$\overline{|M_{fi}|^2_{\mathrm{cm}}} = \frac{2(4\pi)^2 e^4}{m_0^2 q^4}\left[E_i^2 \boldsymbol{p}_f^2 - (\boldsymbol{p}_i \boldsymbol{p}_f)^2\right] = \frac{2\pi^2 e^4 \boldsymbol{p}_f^2}{m_0^2 E_i^4}\left(E_i^2 - \boldsymbol{p}_i^2 \cos^2\theta\right) .$$

Theorem 3.21: Pion production via electrons to leading order

The scattering amplitude for pion-antipion production via electron-positron annihilation to leading order is ($f \neq i$, m_0 =electron mass)

$$S_{fi} = \frac{(2\pi)^4 \delta(p_f + \bar{p}_f - p_i - \bar{p}_i)}{V^2} \sqrt{\frac{m_0^2}{E_i \bar{E}_i}} \sqrt{\frac{1}{4E_f \bar{E}_f}} M_{fi} \ ,$$

with the Lorentz-invariant amplitude ($q = p_i + \bar{p}_i = p_f + \bar{p}_f$)

$$M_{fi} = \frac{4\pi \mathrm{i} e^2}{q^2 + \mathrm{i}\epsilon} \bar{v}(\bar{p}_i, \bar{s}_i)(\not{p}_f - \bar{\not{p}}_f)u(p_i, s_i) .$$

The differential cross section follows as

▷

$$
\begin{aligned}
\mathrm{d}\sigma &= \frac{m_0^2}{\sqrt{(p_i \cdot \bar{p}_i)^2 - m_0^4}} |M_{fi}|^2 (2\pi)^4 \delta(p_f + \bar{p}_f - p_i - \bar{p}_i) \\
&\quad \times \frac{\mathrm{d}^3 p_f}{2E_f (2\pi)^3} \frac{\mathrm{d}^3 \bar{p}_f}{2\bar{E}_f (2\pi)^3}
\end{aligned}
$$

and, particularly in the center of mass system,

$$
\left(\frac{\mathrm{d}\sigma}{\mathrm{d}\Omega}\right)_{\mathrm{cm}} = \frac{m_0^2 |\boldsymbol{p}_f|}{16(2\pi)^2 E_i^2 |\boldsymbol{p}_i|} |M_{fi}|^2_{\mathrm{cm}} ,
$$

where all pionic scattering momenta $\boldsymbol{p}_f$ toward $\mathrm{d}\Omega$ and all antipionic scattering momenta $\bar{\boldsymbol{p}}_f$ have been integrated out in the last equation. If electron polarizations are ignored, the unpolarized amplitude square becomes

$$
\overline{|M_{fi}|^2_{\mathrm{cm}}} = \frac{2\pi^2 e^4 \boldsymbol{p}_f^2}{m_0^2 E_i^4} \left(E_i^2 - \boldsymbol{p}_i^2 \cos^2\theta\right) .
$$

Total cross section. In order to exemplify the limited application range of the bosonic scattering theory, we now calculate the total cross section of pion production via electrons and compare it with the experimental result

$$
E_i = 385 \text{ MeV} \Longrightarrow \sigma_{\exp} = 1.4 \cdot 10^{-30} \text{ cm}^2 .
$$

According to the above theorem, we have

$$
\begin{aligned}
\sigma &= \int_0^{2\pi} \mathrm{d}\varphi \int_{-1}^{1} \mathrm{d}\cos\theta \left(\frac{\mathrm{d}\sigma}{\mathrm{d}\Omega}\right)_{\mathrm{cm}} = 2\pi \frac{e^4 |\boldsymbol{p}_f|^3}{32 E_i^6 |\boldsymbol{p}_i|} \int_{-1}^{1} \mathrm{d}\cos\theta \left(E_i^2 - \boldsymbol{p}_i^2 \cos^2\theta\right) \\
&= \frac{\pi e^4 |\boldsymbol{p}_f|^3}{16 E_i^6 |\boldsymbol{p}_i|} \left(2E_i^2 - \frac{2}{3}\boldsymbol{p}_i^2\right) \approx \frac{\pi e^4 |\boldsymbol{p}_f|^3}{12 E_i^5} ,
\end{aligned}
$$

where the approximation $\boldsymbol{p}_i^2 = E_i^2 - m_0^2 \approx E_i^2$ has been used in the last step due to $E_i > M_0 \gg m_0$. Thus, at the experimental measurement point, we obtain in natural units

$$
\left.
\begin{aligned}
&E_i = 385 \text{ MeV} \\
&M_0 = 139.6 \text{ MeV} \\
&|\boldsymbol{p}_f| = \sqrt{E_i^2 - M_0^2} = 358.8 \text{ MeV} \\
&e^2 = \alpha_e = 1/137
\end{aligned}
\right\} \Longrightarrow \sigma = 7.6 \cdot 10^{-11} \frac{1}{\text{MeV}^2}
$$

or, after translation to MKS units [see (3.69)],

$$
\frac{1}{\text{MeV}} = 1.973 \cdot 10^{-11} \text{ cm} \Longrightarrow \sigma \approx 3 \cdot 10^{-32} \text{ cm}^2 .
$$

Obviously, this value is smaller by about a factor of 50 than the experimentally measured pion production cross section. This shows very clearly that the assumption of structureless bosons interacting solely electromagnetically is not justified. In truth, the virtual photon initially produces other short-lived mesons that, in turn, decay into pions or take part in other strong interaction processes.

3.5.6 Compton Scattering against Pions

Now we turn to the description of processes with real photons and discuss the Compton scattering against pions to lowest order.[31] Another process of this kind, the pion-antipion annihilation, is discussed in Exercise 44. The leading order of the Compton scattering amplitude is $\mathcal{O}\left(e^2\right)$. Hence, in Theorem 3.18, we have to take into account the terms with number one *and* two where the relevant parts of the modified potential V up to order $\mathcal{O}\left(e^2\right)$ are to be inserted:

$$\left.\begin{aligned} S_{fi} &= S_{fi}^{(a)} + S_{fi}^{(b)} \\ S_{fi}^{(a)} &= -\mathrm{i}e^2 \int \mathrm{d}^4x \int \mathrm{d}^4y \Phi_f^*(x) \left[\mathrm{i}\frac{\partial}{\partial x^\mu} A^\mu(x) + A^\mu(x)\mathrm{i}\frac{\partial}{\partial x^\mu}\right] \\ &\quad \times \Delta_{\mathrm{F}}^{(0)}(x-y) \left[\mathrm{i}\frac{\partial}{\partial y^\nu} A^\nu(y) + A^\nu(y)\mathrm{i}\frac{\partial}{\partial y^\nu}\right] \Phi_i(y) \\ S_{fi}^{(b)} &= \mathrm{i}e^2 \int \mathrm{d}^4x \Phi_f^*(x)\Phi_i(x) A_\mu(x) A^\mu(x) \;. \end{aligned}\right\} \quad (3.171)$$

$S_{fi}^{(a)}$ corresponds to the Dirac Compton scattering amplitude in Subsection 3.3.7 and consists of direct and exchange contributions. By contrast, the *seagull scattering amplitude*[32] $S_{fi}^{(b)}$ is purely boson specific.

Direct and exchange scattering amplitude. Let us first consider the direct scattering contribution to $S_{fi}^{(a)}$. Using partial integration, this can be transformed into

$$\begin{aligned} S_{fi}^{(a)}(\mathrm{dir}) &= -\mathrm{i}e^2 \int \mathrm{d}^4x \int \mathrm{d}^4y \left\{\Phi_f^*(x)\mathrm{i}\frac{\overleftrightarrow{\partial}}{\partial x^\mu}\left[\Delta_{\mathrm{F}}^{(0)}(x-y)\mathrm{i}\frac{\overleftrightarrow{\partial}}{\partial y^\nu}\Phi_i(y)\right]\right\} \\ &\quad \times A_f^\mu(x) A_i^\nu(y) \;, \end{aligned}$$

where, referencing Subsection 3.3.7, the first A^μ-field is identified with the outgoing and the second one with the incoming photon. The correct ansatz for these fields is

$$A_i^\nu(y) = \sqrt{\frac{2\pi}{\omega_i V}}\epsilon^\nu(k_i,\lambda_i)\mathrm{e}^{-\mathrm{i}k_i\cdot y} \;,\; A_f^\mu(x) = \sqrt{\frac{2\pi}{\omega_f V}}\epsilon^\mu(k_f,\lambda_f)\mathrm{e}^{+\mathrm{i}k_f\cdot x} \quad (3.172)$$

since, due to the same line of argument as in Subsection 3.3.7 and Exercise 37, all other photon contributions lead to different or not realizable kinematic constellations. Thus, along with

---

[31] See footnote 20 on page 267.

[32] For clarification of this term, see Figure 3.48.

$$\Phi_i(y) = \frac{1}{\sqrt{2E_i V}} \mathrm{e}^{-\mathrm{i}p_i\cdot y} \;,\; \Phi_f(x) = \frac{1}{\sqrt{2E_f V}} \mathrm{e}^{-\mathrm{i}p_f\cdot x} \tag{3.173}$$

and

$$\Delta_{\mathrm{F}}^{(0)}(x-y) = \int \frac{\mathrm{d}^4 p}{(2\pi)^4} \frac{\mathrm{e}^{-\mathrm{i}p\cdot(x-y)}}{p^2 - m_0^2 + \mathrm{i}\epsilon}$$

the direct scattering amplitude follows as

$$\begin{aligned} S_{fi}^{(a)}(\mathrm{dir}) &= -\frac{\mathrm{i}e^2}{V^2}\frac{1}{\sqrt{4E_iE_f}}\sqrt{\frac{(2\pi)^2}{\omega_i\omega_f}}\int \mathrm{d}^4x \int \mathrm{d}^4 y \int \frac{\mathrm{d}^4 p}{(2\pi)^4} \\ &\quad \times \epsilon^\mu(k_f,\lambda_f)(p+p_f)_\mu \frac{1}{p^2 - m_0^2 + \mathrm{i}\epsilon}\epsilon^\nu(k_i,\lambda_i)(p_i+p)_\nu \\ &\quad \times \mathrm{e}^{\mathrm{i}p_f\cdot x}\mathrm{e}^{-\mathrm{i}p\cdot(x-y)}\mathrm{e}^{-\mathrm{i}p_i\cdot y}\mathrm{e}^{\mathrm{i}k_f\cdot x}\mathrm{e}^{-\mathrm{i}k_i\cdot y} \end{aligned}$$

and, after carrying out the coordinate and momentum integrations,

$$\left.\begin{aligned} S_{fi}^{(a)}(\mathrm{dir}) &= \frac{(2\pi)^4\delta(p_f+k_f-p_i-k_i)}{V^2}\frac{1}{\sqrt{4E_iE_f}}\sqrt{\frac{(2\pi)^2}{\omega_i\omega_f}}M_{fi}^{(a)}(\mathrm{dir}) \\ M_{fi}^{(a)}(\mathrm{dir}) &= \epsilon^\mu(k_f,\lambda_f)(-\mathrm{i}e)(2p_f+k_f)_\mu \frac{+\mathrm{i}}{(p_i+k_i)^2 - m_0^2 + \mathrm{i}\epsilon} \\ &\quad \times \epsilon^\nu(k_i,\lambda_i)(-\mathrm{i}e)(2p_i+k_i)_\nu \;. \end{aligned}\right\} \tag{3.174}$$

As expected, this expression and its graphical representation in Figure 3.47**a** coincide with our extended Feynman rules.

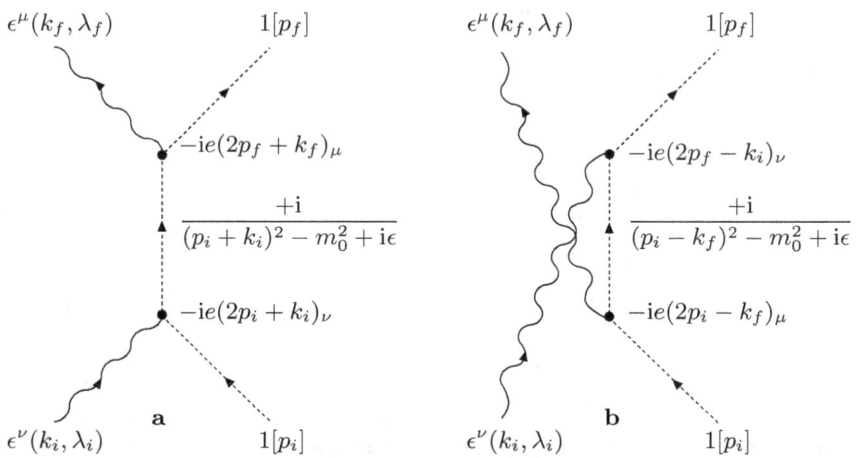

Fig. 3.47. Feynman diagrams of the direct $\mathcal{O}\left(e^2\right)$-scattering amplitude (**a**) and the exchange $\mathcal{O}\left(e^2\right)$-scattering amplitude (**b**) for the Compton scattering against pions in momentum space (compare to Figure 3.22). Analogously to the spin-1/2 scattering theory, internal photon lines are assigned the i-fold of the boson propagator. Energy and momentum are conserved at each vortex.

The exchange amplitude is obtained by swapping the photons' coupling points or by the replacement $k_i \leftrightarrow -k_f$ in (3.174) as (see Figure 3.47**b**)

$$\begin{aligned} S_{fi}^{(a)}(\mathrm{ex}) &= \frac{(2\pi)^4\delta(p_f+k_f-p_i-k_i)}{V^2}\frac{1}{\sqrt{4E_iE_f}}\sqrt{\frac{(2\pi)^2}{\omega_i\omega_f}}M_{fi}^{(a)}(\mathrm{ex}) \\ M_{fi}^{(a)}(\mathrm{ex}) &= \epsilon^\mu(k_f,\lambda_f)(-\mathrm{i}e)(2p_i-k_f)_\mu\frac{+\mathrm{i}}{(p_i-k_f)^2-m_0^2+\mathrm{i}\epsilon} \\ &\quad\times\epsilon^\nu(k_i,\lambda_i)(-\mathrm{i}e)(2p_f-k_i)_\nu \ . \end{aligned}$$

Seagull scattering amplitude. In the scattering amplitude $S_{fi}^{(b)}$ in (3.171), the incoming and outgoing photons join at the same point. Accordingly, there exist two mathematically identical ways of assigning the A^μ-fields: either the first field is considered to be the incoming and the second one the outgoing photon or vice versa. At the insertion of (3.172) and (3.173) in (3.171), we therefore have to take into account an additional factor of 2 so that, after the coordinate and momentum integrations have been performed, we obtain

$$\begin{aligned} S_{fi}^{(b)} &= \frac{(2\pi)^4\delta(p_f+k_f-p_i-k_i)}{V^2}\frac{1}{\sqrt{4E_iE_f}}\sqrt{\frac{(2\pi)^2}{\omega_i\omega_f}}M_{fi}^{(b)} \\ M_{fi}^{(b)} &= 2\mathrm{i}e^2\epsilon_\mu(k_f,\lambda_f)\epsilon^\mu(k_i,\lambda_i) \ . \end{aligned}$$

The Feynman diagram corresponding to $M_{fi}^{(b)}$ is shown in Figure 3.48. It obviously exhibits a particularity that does not exist in the fermion case, namely a

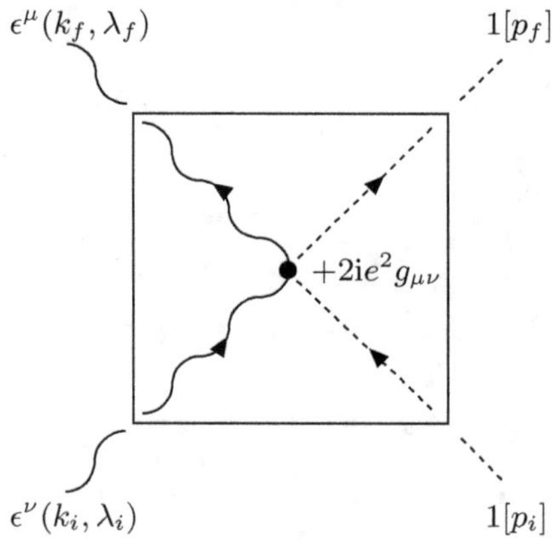

Fig. 3.48. Feynman diagram of the $\mathcal{O}(e^2)$-scattering amplitude $M_{fi}^{(b)}$ for the Compton scattering against pions in momentum space. Inside the box is the two-photon vortex of the bosonic scattering theory. The indices μ and ν of the vortex factor are contracted with those of the photon lines. Because of its shape, the two-photon vortex is also called "seagull vortex" and, correspondingly, $M_{fi}^{(b)}$ the "seagull amplitude".

two-photon vortex (*seagull vortex*) which, as already mentioned, results from the $\mathcal{O}(e^2)$-contribution of the modified potential V from Theorem 3.17.

Bringing all three contributions, $S_{fi}^{(a)}(\mathrm{dir})$, $S_{fi}^{(a)}(\mathrm{ex})$, and $S_{fi}^{(b)}$, together, the scattering amplitude of the Compton scattering against pions to lowest order follows as

$$\left.\begin{aligned}
S_{fi} &= \frac{(2\pi)^4\delta(p_f+k_f-p_i-k_i)}{V^2}\frac{1}{\sqrt{4E_iE_f}}\sqrt{\frac{(2\pi)^2}{\omega_i\omega_f}}M_{fi}\\
M_{fi} &= M_{fi}^{(a)}(\mathrm{dir})+M_{fi}^{(a)}(\mathrm{ex})+M_{fi}^{(b)}\\
&= -\mathrm{i}e^2\epsilon_\mu(k_f,\lambda_f)T^{\mu\nu}\epsilon_\nu(k_i,\lambda_i)\\
T^{\mu\nu} &= \frac{(2p_f+k_f)^\mu(2p_i+k_i)^\nu}{(p_i+k_i)^2-m_0^2+\mathrm{i}\epsilon}+\frac{(2p_i-k_f)^\mu(2p_f-k_i)^\nu}{(p_i-k_f)^2-m_0^2+\mathrm{i}\epsilon}\\
&\quad -2g^{\mu\nu}\;,
\end{aligned}\right\} \tag{3.175}$$

where $T^{\mu\nu}$ denotes the *Compton tensor*. Incidentally, a more formal justification for the factor 2 within the two-photon vortex is that it is necessary to ensure the invariance of M_{fi} under gauge transformations of the four-potential:

$$\epsilon_\nu(k,\lambda)\to\epsilon_\nu(k,\lambda)+k_\nu\Lambda(k)\Longrightarrow T^{\mu\nu}k_\nu=0\;,\;k_\mu T^{\mu\nu}=0\;.$$

Cross section. In the laboratory system where the pion is initially at rest, $p_i=(m_0,\mathbf{0})$, the differential cross section is (as regards the kinematic situation, see Figure 3.21)

$$\mathrm{d}\sigma=\frac{2\pi}{2\sqrt{m_0^2\omega_i^2}}|M_{fi}|^2(2\pi)^4\delta(p_f+k_f-p_i-k_i)\frac{\mathrm{d}^3p_f}{2E_f(2\pi)^3}\frac{2\pi\mathrm{d}^3k_f}{(2\pi)^3\omega_f}\;.$$

Using

$$\mathrm{d}^3k_f=\omega_f^2\mathrm{d}\omega_f\mathrm{d}\Omega\;,\;\frac{\mathrm{d}^3p_f}{2E_f}=\int\mathrm{d}^4p_f\delta(p_f^2-m_0^2)\Theta(p_f^0)\;,$$

this turns into

$$\begin{aligned}
\frac{\mathrm{d}\sigma}{\mathrm{d}\Omega} &= \frac{1}{2m_0\omega_i}\int\mathrm{d}\omega_f\omega_f\int\mathrm{d}^4p_f|M_{fi}|^2\delta(p_f+k_f-p_i-k_i)\\
&\quad\times\delta(p_f^2-m_0^2)\Theta(p_f^0)\;.
\end{aligned}$$

A further evaluation can again be related back to the kinematically equivalent calculation for the Compton scattering against electrons in Subsection 3.3.7 by noting that the above expression is formally identical to the first row in (3.118) multiplied by $1/(4m_0^2)$. Thus, from the same equation it follows for the case in hand that

$$\frac{\mathrm{d}\sigma}{\mathrm{d}\Omega}=\frac{\omega_f^2}{4m_0^2\omega_i^2}|M_{fi}|^2_{\mathrm{co}}\;,\;|M_{fi}|^2_{\mathrm{co}}=|M_{fi}|^2_{p_f=p_i+k_i-k_f}\;,$$

with the secondary condition

$$\omega_f = \frac{\omega_i}{1 + \frac{\omega_i}{m_0}(1 - \cos\theta)} \Longleftrightarrow \lambda_f = \lambda_i + \frac{2\pi}{m_0}(1 - \cos\theta) \ .$$

Amplitude square. In view of the evaluation of $|M_{fi}|^2$ and due to $\epsilon(k,\lambda) \cdot k = 0$, the Compton tensor in (3.175) is initially simplified to

$$T^{\mu\nu} = \frac{4p_f^\mu p_i^\nu}{(p_i + k_i)^2 - m_0^2 + \mathrm{i}\epsilon} + \frac{4p_i^\mu p_f^\nu}{(p_i - k_f)^2 - m_0^2 + \mathrm{i}\epsilon} - 2g^{\mu\nu} \ .$$

Moreover, if we move to the radiation gauge $[\epsilon^\mu(k,\lambda)] = [0, \boldsymbol{\epsilon}(\boldsymbol{k},\lambda)]$, the first two terms of $T^{\mu\nu}$ vanish because of $p_i = (m_0, \boldsymbol{0})$ so that

$$T^{\mu\nu} = -2g^{\mu\nu} \ .$$

In other words, in the radiation gauge and in the laboratory system, the Compton cross section is solely determined by the two-photon vortex. Disregarding any photon polarizations (averaging over the initial and summing over the final polarizations), the amplitude square finally follows as

$$\begin{aligned} \overline{|M_{fi}|^2_{\mathrm{CO}}} &= \frac{e^4}{2} \sum_{\lambda_i,\lambda_f} |\epsilon_\mu(k_f,\lambda_f) T^{\mu\nu} \epsilon_\nu(k_i,\lambda_i)|^2 \\ &= 2e^4 \sum_{\lambda_i,\lambda_f} [\boldsymbol{\epsilon}(\boldsymbol{k}_f,\lambda_f)\boldsymbol{\epsilon}(\boldsymbol{k}_i,\lambda_i)]^2 = 2e^4\left(1 + \cos^2\theta\right) \ , \end{aligned} \tag{3.176}$$

where θ denotes the angle between $\boldsymbol{k}_i$ and $\boldsymbol{k}_f$ [see (3.123)].

Theorem 3.22: Compton scattering against pions to leading order

The scattering amplitude of the Compton scattering against pions to leading order is $(f \neq i)$

$$S_{fi} = \frac{(2\pi)^4 \delta(p_f + k_f - p_i - k_i)}{V^2} \frac{1}{\sqrt{4E_i E_f}} \sqrt{\frac{(2\pi)^2}{\omega_i \omega_f}} M_{fi} \ ,$$

with the Lorentz-invariant amplitude

$$\begin{aligned} M_{fi} &= M_{fi}^{(a)}(\mathrm{dir}) + M_{fi}^{(a)}(\mathrm{ex}) + M_{fi}^{(b)} \\ M_{fi}^{(a)}(\mathrm{dir}) &= -\mathrm{i}e^2 \epsilon_\mu(k_f,\lambda_f) \frac{(2p_f + k_f)^\mu (2p_i + k_i)^\nu}{(p_i + k_i)^2 - m_0^2 + \mathrm{i}\epsilon} \epsilon_\nu(k_i,\lambda_i) \\ M_{fi}^{(a)}(\mathrm{ex}) &= -\mathrm{i}e^2 \epsilon_\mu(k_f,\lambda_f) \frac{(2p_i - k_f)^\mu (2p_f - k_i)^\nu}{(p_i - k_f)^2 - m_0^2 + \mathrm{i}\epsilon} \epsilon_\nu(k_i,\lambda_i) \\ M_{fi}^{(b)} &= 2\mathrm{i}e^2 \epsilon_\mu(k_f,\lambda_f) \epsilon^\mu(k_i,\lambda_i) \ . \end{aligned}$$

The differential cross section follows as ▷

$$\mathrm{d}\sigma = \frac{\pi}{\sqrt{(p_i \cdot k_i)^2}} |M_{fi}|^2 (2\pi)^4 \delta(p_f + k_f - p_i - k_i) \frac{\mathrm{d}^3 p_f}{2E_f (2\pi)^3} \frac{2\pi \mathrm{d}^3 k_f}{(2\pi)^3 \omega_f}$$

and, particularly in the laboratory system where the pion is initially at rest,

$$\frac{\mathrm{d}\sigma}{\mathrm{d}\Omega} = \frac{\omega_f^2}{4m_0^2 \omega_i^2} |M_{fi}|^2_{\mathrm{co}} \ , \ |M_{fi}|^2_{\mathrm{co}} = |M_{fi}|^2_{p_f = p_i + k_i - k_f}$$

$$\omega_f = \frac{\omega_i}{1 + \frac{\omega_i}{m_0}(1 - \cos\theta)} \ ,$$

where all photonic scattering momenta $\boldsymbol{k}_f$ toward $\mathrm{d}\Omega$ and all pionic scattering momenta $\boldsymbol{p}_f$ have been integrated out in $\mathrm{d}\sigma/\mathrm{d}\Omega$. Disregarding photon polarizations, the unpolarized amplitude square becomes

$$\overline{|M_{fi}|^2_{\mathrm{co}}} = 2e^4 \left(1 + \cos^2\theta\right) \ .$$

3.5.7 Conclusion: Enhanced Feynman Rules in Momentum Space

As we see in the preceding examples, the scattering amplitudes of bosonic scattering processes can be constructed very similarly to those of the fermion case. In fact, we only need some further extensions to the spin-1/2 Feynman rules in Subsection 3.3.9 for them to cover also the spin-0 case. In the following we recapitulate these extensions (marked by E) once again.

E1. One boson factor $N_{i,f}^{(\ldots)} = 1/2$ for each (anti)boson.

E3. If bosons are present, the construction of M_{fi} involves Feynman diagrams with one-photon and two-photon vortices as well as internal and external boson lines inasmuch as they lead to topological constellations that are compatible with the considered scattering process.

E4. All vortices and lines within Feynman diagrams belonging to the bosonic sector are supplied by the factors in Figure 3.49. Furthermore, there is

  i) no relative sign between Feynman diagrams differing only by the exchange of two boson lines.

  ii) a factor of 1/2 for each closed photon loop.

To E4. Contrary to rule 4.i) of the fermion case, rule E4.i) accommodates the Bose-Einstein statistics, according to which the whole scattering amplitude must be symmetric under the exchange of two boson lines (of the same type). The factor 1/2 in rule E4.ii) compensates for the double counting of alternatives to connect both vortices of a photon loop – actually two-photon vortices

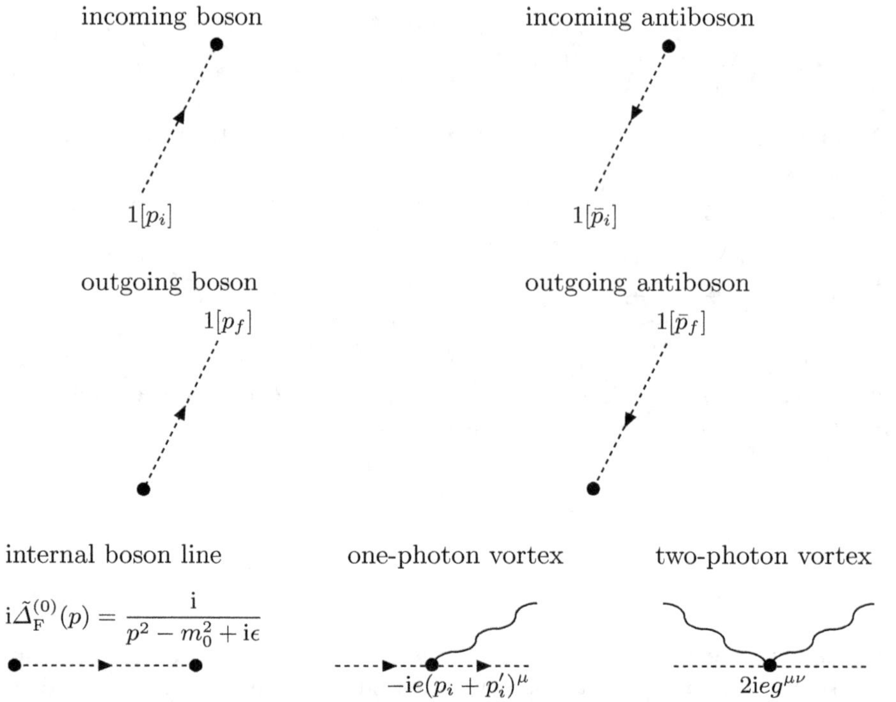

Fig. 3.49. Supplementary Feynman diagram elements and characteristic factors for the bosonic sector in momentum space. The vortex indices are contracted with those of the photon lines.

each assigned a factor of $2\mathrm{i}eg^{\mu\nu}$ – by photon lines. In this context, recall our discussion of the seagull scattering amplitude in the preceding subsection as well as the motivation for the factor 2 within the two-photon vortex.

Note that, as in the fermion case, these rules also allow the construction of loop diagrams in higher orders whose mathematical divergences can be handled using the renormalization methods discussed in Section 3.4. As before, those loop diagrams are of pure quantum field theoretical nature. However, before tackling the concrete calculation of higher order corrections, one should bear in mind that electromagnetic forces are normally superposed by the much stronger interactions between bosons themselves as well as with nucleons that do not play any role in our formalism. To this extent, one should not necessarily expect a better coincidence with experiments by taking higher electromagnetic corrections into account.

Summary

- The Klein-Gordon equation can be solved approximately using the same methods as in the Dirac case.

- Analogously to the fermion propagator, the **Feynman boson propagator** is defined in such a way that it describes the temporally forward [backward] oriented evolution of the positive [negative] parts of a wave function known at a particular time (including its derivative). With this and the Feynman-Stückelberg interpretation, the positive [negative] parts can be interpreted as temporally forward directed particle [antiparticle] propagation.
- Using the Feynman propagator formalism, the amplitude of bosonic scattering processes can be expanded in a series of multiple scatterings. Contrary to the fermion case, each series term contains two different orders of e due to the form of the **modified potential**.
- A practical calculation of spin-0 scattering processes is based on the same prerequisites (plane waves, adiabatic approximation, etc.) and steps (constructing S_{fi}, squaring, inserting in $\mathrm{d}\sigma$, and so on) as the spin-1/2 case.
- The corresponding directives can be easily integrated into the spinor Feynman rules by adding some bosonic extensions.
- With this extended version, Feynman rules allow the study of scattering processes with an involvement of fermionic, bosonic or photonic initial or final states. Scattering processes with purely bosonic initial and final states are the **Coulomb scattering**, **pion-pion scattering**, and **pion-antipion scattering**. A fermionic-photonic process is the **pion-antipion production via electrons** and photonic-bosonic processes are the **Compton scattering** and **pion-antipion annihilation**.
- Compared to the spin-1/2 scattering theory, the validity range of the spin-0 scattering formalism is highly restricted since spin-0 particles are not elementary but consist of quarks that, in turn, are subject to the strong interaction.

Exercises

42. Causality principle of $\Delta_{\mathrm{F}}^{(0)}$. Show with the help of (3.162) the causal relationships mentioned in Theorem 3.17 for the free case.

Solution. The calculation proceeds analogously to that of Exercise 36. Let

$$\phi(x) = \phi^{(+)}(x) + \phi^{(-)}(x) = \int \mathrm{d}^3p' \sum_{r=1}^{2} a^{(r)}(\boldsymbol{p}')\phi_{\boldsymbol{p}'}^{(r)}(x)$$

be an arbitrary Klein-Gordon wave packet. Then,

$$
\begin{aligned}
&\int \mathrm{d}^3 x \Delta_{\mathrm{F}}^{(0)}(x'-x)\mathrm{i}\overleftrightarrow{\partial_0}\phi(x) \\
&\quad = -\mathrm{i}\Theta(x'^0 - x^0)\int \mathrm{d}^3 x \int \mathrm{d}^3 p \int \mathrm{d}^3 p' \\
&\qquad \times \sum_{r=1}^{2} \phi_{\boldsymbol{p}}^{(1)}(x')\phi_{\boldsymbol{p}}^{(1)*}(x)\mathrm{i}\overleftrightarrow{\partial_0}\phi_{\boldsymbol{p}'}^{(r)}(x)a^{(r)}(\boldsymbol{p}') \\
&\qquad -\mathrm{i}\Theta(x^0 - x'^0)\int \mathrm{d}^3 x \int \mathrm{d}^3 p \int \mathrm{d}^3 p' \\
&\qquad \times \sum_{r=1}^{2} \phi_{\boldsymbol{p}}^{(2)}(x')\phi_{\boldsymbol{p}}^{(2)*}(x)\mathrm{i}\overleftrightarrow{\partial_0}\phi_{\boldsymbol{p}'}^{(r)}(x)a^{(r)}(\boldsymbol{p}') \\
&\quad = -\mathrm{i}\Theta(x'^0 - x^0)\int \mathrm{d}^3 p \int \mathrm{d}^3 p' \sum_{r=1}^{2} \epsilon_r \delta_{r1}\delta(\boldsymbol{p}-\boldsymbol{p}')\phi_{\boldsymbol{p}}^{(1)}(x')a^{(r)}(\boldsymbol{p}') \\
&\qquad -\mathrm{i}\Theta(x^0 - x'^0)\int \mathrm{d}^3 p \int \mathrm{d}^3 p' \sum_{r=1}^{2} \epsilon_r \delta_{r2}\delta(\boldsymbol{p}-\boldsymbol{p}')\phi_{\boldsymbol{p}}^{(2)}(x')a^{(r)}(\boldsymbol{p}') \\
&\quad = -\mathrm{i}\Theta(x'^0 - x^0)\int \mathrm{d}^3 p \phi_{\boldsymbol{p}}^{(1)}(x')a^{(1)}(\boldsymbol{p}) \\
&\qquad +\mathrm{i}\Theta(x^0 - x'^0)\int \mathrm{d}^3 p \phi_{\boldsymbol{p}}^{(2)}(x')a^{(2)}(\boldsymbol{p}) \\
&\quad = -\mathrm{i}\Theta(x'^0 - x^0)\phi^{(+)}(x) + \mathrm{i}\Theta(x^0 - x'^0)\phi^{(-)}(x)\ .
\end{aligned}
$$

And, likewise:

$$
\begin{aligned}
&\int \mathrm{d}^3 x \phi^*(x)\mathrm{i}\overleftrightarrow{\partial_0}\Delta_{\mathrm{F}}^{(0)}(x-x') \\
&\quad = -\mathrm{i}\Theta(x^0 - x'^0)\int \mathrm{d}^3 x \int \mathrm{d}^3 p' \int \mathrm{d}^3 p \\
&\qquad \times \sum_{r=1}^{2} a^{(r)*}(\boldsymbol{p}')\phi_{\boldsymbol{p}'}^{(r)}(x)\mathrm{i}\overleftrightarrow{\partial_0}\phi_{\boldsymbol{p}}^{(1)}(x)\phi_{\boldsymbol{p}}^{(1)*}(x') \\
&\qquad -\mathrm{i}\Theta(x'^0 - x^0)\int \mathrm{d}^3 x \int \mathrm{d}^3 p' \int \mathrm{d}^3 p \\
&\qquad \times \sum_{r=1}^{2} a^{(r)*}(\boldsymbol{p}')\phi_{\boldsymbol{p}'}^{(r)}(x)\mathrm{i}\overleftrightarrow{\partial_0}\phi_{\boldsymbol{p}}^{(2)}(x)\phi_{\boldsymbol{p}}^{(2)*}(x') \\
&\quad = -\mathrm{i}\Theta(x^0 - x'^0)\int \mathrm{d}^3 p' \int \mathrm{d}^3 p \sum_{r=1}^{2} a^{(r)*}(\boldsymbol{p}')\phi_{\boldsymbol{p}}^{(1)*}(x')\epsilon_r \delta_{r1}\delta(\boldsymbol{p}-\boldsymbol{p}') \\
&\qquad -\mathrm{i}\Theta(x'^0 - x^0)\int \mathrm{d}^3 p' \int \mathrm{d}^3 p \sum_{r=1}^{2} a^{(r)*}(\boldsymbol{p}')\phi_{\boldsymbol{p}}^{(2)*}(x')\epsilon_r \delta_{r1}\delta(\boldsymbol{p}-\boldsymbol{p}')
\end{aligned}
$$

$$= -\mathrm{i}\Theta(x^0 - x'^0)\int \mathrm{d}^3p a^{(1)*}(\boldsymbol{p})\phi_{\boldsymbol{p}}^{(1)*}(x')$$
$$+\mathrm{i}\Theta(x'^0 - x^0)\int \mathrm{d}^3p a^{(2)*}(\boldsymbol{p})\phi_{\boldsymbol{p}}^{(2)*}(x')$$
$$= -\mathrm{i}\Theta(x^0 - x'^0)\phi^{(+)}(x') + \mathrm{i}\Theta(x'^0 - x^0)\phi^{(-)}(x') \ .$$

43. Pion-antipion scattering in the center of mass system. Calculate the differential cross section of pion-antipion scattering to leading order in the center of mass system using the results from Subsection 3.5.4.

Solution. Similarly to the electron-positron scattering in Subsection 3.3.6, the scattering amplitude for the pion-antipion scattering to leading order encompasses a direct and an exchange or annihilation part whose Feynman diagrams are depicted in Figure 3.50. One obtains from this the scattering

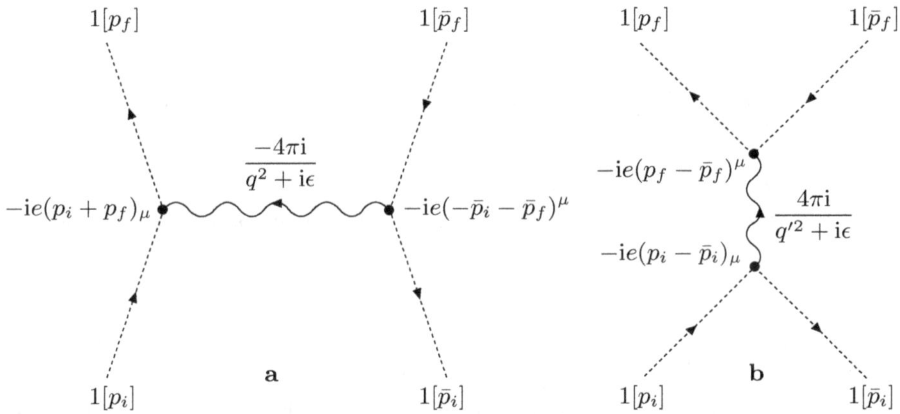

Fig. 3.50. Feynman diagrams of the direct $\mathcal{O}\left(e^2\right)$-scattering amplitude (**a**) and the exchange or annihilation $\mathcal{O}\left(e^2\right)$-scattering amplitude (**b**) for pion-antipion scattering in momentum space (compare to Figures 3.18 and 3.19). Energy and momentum are conserved at each vortex. Therefore, the four-momentum transfers are $q = p_f - p_i = -(\bar{p}_f - \bar{p}_i)$, $q' = p_i + \bar{p}_i = p_f + \bar{p}_f$.

amplitude as

$$\begin{aligned}
S_{fi} &= \frac{(2\pi)^4\delta(p_f + \bar{p}_f - p_i - \bar{p}_i)}{V^2}\frac{1}{\sqrt{4E_iE_f}}\frac{1}{\sqrt{4\bar{E}_i\bar{E}_f}}M_{fi} \\
M_{fi} &= M_{fi}(\mathrm{dir}) + M_{fi}(\mathrm{ex}) \\
M_{fi}(\mathrm{dir}) &= (p_i + p_f)_\mu \frac{-4\pi\mathrm{i}e^2}{q^2 + \mathrm{i}\epsilon}(\bar{p}_i + \bar{p}_f)^\mu \ , \ q = p_f - p_i \\
M_{fi}(\mathrm{ex}) &= (p_i - \bar{p}_i)_\mu \frac{4\pi\mathrm{i}e^2}{q'^2 + \mathrm{i}\epsilon}(p_f - \bar{p}_f)^\mu \ , \ q' = p_i + \bar{p}_i \ .
\end{aligned}$$

A comparison with the scattering amplitude of pion-pion scattering in Theorem 3.20 shows that, also within the bosonic scattering theory, there exist crossing symmetries between processes where an incoming particle is permuted with the corresponding antiparticle and vice versa (see Figure 3.51).

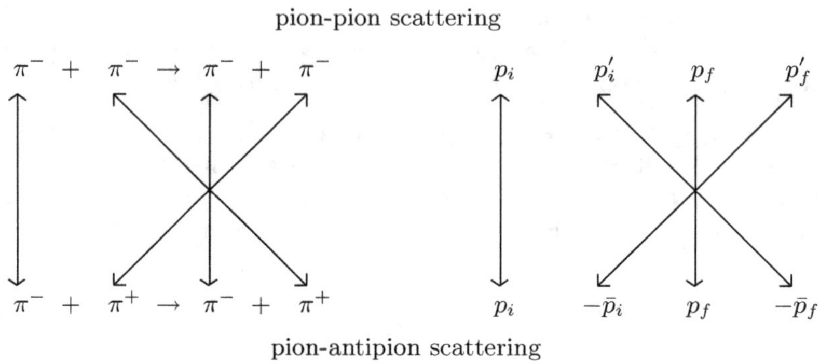

Fig. 3.51. Crossing symmetry between pion-pion and pion-antipion scatterings (compare to Figure 3.20).

Due to the kinematics being equivalent to that of the pion-pion scattering, we can take the differential cross section in Theorem 3.20 for the process in hand, i.e.

$$\left(\frac{\mathrm{d}\sigma}{\mathrm{d}\Omega}\right)_{\mathrm{cm}} = \frac{1}{64(2\pi)^2 E_i^2}|M_{fi}|^2_{\mathrm{cm}} \; .$$

If we finally perform the replacements in (3.169) as indicated in Figure 3.51 and subsequently evaluate the scalar products in the center of mass system, the amplitude square follows as

$$\begin{aligned} |M_{fi}|^2_{\mathrm{cm}} = (4\pi)^2 e^4 & \left[\frac{\left(E_i^2 + \boldsymbol{p}_i^2 \cos^2\frac{\theta}{2}\right)^2}{\boldsymbol{p}_i^4 \sin^4\frac{\theta}{2}} + \frac{\boldsymbol{p}_i^4 \cos^2\theta}{E_i^4}\right. \\ & \left. - \frac{\cos\theta \left(E_i^2 + \boldsymbol{p}_i^2 \cos^2\frac{\theta}{2}\right)}{E_i^2 \sin^2\frac{\theta}{2}}\right] . \end{aligned}$$

44. Pion-antipion annihilation in the center of mass system. Use the results from Exercise 38 and Subsection 3.5.6 to calculate the unpolarized differential cross section of pion-antipion annihilation into two photons to leading order in the center of mass system.

Solution. As regards the differential cross section, we can resort to the results of the electron-positron annihilation in Exercise 38 (see also Figure 3.29) due to the kinematically equivalent situations. Taking into account

$$\frac{1}{4\sqrt{(p_i \cdot \bar{p}_i)^2 - m_0^4}} = \frac{1}{8E_i|\boldsymbol{p}_i|}$$

$$\mathrm{d}^3k_f = \omega_f^2 \mathrm{d}\omega_f \mathrm{d}\Omega \ , \quad \frac{\mathrm{d}^3k'_f}{\omega'_f} = 2\int \mathrm{d}^4k'_f \delta(k'^2_f)\Theta(k'^0_f) \ ,$$

we have

$$\mathrm{d}\sigma = \frac{1}{8E_i|\boldsymbol{p}_i|}|M_{fi}|^2(2\pi)^4\delta(k_f + k'_f - p_i - \bar{p}_i)\frac{2\pi\mathrm{d}^3k_f}{(2\pi)^3\omega_f}\frac{2\pi\mathrm{d}^3k_f}{(2\pi)^3\omega'_f}$$

$$\Longrightarrow \left(\overline{\frac{\mathrm{d}\sigma}{\mathrm{d}\Omega}}\right)_{\mathrm{cm}} = \frac{1}{4E_i|\boldsymbol{p}_i|}\int \mathrm{d}\omega_f\omega_f \int \mathrm{d}^4k'_f\overline{|M_{fi}|^2}\delta(k_f + k'_f - p_i - \bar{p}_i) \times\delta(k'^2_f)\Theta(k'^0_f) \ .$$

Obviously, this equation is formally identical to the first row of (3.129) multiplied by $1/(4m_0^2)$. Thus, for the case in hand, it follows immediately from the last row of (3.129) that

$$\left(\overline{\frac{\mathrm{d}\sigma}{\mathrm{d}\Omega}}\right)_{\mathrm{cm}} = \frac{1}{16E_i|\boldsymbol{p}_i|}\overline{|M_{fi}|^2_{\mathrm{cm}}} \ .$$

Exploiting the relationship of the crossing symmetry between Compton scattering in Subsection 3.5.6 and pion-antipion annihilation (see Figure 3.52), we

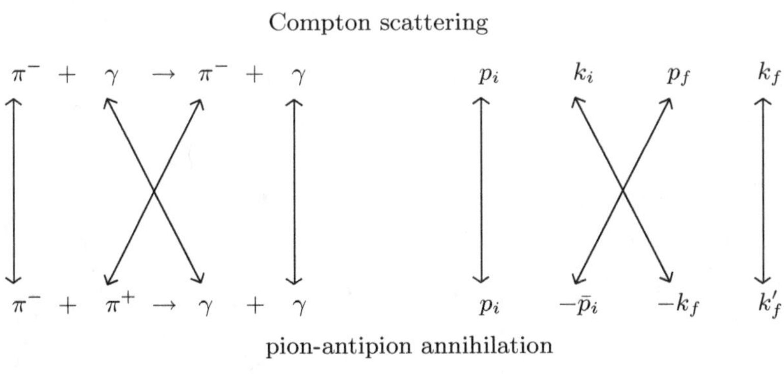

Fig. 3.52. Crossing symmetry between Compton scattering and pion-antipion annihilation (compare to Figure 3.26).

obtain from (3.176) with the corresponding replacements for the unpolarized amplitude square (summation over the photon's final polarizations)

$$
\begin{aligned}
\overline{|M_{fi}|^2_{\mathrm{cm}}} &= e^4 \sum_{\lambda_f,\lambda'_f} \left|\epsilon_\mu(k_f,\lambda_f) T^{\mu\nu} \epsilon_\nu(k'_f,\lambda'_f)\right|^2 \ , \quad T^{\mu\nu} = -2g^{\mu\nu} \\
&= 4e^4 \sum_{\lambda_f,\lambda'_f} \left|\epsilon(k_f,\lambda_f) \cdot \epsilon(k'_f,\lambda'_f)\right|^2 \ .
\end{aligned}
$$

Here we have to bear in mind that this equation only holds under the constraint $\epsilon_f \cdot p_i = \epsilon'_f \cdot p_i = 0$ which was achieved in the laboratory system by using the radiation gauge. Thus, in the center of mass system, the gauge transformation (3.131) from Exercise 38 is appropriate as there the polarization sum turns into (3.132). The remaining evaluation proceeds similarly to that in Exercise 38 and finally leads to

$$
\begin{aligned}
\overline{|M_{fi}|^2_{\mathrm{cm}}} &= 4e^4 \left[2 - \frac{4m_0^2}{E_i^2(1-v_i^2\cos^2\theta)} + \frac{4m_0^4}{E_i^4(1-v_i^2\cos^2\theta)^2}\right] \\
&= 4e^4 \frac{2 - 4v_i^2(1-v_i^2) - 4v_i^4\cos^2\theta + 2v_i^4\cos^4\theta}{(1-v_i^2\cos^2\theta)^2} \ .
\end{aligned}
$$

A. Appendix

A.1 Theory of Special Relativity

The theory of special relativity is based on the following axioms:

1. Constancy of the velocity of light: the vacuum velocity of light is the same in all uniformly moving systems, namely $c \approx 3 \cdot 10^8$ m/s. No information propagates faster than light.

2. Relativity principle: physical laws are equally valid in all inertial systems, i.e. there is no preferred reference frame.

The relativity principle encompasses the homogeneity of space and time as well as the isotropy of space according to which there is no preferred point and direction in space and time.

Due to these axioms, relativistic physical events are described mathematically within a fourdimensional space where the product of the speed of light and time, $x^0 = ct$, appears as an additional dimension on an equal footing with the three space dimensions x^1, x^2, x^3.

Minkowski space. The Minkowski space is a fourdimensional linear vector space over the body of real numbers. Its elements x^μ are represented by four-component coordinate vectors, also called four-vectors,

$$[x^\mu(t)] = \begin{pmatrix} x^0(t) \\ x^1(t) \\ x^2(t) \\ x^3(t) \end{pmatrix} \; , \; x^0(t) = ct \; .$$

The scalar product of two four-vectors is defined as

$$(x^\mu) \cdot (y^\mu) = x^\mu g_{\mu\nu} y^\nu = x^\mu y_\mu = x_\mu g^{\mu\nu} y_\nu = (x_\mu) \cdot (y_\nu) \; ,$$

with the non-Euclidean metric tensor (1. index=row index, 2. index=column index)

$$(g_{\mu\nu}) = (g^{\mu\nu}) = \begin{pmatrix} 1 & 0 & 0 & 0 \\ 0 & -1 & 0 & 0 \\ 0 & 0 & -1 & 0 \\ 0 & 0 & 0 & -1 \end{pmatrix} \; , \; g^{\mu\alpha} g_{\alpha\nu} = g^\mu{}_\nu = \delta^\mu_\nu \; .$$

A. Wachter, Appendix. In: A. Wachter, Relativistic Quantum Mechanics, Theoretical and Mathematical Physics, pp. 349–361 (2011)
DOI 10.1007/978-90-481-3645-2

Notation. Additionally, there exist the following conventions:

- Vectors with an upper index are called contravariant, and those with a lower index are called covariant. This can be transferred to tensors of higher rank. For example, $T^{\mu\nu}{}_{\rho}$ is a twofold contravariant and onefold covariant tensor of rank 3.
- Two indices, one of which is an upper and the other a lower index, are summed over so that the summation sign can be dropped (Einstein sum convention).
- The metric tensor can be used to pull upper indices down and vice versa. For example:

$$x_\mu = g_{\mu\nu}x^\nu \ , \ T^{\mu\nu\rho} = g^{\mu\alpha}T_\alpha{}^{\nu\rho} = g^{\mu\alpha}g^{\nu\beta}T_{\alpha\beta}{}^\gamma = g^{\mu\alpha}g^{\nu\beta}g^{\rho\gamma}T_{\alpha\beta\gamma} \text{ etc.}$$

  Accordingly, the co- and contravariant vectors x_μ and x^μ differ solely by their spatial components.

Contrary to the threedimensional Euclidean case, the norm of a four-vector is not positive definite. Instead, the following cases can occur:

$$x_\mu x^\mu = x_0^2 - \boldsymbol{x}^2 = c^2t^2 - \boldsymbol{x}^2 \begin{cases} > 0 \text{ (time-like)} \\ = 0 \text{ (light-like)} \\ < 0 \text{ (space-like)} \ , \end{cases}$$

where the classification corresponds to the vector's position relative to the light cone $x_\mu x^\mu = 0$.

Lorentz transformations. The axioms of the theory of special relativity imply that the fourdimensional "distance" between two four-vectors x^μ and y^μ is preserved in any inertial system:

$$(x-y)_\mu(x-y)^\mu = (x'-y')_\mu(x'-y')^\mu \ .$$

Lorentz transformations describe the relativistic transition from one inertial system to another. They are defined by the linear transformational equation

$$x^\mu \to x'^\mu = \Lambda^\mu{}_\nu x^\nu + a^\mu \ , \ (a^\mu) = \text{space-time translation} \tag{A.1}$$

of contravariant vectors. Along with the conservation of distance, one obtains from this the conditional equation

$$\Lambda^\mu{}_\alpha g_{\mu\nu}\Lambda^\nu{}_\beta = g_{\alpha\beta} \tag{A.2}$$

as well as the corresponding transformational behavior of covariant vectors

$$x'_\mu = g_{\mu\nu}x'^\nu = g_{\mu\nu}\left(\Lambda^\nu{}_\alpha x^\alpha + a^\nu\right) = x_\beta[\Lambda^{-1}]^\beta{}_\mu + a_\mu \ ,$$

with the inverse transformation

$$[\Lambda^{-1}]^\beta{}_\mu = g_{\mu\nu}\Lambda^\nu{}_\alpha g^{\alpha\beta} = \Lambda_\mu{}^\beta \ , \ [\Lambda^{-1}]^\beta{}_\mu\Lambda^\mu{}_\gamma = g_{\mu\nu}\Lambda^\nu{}_\alpha g^{\alpha\beta}\Lambda^\mu{}_\gamma = \delta^\beta_\gamma . \tag{A.3}$$

In matrix notation the condition (A.2) reads $\Lambda^T g \Lambda = g$ and corresponds to the property $R^T R = 1$ of rotational matrices of threedimensional Euclidean geometry. Lorentz transformations with $a^\mu = 0$ constitute the homogeneous Lorentz group. Besides the distance, they also leave the scalar product of two four-vectors unchanged: $x^\mu y_\mu = x'^\mu y'_\mu$. For the general case, $(a^\mu) \neq 0$, one obtains the inhomogeneous Lorentz group or Poincaré group which is not considered here.

The homogeneous Lorentz group can be classified in the following way:

symbol	$\det(\Lambda)$	$\Lambda^0{}_0$	group name
$\mathcal{L}$	$+1$	>0	proper
$P \cdot \mathcal{L}$	-1	>0	orthochronous
$R \cdot \mathcal{L}$	-1	<0	homogeneous
$P \cdot R \cdot \mathcal{L}$	$+1$	<0	

Contrary to all others, the transformations of the proper Lorentz group $\mathcal{L}$ can be interpreted as a series of infinitesimal transformations. Noninfinitesimal or discrete transformations are, for example, the parity transformation (space reflection)

$$P: \quad x^0 \to x'^0 = x^0 \;, \; x^k \to x'^k = -x^k$$

and Racah time reflection

$$R: \quad x^0 \to x'^0 = -x^0 \;, \; x^k \to x'^k = x^k \;.$$

The orthochronous group consists of the proper transformations $\mathcal{L}$, the space reflection P, and their products $P \cdot \mathcal{L}$. The homogeneous group is composed of the transformations $\mathcal{L}$, P, R, and their products.

Transformational behavior of differential operators. Taking into account $x^\nu = [\Lambda^{-1}]^\nu{}_\mu x'^\mu$ [see (A.1)], we have

$$\frac{\partial x^\nu}{\partial x'^\mu} = [\Lambda^{-1}]^\nu{}_\mu \Longrightarrow \frac{\partial}{\partial x'^\mu} = \frac{\partial}{\partial x^\nu}\frac{\partial x^\nu}{\partial x'^\mu} = \frac{\partial}{\partial x^\nu}[\Lambda^{-1}]^\nu{}_\mu \;.$$

Therefore,

$$\frac{\partial}{\partial x^\mu} = \partial_\mu$$

transforms as a covariant vector and, accordingly,

$$\frac{\partial}{\partial x_\mu} = \partial^\mu$$

as a contravariant vector. Furthermore, it follows that the d'Alembert operator

$$\partial_\mu \partial^\mu = \frac{1}{c^2}\frac{\partial^2}{\partial t^2} - \boldsymbol{\nabla}^2$$

is Lorentz-invariant, i.e. a Lorentz scalar.

Relativistic kinematics. In order to write down the equations of motions of relativistic mechanics Lorentz-covariantly[1] (form invariantly), we must reformulate the Newtonian quantities of velocity, momentum, and force in such a way that they have a defined transformational behavior under Lorentz transformations. Contrary to $\mathrm{d}t$, the eigentime differential

$$\mathrm{d}\tau = \mathrm{d}t\sqrt{1 - \frac{1}{c^2}\left(\frac{\mathrm{d}\boldsymbol{x}}{\mathrm{d}t}\right)^2}$$

is a Lorentz scalar since, due to the conservation of the scalar product, we have

$$\mathrm{d}\tau^2 = \mathrm{d}t^2 - \frac{1}{c^2}\mathrm{d}\boldsymbol{x}^2 = \mathrm{d}t'^2 - \frac{1}{c^2}\mathrm{d}\boldsymbol{x}'^2 \quad (\mathrm{d}\tau^2 > 0 \text{ for time-like vectors}) .$$

With this, the following quantities can be defined, of which the four-versions transform as x^μ:

- Four-velocity u^μ:

$$u^\mu = \frac{\mathrm{d}x^\mu}{\mathrm{d}\tau} = \frac{\mathrm{d}t}{\mathrm{d}\tau}\frac{\mathrm{d}x^\mu}{\mathrm{d}t} \ , \ (u^\mu) = \frac{1}{\sqrt{1-\frac{\boldsymbol{v}^2}{c^2}}}\begin{pmatrix} c \\ \boldsymbol{v} \end{pmatrix}$$

$$\boldsymbol{v} = \frac{\mathrm{d}\boldsymbol{x}}{\mathrm{d}t} = \text{physical velocity.}$$

- Four-momentum p^μ:

$$p^\mu = m_0 u^\mu \ , \ (p^\mu) = \begin{pmatrix} cm \\ \boldsymbol{p} \end{pmatrix} \ , \ m = \frac{m_0}{\sqrt{1-\frac{\boldsymbol{v}^2}{c^2}}} \ , \ m_0 = \text{rest mass}$$

$$\boldsymbol{p} = m\boldsymbol{v} = \text{physical momentum.}$$

- Four-force F^μ:

$$F^\mu = \frac{\mathrm{d}p^\mu}{\mathrm{d}\tau} = \frac{1}{\sqrt{1-\frac{\boldsymbol{v}^2}{c^2}}}\frac{\mathrm{d}p^\mu}{\mathrm{d}t} \ , \ (F^\mu) = \frac{1}{\sqrt{1-\frac{\boldsymbol{v}^2}{c^2}}}\begin{pmatrix} c\frac{\mathrm{d}m}{\mathrm{d}t} \\ \boldsymbol{F} \end{pmatrix} \qquad \text{(A.4)}$$

$$\boldsymbol{F} = \frac{\mathrm{d}\boldsymbol{p}}{\mathrm{d}t} = \text{physical force.}$$

[1] In this book, as in many other textbooks, the transformational behavior of relativistic four-quantities is called "Lorentz-covariant" regardless of the subtle difference between "covariant" and "contravariant". The same holds for relativistic equations whose form remains unchanged under Lorentz transformations. Quantities and equations are called "Lorentz-invariant" if their value is unchanged by Lorentz transformations (Lorentz scalar).

The first force equation also represents the Lorentz-covariant equation of motion of relativistic mechanics, whereas the second equation is the threedimensional relativistic analogon of Newton's equation. In the case of a conservative force field $\boldsymbol{F} = -\boldsymbol{\nabla} V(\boldsymbol{x})$, the latter leads to the energy conservation

$$E = mc^2 + V(\boldsymbol{x}) = \text{const} \Longrightarrow E = mc^2 \text{ for } V = 0$$

and, together with the definition of p^μ, the energy-momentum relation for free particles

$$p_\mu p^\mu = p_0^2 - \boldsymbol{p}^2 = m_0^2 c^2 \Longleftrightarrow E^2 = \boldsymbol{p}^2 c^2 + m_0^2 c^4 \ .$$

It is not always possible to formulate a given mechanical problem covariantly via (A.4) since not all types of forces allow a fourdimensional generalization. One example of this kind is the Coulomb force. As a static long range force, it requires an infinitely high propagational velocity, thus being in contradiction to the first axiom of special relativity.

Lorentz-covariant electrodynamics. The basic equations of classical electrodynamics are the four Maxwell equations

$$\left.\begin{aligned} &\boldsymbol{\nabla}\boldsymbol{E}(\boldsymbol{x},t) = 4\pi\rho(\boldsymbol{x},t) \\ &\boldsymbol{\nabla}\times\boldsymbol{E}(\boldsymbol{x},t) + \frac{1}{c}\frac{\partial \boldsymbol{B}(\boldsymbol{x},t)}{\partial t} = 0 \\ &\boldsymbol{\nabla}\boldsymbol{B}(\boldsymbol{x},t) = 0 \\ &\boldsymbol{\nabla}\times\boldsymbol{B}(\boldsymbol{x},t) - \frac{1}{c}\frac{\partial \boldsymbol{E}(\boldsymbol{x},t)}{\partial t} = \frac{4\pi}{c}\boldsymbol{j}(\boldsymbol{x},t) \end{aligned}\right\} \quad \text{(A.5)}$$

(in the Gaussian unit system), the continuity equation

$$\frac{\partial\rho(\boldsymbol{x},t)}{\partial t} + \boldsymbol{\nabla}\boldsymbol{j}(\boldsymbol{x},t) = 0$$

that follows from the first and fourth Maxwell equation, as well as the Lorentz force

$$\boldsymbol{F}_{\rm L}(\boldsymbol{x},t) = q\left[\boldsymbol{E}(\boldsymbol{x},t) + \frac{\boldsymbol{v}}{c}\times\boldsymbol{B}(\boldsymbol{x},t)\right] \ , \quad \text{(A.6)}$$

describing the force acting on a particle with charge q due to its movement through the fields $\boldsymbol{E}$ and $\boldsymbol{B}$.

That these equations can be formulated Lorentz-covariantly, i.e. that they are in accordance with special relativity, can be shown as follows:

- First, the continuity equation can be immediately brought into the form

  $$\partial_\mu j^\mu(x) = 0 \ , \ (j^\mu) = \begin{pmatrix} c\rho \\ \boldsymbol{j} \end{pmatrix}$$

  showing explicitly its Lorentz covariance, provided that j^μ is a four-vector. However, this follows from the experimental fact that the charge q is a

Lorentz scalar: $\mathrm{d}q = \mathrm{d}^3x\rho = \mathrm{d}^3x'\rho'$. Thus, $\mathrm{d}q$ transforms as a fourdimensional volume element $\mathrm{d}^3x\mathrm{d}x^0 = \mathrm{d}^3x'\mathrm{d}x'^0$ and, consequently, ρ as the 0-th component of a four-vector.

- Introducing the scalar potential ϕ and the vector potential $\boldsymbol{A}$,

$$\boldsymbol{B}(\boldsymbol{x},t) = \boldsymbol{\nabla} \times \boldsymbol{A}(\boldsymbol{x},t) \ , \ \boldsymbol{E}(\boldsymbol{x},t) + \frac{1}{c}\frac{\partial \boldsymbol{A}(\boldsymbol{x},t)}{\partial t} = -\boldsymbol{\nabla}\phi(\boldsymbol{x},t) \ , \tag{A.7}$$

Maxwell's equations can be rewritten as

$$\boldsymbol{\nabla}^2\phi + \frac{1}{c}\frac{\partial}{\partial t}\boldsymbol{\nabla}\boldsymbol{A} = -4\pi\rho$$
$$\left(\boldsymbol{\nabla}^2\boldsymbol{A} - \frac{1}{c^2}\frac{\partial^2 \boldsymbol{A}}{\partial t^2}\right) - \boldsymbol{\nabla}\left(\boldsymbol{\nabla}\boldsymbol{A} + \frac{1}{c}\frac{\partial\phi}{\partial t}\right) = -\frac{4\pi}{c}\boldsymbol{j} \ .$$

From here and choosing the gauge

$$\boldsymbol{\nabla}\boldsymbol{A} = -\frac{1}{c}\frac{\partial\phi}{\partial t} \quad \text{(Lorentz gauge)} \ ,$$

one obtains the disentangled inhomogeneous wave equations

$$\left(\boldsymbol{\nabla}^2 - \frac{1}{c^2}\frac{\partial^2}{\partial t^2}\right)\phi = -4\pi\rho \iff \partial_\mu\partial^\mu\phi = 4\pi\rho$$
$$\left(\boldsymbol{\nabla}^2 - \frac{1}{c^2}\frac{\partial^2}{\partial t^2}\right)\boldsymbol{A} = -\frac{4\pi}{c}\boldsymbol{j} \iff \partial_\mu\partial^\mu\boldsymbol{A} = \frac{4\pi}{c}\boldsymbol{j}$$

that are symmetric in $\boldsymbol{A}$ and ϕ. Since ρ and $\boldsymbol{j}$ are the components of a four-vector and $\partial_\mu\partial^\mu$ is a Lorentz scalar, the last three equations can also be cast into a manifestly Lorentz-covariant form:

$$\partial_\mu A^\mu = 0 \ , \ \partial_\mu\partial^\mu A^\nu = \frac{4\pi}{c}j^\nu \ , \ (A^\mu) = \begin{pmatrix}\phi \\ \boldsymbol{A}\end{pmatrix} \ .$$

- Knowing that j^μ and A^μ are four-vectors, the Lorentz covariance of Maxwell's equations themselves can be shown in the following way: first, we rewrite the defining equations (A.7) as

$$F^{\mu\nu} = \partial^\mu A^\nu - \partial^\nu A^\mu \ , \ (F^{\mu\nu}) = \begin{pmatrix} 0 & -E_x & -E_y & -E_z \\ E_x & 0 & -B_z & B_y \\ E_y & B_z & 0 & -B_x \\ E_z & -B_y & B_x & 0 \end{pmatrix} \ ,$$

where the field strength tensor $F^{\mu\nu}$ must be a twofold contravariant (antisymmetric) tensor transforming as

$$F'^{\mu\nu} = \Lambda^\mu{}_\alpha \Lambda^\nu{}_\beta F^{\alpha\beta} \ .$$

With this, the two homogeneous Maxwell equations in (A.5) turn into the Lorentz-covariant equation

$$\partial_\mu F^{\mu\nu} = \frac{4\pi}{c}j^\nu \ .$$

Both sides transform as onefold contravariants. With the help of the dual field strength tensor

$$G^{\mu\nu} = \frac{1}{2}\epsilon^{\mu\nu\alpha\beta}F_{\alpha\beta} = F_{\mu\nu}(\boldsymbol{E}\to\boldsymbol{B}\ ,\ \boldsymbol{B}\to-\boldsymbol{E})$$

$$\epsilon^{\mu\nu\alpha\beta} = \begin{cases} +1 \text{ if } (\mu\nu\alpha\beta) \text{ even permutation of } (0123) \\ -1 \text{ if } (\mu\nu\alpha\beta) \text{ odd permutation of } (0123) \\ 0 \quad \text{else}\ , \end{cases}$$

the remaining inhomogeneous Maxwell equations (A.5) acquire the covariant form

$$\partial_\mu G^{\mu\nu} = 0\ .$$

- Contrary to the Coulomb force, the Lorentz force (A.6) allows a relativistic generalization to a four-vector $F_{\rm L}^\mu$ which is

  $$F_{\rm L}^\mu = \frac{q}{c}F^{\mu\nu}u_\nu = \frac{q}{c}F^{\mu\nu}\frac{\mathrm{d}x_\nu}{\mathrm{d}\tau}\ .$$

  According to (A.4), we therefore have the equations

  $$\frac{\mathrm{d}p^\mu}{\mathrm{d}\tau} = F_{\rm L}^\mu \Longrightarrow \begin{cases} \mu=0: \ \dfrac{\mathrm{d}}{\mathrm{d}t}mc^2 = q\boldsymbol{E}\boldsymbol{v} \\ \mu=i: \ \dfrac{\mathrm{d}\boldsymbol{p}}{\mathrm{d}t} = q\left(\boldsymbol{E}+\dfrac{\boldsymbol{v}}{c}\times\boldsymbol{B}\right) = \boldsymbol{F}_{\rm L}\ . \end{cases}$$

A.2 Bessel Functions, Spherical Bessel Functions

Bessel functions. The Bessel differential equation is

$$\left[\frac{\mathrm{d}^2}{\mathrm{d}x^2} + \frac{1}{x}\frac{\mathrm{d}}{\mathrm{d}x} + \left(1-\frac{m^2}{x^2}\right)\right]f(x) = 0\ ,\ m\in\mathbb{R}.$$

Its solutions are the Bessel functions J_m and J_{-m} with

$$J_m(x) = \left(\frac{x}{2}\right)^m \sum_{i=0}^\infty \frac{(-1)^i}{i!\Gamma(m+i+1)}\left(\frac{x}{2}\right)^{2i}\ .$$

If m is an integer, then

$$J_m(x) = \left(\frac{x}{2}\right)^m \sum_{i=0}^\infty \frac{(-1)^i}{i!(m+i)!}\left(\frac{x}{2}\right)^{2i}\ ,\ J_{-m}(x) = (-1)^m J_m(x)\ .$$

Spherical Bessel functions. The spherical Bessel differential equation is

$$\left[\frac{\mathrm{d}^2}{\mathrm{d}x^2} + \frac{2}{x}\frac{\mathrm{d}}{\mathrm{d}x} + 1 - \frac{l(l+1)}{x^2}\right] f(x) = 0 \ , \ l = 0, 1, 2, \ldots \ .$$

Its solutions are the spherical Bessel functions j_l, n_l (the latter are also called Neumann functions) and, therefore, also the Hankel functions $h_l^{(\pm)}$:

$$\begin{aligned} j_l(x) &= \left(\frac{\pi}{2x}\right)^{1/2} J_{l+1/2}(x) \\ n_l(x) &= (-1)^l \left(\frac{\pi}{2x}\right)^{1/2} J_{-l-1/2}(x) \\ h_l^{(\pm)}(x) &= n_l(x) \pm \mathrm{i} j_l(x) \ . \end{aligned}$$

Their explicit forms are

$$\left.\begin{aligned} j_l(x) &= R_l(x)\frac{\sin x}{x} + S_l(x)\frac{\cos x}{x} \\ n_l(x) &= R_l(x)\frac{\cos x}{x} - S_l(x)\frac{\sin x}{x} \\ h_l^{(\pm)}(x) &= [R_l(x) \pm \mathrm{i}S_l(x)]\frac{\mathrm{e}^{\pm \mathrm{i}x}}{x} \ , \end{aligned}\right\} \tag{A.8}$$

with

$$R_l(x) + \mathrm{i}S_l(x) = \sum_{s=0}^{l} \frac{\mathrm{i}^{s-l}}{2^s s!}\frac{(l+s)!}{(l-s)!} x^{-s} \ , \ R_l, S_l \in \mathbb{R} \ .$$

R_l and S_l are polynomials in $1/x$ of order l with real coefficients and parity $(-1)^l$ and $-(-1)^l$ respectively. For any linear combination $f_l = aj_l + bn_l$, a, b fixed, we have the recursion formulae

$$\begin{aligned} (2l+1)f_l(x) &= x\left[f_{l+1}(x) + f_{l-1}(x)\right] \\ f_{l-1} &= \left(\frac{\mathrm{d}}{\mathrm{d}x} + \frac{l+1}{x}\right) f_l = \frac{1}{x^{l+1}}\frac{\mathrm{d}}{\mathrm{d}x}\left(x^{l+1} f_l\right) \\ f_l &= \left(-\frac{\mathrm{d}}{\mathrm{d}x} + \frac{l-1}{x}\right) f_{l-1} = -x^{l-1}\frac{\mathrm{d}}{\mathrm{d}x}\left(\frac{f_{l-1}}{x^{l-1}}\right) \ , \end{aligned}$$

which imply that

$$f_l = \left[x^l \left(-\frac{1}{x}\frac{\mathrm{d}}{\mathrm{d}x}\right)^l\right] f_0 \ .$$

The first spherical functions are obtained from (A.8) as

$$j_0(x) = \frac{\sin x}{x} \ , \ j_1(x) = \frac{\sin x}{x^2} - \frac{\cos x}{x}$$

$$n_0(x) = \frac{\cos x}{x} \ , \ n_1(x) = \frac{\cos x}{x^2} + \frac{\sin x}{x}$$

$$h_0^{(\pm)}(x) = \frac{\mathrm{e}^{\pm \mathrm{i}x}}{x} \; , \; h_1^{(\pm)}(x) = \left(\frac{1}{x^2} \mp \frac{\mathrm{i}}{x}\right) \frac{\mathrm{e}^{\pm \mathrm{i}x}}{x} \; .$$

A.3 Legendre Functions, Legendre Polynomials, Spherical Harmonics

Legendre functions. The Legendre differential equation is

$$\left[(1-x^2)\frac{\mathrm{d}^2}{\mathrm{d}x^2} - 2x\frac{\mathrm{d}}{\mathrm{d}x} + l(l+1) - \frac{m^2}{1-x^2}\right] f(x) = 0 \; ,$$

with $l = 0, 1, 2, \ldots$, $m = 0, \ldots, \pm l$. Its limited solutions within the interval $[-1:1]$ are the Legendre functions

$$P_{l,m}(x) = \frac{(1-x^2)^{m/2}}{2^l l!} \frac{\mathrm{d}^{l+m}}{\mathrm{d}x^{l+m}} (x^2-1)^l \; . \tag{A.9}$$

$P_{l,m}$ is the product of $(1-x)^{m/2}$ with a polynomial of order $l-m$ and parity $(-1)^{l-m}$, and it has $l-m$ zeros within the interval $[-1:1]$. We have the following recursion formulae ($P_{-1,\ldots} = 0$):

$$\begin{aligned}
(2l+1)xP_{l,m} &= (l+1-m)P_{l+1,m} + (l+m)P_{l-1,m} \\
(1-x^2)\frac{\mathrm{d}}{\mathrm{d}x}P_{l,m} &= -lxP_{l,m} + (l+m)P_{l-1,m} \\
&= (l+1)xP_{l,m} - (l+1-m)P_{l+1,m}
\end{aligned}$$

as well as the orthonormality relations

$$\int_{-1}^{1} \mathrm{d}x P_{l,m}(x) P_{l',m}(x) = \frac{2}{2l+1} \frac{(l+m)!}{(l-m)!} \delta_{ll'} \; .$$

Legendre polynomials. In the case of $m = 0$ the Legendre polynomials follow from (A.9) as

$$P_l(x) = P_{l,0}(x) = \frac{1}{2^l l!} \frac{\mathrm{d}^l}{\mathrm{d}x^l} (x^2-1)^l \; .$$

P_l is a polynomial of order l with parity $(-1)^l$ and possesses l zeros within the interval $[-1:1]$. The Legendre polynomials can be obtained by expanding the functions $(1-2xy+y^2)^{-1/2}$ in powers of y:

$$\frac{1}{\sqrt{1-2xy+y^2}} = \sum_{l=0}^{\infty} y^l P_l(x) \; , \; |y| < 1 \; . \tag{A.10}$$

The first five Legendre polynomials are

$$P_0(x) = 1 \; , \; P_1(x) = x \; , \; P_2(x) = \frac{1}{2}(3x^2-1)$$

$$P_3(x) = \frac{1}{2}(5x^3 - 3x) \; , \; P_4(x) = \frac{1}{8}(35x^4 - 30x^2 + 3) \; .$$

Spherical harmonics. The spherical harmonics $Y_{l,m}$ are defined as the eigenfunctions of the quantum mechanical angular momentum operators $\boldsymbol{L}^2$ and L_z:

$$\begin{aligned}\boldsymbol{L}^2 Y_{l,m} &= \hbar^2 l(l+1) Y_{l,m} \ , \ l = 0, 1, 2, \ldots \\ L_z Y_{l,m} &= \hbar m Y_{l,m} \ , \ m = 0, \ldots, \pm l \ .\end{aligned}$$

Their explicit forms are

$$\begin{aligned}Y_{l,m}(\theta,\varphi) &= \frac{(-1)^l}{2^l l!}\sqrt{\frac{(2l+1)!}{4\pi}}\sqrt{\frac{(l+m)!}{(2l)!(l-m)!}} \\ &\quad \times \mathrm{e}^{\mathrm{i}m\varphi}\sin^{-m}\theta \frac{\mathrm{d}^{l-m}}{\mathrm{d}(\cos\theta)^{l-m}}\sin^{2l}\theta \ .\end{aligned}$$

They form a complete orthonormal function system on the unit circle. This means that the following orthonormality and completeness relations hold:

$$\int Y_{l,m}^* Y_{l',m'} \mathrm{d}\Omega = \int\limits_0^{2\pi} \mathrm{d}\varphi \int\limits_0^{\pi} \mathrm{d}\theta \sin\theta Y_{l,m}^*(\theta,\varphi) Y_{l',m'}(\theta,\varphi) = \delta_{ll'}\delta_{mm'}$$

$$\sum_{l=0}^{\infty}\sum_{m=-l}^{l} Y_{l,m}^*(\theta,\varphi) Y_{l,m}(\theta',\varphi') = \frac{\delta(\varphi-\varphi')\delta(\cos\theta-\cos\theta')}{\sin\theta} = \delta(\Omega-\Omega').$$

Further properties are:

- Parity:

$$Y_{l,m}(\pi-\theta,\varphi+\pi) = (-1)^l Y_{l,m}(\theta,\varphi) \ .$$

- Complex conjugation:

$$Y_{l,m}^*(\theta,\varphi) = (-1)^m Y_{l,-m}(\theta,\varphi) \ .$$

- Relationship with Legendre functions:

$$Y_{l,m}(\theta,\varphi) = \sqrt{\frac{2l+1}{4\pi}\frac{(l-m)!}{(l+m)!}} P_{l,m}(\cos\theta)\mathrm{e}^{\mathrm{i}m\varphi} \ , \ m \ge 0 \ .$$

- Addition theorem: using

$$\boldsymbol{x} = r\begin{pmatrix}\cos\varphi\sin\theta \\ \sin\varphi\sin\theta \\ \cos\theta\end{pmatrix} \ , \ \boldsymbol{x}' = r'\begin{pmatrix}\cos\varphi'\sin\theta' \\ \sin\varphi'\sin\theta' \\ \cos\theta'\end{pmatrix}$$

and

$$\boldsymbol{x}\boldsymbol{x}' = rr'\cos\alpha \ , \ \cos\alpha = \sin\theta\sin\theta'\cos(\varphi-\varphi') + \cos\theta\cos\theta' \ ,$$

it follows that

$$P_l(\cos\alpha) = \frac{4\pi}{2l+1}\sum_{m=-l}^{l} Y^*_{l,m}(\theta',\varphi')Y_{l,m}(\theta,\varphi) \ .$$

We obtain from this, in line with (A.10),

$$\frac{1}{|\boldsymbol{x}-\boldsymbol{x}'|} = \frac{1}{r\sqrt{1-2\frac{r'}{r}\cos\alpha+\left(\frac{r'}{r}\right)^2}} = \frac{1}{r}\sum_{l=0}^{\infty}\left(\frac{r'}{r}\right)^l P_l(\cos\alpha)$$
$$= \sum_{l=0}^{\infty}\sum_{m=-l}^{l}\frac{4\pi}{2l+1}\frac{r'^l}{r^{l+1}}Y^*_{l,m}(\theta',\varphi')Y_{l,m}(\theta,\varphi) \ .$$

The first spherical harmonics are

$$Y_{0,0}(\theta,\varphi) = \frac{1}{\sqrt{4\pi}} \ , \ Y_{1,1}(\theta,\varphi) = -\sqrt{\frac{3}{8\pi}}\mathrm{e}^{\mathrm{i}\varphi}\sin\theta$$
$$Y_{1,0}(\theta,\varphi) = \sqrt{\frac{3}{4\pi}}\cos\theta \ , \ Y_{2,2}(\theta,\varphi) = \sqrt{\frac{15}{32\pi}}\mathrm{e}^{2\mathrm{i}\varphi}\sin^2\theta$$
$$Y_{2,1}(\theta,\varphi) = -\sqrt{\frac{15}{8\pi}}\mathrm{e}^{\mathrm{i}\varphi}\sin\theta\cos\theta \ , \ Y_{2,0}(\theta,\varphi) = \sqrt{\frac{5}{16\pi}}\left(3\cos^2\theta-1\right) \ .$$

A.4 Dirac Matrices and Bispinors

The Dirac matrices $\{\alpha_1,\alpha_2,\alpha_3,\beta\}$ and $\{\gamma^0,\gamma^1,\gamma^2,\gamma^3\}$ as well as γ^5 and $\sigma^{\mu\nu}$ are defined representation-independently by

$$\{\alpha_i,\alpha_j\} = 2\delta_{ij} \ , \ \{\alpha_i,\beta\} = 0 \ , \ \alpha_i^2 = \beta^2 = 1 \ , \ \alpha_i = \alpha_i^\dagger \ , \ \beta = \beta^\dagger$$
$$\gamma^0 = \beta \ , \ \gamma^i = \beta\alpha_i \ , \ \gamma_\mu = g_{\mu\nu}\gamma^\nu$$
$$\gamma^5 = \mathrm{i}\gamma^0\gamma^1\gamma^2\gamma^3 = -\mathrm{i}\gamma_3\gamma_2\gamma_1\gamma_0 = \gamma_5 \ , \ \sigma^{\mu\nu} = \frac{\mathrm{i}}{2}[\gamma^\mu,\gamma^\nu] \ .$$

This implies the following identities:

$$\begin{aligned}
\{\gamma^\mu,\gamma^\nu\} &= 2g^{\mu\nu} \quad \text{(Clifford algebra)} \ , \ (\gamma^\mu)^2 = g^{\mu\mu} \\
\gamma^5 &= -\frac{\mathrm{i}}{4!}\epsilon_{\mu\nu\alpha\beta}\gamma^\mu\gamma^\nu\gamma^\alpha\gamma^\beta \ , \ \gamma_5^2 = 1 \\
\{\gamma^5,\gamma^\mu\} &= 0 \\
\gamma^5\sigma^{\mu\nu} &= \frac{\mathrm{i}}{2}\epsilon^{\mu\nu\alpha\beta}\sigma_{\alpha\beta} \\
\left[\gamma^5,\sigma^{\mu\nu}\right] &= 0 \\
\gamma^\mu\gamma^\nu &= g^{\mu\nu} - \mathrm{i}\sigma^{\mu\nu} \\
\gamma_\mu\gamma^\mu &= 4
\end{aligned}$$

$$\begin{aligned}
\gamma^\mu\gamma^\nu\gamma_\mu &= -2\gamma^\nu \\
\gamma^\mu\gamma^\nu\gamma^\alpha\gamma_\mu &= 4g^{\nu\alpha} \\
\gamma^\mu\gamma^\nu\gamma^\alpha\gamma^\beta\gamma_\mu &= -2\gamma^\beta\gamma^\alpha\gamma^\nu \\
\gamma^\mu\gamma^\nu\gamma^\alpha\gamma^\beta\gamma^\rho\gamma_\mu &= 2\left(\gamma^\rho\gamma^\nu\gamma^\alpha\gamma^\beta - \gamma^\beta\gamma^\alpha\gamma^\nu\gamma^\rho\right) \\
\gamma^\mu\sigma^{\alpha\beta}\gamma_\mu &= 0 \\
\gamma^\mu\sigma^{\alpha\beta}\gamma^\rho\gamma_\mu &= 2\gamma^\rho\sigma^{\alpha\beta} \; .
\end{aligned}$$

Traces:

$$\begin{aligned}
\mathrm{tr}(\gamma^\mu) &= \mathrm{tr}(\gamma^5) = 0 \\
\mathrm{tr}(\gamma^\mu\gamma^\nu) &= 4g^{\mu\nu} \\
\mathrm{tr}(\sigma^{\mu\nu}) &= 0 \\
\mathrm{tr}(\gamma^\mu\gamma^\nu\gamma^5) &= 0 \\
\mathrm{tr}(\gamma^\mu\gamma^\nu\gamma^\alpha\gamma^\beta) &= 4(g^{\mu\nu}g^{\alpha\beta} - g^{\mu\alpha}g^{\nu\beta} + g^{\mu\beta}g^{\nu\alpha}) \\
\mathrm{tr}(\gamma^5\gamma^\mu\gamma^\nu\gamma^\alpha\gamma^\beta) &= -4\mathrm{i}\epsilon^{\mu\nu\alpha\beta} = 4\mathrm{i}\epsilon_{\mu\nu\alpha\beta} \; .
\end{aligned}$$

Hermitean conjugation:

$$\begin{aligned}
\gamma^{0\dagger} &= \gamma^0 \; , \; \gamma^{i\dagger} = -\gamma^i \; , \; \gamma^{5\dagger} = \gamma^5 \\
\gamma^0\gamma^\mu\gamma^0 &= \gamma^{\mu\dagger} \; , \; \gamma^0\gamma^5\gamma^0 = -\gamma^{5\dagger} \\
\gamma^0\gamma^5\gamma^\mu\gamma^0 &= (\gamma^5\gamma^\mu)^\dagger \\
\gamma^0\sigma^{\mu\nu}\gamma^0 &= \sigma^{\mu\nu\dagger} \; .
\end{aligned}$$

Fourdimensional representations of the γ-matrices.

Dirac representation:

$$\gamma^0 = \begin{pmatrix} 1 & 0 \\ 0 & -1 \end{pmatrix} , \; \gamma^i = \begin{pmatrix} 0 & \sigma_i \\ -\sigma_i & 0 \end{pmatrix} , \; \gamma^5 = \begin{pmatrix} 0 & 1 \\ 1 & 0 \end{pmatrix}$$

$$\sigma^{0i} = \mathrm{i}\begin{pmatrix} 0 & \sigma_i \\ \sigma_i & 0 \end{pmatrix} , \; \sigma^{ij} = \epsilon_{ijk}\begin{pmatrix} \sigma_k & 0 \\ 0 & \sigma_k \end{pmatrix} , \; C = \mathrm{i}\gamma^2 \; .$$

Weyl representation:

$$\gamma^0 = \begin{pmatrix} 0 & -1 \\ -1 & 0 \end{pmatrix} , \; \gamma^i = \begin{pmatrix} 0 & \sigma_i \\ -\sigma_i & 0 \end{pmatrix} , \; \gamma^5 = \begin{pmatrix} 1 & 0 \\ 0 & -1 \end{pmatrix}$$

$$\sigma^{0i} = \mathrm{i}\begin{pmatrix} \sigma_i & 0 \\ 0 & -\sigma_i \end{pmatrix} , \; \sigma^{ij} = \epsilon_{ijk}\begin{pmatrix} \sigma_k & 0 \\ 0 & \sigma_k \end{pmatrix} , \; C = \mathrm{i}\gamma^2$$

$$\gamma^\mu_{\mathrm{Weyl}} = U^\dagger\gamma^\mu_{\mathrm{Dirac}}U \; , \; U = \frac{1}{\sqrt{2}}\begin{pmatrix} 1 & -1 \\ 1 & 1 \end{pmatrix} \; .$$

Majorana representation:

$$\gamma^0 = \begin{pmatrix} 0 & \sigma_2 \\ \sigma_2 & 0 \end{pmatrix} , \; \gamma^1 = \mathrm{i}\begin{pmatrix} \sigma_3 & 0 \\ 0 & \sigma_3 \end{pmatrix} , \; \gamma^2 = \begin{pmatrix} 0 & -\sigma_2 \\ \sigma_2 & 0 \end{pmatrix}$$

$$\gamma^3 = \mathrm{i}\begin{pmatrix} -\sigma_1 & 0 \\ 0 & -\sigma_1 \end{pmatrix} , \; \gamma^5 = \begin{pmatrix} \sigma_2 & 0 \\ 0 & -\sigma_2 \end{pmatrix} , \; C = 1$$

$$\gamma^\mu_{\mathrm{Majorana}} = U^\dagger \gamma^\mu_{\mathrm{Dirac}} U \ , \ U = \frac{1}{\sqrt{2}} \begin{pmatrix} 1 & \sigma_2 \\ \sigma_2 & -1 \end{pmatrix} .$$

Here σ_i denote the Pauli matrices:

$$\sigma_1 = \begin{pmatrix} 0 & 1 \\ 1 & 0 \end{pmatrix} \ , \ \sigma_2 = \begin{pmatrix} 0 & -\mathrm{i} \\ \mathrm{i} & 0 \end{pmatrix} \ , \ \sigma_3 = \begin{pmatrix} 1 & 0 \\ 0 & -1 \end{pmatrix} .$$

Dirac bispinors. The bispinors $u(p,s)$, $v(p,s)$, as well as their adjoints $\bar{u}(p,s) = u^\dagger(p,s)\gamma^0$, $\bar{v}(p,s) = v^\dagger(p,s)\gamma^0$, fulfill the Dirac equations in momentum space ($\hbar = c = 1$, $p_0 = \sqrt{\boldsymbol{p}^2 + m_0^2}$):

$$(\not{p} - m_0)u(p,s) = 0 \ , \ (\not{p} + m_0)v(p,s) = 0$$

$$\bar{u}(p,s)(\not{p} - m_0) = 0 \ , \ \bar{v}(p,s)(\not{p} + m_0) = 0 .$$

Normalization:

$$\bar{u}(p,s)u(p,s) = 1 \ , \ \bar{v}(p,s)v(p,s) = -1$$

$$\bar{u}(p,s)v(p,s) = \bar{v}(p,s)u(p,s) = 0 .$$

Completeness relation:

$$\sum_s u_\alpha(p,s)\bar{u}_\beta(p,s) - v_\alpha(p,s)\bar{v}_\beta(p,s) = \delta_{\alpha\beta} .$$

Projection operators:

$$\begin{aligned}
\sum_s u_\alpha(p,s)\bar{u}_\beta(p,s) &= \left(\frac{\not{p} + m_0}{2m_0}\right)_{\alpha\beta} = [\Lambda_+(p)]_{\alpha\beta} \\
-\sum_s v_\alpha(p,s)\bar{v}_\beta(p,s) &= \left(\frac{-\not{p} + m_0}{2m_0}\right)_{\alpha\beta} = [\Lambda_-(p)]_{\alpha\beta} \\
u_\alpha(p,s)\bar{u}_\beta(p,s) &= \left(\frac{\not{p} + m_0}{2m_0}\frac{1 + \gamma^5\not{s}}{2}\right)_{\alpha\beta} = [\Lambda_+(p)\Sigma(s)]_{\alpha\beta} \\
-v_\alpha(p,s)\bar{v}_\beta(p,s) &= \left(\frac{-\not{p} + m_0}{2m_0}\frac{1 + \gamma^5\not{s}}{2}\right)_{\alpha\beta} = [\Lambda_-(p)\Sigma(s)]_{\alpha\beta} .
\end{aligned}$$

Gordon decompositions:

$$\begin{aligned}
\bar{u}(p',s')\gamma^\mu u(p,s) &= \frac{1}{2m_0}\bar{u}(p',s')\left[(p'+p)^\mu + \mathrm{i}\sigma^{\mu\nu}(p'-p)_\nu\right]u(p,s) \\
\bar{v}(p',s')\gamma^\mu v(p,s) &= -\frac{1}{2m_0}\bar{v}(p',s')\left[(p'+p)^\mu + \mathrm{i}\sigma^{\mu\nu}(p'-p)_\nu\right]v(p,s) \\
\bar{u}(p',s')\gamma^\mu v(p,s) &= \frac{1}{2m_0}\bar{u}(p',s')\left[(p'-p)^\mu + \mathrm{i}\sigma^{\mu\nu}(p'+p)_\nu\right]v(p,s) \\
\bar{v}(p',s')\gamma^\mu u(p,s) &= -\frac{1}{2m_0}\bar{v}(p',s')\left[(p'-p)^\mu + \mathrm{i}\sigma^{\mu\nu}(p'+p)_\nu\right]u(p,s) .
\end{aligned}$$

Index

www.ingramcontent.com/pod-product-compliance
Lightning Source LLC
Chambersburg PA
CBHW071726290626
47387CB00004B/570
* 9 7 8 9 0 4 8 1 3 6 4 4 5 *